Gums and Stabilisers for the Food Industry 16

Gums and Stabilisers for the Food Industry 16

Edited by

Peter A. Williams
Centre of Water Soluble Polymers, Glyndwr University, Wrexham, UK
Email : p.a.williams@glyndwr.ac.uk

Glynn O. Phillips
Phillips Hyrdocolloids Research Ltd, London , UK

RSC Publishing

The proceedings of the 16th Gums and Stabilisers for the Food Industry Conference held on 28 June - 1 July 2011 in Wageningen, The Netherlands.

Special Publication No. 335

ISBN: 978-1-84973-358-8

A catalogue record for this book is available from the British Library

Published by The Royal Society of Chemistry,
Thomas Graham House, Science Park, Milton Road,
Cambridge CB4 0WF, UK

Registered Charity Number 207890

For further information see our web site at www.rsc.org

Printed and bound in Great Britain by CPI Group (UK) Ltd, Croydon, CR0 4YY, UK

Preface

This volume presents the most successfully reviewed papers from the 16th International Gums and Stabilisers for the Food Industry Conference. The first book in this series appeared in 1982 and in the subsequent years the readership has grown continuously. The volumes have now established themselves as the most widely referenced publications in the food hydrocolloids sector. It was these publications which inspired the parallel Journal *"Food Hydrocolloids"* which during the 16th conference celebrated its 25th Anniversary. The review by Dennis Seisun shows that the subject is continuing to grow at a rapid rate with the applications calling for new and innovating materials and processes. The publications presented here show how imaginatively this challenge is being met.

For this conference the venue was changed from Wales (at the North East Wales Institute (now Glyndwr University), to the Netherlands. I must thank our friends in Wageningen for their hospitality. We thought that a change of venue would give us a fresh impetus. And where better could we go than to Wageningen University with its rich tradition in food and polysaccharide research. Professor Pilnic would have certainly approved and his memory was honoured at the conference with Professor Bill Williams giving the now established Pilnik Lecture. Our valued committee member Professor Fons Voragen was the catalyst in our fixing on Wageningen for our "sabbatical" and we thank him and his colleagues for their support and welcome.

The book illustrates the breadth of materials, their interaction, methods, functionality, and processing which need to be combined to give innovative and creative solutions. Polysaccharides have been the traditional foundation of this subject and their number and type continue to grow.

Included now in the volume are: xanthan, beta glucan, carrageenan, succinoglycan, chitosan, sago and wheat starch, native and high amylose starch, citrus and sugar beet pectins, CMC, gum arabic, methyl and other cellulose ethers, corn fibre gum and konjac glucomannan.

We wonder now why proteins were not immediately classed as food hydrocolloids but their involvement has now been found to enhance and support the polysaccharide contribution to the product properties.

The following proteins are shown in the book to be able to function alone or in associated complex interactions: fish and other gelatines, beta lactoglobulin, sodium caseinate, papein, cray fish protein isolate, potato protein. Further details of the major role of proteins are given in a recent publication *(Handbook of Food Proteins, Ed.*

G.O.Phillips and P.A.Williams, Woodhead Publishing, ISBN 978-1-84569-7, 2011, 432pp)

The use of the term "product design" is now increasing and justifiably so. Texture is being increasingly planned to deal with specific medical conditions, for example, dysphphagia - the inability for older people to masticate and swallow food correctly. The papers describe how the various materials can be engineered to give the desired physical, mechanical and release properties, whether emulsions, gels or films.

For such controlled modification and still retain the natural character and status the associated rheological and other methodology is necessary as the book testifies. The titles alone are fascinating:

- electrospinning of fish gelatine;
- chitosan films prepared in electrical fields;
- gelatine from black tilapia *(Oreschromis mossambicus)* skins;
- Pectin – zein hydrogels;
- Effect of palm oil incorporation on the properties of biodegradable cobia (*Rachycentron canadum*) skin gelatine;
- Interfacial behaviour of chitosan and crayfish protein isolate mixed films;
- Encapsulation by electrospinning of live bacteria

Drawing attention to these exotic subject areas should not reflect adversely on the traditional structure-function studies which give authority to the subject. The move into nutrition to ensure improved satiety, sensory and health continues in this volume also. Indeed, there is a wealth of fascinating material now available in the book for academic researchers, ingredient providers and users and industrial R&D scientists of hydrocolloids. Please enjoy.

Glyn O. Phillips

Chairman, Gums and Stabilisers for the Food Industry Organising Committee

Gums and Stabilisers for the Food Industry Conferences

The international conference on 'Gums and Stabilisers for the Food Industry' was initiated at the North East Wales Institute, Wrexham, in 1981 and has been held biennially since then. The proceedings of all of the Conferences have been published and details can be found below.

1. Prog. Fd. Nutr. Sci., "Gums and Stabilisers for the Food Industry" (Eds. G. O. Phillips, D. J. Wedlock and P. A. Williams). Pergamon Press Ltd., Oxford Vol 6 (1982).
2. "Gums and Stabilisers for the Food Industry 2" (Eds., G. O. Phillips, D. J. Wedlock and P. A. Williams), Pergamon Press Ltd., Oxford (1984).
3. "Gums and Stabilisers for the Food Industry 3" (Eds., G. O. Phillips, D. J. Wedlock and P. A. Williams), Elsevier Applied Science Publishers (1986).
4. "Gums and Stabilisers for the Food Industry 4" (Eds., G. O. Phillips, D. J. Wedlock and P. A. Williams), IRL Press, Oxford, (1988).
5. "Gums and Stabilisers for the Food Industry 5 (G. O. Phillips, D. J. Wedlock and P. A. Williams, Oxford University Press Ltd., Oxford, (1990).
6. "Gums and Stabilisers for the Food Industry 6" (eds Phillips G. O., Williams P. A. and Wedlock D. J.) Oxford University Press Ltd, Oxford (1992).
7. "Gums and Stabilisers for the Food Industry 7" (eds Phillips G. O., Williams, P. A. and Wedlock D. J.) Oxford University Press, Oxford (1994).
8. "Gums and Stabilisers for the Food Industry 8" (eds Phillips G. O., Williams, P. A. and Wedlock D. J.) Oxford University Press (1996).
9. "Gums and Stabilisers for the Food Industry 9" (eds.Williams, P. A. aand Phillips, G.O.), Royal Society of Chemistry, Cambridge, UK (1998).
10. "Gums and Stabilisers for the Food Industry 10" (eds Williams, P. A. and Phillips G.O.) Royal Society of Chemistry, Cambridge, UK (2000).
11. "Gums and Stabilisers for the Food Industry 11" (eds Williams, P. A. and Phillips G.O.) Royal Society of Chemistry, Cambridge, UK (2002).
12. "Gums and Stabilisers for the Food Industry 12" (eds Williams, P.A. and Phillips, G.O.) Royal Society of Chemistry, Cambridge, UK (2004)
13. "Gums and Stabilisers for the Food Industry 13" (eds Williams, P.A. and phillips, G.O.) Royal Society of Chemistry, Cambridge, UK (2006)
14. "Gums and Stabilisers for the Food Industry 14" (eds Williams, P.A. and Phillips, G.O.) Royal Society of Chemistry, Cambridge, UK (2008)
15. "Gums and Stabilisers for the Food Industry 15" (eds Williams, P.A. and Phillips, G.O.) Royal Society of Chemistry, Cambridge, UK (2009)

Acknowledgements

The Food Hydrocolloids Trust is indebted to the members of the conference organising committee.

Dr P. Boulenguer, Cargill France SAS
Dr M. Capelle, Nestle, France
Professor E. A. Foegeding, North Caroloina State University, USA
Dr T. J. Foster, ***(Treasurer),*** University of Nottingham, UK
Mr H. Hughes *(Administrative Secretary),* Glyndwr University
Dr A. Imeson, FMC Corporation
Dr A. Koliandris, Mars GmbH, Germany
Dr R.G. Morley, Delphi Consultant Services Inc., USA
Profesoor J.R. Mitchell, University of Nottingham
Professor E. R. Morris, University College Cork, Ireland
Professor K. Nishinari, Osaka City University
Professor G.O.Phillips *(Chairman*), Phillips Hydrocolloids Research Ltd
Dr K. Philp, CyberColloids Ltd, Ireland
Dr C. Rolin, CP Kelco, Denmark
Dr C. Schorsch, Danone, France
Dr C. Speirs, CCFRA, UK
Dr G. Sworn, Danisco, France
Professor A.G.J. Voragen, University of Wageningen, The Netherlands
Professor P.A. Williams *(Scientific Secretary*), Glyndwr University

and also acknowledges financial support from

Biopolymer Solutions
Campden BRI
Cargill
CP Kelco
Cyber Colloids
Danisco
Elsevier
Kerry Food and Beverage
FMC
Mars
National Starch and Chenical
Pepsico
Phillips Hydrocolloids Research Centre
Unilever
Woodhead Publishing Ltd

CONTENTS

Chapter 1	**Market overview**	1
D. Seisun	Over view of the food hydrocolloids market	3
	Isolation, characterisation and modification	
N. Wellner, M.L. Parker, D.M.R. Georget and V.J. Morris	Raman and FTIR microscopy of native and high amylose maize starch *in situ*	11
W. Helbert	Analysis and modification of carrageenan structure using enzymes	19
L.M. Nwokocha and P.A. Williams	Evaluating the potential of Nigerian plants as a source of industrial hydrocolloids	27
A.Sullo, Y.Wang, J.R. Mitchell, A. Koshelle, T. Heinze and T.J. Foster	New regioselective substituted cellulose ethers: A thermo-rheological study	45
M.P. Yadav, R.A. Moreau and K.B. Hicks	Production of corn fibre gum under conditions that retain its functional components	59
F. Hajji, J.R. Mitchell and T.J. Foster	Relationship between the conformational structure and processability of hydrocolloids	67
T.J. Foster and J.R. Mitchell	Physical modification of xanthan gum	77
G.Sworn and E. Kerdavid	A comparison of the rheological properties of succinoglycan with xanthan gum	89
L.M. Perez-Mosqueda, P. Ramirez, M.C. Alfaro, J. Munoz, F. Rincon and G. Leon de Pinto	Bulk and surface rheology characterisation of *SterculiaApetala* exudate polysaccharide solutions	99
M. Gupta, B.K. Tiwari, C.S. Brennan, P.J. Cullen and C. O'Donnell	Effect of ozonation on the physicochemical, thermal and structural properties of wheat starch samples	107
M. Majzoobi , E. Abedi, A. Farahnaky and M. Aminlari	Functional properties of acetylated glutenin at varying pH values	115
P. Lai, Y.-T. Huang and K.-Y. Jiang	Optimisation for rapid separation of immunostimulatory*Chlorella* biopolymers	123

Chapter 2	**Mixed hydrocolloid systems**	
J.K. Agbenorhevi, V.Kontogiorgos and S. Kasapis	Phase behaviour of beta glucan / sodium caseinate mixtures varying in molecular weight	133
XuanThang Le and S.L. Turgeon	Structure as related to rheological, textural properties and water binding of β – lactoglobulin / xanthan gum electrostatic gels in associative conditions	141
A. H. Abbaszadeh and T.J. Foster	The effect of polymer fine structure on synergistic interactions of xanthan with konjacglucomannan	151
Chapter 3	**Hydrocolloid gels**	
H.M. Shewan and J.R. Stokes	Biopolymer microgel suspension rheology as a function of particle modulus and effective phase volume	165
M. Takigami, N. Nagasawa, A. Hiroki, M. Taguchi and S. Takigami	Preparation of stable CMC - acid gel	175
D.A. Garrec and I.T. Norton	On the lubricating behaviour of kappa carrageenan gelled networks	183
H. Kastner, U. Einhorn-Stoll and B. Senge	New parameters for the examination of the pectin gelation process	191
U. Einhorn-Stoll, H. Kastner and B. Senge	Comparison of molecular parameters: material properties and gelling behaviour of commercial citrus pectins	199
Chapter 4	**Emulsions**	
V.J. Morris, A.Gromer, A.R. Kirby, R. Penfold and A.P. Gunning	Atomic force microscopy and force spectroscopy studies of the emulsifying action of sugar beet pectin	209
N.C. Woodward, C. Fernandez Fraguas, P.J. Wilde and V.J. Morris	Enhancing protein-protein interactions at the emulsion interface to resist displacement by biosurfactants	219
N. Calero, J. Munoz and A. Guerro	Effect of pH on O/W emulsions formulated with potato protein and chitosan	227
Chapter 5	**Fibres and films**	
M.H. Norziah, H.L Farm, A.B. Khairulnizam and K. Sudesh	Electrospinning of fish gelatine from black tilapia *(Oreschromismossambicus)* skins	237

R. Hans Tromp, C. Vink and A. Stijnman	Encapsulation by electrospinning of live bacteria used in the food industry	247
M.H. Norziah and J.L. Wong	Effects of sucrose palmitate on physical and mechanical properties of sago starch – gelatine edible films	257
M.A. Amiza and Y.P. Wong	Effect of palm oil incorporation on the properties of biodegradable cobia (*Rachycentroncanadum*) skin gelatin	269
M. Marudova,T. Yovcheva, S. Zhelyazkov, G. Zsivanovits and A. Viraneva	Controlled release of water soluble vitamins from chitosan films prepared in electric field	279
A.Romero, T. Werwijlen, A. Guerrero and J. Vermant	Interfacial behaviour of chitosan and crayfish protein isolate mixed films	287
L. Zarate-Ramirez, A. Romero, C. Bengoechea, P. Partel and A. Guerrero	Influence of the addition of polysaccharide to protein-based biodegradable polymeric materials processed by a thermomechanical procedure	295
Chapter 6	**Microstructure and texture**	
M. Peleg and M.G. Corradini	Potential uses of hydrocolloids and gums in food texture studies	305
P. Lopez-Sanchez, L. Bialek, R. Farr, S. Schumm and M. Langton	Structuring properties of plant particles as a result of physical treatments	317
K. Kohyama, F. Hayakawa and T. Sasaki	Relations between mastication variables, mechanical properties and sensory scores of cheese samples with various textures	331
L. Knarreborg and H.L. Nielsen	The texture of low fat yoghurt predicted from its content of starch, pectin and gelatine	339
Chapter 7	**Food applications**	
R. Abson , B. Wolf, S.Hill and T. Foster	Contributions of the particulates and soluble materials to the viscosity behaviour of tomato puree	351
M. Majzoobi , B. Layegh and A. Farahnaky	Inclusion of oat flour in the formulation of regular salted dried noodles and its effects on dough and noodle properties	359
S. Martinez-Cervera, T.Sanz and A. Salvador	Functionality of polyols as sucrose replacers in Spanish muffins	369
S. Martinez-Cervera, E. de la Hera, T. Sanz, M. Gomez and A. Salvador	Replacing sucrose in Spanish muffins with erythritol and xanthan gum	375

L. Laguna, M.J. Hernandez, A. Salvador and T. Sanz	Linear viscoelastic properties of short dough enriched with resistant starch	385
Chapter 8	**Health - related aspects**	
G.H. Zhao, N. Kapur and J.T. Guthrie	The application of cellulose ethers as rheological modifiers in a controlled droplet delivery system	391
Z.K. Muhidinov, Kh. I. Teshaev, G.F. Kasimova, A.S. Nasridinov and L.S. Liu	Pectin – zein hydrogels for the delivery of drugs and nutrients	401
T. Funami, S. Ishihara and M. Nakauma	Texture design of dysphagia foods using polysaccharides	407
V.J. Morris, R.J.M. Bongaerts and A.P. Gunning	Probing specific interactions between pectin-derived galacto-oligosaccharides and the mammalian lectingalectin 3 – testing a model for pectin bioactivity	415
R. Adden, K. Anderson, W. Calame, S. Clegg, B. Hubner-Keese, M. Knarr, R. Re and M. Wickham	Satiety induction with novel food grade methyl cellulose	423
Subject index		425

Market overview

OVERVIEW OF THE FOOD HYDROCOLLOIDS MARKET

Dennis Seisun

IMR International, San Diego CA, USA
[www.hydrocolloids.com; email dseisun:hydrocolloid.com]

This paper focuses on the various definitions and classifications of hydrocolloids and presents a global overview of hydrocolloids from both a numerical and perceptual point of view. Hydrocolloids for all the studying and analyzing conducted on them, remain a nebulous subject. Ask ten scientists for an exact definition and one is likely to receive ten somewhat different answers. In general, hydrocolloids 'work with water', they thicken, suspend, gel or stabilize a solution. As colloids they fall between a true solution and solid suspension. A range of novel, nutraceutical properties are being discovered and established for hydrocolloids. A key property of some is that of providing fiber, both soluble and/or insoluble.

Hydrocolloids are used in three key segments, industrial, food and oilfield applications. The relative importance of each segment in terms of hydrocolloid value and volume is given in the table below:

Hydrocolloid Market All Applications

Application	'000 TPA	$ Million	% Vol	% Val
Industrial	5,120	6,680	54	42
Food	1,890	5,010	20	32
Oilfield	2,420	4,090	26	26
Total	9,430	15,780	100	100

Source :IMR International
Note : Excluding China

The individual hydrocolloids in food can be roughly divided into three major segments, based on overall value. Starches and gelatin are the giant category with over $1 billion in sales each. The second group includes five hydrocolloids with sales between $200-700 million and the remainder are those with sales less than $200 million as indicated in the table below:

Food Hydrocolloid Market 2010

Hydrocolloid	$ Million	% Total
Starches	1,275	25
Gelatin	1,030	21
Pectin	658	13
Carrageenan	561	11
Xanthan	277	6
Alginates	272	5
Agar	219	4
CMC	169	3
Arabic	147	3
Guar	94	2
MCC	91	2
LBG	85	2
MC/HPMC	80	2
Other	51	1
Total	**5,008**	**100**

Source : IMR International
Note : Excluding China

Unfortunately, data for Chinese production and consumption of hydrocolloids is sporadic and unreliable. The above estimates therefore, do not include China. There is no doubt, however, that China represents a significant additional volume. In the case of CMC for food applications in China, reliable data has been obtained. Chinese consumption of food grade CMC is thought to be twice the volume of CMC consumption in all other parts of the world. Chinese food CMC specifications require a purity level of 95% whereas other parts of the world specify purity above 99%. IMR is gathering estimates of Chinese hydrocolloid consumption as an on-going effort. Results of these efforts will be published in future issues of *The Quarterly Review of Food Hydrocolloids* as they are obtained.

Prices of hydrocolloids vary widely, with the costliest, pure gellan gum (approx $60.00/kg), being as much as 60 times the price of native starch ($1.00/kg). Even within a hydrocolloid category, there is a wide range of different grades, eg gum arabic of the emulsifying *Acacia senegal* grade may cost double the price of non-emulsifying *Acacia seyal* grade. Despite these variations, IMR has established a rough average for each hydrocolloid category as indicated in the below table:

Hydrocolloid Prices - 2010 Average

Hydrocolloid	$/Lb	$/Kg
Starches	0.42	0.92
Gelatin	2.60	5.72
Pectin	6.50	14.30
Carrageenan	5.60	12.32
Xanthan	2.35	5.17
Agar	9.30	20.46
Arabic	1.65	3.63
CMC	2.40	5.28
LBG	3.90	8.58
Alginates	9.15	20.13
Guar	0.83	1.83
MCC	3.85	8.47
MC/HPMC	5.35	11.77

Source : IMR International Quarterly Hydrocolloid Review

The above are averages based on the four quarterly prices published in IMR's *Quarterly Review of Food Hydrocolloids* for 2010. There is much volatility in prices and readers are encouraged to obtain up to date price information, from IMR or elsewhere. Guar gum for example, has more than doubled in price since the above table was produced. Xanthan prices vary widely depending on country of origin. Most prices are directly linked to upstream raw material costs which have increased significantly for most hydrocolloids in 2010-2011.

The rapidly changing world of commerce has wrought many changes on the hydrocolloid industry. Channels of distribution have changed as have relationships between producers themselves with strategic alliances being made. Distributors are more widely used even by the large multinational hydrocolloid producers. Technical service and formulation development are also being offered by some distributors in an effort to differentiate themselves. Overall channels, however, remain the same as indicated in the below diagram established decades ago by IMR:

Hydrocolloid Distribution Channels

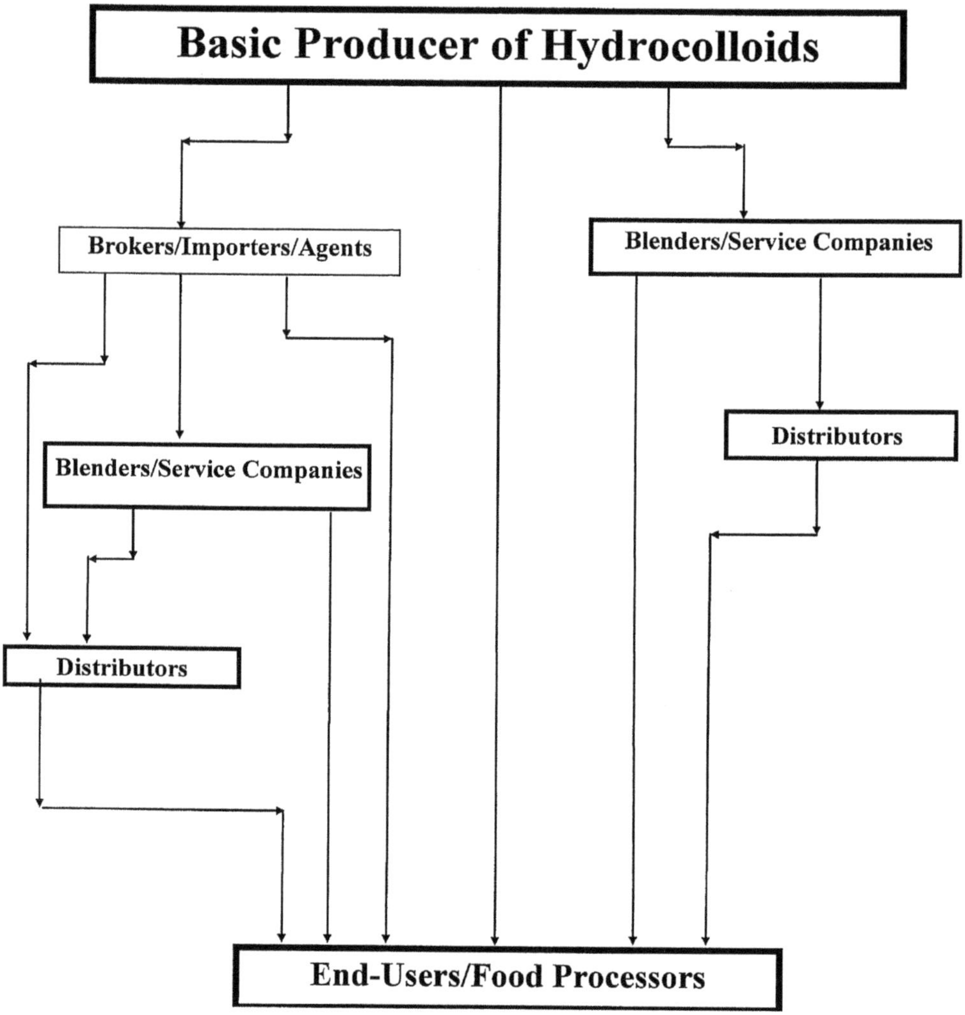

Crystal ball gazing and forecasting in any industry is an inexact science if a science at all. Forecasting the future of hydrocolloids is no easier or more accurate. Nevertheless with a knowledge of properties, applications, consumer trends and raw material tendencies an 'informed guess' can be made. The table below is presented on a best effort basis:

Forecast Hydrocolloid Market

Hydrocolloid	$ MM		AGR
	2010	2015	
Starches	1,275	1,422	2.2
Gelatin	1,030	1,126	1.8
Pectin	658	856	5.4
Carrageenan	561	653	3.1
Xanthan	277	351	4.9
Alginates	272	312	2.8
Agar	219	250	2.7
CMC	169	193	2.7
Arabic	147	169	2.8
Guar	94	108	2.9
MCC	91	112	4.2
LBG	85	98	2.9
MC/HPMC	80	108	6.2
Other	51	60	3.0
Total	**5,008**	**5,818**	**2.4**

Note Forecast based on constant price. Excludes China
Source : IMR International

Pectin, xanthan and MC/HPMC are expected to be the most rapidly growing hydrocolloids each for their specific reasons. Pectin is the most label and consumer friendly of all hydrocolloids. Consumer concerns and label preferences are paramount in hydrocolloid selection and likely to remain so in the foreseeable future. Pectin is a 'no problem' hydrocolloid on a food label. Xanthan gum has become the 'ubiquitous' hydrocolloid which is appearing on an increasing number of labels worldwide. Xanthan in many cases, has gained the reputation of 'try this hydrocolloid first' in a formulator's product development efforts. If xanthan works it is likely to be used. Only if xanthan fails, will efforts then focus on alternative hydrocolloids. The group of cellulosics in the MC/HPMC category are rapidly growing for two primary reasons, a) they are currently a small volume item and high growth more easily achieved and b) formulators are discovering properties well beyond the thermal gelation for which they were initially known.

Of course the above are generalizations each subject to their own limitations. Also, what is applicable now may not apply in the future. Gelatin for example was, along with pectin, amongst the most consumer friendly items on a food label. And then, mad cow disease appeared, the number of vegetarians grew exponentially and kosher/halal requirements became more widespread. Change is the only steady factor.

A large part of the future of hydrocolloids will remain in thickening, stabilizing, gelling and suspending. A growing value however is found in nutritional and nutraceutical aspects which had never before been considered. Hydrocolloids as sources of soluble and/or insoluble fiber is only one such aspect. Satiety control and even nutraceutical benefits are all in the pipeline of evaluation and certification.

Isolation, characterisation and modification

RAMAN AND FT-IR MICROSCOPY OF NATIVE AND HIGH-AMYLOSE MAIZE STARCH *IN SITU*

N. Wellner[1], M.L. Parker[1], D.M.R Georget[2] and V.J. Morris[1]

[1]Institute of Food Research, Norwich Research Park, NR4 7UA, UK
[2]School of Pharmacy, University of East Anglia, Norwich Research Park, NR4 7TJ, UK

1 INTRODUCTION

Knowledge of the structure of starch granules has advanced considerably over the last 30 years[1]. For wild-type starches the granule structures are homogeneous and the properties of individual granules can be deduced from studies on populations of isolated granules. Natural or induced mutations in the biosynthesis of starch can be used to generate so-called high-amylose or low-amylose starches. Despite this nomenclature it is clear that such mutations actually lead to significant modifications of the starch granule structure rather than simply altering the levels of amylose and amylopectin which can be isolated from the granules. Even in the case of isogenic mutants such as *amylose extender ae* (high-amylose) maize, studies on individual granules reveal substantial heterogeneity in the structure for isolated populations of starch granules[2-4]. In order to understand the effects of mutations in biosynthesis on granule structure it is necessary to map the chemical and physical structures of individual granules. Based on a previously described method[5-6] for *in situ* imaging of starch in dry seeds, it is shown in this article that Raman microscopy can be used for mapping the chemical and physical structures of starch granules within the dry seeds of wildtype and *ae* maize, an isogenic mutant resulting from the suppression of starch branching enzyme IIb (SBEIIb) during biosynthesis.

2 MATERIALS AND METHODS

2.1 Maize kernel samples

Kernels of wild-type maize (cultivated Oh 43) and the high-amylose *amylose extender* (ae) mutant in an Oh 43 background were provided by Dr K. Denyer (John Innes Centre, Norwich, UK). Thick (2 mm) transverse slices were cut from the wild-type and mutant maize samples approximately 2 mm from the crown end of the kernels using a fine-toothed saw. These slices were cut in half across the long axis to obtain blocks including the pericarp, aleurone and outer endosperm. The blocks were glued onto aluminium scanning electron microscopy stubs with the transverse orientation uppermost. An ultra-microtome (Reichert-Jung Ultracut E) and a glass knife were used to polish the face of these blocks

(Figure 1a). Sections of endosperm tissue (1.5-2.0 μm) were cut from these polished blocks with a glass knife (Figure 1b).

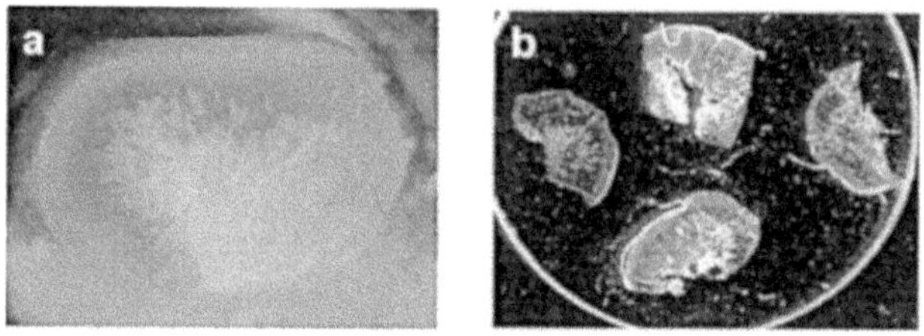

Figure 1 *Polished face (a) and (b)1-2 μm thick endosperm sections*

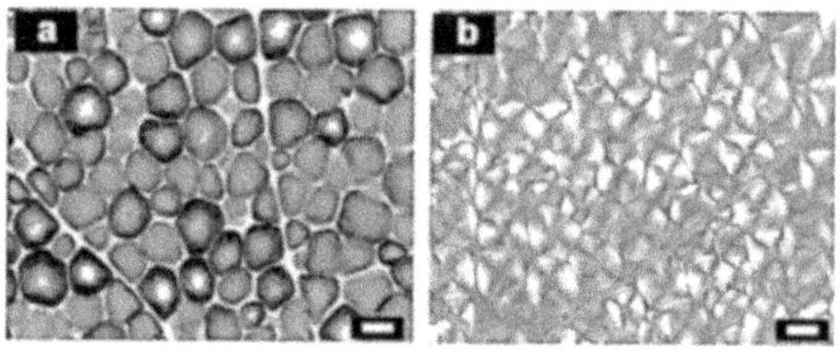

Figure 2 *Representative optical images of wild-type endosperm sections. (a) KI/I_2 staining and (b) polarised light microscopy. Scale bar10 μm*

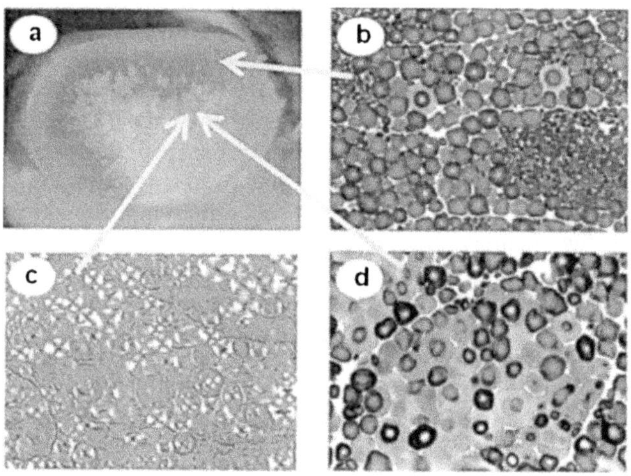

Figure 3 *Representative images of KI/I_2 stained ae maize sections from the (b) outer and (d) inner endosperm and (c) polarised light microscopy image of the inner endosperm*

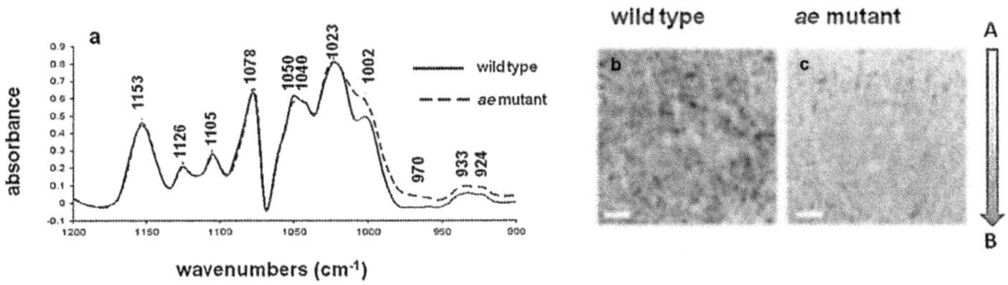

Figure 4 *FT-IR data. (a) Average FT-IR spectra for wild type and ae maize showing the differences in the starch bands. Maps of crystal type for (b) wild type and (c) ae maize. Scale bar is 100 μm.*

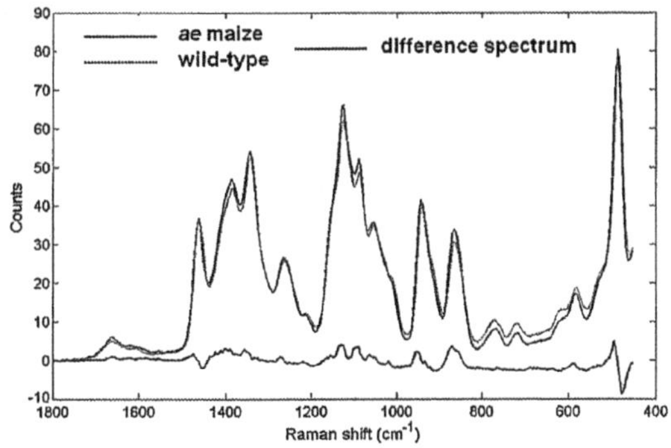

Figure 5 *Averaged Raman spectra for wild-type and ae maize and difference spectra*

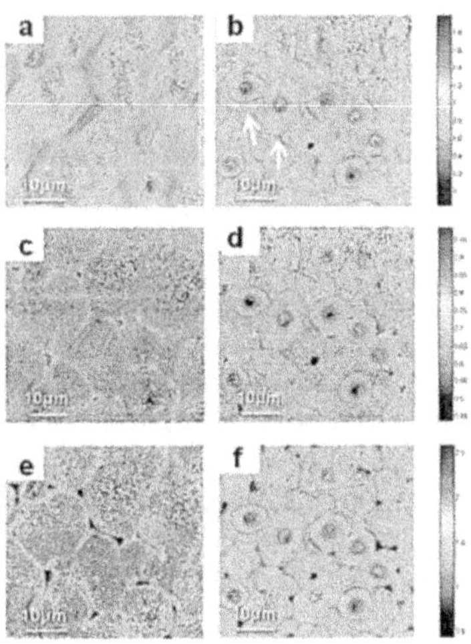

Figure 6 *Raman maps of inner endosperm sections showing (1) crystal orientation in (a) wild type and (b)ae mutant(2) the linear/branched ratio in (c) wild type and (d) ae mutant, (3) crystallinity in (e) wild type and (f) ae mutant. Colouration in the boundary regions between granules is an artefact due to the absence of starch in these regions. Scale bars 10 μm.*

2.2 Conventional light microscopy

For polarised light microscopy, sections were mounted in water and examined using an Olympus BX60 (Olympus, Japan). Sections stained with 0.2% iodine in 2% potassium iodide diluted 2:1 with distilled water were viewed by bright field optics.

Sections for FT-IR and Raman microscopy/spectroscopy were floated onto a small drop of distilled water or 30% ethanol on infra-red transparent BaF_2 discs (Figure 1b). Any excess liquid was removed with a strip of filter paper and the samples were allowed to dry under ambient conditions.

2.3 Raman microscopy

Raman images were obtained using an upgraded WITec CRM200 (Witec GmbH, Ulm, Germany) confocal microscope using a frequency doubled Nd YAG laser (532 nm) and a 100x objective. Raman data were collected using a 600 nm grating in the spectral range -30 to 3980 cm^{-1}. A 50 x 50 mm^2 areal region of the sample was scanned in 0.39 mm steps, producing a map containing 128 x 128 pixels. An initial scan (100 ms integration time) was used to determine the quality of the sample in the chosen area before collecting a full scan at 1 s/pixel in order to obtain spectra with a reasonable signal to noise ratio. The Raman maps were analysed with Matlab 7.9 (The MathWorks, Inc, USA). Spectra were baseline corrected by subtraction of the value at 1800 cm^{-1} as a linear offset: the background was approximately linear in the starch-rich areas but contained broad fluorescence bands in the carotene-rich areas. Areas of bands were calculated by integrating the three values closest to the peak maximum, assuming a 12 cm^{-1} band width, in order to avoid overlap of neighbouring bands. Several sections from different maize kernels were examined, and images were recorded from different areas in each section. All gave consistent results, although the extent of the observed differences varied because of the variability of kernels and locations. The data shown are typical of the observed variations in the wild-type and *ae* maize samples.

2.4 FT-IR microscopy

FT-IR data were acquired with a Varian UMA600 Microscope interfaced to a FTS6000 FT-IR spectrometer (Digilab/Varian). Average spectra were collected with a MCT detector (128 scans at 2 cm^{-1} resolution) from a 1 mm diameter spot. Microscopic FT-IR images of the sections were acquired with a 128 x 128 pixel FPA detector. The image area was 0.64 x 0.64 mm^2. Using a 15x objective, the nominal spatial resolution was 5 x 5 μm per pixel, however, the effective spatial resolution was diffraction limited by the IR wavelength (2-10 μm). 128 scans at 8 cm^{-1} resolution were co-added in transmission and referenced against a background from an empty area of the BaF_2 disk FT-IR maps were imported into ENVI 4.6 (ITT Visual Information Solutions). The spectra were baseline corrected in the region 1800-900 cm^{-1} with the inbuilt continuum removal function, and then classified according to the shapes of the carbohydrate band regions.

3 RESULTS AND DISCUSSION

Optical microscopy reveals marked changes in the structure of the starch granules in the wild-type and *ae* mutants (Figures 2 & 3). For the wild type the granules are similar in shape and size, and stain uniformly blue with KI/I_2 in cells across the endosperm (Figure

2a). Images obtained using polarising optics reveal characteristic Maltese cross patterns indicating radial orientation of the crystalline regions within the granules (Figure 2b): at higher magnification the granules reveal growth ring structures indicating a periodic radial variation in the crystalline/amorphous ratio within the granules. The granules structures are homogeneous and the population of granules within the seeds is homogeneous.

For the *ae* mutant, the optical microscopy reveals gross heterogeneity of the starch granule structure within individual granules, between granules within cells, and between cells across the starchy endosperm (Figure 3). Figures 3b and d show representative regions of the KI/I_2 stained sections from the outer and inner regions of the starchy endosperm. The spheroidal starch granules tend to be smaller than the wild-type granules, with evidence for the presence of elongated and dimpled granules, reported in previous studies on isolated *ae* mutant maize starch[2-4]. In the outer endosperm regions, the granules tend to stain blue in KI/I_2 with occasional granules showing central regions which stain blue and outer regions which stain pink: the central hilum region enlarges when the starch granule swells in the aqueous mounting conditions and remains unstained (Figure 3b). The heterogeneity of the starch granule structure can vary markedly between neighbouring cells and, in general, becomes more heterogeneous as sampling progresses from the outer to the inner endosperm. In the inner endosperm, there is a wider variation between blue and pink stained granules with a greater propensity for pink stained regions at the periphery of granules (Figure 3d). The difference between the pink and blue stained granules, or regions within granules, is well illustrated if the granules are exposed to KI/I_2 or iodine vapour. However, prolonged exposure results in the pink regions eventually staining blue: this suggests that the difference in colour is due to differences in the structure which influences the rate of diffusion of the iodine into the structure, rather than the length of the ordered helical regions within the structure. Polarised light microscopy of regions within the inner endosperm also emphasise the structural heterogeneity (Figure 3c). These images reveal some granules in which the crystal structures are radially aligned and homogeneous, some granules that show no radial alignment, and further granules with various intermediate structures showing one or more radially aligned regions within the granules, and a peripheral region with no radial alignment (Figure 3c). These images show that a single mutation in an isogenic line does not result in a simple change in the ratio of amylose and amylopectin within the granules, but rather results in a complete and variable heterogeneous population of granular structures. FT-IR and Raman chemical mapping of the starch structures within seeds provides a basis for defining the chemical and physical structures within and across the endosperm, within and across individual cells and within individual granules.

The FT-IR absorption spectra provide information on the composition and structure of the samples which can be mapped in the FT-IR images for the wild-type and *ae* mutant maize samples (Figure 4). The FT-IR spectra of maize showed strong absorption in the protein and carbohydrate regions. The carbohydrate region is dominated by the starch content with only a minor contribution from the non-starch cell wall components. Fourier deconvolution of the spectra resolved bands at 1153, 1126, 1105, 1078, 1050, 1044, 1023, 1002, 933 and 924 cm^{-1} in the carbohydrate region. Comparison of the FT-IR spectra for the wild-type and *ae* mutant show similar protein bands at 1650 cm^{-1} but differ for the starch bands, particularly in the absorption region sensitive to the crystal structure of the starch (Figure 4a). The *ae* maize spectra showed a large increase in the 1002 cm^{-1} band and also some broadening, which resulted in greater overlap with the main 1023 cm^{-1} band, and a broader wing around 970 cm^{-1}. In addition, the band at 1050 cm^{-1} was slightly smaller than in the equivalent band for the wild-type samples. The band at 1022 cm^{-1} has been assigned to the amorphous starch phase, whereas lintnerised starches and granular starches are characterised by a band at 1000 cm^{-1}. However, the ratio of these two bands is also highly influenced by water content and, in this case, indicates a change in the short-range order of amylosic helices within the crystal structure[7], because the 1047 cm^{-1} band, which has been linked to starch crystallinity[8], is slightly lower in the *ae* maize samples. The differences in the two bands at 1050 and 1002 cm^{-1} in the *ae* maize when compared to the data for the *ae* mutant indicates that the starch structure in the maize mutant was significantly different from that in the wild type. These data are consistent with the well documented observations from x-ray diffraction data that the mutation in the biosynthesis leads to a progressive transition from A- to B-type crystals.

The spectral data can be used to generate low-resolution maps of the crystal structure within the sections. The images shown in Figure 4b-c, taken from regions within the inner endosperm where the optical images show marked variability, are colour coded to indicate a transition from A-type crystals (blue) to B-type crystals (red). Such chemical imaging revealed differences in the spatial distribution of the crystal types across the starchy endosperm for the *ae* mutant. Whereas in the wild type, the starch structure was homogeneous across the endosperm, there were local spatial variations observed within the mutant sections with regions nearer the inner endosperm showing a higher level of B-type crystal structure.

The Raman microscopy provides a basis for higher resolution mapping of the structural changes within the starch granules for the mutant starches. Figure 5 shows the averaged Raman spectra for the wild-type and *ae* mutant starches. The bands in the Raman spectra have been assigned[6] and the difference spectrum (Figure 5, blue line) indicates differences between the chemical and physical structures of the granules due to the mutation in the biosynthesis: the differences can be used to map changes in the physical and chemical properties of the granules (Figure 6).

The normal mode of the 865 cm^{-1} band, which is assigned to ν_s C-O-C/ring breathing mode, appears to result from C-O-C groups in the glucose rings aligned along the axes of starch double helices. The laser source used in the Raman microscope is vertically polarised and the ratio of the 865/042 bands shows dichroic effects which can used to map orientation of ordered (crystalline) structures, akin to the Maltese cross patterns obtained by conventional polarised light microscopy (Figures 2b, 3c,). For the wild-type maize samples, the granule structures are homogeneous and the crystalline structures are radially orientated (Figure 6a). For the *ae* maize, Raman maps constructed for samples in the inner endosperm region (Figure 6b) show heterogeneity equivalent to that observed by polarised light microscopy and iodine staining (Figure 3c and d). Thus these Raman maps allow one to locate regions within and between granules which are associated with the blue and pink stained regions. Arrows in Figure 6b indicate examples of granules containing blue central regions and pink-staining extremities. It is then possible to use other bands in the Raman spectra to map chemical and physical differences in the structures within these regions.

The *ae* mutation results from the suppression of the activity of the starch branching enzyme IIb (SBEIIb) and should result in a lower level of branching in the mutant starches. The band at 942 cm^{-1} is sensitive to the level of branching of the starch polysaccharides. It has been shown by Liu and co-workers[9] using 2D Raman measurements that this band can be de-convoluted into two components, a band at 954 cm^{-1} due to linear α(1-4) amylosic chain linkages and a band at 938 cm^{-1} due to α(1-4,6) linkages. The presence of a positive band in the difference spectra (Figure 5) at 952 cm^{-1} indicates that the level of branching is decreased for the *ae* mutant with respect to the wild type sample. The ratio of the bands at 952 cm^{-1} and 942 cm^{-1} can be used to map the ratio of linear/branched residues (Figure 6c-d), with the associated colour table (blue-green-yellow-red) indicating an increase in the linear to branched ratio. The ratio is constant for granules in the wild-type samples (Figure 6c). However, for the *ae* mutant the maps show heterogeneity across the endosperm between different cells and within individual granules (Figure 6d). For the granules shown in Figure 6d, the linear/branch ratio is lower in the central (blue staining) regions than in the outer (pink staining) periphery. The band at 486 cm^{-1} normalised with respect to the band at 942 cm^{-1} can be used to map local crystallinity (Figure 6e-f): decreasing crystallinity is indicated in the colour table by the transition red-yellow-green-blue. For the wild-type samples the level of crystallinity is constant for all the granules (Figure 6e). It can be seen that the level of crystallinity in the *ae* mutant is lower than in the wild type (Figure 6f), but the variation between granules in the inner endosperm is relatively small, although for individual granules the outer pink staining regions appear more crystalline than the central blue staining regions. These studies support the concept of crystalline amylose chains of variable length within *ae* maize granules[10] and suggest they exist in the pink staining regions of the granules.

4 CONCLUSIONS

Comparative studies on wild-type and *ae* maize mutant samples suggest that the mutation induced a heterogeneous population of granules: the level of heterogeneity varies across the starchy endosperm, between individual cells, between granules and within individual granules. The heterogeneity observed within individual seeds appear to be more marked than that reported in the literature for isolated starch granules, suggesting that isolation may select certain sub-populations of the granular structures present in the kernels. Infra-red microscopy provides low-resolution mapping of the type of crystal structure within the seeds. Raman microscopy provides higher-resolution maps of the chemical composition, level of crystallinity and crystal orientation within individual granules. The results are consistent with the *ae* mutations inducing defects in a self-assembly process that accumulate to form increasingly random physical structures in the granules. This disruption could result in crystals in the pink staining regions formed from amylose chains of variable length; accounting for the broadened gelatinisation behaviour of these starches.

ACKNOWLEDGEMENT The research at IFR was funded through the BBSRC core strategic grant to the Institute.

References

1 S. Pérez and E. Bertoft, *Starch-Stärke*, 2010, **62**, 389.
2 P. Chen, L. Yu, L. Chen and X. Li, *Starch-Stärke*, 2009, **56**, 611.
3 A. Evans, N. McNish and D.B. Thompson, *Starch-Stärke*, 2003, **55**, 250.
4 F.R. Katz,S.L. Furesik, F.L. Tenbarge, R.J. Hauber and R.B. Friedman, *Carbohydr. Polym.*, 1993, **21**, 133.
5 M.L.Parker, A.R. Kirby and V.J. Morris, *Food Biophys.*, 2008, **3**, 66.
6 N. Wellner, D.M.R. Georget, M.L. Parker and V.J. Morris, *Starch-Stärke*, 2010, **63**, 128.
7 I. Capron, P. Robert, P. Colonna, M. Brogly and V. Planchot, *Carbohydr. Polym.*, 2007, **68**, 24.

8 J.J.G. vanSoest, H. Tournois, D. deWit and J.F.G. Vliegenthart, *Carbohydr. Res.* 1995, **279**, 201.
9 Y. Liu, D.S. Himmelsbach and F.E. Barton, *Appl. Spectrosco.*, 2004, **58**, 745.
10 H. Jiang, M. Campbell, M. Blanco and J-L. Jane, *Carbohydr. Polym.* 2010, **80**, 1.

Footnote

Copies of this article which includes figures in colour can be obtained directly from the authors

ANALYSIS AND MODIFICATION OF CARRAGEENAN STRUCTURE USING ENZYMES

W. Helbert

Centre National de la Recherche Scientifique; Université Pierre et Marie Curie-Paris 6, Unité Mixte de Recherche 7139 "Marine Plants and Biomolecules", Station Biologique, F-29682 Roscoff Cedex, Bretagne, France

1 INTRODUCTION

Carrageenans represent a class of sulphated galactans that have a wide spectrum of chemical structures. This class of macromolecules, occurring exclusively in the cell walls of red algae, shares a common galactan backbone of alternating 3-linked-β-D-galactopyranoses (G) and 4-linked-α-D-galactopyranoses (D). The disaccharide repetition moieties or carrabiose units are classified according to the number and position of sulphate ester groups (S) and the occurrence of 3,6-anhydro-D-galactose (DA) on the 1,4-linked residue found in gelling carrageenans. For example, the three most widely used carrageenans in industry — kappa- (κ, DA-G4S), iota- (ι, DA2S-G4S) and lambda- (λ, D2S6S-G2S) carrageenans — are differentiated by the presence of one, two and three ester sulphate groups, respectively, on each repeating disaccharide unit (Figure 1). Other derivatives such as methyl (M) or pyruvate (P) groups have been also observed. The number of possible carrabiose structures is therefore very large.

Because carrageenans are the products of a complex biosynthetic pathway, they are not homopolymers, but heteropolymers or hybrid structures composed of sequences of several types of moieties. As a result, to illustrate just one case, the so-called κ-carrageenan defines a family of polysaccharides composed mainly — but not only — of κ-carrabiose units. The most typical hybrid structures of carrageenan are those found in native or unprocessed κ- and ι-carrageenan chains. They usually contain fractions of their biosynthetic precursors named mu- (μ, D6S-G4S) and nu- (ν, D2S6S-G4S) carrageenans, respectively[1,2]. Other carrabiose combinations have also been reported, such as hybrid κ-/ι-carrageenans (DA-G4S/DA2SG4S) in several species of Gigartinales[3,4] and κ–/β–carrageenans (DA-G4S/AG) in *Furcellaria lubricallis*[5].

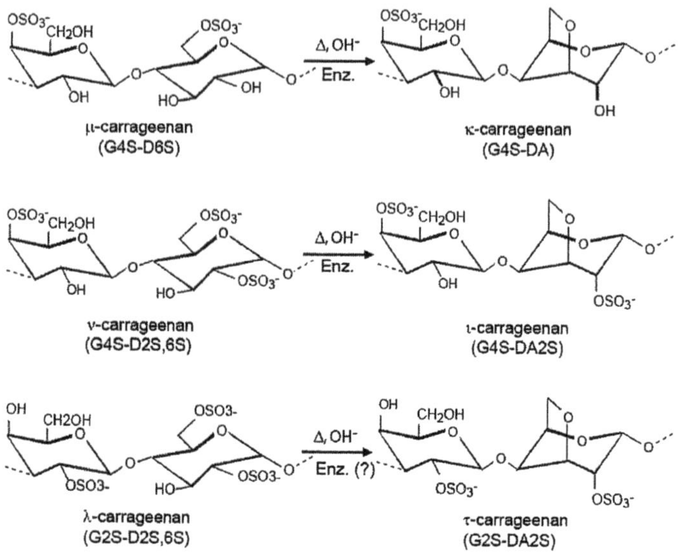

Figure 1: Chemical structure of disaccharide units found in the most widely used carrageenans. The μ- and ν-carrageenans are converted into κ- and ι-carrageenans, respectively, *in vitro* by hot alkaline treatment or *in vivo* by the action the galactose-6-sulphurylases. λ-Carrageenan can also be converted to τ-carrageenan in alkaline solutions; however, no enzyme has yet been identified.

The physico-chemical properties of carrageenans that are widely used in industry are directly linked to their chemical structure. It is well-known that the amount of biosynthetic precursors, i.e. μ- and ν-carrabiose units, occurring in κ- and ι-carrageenan chains respectively, strongly affects the ability of these macromolecules to form gels. Alkaline treatments induce cyclisation of μ- and ν-carrabiose leading to κ- and ι-carrabiose, respectively, and strongly enhance the gel strength of carrageenans. Other hybrid structures of economic interest are κ-/ι-carrageenans, which are characterised by a wide range of κ-carrabiose:ι-carrabiose ratios[6,7].

The first part of this article covers recent progress in analysing carrageenan hybridity. Enzymes can be used to describe the distribution of carrabiose moieties along the polysaccharide chain. Knowledge of the carrabiose composition and distribution in carrageenans allows their physico-chemical properties to be predicted. Current strategies for monitoring the physico-chemical properties of carrageenans involve mixing various carrageenans and/or adding other polysaccharides or additives. Enzymes could therefore constitute an interesting alternative for determining chemical structure. In the second part of this article, advances in the identification and characterisation of various carrageenan-modifying enzymes are presented.

2 STRUCTURE ANALYSIS USING CARRAGEENASES

2.1. A brief introduction to carrageenases

Carrageenases are enzymes that catalyse the hydrolysis of the glycosidic bonds in carrageenans. They have been isolated from marine bacteria and some of them have been produced in heterologous expression systems (i.e. *E. coli*). Three classes of carrageenases have been reported: κ-, ι-, and λ-carrageenases, which specifically cleave the glycosidic linkage between two κ-, ι- and λ-carrabioses, respectively. κ-Carrageenases (EC 3.2.1.83) belong to the GH16 (glycoside hydrolase 16) family, a polyspecific family whose members show at least eight different enzymatic activities, such as, notably, β-agarases[8]. Like other members of the GH16 family, κ-carrageenases hydrolyse β-(1→4) glycosidic linkages with retention of the anomeric configuration[9,10]. The end-products of the *Pseudoalteromonas carrageenovora* κ-carrageenase are neo-κ-carratetraose and neo-κ-carrabiose. The ι-carrageenases from the marine bacteria *Alteromonas fortis* and *Zobellia galactanivorans* have homologous gene sequences and belong to the monospecific GH82 family. They cleave β-(1→4) glycosidic linkages according to a mechanism that inverts the anomeric configuration and leads to the production of neo-ι-carratetraose and neo-ι-carrahexaose as end-products[11]. Recently, a novel ι-carrageenase was observed in the marine bacterium *Microbulbifer thermotolerans* and seems to have a different ι-carrageenan recognition site from that known for GH82 ι-carrageenases[12]. λ-Carrageenases have been isolated from *P. carrageenovora*[13] and from a deep-sea *Pseudoalteromonas* strain CL19[14]. These enzymes have been purified, cloned and sequenced. Sequences of two of these λ-carrageenases are highly homologous and are likely the two first members of a new glycoside hydrolase family unrelated to the GH16 (κ-carrageenase) or GH82 (ι-carrageenase) families. The enzyme proceeds according to an endolytic mode of action and inverts the anomeric configuration[13]. The main degradation products have been purified, characterised and identified as neo-λ-carratetraose and neo-λ-carrahexaose.

2.2 Active site mapping of carrageenases

Using enzymes to analyse carrageenan structures takes advantage of the enzyme's cleavage-site specificity. The active sites of carrageenases can be mapped through detailed analyses of carrageenase end-products. Accordingly, the active sites of κ-carrageenase have been mapped based on a structural analysis of end-products obtained after degradation of κ-/ι-, κ-/μ-, κ-/β-carrageenans[5,15,16]. In the case *of A. fortis* ι-carrageenase, κ-/ι- and ι-/ν-carrageenans were used [as a substrate?][15,17]. Only *Gigartina skottsbergi* λ-carrageenans have been used as substrate of λ-carrageenase[13]. All the resulting series of hybrid oligosaccharides have been purified and fully analysed using NMR and mass spectrometry. Based on oligosaccharide structure, active site topology was deduced and exact recognition specificities could be determined for these carrageenases. These results have been confirmed by protein crystallography analysis.

The crystallographic structure of *P. carrageenovora* κ-carrageenase revealed a six sub-site topology for the active site. The degradation products indicate that positions -4 and +2 of the active site are able to accommodate residues other than κ-carrabiose, conferring some flexibility in substrate recognition. For example, the degradation of κ-/β-carrageenan resulted in β-carrabiose at the non-reducing end. *A. fortis* ι-carrageenase has also been crystallised and its active site is made of eight sub-sites. In contrast to κ-carrageenase, its specificity is very conservative and all eight sub-sites must be occupied by the same species of substrate for degradation to occur. Therefore, stretches of four contiguous ι-carrabioses must be present to observe one cleavage and the production of one tetrasaccharide requires six contiguous ι-carrabioses (two cleavage sites). λ-Carrageenase has not been crystallised and the active site has only been mapped according to the degradation products of purified oligosaccharides. The active site is composed of at least

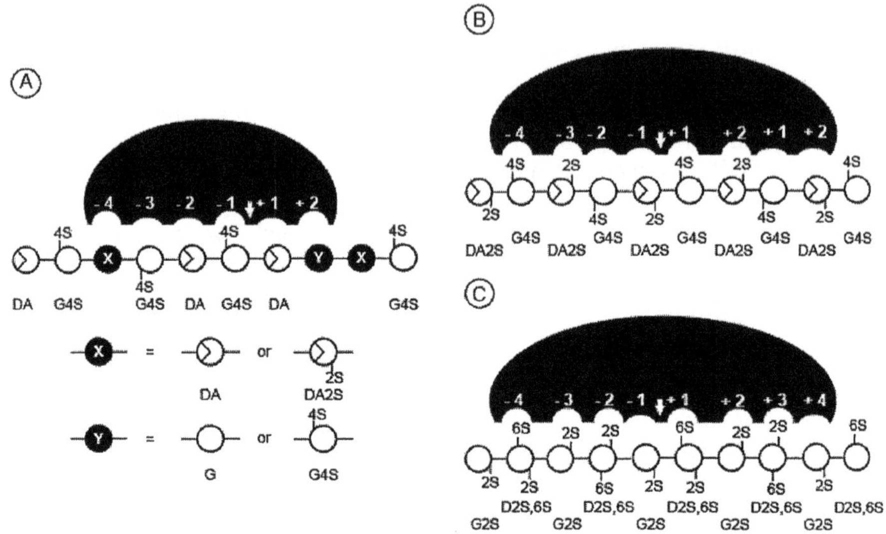

Figure 2: Active site organisation in carrageenases: **A**) *P. carrageenovora* κ-carrageenase; **B**) *A. fortis* ι-carrageenase and **C**) *P. carrageenovora* λ-carrageenase. White arrows indicate the site of cleavage.

eight sub-sites but cleavage can occur if six sub-sites are filled, as deduced from the degradation of neo-λ-carrahexose to di- and tetra-saccharides.

2.3 Example: Distribution of μ-carrabiose in hybrid κ-/μ-carrageenans

Hybrid κ-/μ-carrageenan water-extracted from *Kappaphycus alvarezzi* has been incubated with *P. carrageenovora* κ-carrageenase[17]. The degradation products are composed of the standard neo-κ-carrabiose (DP2 and DP4), as well as five abundant hybrid κ-/μ-carrageenan oligosaccharides. These hybrid oligosaccharides have been purified and their structures determined by NMR and mass spectrometry. Since the enzyme-resistant fraction is very small, these five oligosaccharides are representative of the distribution of μ-carrabiose moieties along the carrageenan chain. Three main distributions of μ-carrabiose have been found: -μ-, -μ-μ- and -μ-μ-μ-. Interestingly, the chemical shifts of the anomeric proton and carbon are strongly influenced by the identity of the adjacent carrabiose, which explains the asymmetrical D6S-H1 NMR signal. Although μ-carrabiose chemical shifts are usually attributed to blocks of μ-carrabiose as expected in ideal μ-carrageenan, the study by Jouanneau et al.[16] has demonstrated that they are in fact due to μ-carrabiose units positioned between κ-carrabiose units.

In industry, alkaline conversion of μ-carrabiose into κ-carrabiose is routinely carried out to improve carrageenan gel strength. By describing the distribution of μ-carrabiose, the reactivity of μ-carrabiose to alkaline treatments can be investigated. The kinetics of conversion followed by chromatography and NMR have revealed that the rate of conversion of μ-carrabioses is independent of neighbouring carrabioses. This means that one μ-carrabiose positioned between two μ-carrabioses is converted at the same rate as a μ-carrabiose located between two κ-carrabioses.

3 MONITORING THE CHEMICAL STRUCTURE OF CARRAGEENANS

3.1 Enzymes involved in biosynthesis

The biosynthetic pathways of carrageenans are currently only hypothetical. However, based on the structure of carrabiose moieties that co-occur in polysaccharide chains, the proposed sequences of catalytic events accounting for the chemical structure of carrageenans usually assume that three classes of enzymes are involved: galactosyl-transferases, sulphotransferases and galactose-6-sulphurylases, also referred to as "sulphohydrolases" in the case of partially purified protein extracts[18]. The α- and β-galactosyl transferases catalyse the polymerisation of the galactose backbone. Sulphotransferases decorate the neutral galactose chain with sulphate side groups. These reactions have been shown to take place in the Golgi apparatus, but none of these enzymes have been isolated *in vitro* [19,20].

Only the final step of biosynthesis has been unambiguously demonstrated by characterising the enzymatic activity of galactose-6-sulphurylases which catalyse the formation of 3,6-anhydro rings in agarose[21,22]. The formation of κ- and ι-carrageenan anhydro rings has also been observed after incubation of μ- and ν-carrageenans, respectively, with red algal protein extracts that had been named "sulphohydrolases"[18,23]. In contrast to sulphurylases, sulpho- and galactosyl-transferases require phosphoadenylyl sulphate (PAPS) or adenosine triphosphate (ATP) respectively, as their source of energy to catalyse the reaction. Therefore, sulphurylases, which make anhydro rings without costly co-factors, are better candidates for modulating carrageenan properties using biotechnology.

Two galactose-6-sulphurylases catalysing the conversion of ν- to ι-carrageenan have been purified to electrophoretic homogeneity[24]. These two enzymes have similar biochemical (pH and temperature optimum) and kinetic (Km, Vm, kcat) parameters. However, they have different molecular weights: 65 kD (sulphurylase I) and 32 kD (sulphurylase II), and their amino acid sequences show no homology[25]. The conversion of ν- to ι-carrageenan by sulphurylase is accompanied by an increase in the viscosity of the medium followed by gelation of the polysaccharide. However, at equal concentrations of

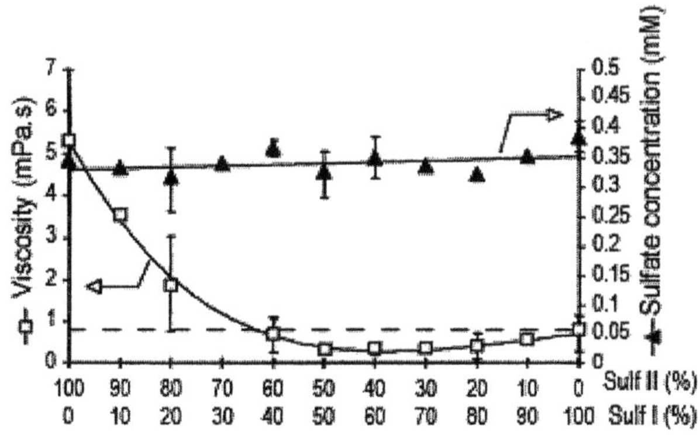

Figure 3: Viscosity (□) and concentration of released sulphate (▲) after 12 hours of incubation of ι-/ν-carrageenan with increasing sulphurylase I:sulphurylase II ratios. Bars correspond to the standard deviation determined from three independent experiments.

released sulphate, viscosities are not the same when the conversion is catalysed by sulphurylase I or II. This is clearly illustrated in Figure 3 which shows that sulphurylase II induces greater viscosity at lower concentrations of sulphate than sulphurylase I.

3.2 Enzymes involved in biodegradation

The biodegradation of carrageenan involves carrageenases, which are very useful for the structural analysis of carrageenans. However, because their activity leads to a decrease in carrageenan molecular mass and, as a consequence, to the loss of their useful functional properties, carrageenases do not seem to be interesting tools for controlling the structure and the properties of carrageenans. A second class of enzymes are sulphatases, which catalyse the removal of sulphate. Weigl and Yaphe[26] and then McLean and Williamson[27] have demonstrated the occurrence of enzymes capable of desulphating the end-products of k-carrageenases: neo-κ-carratetraose and neo-κ-carrabiose. NMR analysis of the end-products have shown that the sulphatases are exo-enzymes acting at the non-reducing end of oligosaccharides, releasing a neutral disaccharide, which is further degraded by a specific glycoside hydrolase[27,28]. These observations suggest a sequential carrageenan degradation scheme involving sulphatases that act on low molecular weight oligo-carrageenans.

Literature dealing with polysaccharide-acting sulphatases involves mainly glycoaminoglycans (*i.e.* heparan, dermatan, chondroitin). It has been shown that these enzymes have very specific desulphatation sites and both exo- and endo-acting sulphatases have been encountered. We have therefore screened extracts prepared from well-known carrageenolytic bacteria (i.e. *P. carrageenovora*) for the possible occurrence the sulphatase activities. Preliminary results demonstrate that bacteria can carry out the desulphatation of undegraded carrageenans in its polymeric form. Consequently, as found in glycoamoniglycans, both endo- and exo-acting carrageenan sulphatases do indeed occur *in vivo*.

4 CONCLUSION

As found in glycosaminoglycans, pectins and alginates, the use of enzymes has been instrumental in deciphering the structure of complex polysaccharide structures. However, in-depth analysis of these complex polysaccharides requires several enzymes having different specificities regarding the structure and length of the recognised moieties. Similarly, improvements in the description of carrageenan structure require the discovery of new carrageenases. For example, novel κ- and ι-carrageenases would greatly enhance understanding of hybrid κ-/ι-carrageenans. Novel enzymes would also be very helpful to better describe the amount and size of carrabiose blocks, which would allow the introduction of the concept of "blockiness", as has been developed for pectins.

Structural analysis should lead to the prediction of functional properties and thereby, at the very least, facilitate polysaccharide formulation for industry. Using enzymes to modify polysaccharide structure is an alternative strategy for monitoring the physico-chemical properties of carrageenans. Although sulphurylases are far from being well characterised, they are the only enzymes that have been found to control the viscosity of carrageenan solutions. Other enzymes such as sulphatases have very promising properties.

For example, sulphatases may be useful for converting soft ι-carrageenan gels into rigid κ-carrageenan gels.

REFERENCES

1 Bellion, C.; Brigand, G.; Prome, J.-C.; Welti, D.; Bociek, S., *Carbohydr. Res.*, 1983, **119**, 31.
2 van de Velde, F.; Rollema, H.; Grinberg, N.; Burova, T.; Grinberg, V.; Tromp, R., *Biopolymers*, 2002, **65**, 299.
3 Greer, C.; Yaphe, W., *Bot. Mar.*, 1984, **27**, 479.
4 Bixler, H., *Hydrobiologia* 1996, **326/327**, 35.
5 Knutsen, S.; Grasdalen, H., *Carbohydr. Polym.*, 1992, **19**, 199.
6 Bixler, H.; Johndro, K.; Falshaw, R., *Food Hydrocoll.*, 2001, **15**, 619.
7 van de Velde, F.; Antipova, A. S.; Rollema, H. S.; Burova, T. V.; Grinberg, N. V.; Pereira,L.; Gilsenan, P. M.; Tromp, R. H.; Rudolph, B.; Grinberg, V. Y., *Carbohydr. Res.* 2005, **340**, 1113.
8 Barbeyron, T.; Gerard, A.; Potin, P.; Henrissat, B.; Kloareg, B., *Mol. Biol. Evol.* 1998, **15**, 528
9 Potin, P.; Richard, C.; Barbeyron, T.; Henrissat, B.; Gey, C.; Petillot, Y.; Forest, E.; Dideberg O., Rochas C. and Kloareg B., *Eur. J. Biochem.*, 1995, **228**, 971
10 Jam, M.; Flament, D.; Allouch, J.; Potin, P.; Thion, L.; Kloareg, B.; Czjzek, M.; Helbert, W.; *Biochemical J.*, 2005, **385**, 703
11 Barbeyron, T.; Michel, G.; Potin, P.; Henrissat, B.; Kloareg, B., *J. Biol. Chem.* 2000, **275**, 35499
12 Hatada Y., Mizuno M. and Ohta Y.. *Mar Biotechnol*, 2010, DOI 10.1007/s10126-010-9312-0
13 Guibet, M.; Colin, S.; Barbeyron, T.; Génicot, S.; Kloareg, B.; Michel, G.; Helbert, W., *Biochem. J.*, 2007, **404**, 105
14 Ohta Y.and Hatada U, *J. Biochem.*, 2006. **140**, 475
15 Guibet, M.; Boulenguer, P.; Mazoyer, J.; Kervarec, N.; Antonopoulos, A.; Lafosse, M.; Helbert, W., *Biomacromolecules*, 2008, **9**, 408
16 Jouanneau, D.; Guibet, M.; Boulenguer, P.; Mazoyer, J.; Smietana, M.; Helbert, W., *Food Hydrocoll*, 2010, **24**, 452
17 Jouanneau, D.; Boulenguer, P.; Mazoyer, J.; Helbert, W., Carbohydr. Res. 2010b, 345, 934
18 Wong, K.F. and Craigie, J.S., *Plant Physiol.*, 1978, **61**, 663
19 Tveter-Gallagher, E.; Cogburn J.N.; Mathieson, A.C.; Schiff, J.A., *Hydrobiologia*, 1984, **116/117**, 488
20 Gretz, M.R.; Wu, Y., *Abstracts of the annual meeting of the Phycological Society of America. J. Phycol.* 1990, **2s**, 4
21 Rees, D.A., *Biochem. J.* 1961, **80**, 449
22 Rees, D.A., *Biochem. J.* 1961, **81**, 347
23 Zinoun, M.; Diouris, M.; Potin, P.; Floc'h J.Y.; Deslandes, E., Bot Mar 1997, 40, 49-53.
24 Genicot-Joncour, S.; Poinas, A.; Richard, O.; Potin, P.; Rudolph, B.; Kloareg, B.; Helbert, W.; *Plant Physiol.* 2009, **151**, 1609
25 Genicot, S.; Kloareg, B.; De Ruiter, G.; Penninkhof, B.; Potin, P.; Richard, O.; Rudolf, B.; 2004 Patent number US-6,620,604
26 Weigl, J.; Yaphe, W., *Can. J. Microbiol.* 1966, **12**, 939
27 McLean M.W. and Willamson F.B., *Eur. J. Biochem.*, 1979, **101** 497-505
28 McLean M.W. and Willamson F.B., *Eur. J. Biochem.*, 1981, **113**, 447-456

EVALUATING THE POTENTIAL OF NIGERIAN PLANTS AS A SOURCE OF INDUSTRIAL HYDROCOLLOIDS

Louis M. Nwokocha[1] and Peter A. Williams[2]

[1]University of Ibadan, Ibadan, Nigeria
[2]Glyndwr University, Wrexham, United Kingdom

Abstract

Plant hydrocolloids dominate the food industry as thickeners, gelling agents and viscosity modifiers. These hydrocolloids are mostly sourced from tropic plants. Nigeria, by virtue of its location in the tropics, is a harbinger of these plants which are distributed across its six ecological zones, from the mangrove rain forest to the arid zone. These hydrocolloid-rich plants are under-utilized having only served as thickeners in local cuisines. Lack of information on the properties of the hydrocolloids from these plants has made them to remain largely commercially unexplored.

Among the plants studied are *Brachystigia eurycoma, Prosopis africana, Afzelia africana, Mucuna flagellipes, Detarium microcarpum, Adansonia digitata, Irvingia gabonensis, Gliricidia sepium, Peltrophorum pterocarpum, Bauhinia monandra,* Cola milleni, *Caesalpinea pulcherrima, Delonix regia and Leucaena leucocephala.*

This study evaluates the potential of these plants as sources of industrial hydrocolloids. We have examined the endosperm-seed ratio, yield, rheological and molecular properties of the hydrocolloids. Our results show that some of these hydrocolloids possess novel properties which can be exploited in the food industry.

Introduction

Plant hydrocolloids are used in the food industry to improve consistency and they function as viscosity modifiers, gelling agents, water binders, stabilizers, ice crystal inhibitors and fibre source. The main commercial hydrocolloids obtained from plants include, starch, cellulose and its derivatives, pectin, guar gum, locust bean gum, tara gum, konjac mannan, gum Arabic, gum karaya and gum tragacanth. However, supply has not matched up with demand for these hydrocolloids and known hydrocolloids do not meet all the versatile properties required in diverse applications hence the need for new sources and new properties. Nigeria is strategically located in the tropics covering about six ecological zones (Olaleye and Ameh,

1999). Many plants in these zones are rich in hydrocolloids and many of these are employed as food thickeners in local cuisines. However, there is scanty information on the properties of hydrocolloids from these plants. A few available reports are mostly centred on the emulsification and functional properties of the hydrocolloids (Onweluzo et al., 1995; Chukwu et al., 1998; Andrade, et al., 1999; Uzomah & Ahiligwo, 1999; Achi & Okolo, 2004; Adetogun & Alebiowu, 2007, Saidou et al, 2011). However, solution rheology and average molecular weight of the polysaccharides from *Afzelia africana, Detarium microcarpum* (Picout et al., 2003; Ren et al., 2004; Ren et al., 2005), *Irvingia gabonensis, Hibiscus esculentus* (Ndjouenkeu et al., 1996), and *Corchorus olitorius* mucilages (Yamazaki et al., 2009) have been reported.

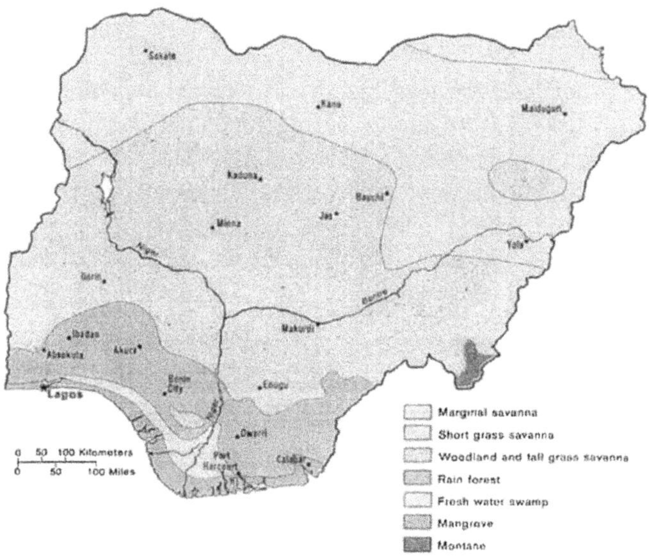

Figure 1. Vegetation zones of Nigeria (Source: University of Texas at Austin)

Objectives

- To isolate and characterize plant hydrocolloids of tropical Nigeria to generate information that may serve as guide to food processors in selecting hydrocolloids with appropriate functionality.
- To provide an inventory of tropical hydrocolloid plants of Nigeria to form a basis for future research reference.
- To provide information on the endosperm-seed ratio, hydrocolloid content, molecular weight, hydrodynamic and rheological properties of these plant hydrocolloids.

Materials and methods

Materials

Polysaccharides were isolated from the plant materials listed in Table 1. Except for *Treculia africana* (fruit pulp) and *Adansonia digitata* (leaves), the polysaccharides were isolated from the seeds.

Table 1. Polysaccharide sources

Botanical name	Family-subfamily	Plant material
Cola milleni	*Sterculiaceae*	seed
Irvingia gabonensis	*Irvingiaceae*	seed
Treculia africana	*Moraceae*	fruit pulp
Adansonia digitata	*Malvaceae- Bombacaceae*	leaves
Mucuna flagellipes	*Leguminosae-Papilionoideae*	seed
Detarium microcarpum	*Leguminosae-Caesalpinioideae*	seed
Afzelia africana	*Leguminosae-Caesalpinioideae*	seed
Leucaena leucocephala	*Leguminosae-Mimosoideae*	seed
Brachystigia eurycoma	*Leguminosae-Caesalpinioideae*	seed
Delonix regia	*Leguminosae-Caesalpinioideae*	seed
Prosopis africana	*Leguminosae-Mimosoideae*	seed
Prosopis glandulosa	*Leguminosae-Mimosoideae*	seed
Caesalpinia pulcherrima	Leguminosae-Caesalpinioideae	seed
Peltophorum pterocarpum	*Leguminosae-Fabaceae*	seed
Bauhinia monandra	*Leguminosae-Fabaceae*	seed

Experimental Methods

Endosperm-seed ratio

The seeds were dehulled by cracking to free the endosperm. In some cases, seeds were swollen in warm water to loosen the endosperm before separating the hull and endosperm and both the hull and endosperm dried to constant weight. These were expressed as a percentage of the whole seed. Where the seed germ constituted $< 0.5\%$, it was considered negligible.

Isolation of polysaccharide

Whole seed or endosperm or other source materials were pulverized and defatted by hexane extraction for 12 hr. The flour was aqueous extracted followed by centrifuging and the polysaccharide solution precipitated with isopropanol. Reconstitution in distilled water and re-precipitation with isopropanol was used to purify the polysaccharide. The purified sample was freeze dried.

Characterization

Intrinsic viscosity: The flow time of the polysaccharide solutions and the solvent were determined using a Canon-Ubbelohde capillary viscometer using serial iso-ionic dilution at 25°C. The flow time measured in this way was converted to relative viscosity (η_r) by dividing with the flow time of the solvent and the intrinsic viscosity, [η], evaluated by applying Fedors equation (eq. (1)).

$$\text{Fedors Equation:} \quad \frac{1}{2(\eta_r^{\frac{1}{2}}-1)} = \frac{1}{[\eta]c} - \frac{1}{c_m[\eta]} \qquad (1)$$

Where c is concentration of polymer and c_m is a concentration factor.

Steady shear and dynamic viscosities: Steady shear viscosity and small angle deformation oscillation measurements were obtained on different concentrations of the polysaccharide solutions at 25°C using cone and plate geometries and recessed end concentric cylinders depending on the concentration. Flow measurements were determined at shear rates of 10^{-3} to 10^{3} s^{-1} while dynamic measurements were carried out by performing frequency sweeps at an angular frequency of 10^{-2} to 628 rad/s in the linear viscoelastic region. The linear viscoelastic region was determined from a stress sweep in the region of 10^{-3} to 200 Pa at an angular frequency of 6.28 rad/s.

The Williamson model (eq. (2)) was used to describe the flow characteristics.

$$\frac{\eta}{\eta_o} = \frac{1}{(1+(\tau\dot{\gamma})^N)} \qquad (2)$$

Where η and η_o are shear and zero shear viscosities, respectively; τ is the relaxation constant, γ is the shear rate and N is the rate index.

Molecular weight: The molecular weight was determined using gel permeation chromatography coupled to multiangle laser light scattering, refractive index and UV detectors. The polysaccharide solutions were filtered through a 0.45μm syringe filter and injected into a 200 μl loop, passed through a combination of Suprema columns (100Å, 3000Å and 30000Å). The flow rate was 0.5 ml/min and the effluent was 0.1M $NaNO_3$ solution containing 0.005% of sodium azide. The chromatograms were analyzed with Astra software with a predetermined dn/dc values.

Results and Discussion

The seed components and gum yields are presented in Table 2

Table 2. Seed components and gum yields (%)

Seeds	Hull	Endosperm	Germ	Gum
Brachystigia eurycoma	30.22±4.19	69.78±4.19	N	44.3±3.3*
Detarium microcarpum	17.99±2.64	82.01±2.64	N	61.58±0.56*
Delonix regia	39.51±3.11	38.20±2.94	22.30±2.23	77.8*
Afzelia africana	18.87±1.69	81.13±1.69	N	41.55±1.2*
Leucaena leucocephala	46.86±2.60	53.14±2.60	N	7.9 (33*)
Caesalpinia pulcherrima	28.85±1.74	34.74±2.34	36.40±0.91	26.53±1.41
Bauhinia monandra	23.4±1.3	37.1±1.6	39.5±2.2	17.8±0.03*
Cola milleni	ND	ND	ND	7.41±0.95*
Peltophorum pterocarpum	42.06±8.05	29.32±1.74	28.62±8.36	20.4±3.1
Prosopis glandulosa	43.24±4.53	56.76±4.53	N	39.03±1.71*
Prosopis africana	ND	ND	ND	53.6*
Mucuna flagellipes	32.85±10.15	67.15±10.15	N	32.6±1.97*
Irvingia gabonensis	17.02±3.81	88.98±3.81	N	95.5*

*based on defatted endosperm , N= Negligible (value < 0.5%) , ND = Not determined

It is noted that seeds that are low in percentage endosperm had high content of germs and thick seed coats. It is the endosperm of the seeds that has thickening characteristics. As a result, the ratio of the endosperm to the whole seed gives an idea of potential of a seed as a source of polysaccharide. Most of the seeds gave high yield of gum per endosperm. Defatted *I. gabonensis* seed endosperm flour contained ~95% gum.

Rheological properties.

The steady shear viscosities (Figure 2 a-c) are presented as a plot of η/η_o versus $\gamma/\gamma_{0.8}$ for the different polysaccharides (where $\gamma_{0.8}$ is the shear rate at $\eta = 0.8\eta_o$). The main observations included:

- Flow curves at different concentrations were reduced to a single flow curve
- η/η_o approached unity as $\gamma/\gamma_{0.8}$ approached zero for all the polysaccharides
- The shapes of the profiles of η/η_o versus $\gamma/\gamma_{0.8}$ were not same for all polysaccharides but depended on the shear thinning properties of the polysaccharides.
- Deviations were observed at low concentrations at high shear rates in *Cola milleni* and *Afzelia africana* polysaccharides. Similar deviations were reported to occur in xanthan gum at transition zones between the power law zone and second Newtonian viscosity (Launay et al, 1983).

In Figure 3, the shear viscosities of the polysaccharides were compared with guar gum at 2% concentration while the flow characteristics are presented in Table 3. Most of the polysaccharides exhibited low shear viscosities compared with guar at 2% concentration except *Delonix regia, Prosopis africana* and *Cola milleni* which had fairly high zero shear viscosities. *Delonix regia* and *Prosopis africana* were more stable to shear thinning than guar indicating they might have a more rigid structure. *Adansonia digitata* (τ =51.82 s), *Cola milleni* (τ =49.58 s) had higher relaxation time (τ) compared with guar (τ = 43.84 s) indicating presence of very weak structures which broke down at very low shear rates. *Treculia africana* polysaccharide solution was Newtonian at 2% concentration.

Figure 4 shows the dependence of storage (G') and loss (G") moduli on angular frequency (ω) and Table 4 presents the parameters of the frequency sweep for 2% concentrations of the polysaccharides compared with guar gum. Guar gum at 2% behaved more like a gel with G' > G" at all ω and both moduli showed only slight dependence on ω. All the other polysaccharides behaved like concentrated polymer solutions at 2% concentration with varying dependence of G' and G" on ω. For *Adansonia digitata,* G' > G" but both moduli increased with increase in ω, a characteristic of a weak gel. *Cola milleni* exhibited two transitions of G' and G" at very low ω but at ω > 0.1 rad/s it exhibited gel-like properties with G' > G". *Peltophorum pterocarpum, Leucaena leucocephela*, and *Delonix regia* exhibited gel-like properties at lower ω, with G' > G" but liquid-like properties at higher ω. On the other hand *Prosopis africana*, *Prosopis grandulosa, Brachystegia eurycoma* and *Afzelia africana* were dominated by liquid-like characteristics at lower ω and gel-like properties at higher ω.

The dependence of the zero shear viscosity on the space occupancy is presented in Figure 5. All the polysaccharides gave a very close superimposition of their curves in the region of log C[η] < 0.6. This is the dilute region below the coil overlap regime. The transition from the dilute region to the semidilute region, in which the polymer coils start to overlap, was smooth rather than sharp to be described by a single slope. The varying degree of molecular interactions (hyper-entanglements) among the polymers of the different polysaccharides resulted in zone broadening in this region. The polymer concentration at log C[η] = 0.6, at which transition from the dilute to the concentration region occurred is called critical coil overlap concentration, C*, and occurs at $C^* \sim 4/[\eta]_o$ for most random coil polymers. The average slope of the curve below C* was estimated to be 1.16 while the average slope above C* was 4.36. The [η] and hydrodynamic properties of the polysaccharides are presented in Table 5.

In most industrial applications, the polysaccharide has to be used at a concentration > C* for its rheological effects to be noticed.

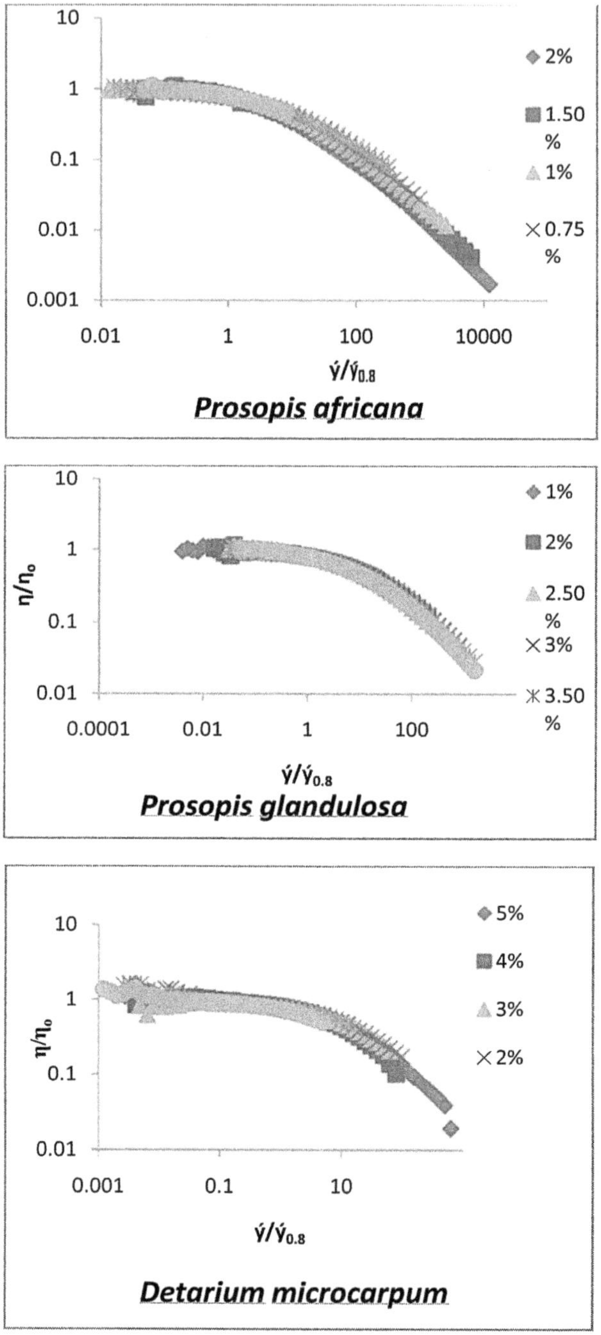

Figure 2a η/η_o versus $\gamma/\gamma_{0.8}$ for different concentration of polysaccharide solutions

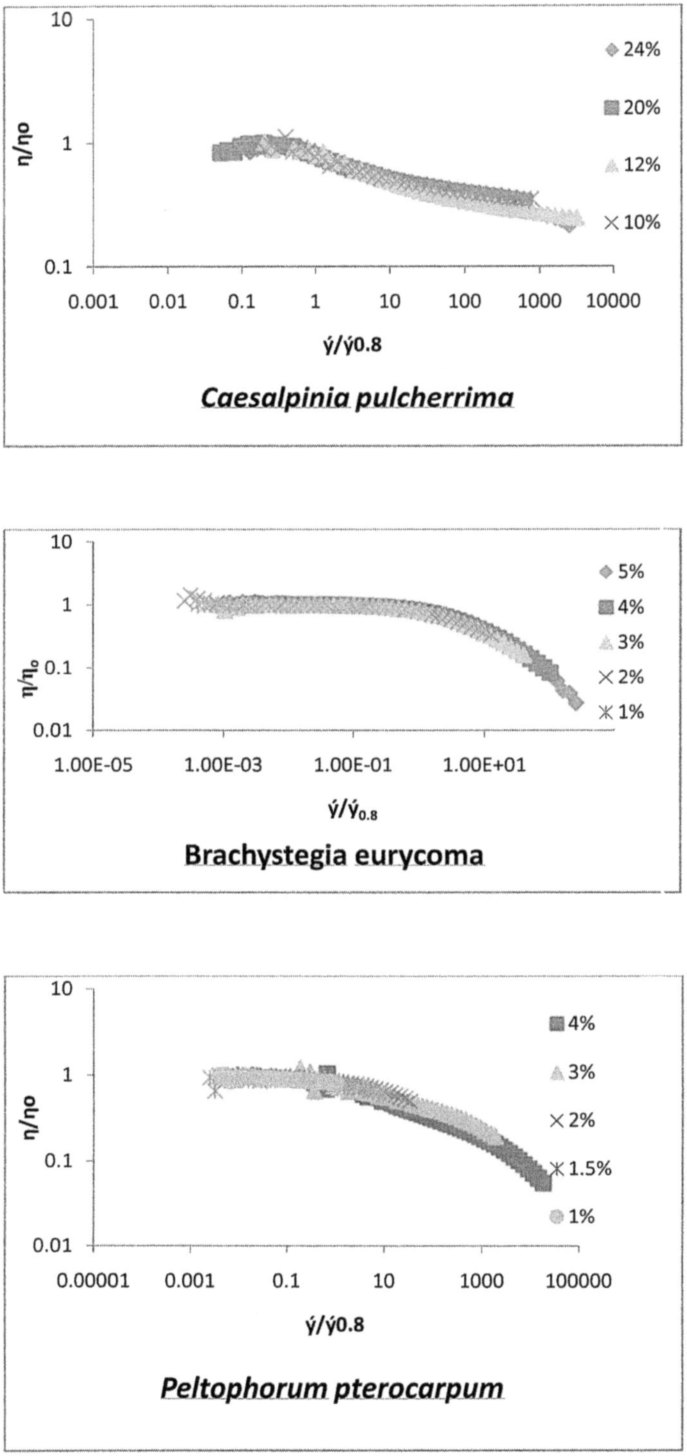

Figure 2b η/η_o versus $\gamma/\gamma_{0.8}$ for different concentration of polysaccharide solutions

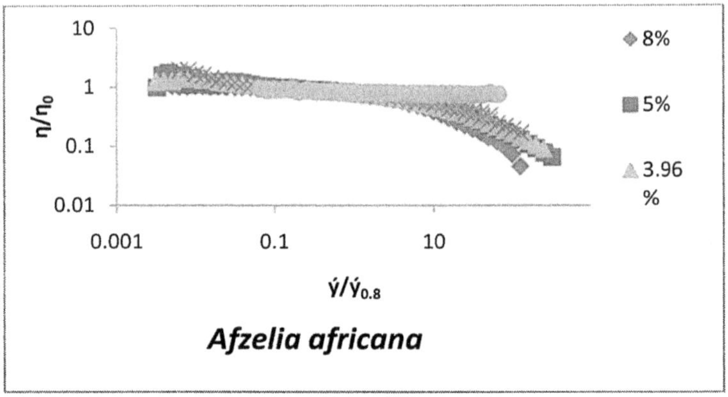

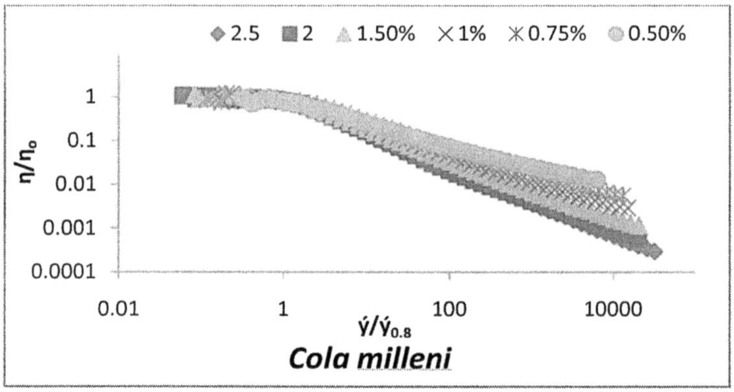

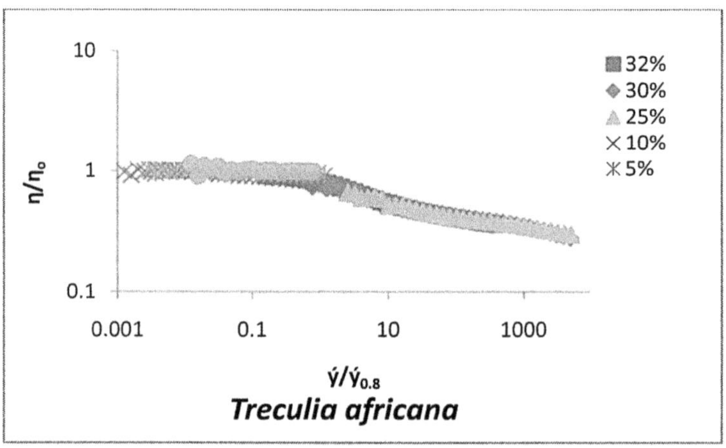

Figure 2c η/η_o versus $\gamma/\gamma_{0.8}$ for different concentration of polysaccharide solutions

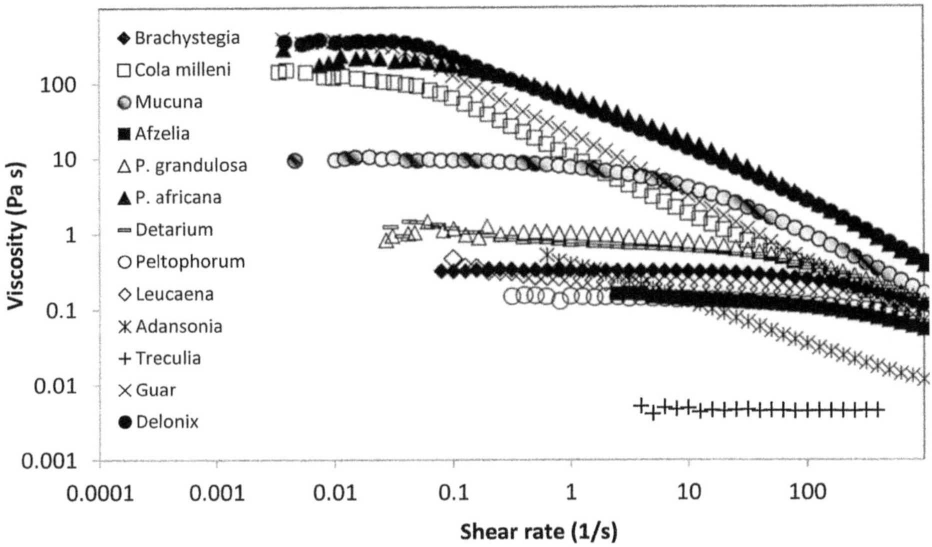

Figure 3. Viscosity - shear rate profiles of 2 % solutions of the different polysaccharides

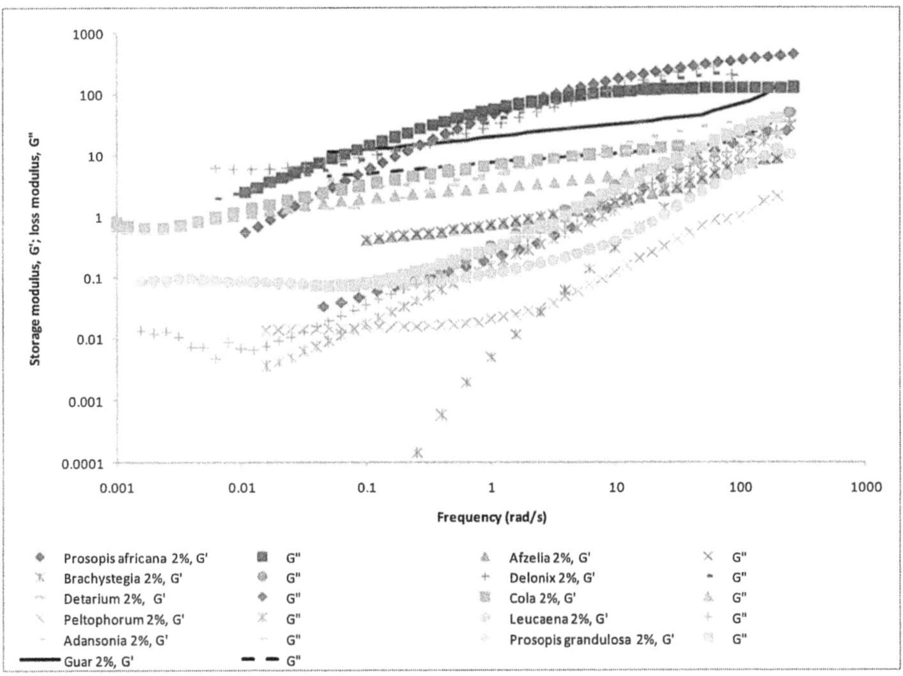

Figure 4. Storage and loss moduli versus angular frequency for 2% polysaccharide solutions

Table 3. Comparison of flow characteristics of 2% solutions of polysaccharides and guar gum

Polysaccharide	η_0 (Pa.s)	τ (s)	N	S.E	Model
Treculia africana	4.488E-3	-	-	2.409	Newtonian
Peltophorum pterocarpum	0.1552	9.808E-4	0.4716	15.15	Williamson
Afzelia africana	0.1512	2.313E-3	0.6540	7.792	Williamson
Leucaena leucocephala	0.2366	3.229E-3	0.5859	9.30	Williamson
Brachystegia eurycoma	0.3334	2.420E-3	0.7861	6.772	Williamson
Detarium microcarpum	0.7554	0.01010	0.6524	15.08	Williamson
Prosopis glandulosa	1.096	0.02193	0.6422	6.103	Williamson
Adansonia digitata	3.968	51.82	0.5434	9.591	Williamson
Mucuna flagellipes	14.73	0.7372	0.6672	19.25	Williamson
Cola milleni	221.3	49.58	0.7473	17.22	Williamson
Prosopis africana	223.9	2.472	0.7956	11.30	Williamson
Delonix regia	503.4	16.03	0.7156	12.40	Williamson
Guar	567.5	43.84	0.8510	9.841	Williamson

η_0 =zero shear viscosity, τ = relaxation time, N =rate index, S.E= standard error

Table 4. Viscoelastic properties obtained by small deformation oscillation measurements

Polysaccharide	No. Crossover points	G' (Pa s)	ω (rad/s)	τ (s)= 1/ω
Prosopis africana	1	72.37	1.994	0.502
Prosopis glandulosa	-	-	-	-
Afzelia africana	1	1.028	2.735	0.3656
Brachystegia	-	-	-	-
Delonix regia	2	113.2	8.840	0.113
		10.14	0.1205	8.30
Detarium microcarpum	-	-	-	-
Cola milleni	2	0.7083	3.047E-3	328.1
		0.6542	1.922E-3	520.29
Luecaena leucocephala	1	0.08261	0.2694	3.7
Adansonia digitata	-	-	-	-
Peltophorum pterocarpum	1	0.01463	0.0846	11.82
Guar	-	-	-	-

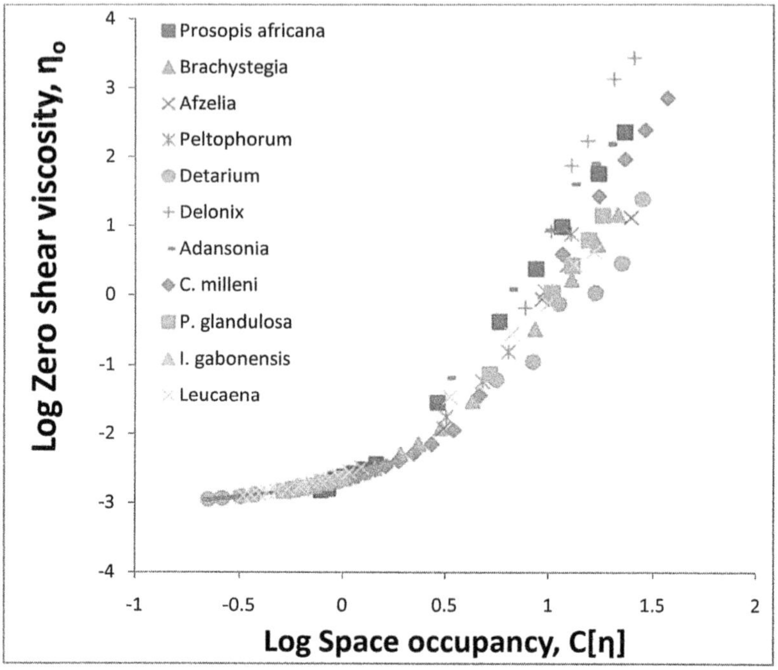

Figure 5. Zero shear viscosity versus space occupancy for the polysaccharide solutions

Molecular weight

The average molecular weights of the polysaccharides are presented in Table 6. Molecular weights of most of the polysaccharides are in the order of half a million. Figure 6 shows the RI [lower curve] and UV [upper curve] profiles for *Afzelia africana* polysaccharide. The presence of a UV peaks indicates the presence of a protein-rich fraction which may provide good emulsifying properties.

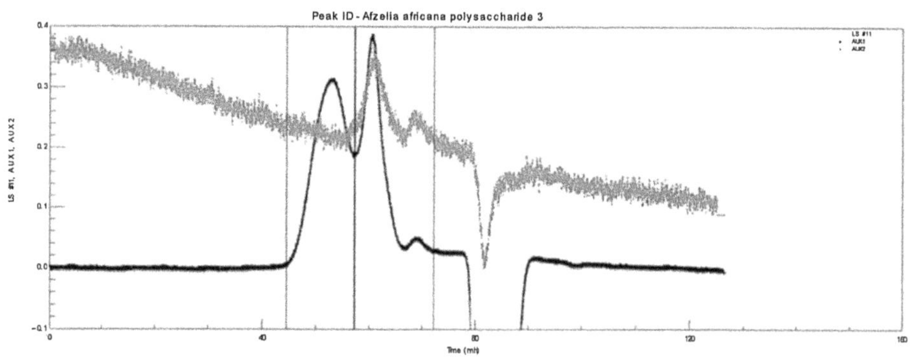

Figure 6. GPC elution profile of *Afzelia africana* polysaccharide

Table 5. Analysis results of Log $\eta_{sp,o}$ versus Log C[η]

	[η]	C*[η]	C* (g/dL)	Slope below C*	Slope above C*
Brachystegia eurycoma	4.3	3.6	0.8	1.4	3.8
Afzelia africana	3.0	3.0	1.0	1.2	3.9
Leucaena leucocephala	3.5	2.7	0.6	1.3	3.0
Adansonia digitata	3.3	0.27	0.07	0.8	4.4
Prosopis africana	11.6	2.2	0.2	1.4	4.3
Peltophorum pterocarpum	3.2	3.4	1.1	1.46	4.3
Mucuna flagellipes	7.9	3.98	0.5	1.0	4.6
Delonix regia	10.3	4.0	0.39	1.2	6.8
Cola milleni	11.65	2.8	0.24	1.3	4.5
Prosopis glandulosa	5.2	3.7	0.7	0.7	4.1
Caesalpinia pulcherrima	0.353	1.6	4.6	1.3	3.8
Treculia africana	1.8	24.4	13.6	0.9	4.8

C = coil overlap concentration*

Table 6. Average molecular weight of some polysaccharides

Polysaccharide	Molecular weight (g/mol)		Polysaccharide	Molecular weight (g/mol)
Brachystigia eurycoma	5.47±0.68E+05		*Prosopis africana*	1.53±0.05E+06
Detarium microcarpum	8.32±0.41E+05		*Prosopis glandulosa*	5. 10±0.04E+05
Delonix regia	7.65±0.14E+05		*Treculia africana*	1.60±0.03E+06
Adansonia digitata	4.01±0.08E+06		*Irvingia gabonensis*	1.51±0.002E+06
Leucaena leucocephala	5.12±0.25E+05		*Cola milleni*	4.66±0.16E+06
Caesalpinia pulcherrima	2.64±0.13E+04		*Afzelia africana*	7.87±0.13E+05 4.55±0.03E+05*
Peltophorum pterocarpum	2.485E+05			

* *Protein-rich fraction*

Conclusions

- We have studied a number of polysaccharides of Nigerian plants especially those employed as food thickeners in local cuisines.
- The yield from most of the seeds is greater than 30% and in the range of values found in commercial seed such as guar and therefore have potential for exploitation.
- Rheological and molecular characteristics of most of the polysaccharides are being reported for the first time especially *Brachystegia eurycoma, Prosopis africana* and *Treculia africana.*
- *Cola milleni* and *Peltophorum pterocarpum* are polysaccharides not previously reported in literature.

- The GPC of *Afzelia africana* polysaccharide shows the presence of a protein-rich fraction indicating potential emulsification properties.

Acknowledgement

This study was supported by a Fellowship for LMN by the Leverhulme Trust Foundation.

References

Adetogun, G. E., Alebiowu, G. (2007). Influence of *Delonix regia* seed gum on the compressional characteristics of paracetamol tablet formulations. Journal of Drug Delivery Science and Technology, 17(6), 443-445.

Andrade, C.T., Azero, E.G., Luciano, L., Goncalves, M.P. (1999). Solution properties of the galactomannans extracted from the seeds of *Caesalpinia pulcherrima* and *Cassia javanica*: comparison with locust bean gum. International Journal of Biological Macromolecules 26(2-3):181-5.

Chukwu, A., Ofoefule, S.I., Ugueze, K. (1998). Application of a polysaccharide derived from *Treculia africana* as a sustained release hydrophilic matrix for Theophylline hydrate tablet. Nigeria Journal of Natural Products and Medicine 2, 37-41.

Launay, B., Cuvelier, G. and Martinez-Reyes, S. (1983). Xanthan gum in various solvent conditions: Intrinsic viscosity and flow properties. In Gums and Stabilizers for the Food Industry 2. Application of hydrocolloids. (Eds. G.O. Phillips, D.J. Wedlock and P.A. Williams) Pergamon Press, Oxford, England. p79-98

Ndjouenkeu, R., Goycoolea, F.M., Morrisa, E.R. and Akingbala, J.O. (1996). Rheology of Okra (*Hibiscus esculenta* L) and dika nut (*Irvingia gabonensis*) polysaccharides. Carbohydrate Polymers 29(3), 263-269.

Nwokocha, L.M., Williams, P.A. (2009). Isolation and rheological characterization of *Mucuna flagellipes* seed gum. Food Hydrocolloids 23 (5), 1394-1397.

Olaleye, O.A., Ameh, C. E. (1999). Forest Resource Situation Assessment of Nigeria. FAO Corporate Document Repository. http://www.fao.org/docrep/004/AB578E/AB578E00.HTM

Onweluzo, J.C., Onuoha, K.C., Obanu, Z.A. (1995). Certain functional properties of gums derived from lesser known tropical legumes (*Afzelia africana*, *Detarium microcarpum* and *Mucuna flagellipes*. Plant Foods for Human Nutrition 48, 55-63.

Picout, D.R., Ross-Murphy, S. B., Errington, N., Harding, S. E. (2003). Pressure cell assisted solubilization of xyloglucans: Tamarind seed polysaccharide and Detarium gum. Biomacromolecules, 4, 799-807.

Ren, Y., Picout, D. R., Ellis, P. R., Ross-Murphy, S.B. (2004). Solution properties of the xyloglucan polymer from *Afzelia africana*. Biomacromolecules, 5, 2384– 2391.

Ren, Y., Picout, D. R., Ellis, P. R., Ross-Murphy, S.B., Grant Reid, J.S. (2005). A novel xyloglucan from seeds of *Afzelia africana* Se. Pers.—extraction, characterization and conformational properties. Carbohydrate Polymers 340 (5), 997-1005.

Saidou, C., Tchatchueng, J.B., Ndjouenkeu, R., and Roux, D.C.D. (2011). Extraction and partial characteristion of hydrocolloid gums of some African Legumes. Journal of Food Engineering Vol. 7 Issue 3, Article 15.

Uzomah, A., Ahiligwo, R. N. (1999). Studies on the rheological properties and functional potentials of achi (*Brachystegea eurycoma*) and ogbono (*Irvingia gabonesis*) seed gums. Food Chemistry 67(3), 217-222.

Yamazaki, E., Kurita, O., Matsumura, Y. (2009). High viscosity of hydrocolloid from leaves of *Corchorus olitorius* L. Food Hydrocolloids 23 (3), 655-660.

NEW REGIOSELECTIVE SUBSTITUED CELLULOSE ETHERS: THERMO-RHEOLOGICAL STUDY

A.Sullo[1], Y.Wang[2], J. R. Mitchell[1], A. Koshella[2], T. Heinze[2] and T.J. Foster[1].

[1]Division of Food Science, University of Nottingham, Sutton Bonington Campus, LE125RD Loughborough, United Kingdom

[2]Center of Excellence for Polysaccharide Research, Friedrich Schiller University of Jena, Humboldtstrasse 10, D-07743 Jena, Germany

Abstract

Aqueous solutions of regioselective substituted 3-O-ethyl-propyl cellulose ethers with measured degrees of substitution close to 1 but varying in the ratio of ethyl to propyl groups have been studied by means of differential scanning calorimetry and oscillatory shear rheology. It was found that the modulus developed on heating and the enthalpies of the thermal transitions increases strongly with increasing propyl content whereas the cloud point temperature decreases. The shapes of the endothermic traces are indicative of an aggregation process which is intimately coupled with phase separation as shown by the appearance of clouding in the same temperature range. The correlation between rheology and thermal analysis clearly demonstrates that aggregation eventually leads to formation of a gel network. In fact the aggregation is accompanied by an increase in the elastic modulus and formation of an opaque gel network. Reversibility was observed on cooling. Thermal and rheological properties are both controlled by the ratio of the two alkyl moieties. Thermal traces were also de-convoluted in several processes and an attempt to explain the physical meaning of those processes is also reported.

1. INTRODUCTION

Cellulose is a linear homopolymer made from glucose molecules linked through (β 1→4) glycosidic bonds. Crystallinity, due to the large number of intermolecular hydrogen bonds between adjacent hydroxyl groups, is mainly responsible for the insolubility in water.[1,2] However, chemical substitution of a certain fraction of hydroxyl groups with hydrophobic moieties enables production of water-soluble cellulose derivatives, widely used in a large number of industrial applications as rheology modifiers.[3-6] In general, the accessible reactive hydroxyl groups for substitution are at the *C*-2, *C*-3, and *C*-6 atoms of the glucose unit.[7,8] It is well established that physico-chemical properties of cellulose derivatives are mainly determined by the nature of the substituents, degree of substitution (DS) and substitution pattern along the cellulose chain.[9-12] It is now possible substitute cellulose regioselectively i.e. exclusively at positions 2, 3 or 6 of the anhydroglucose unit.[13-14] The most significant feature of substitution, along with solubility in water, is the heat-induced self-association, the so-called "Thermogelation". A number of researchers have studied the thermogelation process in water, especially with reference to random substituted methylcellulose (MC) and hydroxypropylmethylcellulose (HPMC).[15-23] Aqueous solutions of MC and HPMC have shown formation of physical gels when the temperature is raised above a critical value (heat setting), in contrast with the majority of the other polysaccharides (cold setting).[11] The mechanism of gelation in water is believed to be mainly caused by the aggregation of hydrophobic moieties on the cellulose backbone leading to cross linking.[23] A well agreed thermodynamic model describing the behaviour of aqueous solutions of the hydrophobic polymer is the following: at low temperature, water molecules, immediately adjacent to the hydrophobic solute, to which they cannot hydrogen bond, are presumed to form an ice-like structure. The enthalpy of mixing is exothermic and indicates an increased number of hydrogen bonds between water molecules surrounding the hydrophobic solute. This phenomenon is called hydrophobic hydration which accounts for the solubility of those polymers at low temperature.[24-25] However extensive water structuring, associated with formation of hydrogen bonds, carries a large negative entropy penalty. At high temperature the ice-like water structures "melt", which causes an increase in entropy. Increase in entropy is responsible for the overall free energy becoming negative ($\Delta G = \Delta H - T\Delta S$), thus it is believed to be the driving force for association. The system undergoes the transition to a more stable state which involves polymer-polymer interaction in order to minimize the contact with water. Eventually aggregates can form a three-dimensional gel network.[26] The sol-gel transition is reversible on cooling because the aggregates formed at elevated temperature disassociate when the temperature is lowered. However there is still a certain degree of controversy regarding the mechanism of gelation which is even more complicated by a range of concomitant phenomena occurring at high temperature such as precipitation, which is due to the inverse temperature–solubility relationship of these polymers.[21] Several authors, using different techniques, proposed explanations for the thermogelation phenomena. Haque *et al.*,[17,18] proposed a model of gelation based on the existent of a residual crystallinity which implies cellulose derivatives molecules are organized in bundles at room temperature. On heating the gelation proceeds in two stages: 1) the exposure of the hydrophobic groups enabled by bundle separation and then, 2) the association of the hydrophobic groups at higher temperature. Ibbett *et al.*,[27] have proposed a different explanation for the two stage phenomena. First, highly substituted regions separate forming micelles. On further heating micelles connect leading eventually to the formation of a network. Kato[28] concluded that highly substituted units, responsible for gelation and junction zones, are "crystalline and consist of trimethyl glucose units". A common observation seems to be the fact that thermally induced "gelation" does not occur monotonically. A recent study on regioselectively substituted cellulose derivatives stressed the importance of the substitution pattern on the aggregation phenomena. More insight into the gelation of these materials can be gained from a study of regioselectively substituted celluloses where the position of the substituent group is known and the structure of the group varied. For example Sun *et al.*,[12] reported a one step gelation for regioselective 3-0-ethylcellulose in contrast with random substituted ethylcellulose derivatives. As a continuation of Sun *et al.*,[12] studies, the thermogelation properties of new regioselective celluloses with varying ratios of ethyl to propyl substitutes were studied by coupling rheological with thermal measurements. This might be relevant not only to further understand the phenomena of thermogelation but also to potential industrial applications.

2. MATERIALS AND METHODS

3-mono-O-alkyl cellulose was synthesized using two different alkyl groups (ethyl and propyl), in three different molar ratios, via 2,6-di-O-thexyldimethylsilyl cellulose. Detailed description of the chemistry involved is reported elsewhere.[29] Figure 1 shows an illustration of the 3-mono-*O*-alkyl cellulose backbone containing two different moieties at position 3.

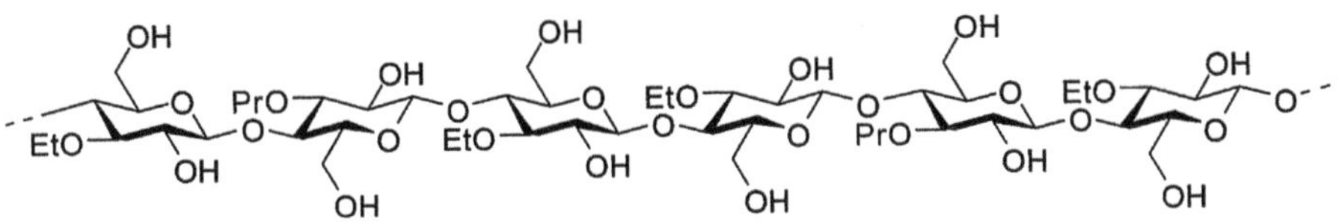

Figure 1: Mixed-type-substituted cellulose backbone.

Degree of substitution (DS) for the three samples studied were calculated using ^{1}H-NMR (table1). Samples with different ratios of the two hydrophobic moieties will be referred to as sample A, B and C as describe in table 1. Nominal degree of polymerisation (DP_n), determined by size exclusion chromatography, was respectively: 172, 252 and 235.

Tab. 1 - DS values of ethyl and n-propyl group

Sample	DS_{total}	DS_{Et}/DS_{Pr}	DS_{Et}	DS_{Pr}
A	1,10	5,8	0,95	0,15
B	1,04	2,7	0,77	0,27
C	0,83	0,8	0,27	0,56

2.1 Sample preparation

Regioselective cellulose derivatives were dissolved in distilled water at temperatures close to 2°C using an ice water bath. Dissolution was achieved after 24h continuous stirring in a cold room (4°C). Solutions were then centrifuged at 14,000 RPM for 30 min. An aliquot from the supernatant was then dried at 105°C overnight to calculate concentrations. All samples were used at a concentration of 6.2% wt.

2.2 Cloud point (visual inspection)

Aqueous solutions of cellulose derivatives at 5°C were sealed in test tubes transferred to a water bath and heated to the desired temperature. The temperature was allowed to stabilize at the desired value and digital images highlighting the state of each MC solution were captured using a digital camera. After the first heating process, samples were allowed to cool down naturally and kept in the refrigerator for 1 h. Then, bath temperature was increased and the process was repeated.

2.3 Micro-Differential Scanning Calorimetry

Thermal characteristics of sample solution were determined using a SETARAM micro DSC-III calorimeter (Caluire, France). About 850 mg of each sample were sealed into DSC cells. An equal weight of water was placed in the reference cell. The water is required to match the heat capacity of the sample. Both the sample and reference cells were allowed to equilibrate at 10 °C for 1h. Temperature was raised from 10 to 78°C at a heating rate of 1°C/min and then lowered immediately to 10 °C at the same rate. Enthalpy values (ΔH), onset temperature T_o, peak temperature T_m and end temperature T_e of the transition were calculated using Setaram software with an interpolated baseline.

2.4 Rheology

Dynamic viscoelastic measurements were carried out on a controlled stress rheometer (MCR 301; Anton Parr, Austria). Parallel plate geometry of 25 mm diameter and 0.5 mm as gap were used. All measurements were performed under oscillatory shear mode at an angular frequency of 10 rad/s and a shear strain ensuring operation within the linear viscoelastic region. To prevent dehydration during the measurements, a thin layer of low viscosity silicone oil was placed on the periphery surface of the solution. Elastic modulus and loss modulus were monitored during the thermal cycle with a heating rate (1°C/min) and temperature range (10 to 78°C) as for the differential scanning calorimeter experiment.

3. RESULT AND DISCUSSION

3.1 Turbidity

Figure 2 shows changes in turbidity (visual inspection) for sample A, B and C, equilibrated at the desired temperature. The turbidity increases dramatically as the solution temperature exceeded the critical onset value, i.e. the cloud point temperature. Although any evaluation of the cloud point in this way would be rather approximate, which reflects the nature of the experiment; correlations with molecule structure of the three samples are still possible. Cloud point temperature is mainly influenced by the ratio of the two hydrophobic moieties present on the cellulose backbone (DS_{Et}/DS_{Pr}). Thus a specific trend was observed: sample A>sample B>sample C, which means the higher the propyl group (lower DS_{Et}/DS_{Pr} - sample C) on the cellulose molecule the lower the cloud point temperature and vice versa. The onset of clouding is normally an indication of phase separation which becomes even more severe as the temperature approached 80°C, as all samples show formation of a "white" precipitate and a clear supernatant.

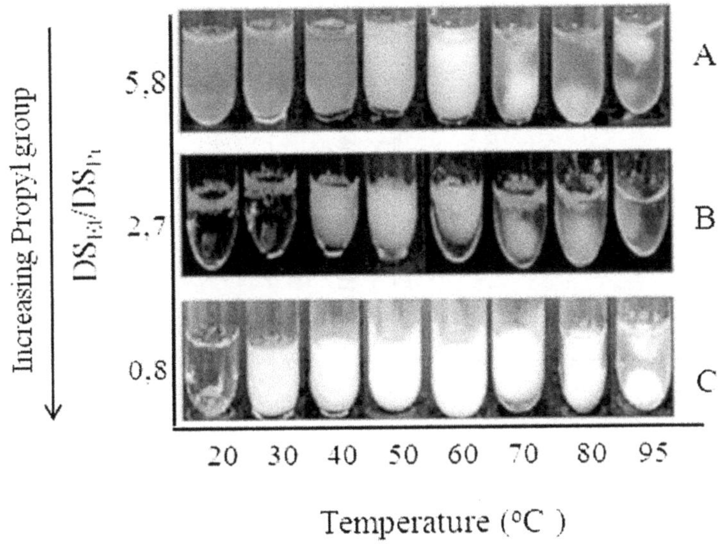

Fig.2- Visual inspection of cloud point for sample A (DS_{Et}/DS_{Pr} = 5.8), sample B (DS_{Et}/DS_{Pr} = 2.7) and sample C (DS_{Et}/DS_{Pr}=0.8).

3.2 Micro-Differential Scanning Calorimetry

Differential Scanning calorimetric (DSC) data, for all the samples studied, on heating and cooling at a scan rate of 1°C/min are reported in figure 3 (a and b). Sample A is reported separately to avoid confusion. On heating, all samples show a broad asymmetric endothermic peak. The endothermic origin of the traces comes from the loss of the water of hydration, in other words the molecule becomes desolvated. The loss of water of hydration, which has a hydrogen-bonded structure (ice-like), must be endothermic which give rise to the signal in the calorimeter. Hindered in this argument, disruption of water structure, there is the self association of cellulose molecules occurring to minimize the contact between hydrophobic moieties and water molecules. It should be noticed that, on heating, the onset of the endotherm, the temperature at which the traces deviate from an imaginary baseline, occurs quite close to the temperature at which the solutions go cloudy (figure1). Turbidity is a clear indication of phase separation, and the shape of the endotherm captures most of the features of the aggregation process. Armstrong *et al.*,[30] described differences in differential scanning calorimetry output between aqueous solution of synthetic polymers which are believed to either aggregate or phase separate. They reported, based on model simulations and experimental data, that the phenomena of phase separation and aggregation give rise to asymmetric endothermic traces. However phase separation is characterised by a "steep leading edge and long gently sloping post transitional tail", with an enthalpy value generally smaller than the one observed for aggregation. In favor of the aggregation argument there is also the similarity in shape with endothermic traces for random substituted cellulose aqueous system for which the aggregation phenomena has been investigated with different techniques.[18, 27] However the gently sloping post transitional tail might, on the other hand, imply phase separation, which also occurs within the same temperature range.

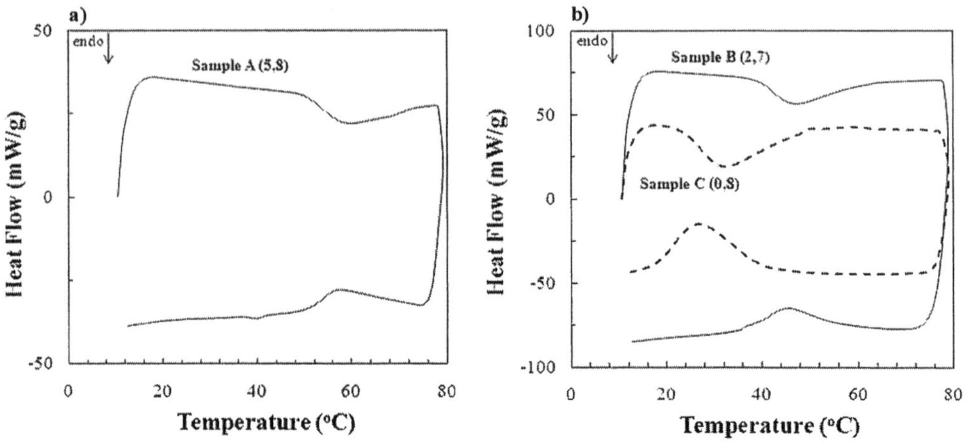

Fig.3- Differential Scanning Calorimetry traces: a) sample A (DS_{Et}/DS_{Pr} = 5.8) and b) sample B (DS_{Et}/DS_{Pr} = 2.7) and sample C (DS_{Et}/DS_{Pr} = 0.8).

This rather qualitative analysis of the DSC traces suggests that aggregation and phase separation are intimately coupled. Characteristic parameters for the transitions of each of the samples studied on heating and cooling are summarised in table 2. T_m, T_o and T_e denote respectively onset, peak and end set temperature; ΔH is the enthalpy change calculated from the area under the peaks. The trend is reasonably obvious and is the same as for the cloud point temperature: the lower the DS_{Et}/DS_{Pr} ratio the lower the temperature of the transition and vice versa. For instance, on heating sample A (DS_{Et}/DS_{Pr} = 5.8), low level of propyl, shows the highest T_m, while sample C (DS_{Et}/DS_{Pr} = 0.8), with high level of propyl, shows the lowest T_m among the sample studied. The enthalpy of transition, as well, exhibits the same significant dependence on DS_{Et}/DS_{Pr} ratio. Differences in temperature and enthalpy of transition bring into play the hydrophobic character of each sample. It might be reasonable to assume that the propyl group, because of its higher molecular size, is more hydrophobic than the ethyl group. Therefore the hydrophobicity of each sample is itself controlled by the presence of the two substituents in different ratio: sample C>sample B>sample A. As anticipated, the endothermic trace on heating arises from the loss of structured water, which accounts for the large entropy penalty. Following this argument sample C might carry higher negative entropy of mixing with water than sample B and A, and therefore the entropy gain arising from hydrophobic interaction tends to be higher. Thus it would be expected that the more hydrophobic the sample is the lower the temperature of the transition and the higher the enthalpy which is, indeed, what was observed in Table 2. The breath of the transition (width at half of the maximum peak intensity) is believed to be due to polydispersity in molecular weight (*M*w/*M*n) (figure 3). In fact the range of molecular dimensions also gives rise to a range of different hydrophobicity within the same sample. For instance the longer the chain length, the higher the hydrophobic character. It might therefore be expected that the more hydrophobic molecules would tend to aggregate first. The similarity of the two signals, heating and cooling, indicates that the process occurring on cooling is the reversal of the one occurring on heating. The two signals carry also the same enthalpy, which indicates the almost total reversibility of the process. Therefore if the hydration/aggregation phenomenon, on heating, is accepted than the reversal is the dis-association and subsequent hydration of the molecules.

Table 2 – Transition parameters for 3-O-ethyl-propyl cellulose ethers.

Heating

Sample	DS_{Et}/DS_{Pr}	T_o	T_m	T_{end}	ΔH(J/g solid)
A	5.8	51.5	58.9	72.3	6.7
B	2.7	39.13	46.99	61.11	11.6
C	0.8	22.8	32.1	50.1	21.83

Cooling

Sample	DS_{Et}/DS_{Pr}	T_o	T_m	T_{end}	ΔH(J/g solid)
A	5.8	64.3	57.1	51.5	-5.1
B	2.7	57.9	45.34	34.8	-10.94
C	0.8	40.35	26.82	18.1	-22.71

The hydration of the molecules implies the formation of the structured water melted on heating, with the creation of hydrogen bond, which is an exothermic event. However the exotherm on cooling, for all the samples, does not occur exactly at the same temperature showing a little shift to a lower temperature (hysteresis). Two possible hypotheses arise from the hysteresis phenomenon. The first case implies that cellulose derivative molecules aggregate specifically, in other words, the most hydrophobic molecule seeks molecules with similar hydrophobicity which might be referred to as a "like-likes-like" concept. Thus it is possible that diffusion of molecules, seeking their similar partner, shifts the transition to higher temperature which otherwise would have occurred in the same temperature range observed on cooling where diffusion is not taken into account. In fact it would be difficult to imply any sort of diffusive transport when the two molecule dis- associates to form near neighbours. The second case involves some sort of structural re- arrangement occurring at high temperature which might increase the activation energy of the reversal process on cooling.

3.3 Rheology

Dynamic viscoelastic measurements for a 6.2 wt% solution of each of the three samples studied, on heating and cooling are shown in figures 4, 5 and 6. The heating rate (1°C/min) and the temperature range (10 to 78°C) were the same as for the differential scanning calorimeter experiment. All samples show the characteristic two stage increase of the elastic modulus (G') as previously reported by many other authors. [17-19, 21-23] The main difference among the samples is the temperature ranges in which the key changes in rheological properties occurs. Figure 4a shows the temperature course, on heating, of G' and G'' for sample A, which was previously argued to be the least hydrophobic.

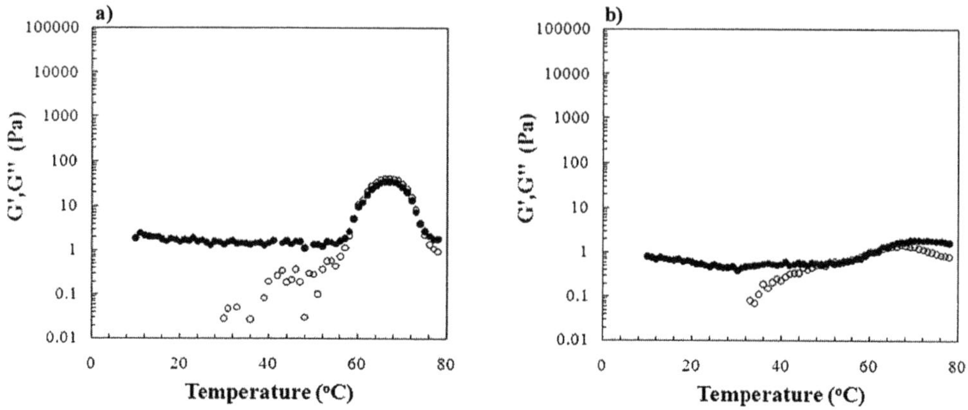

Fig.4 – Temperature course of G' (open symbols) and G" (filled symbol), on heating (a) and cooling (b), for sample A (DS_{Et}/DS_{Pr} = 5.8).

Initial values of G' are too low to be measured and starts to be detectable when temperatures approach 40°C. An increase in temperature is accompanied by the characteristic two step increase of the elastic response, with the onset of the second dramatic increase in modulus around 56°C. On the other hand G'' first decreases with increasing temperature, as expected for a homogeneous polymer solution, and then suddenly increases with an onset around 58°C. Two main "regions" of structure development might be identified. Below 40°C the rheological properties are typical of a solution with the elastic response less than the viscous response (G'<G''). In contrast at temperatures above 56°C the two moduli steeply increase and eventually G′ becomes greater than G'', indicating the "gel-like" character of the system which correlates excellently with the mid-point observed by the micro DSC measurements. However G' never considerably exceeds G'', which, in the absence of a frequency sweep measurement, is an indication of a weak gel. As the temperature increases further, the two moduli reach a maxima before dropping to a lower value, with G'' becoming greater than G' at the end of the heating cycle. The drop of the two moduli is believed to be a consequence of the macroscopic phase separation occurring at high temperature which accounts for the heterogeneity of the sample between the two plates in the rheometer. Indeed separation into polymer rich and polymer poor phases is what was observed visually in the test tube at high temperature (figure2). On cooling both moduli gradually decrease to their original value (figure 4b). The temperature at which the structure reverts to a solution (G''<G') closely matches the exotherm observed on cooling in the DSC experiment. However any conclusion on the cooling cycle must take into account heterogeneity caused by phase separation experienced by the system on heating. Figure 5a shows dynamic viscoelastic measurement results on heating for sample B. The shape of the two moduli is broadly similar to sample A; however the onset of the second step of structure development occurs around 43°C, almost 10 °C lower than sample A, again arguing differences in hydrophobicity being the cause. Above

43°C the solid –like behavior of the material is confirmed by G' becoming greater than G''. The two moduli show a maxima which is one order of magnitude higher than sample A. Moreover, the ratio of elastic over viscous modulus (G'/G'') is an indication of a stronger gel than sample A.

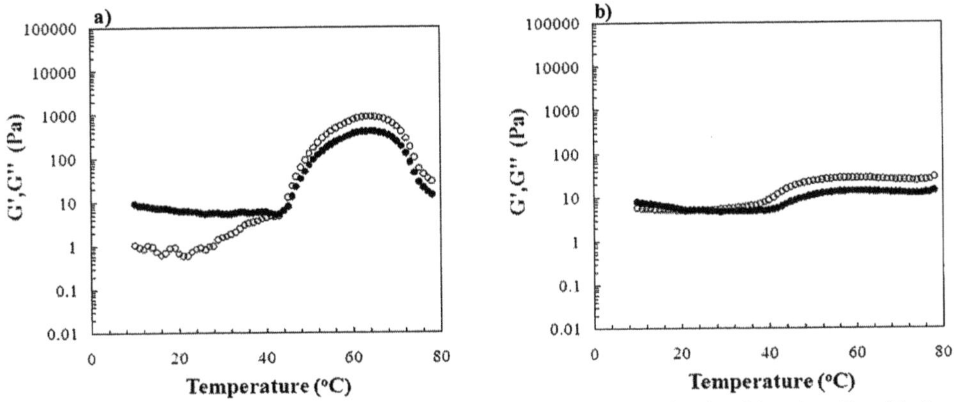

Fig.5 – Temperature course of G' (open symbols) and G" (filled symbol), on heating (a) and cooling (b), for sample B ($DS_{Et}/DS_{Pr} = 2.7$).

Further heating again implies phase separation occurring as a drop in both moduli at the end of the heating cycles. On cooling the two moduli gradually decrease to their original value with a two step shape, with the endset of the first steps matching the end of the exothermic transition observed in DSC. Figure 6 shows changes in rheological properties for sample C.

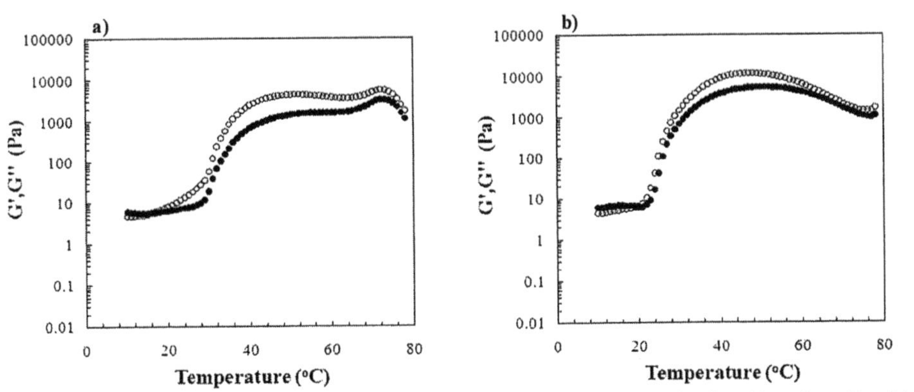

Fig.6 – Temperature course of G' (open symbols) and G" (filled symbol), on heating (a) and cooling (b), for sample C ($DS_{Et}/DS_{Pr} = 0.8$).

On heating (figure 6a) the two moduli increase immediately after the beginning of the experiment, with the onset of the second dramatic increase close to 30°C. The maximum value reached by G' is respectively one and two orders of magnitude higher than sample B and A. On cooling (figure 6b) the two moduli first increase to a value higher than a comparable temperature on heating then revert to their original value with little hysteresis. The temperature course of the rheological properties for the three samples confirm the presence of a specific trend which correlates with the ratio of the two alkyl moieties or the hydrophobicity, with the onset of the structure formation as well as the maximum elasticity reached on heating. For instance, sample A, the least hydrophobic, shows the highest onset temperature of structure formation and weakest gel (low G'/G'' ratio) among the sample studied. Therefore the higher the propyl content/hydrophobic character the higher the gel strength and the lower the onset of structure formation.

4. Deconvolution of thermal traces and further discussion

Deconvolution of the thermal traces was attempted using a similar fitting method to that reported by Sun *et al.*[12] Indication of the presence of more than one step in the aggregation of 3-0-ethyl-propyl cellulose was borne out from the asymmetry of the endothermic traces. Those steps are assumed to not simply superimpose on each other but considered as sequential equilibria. The uncertainty of the base line procedure carries the most part of the error associated with the fittings.

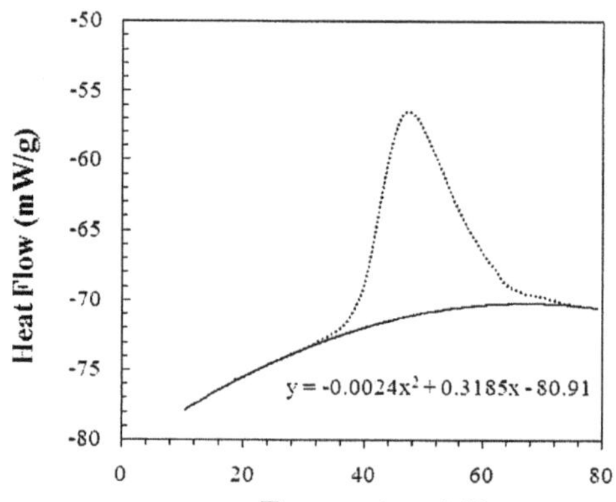

Fig.7 – Example of quadratic polynomial baseline

Different base line procedures were investigated but a robust and consistent one resulted to be a quadratic polynomial function from the leftmost to rightmost of the trace as shown in figure 7. For clarity only sample A and C will be discussed. Analysis of the reduced data, after the base line subtraction, on heating and cooling for sample A and C are reported respectively in figures 8 and 9. It was found, within experimental uncertainty, that the microDSC trace for sample A could be represented as the sum of three Gaussian functions, here referred to as component 1(P1), component 2 (P2) and component 3 (P3). On the other hand, simulation with two component transitions (P1 and P2) gave a good fit for sample C; however using more than three components gave no significant improvement for all the samples. Although the physical meaning of the deconvoluted components is indeed arguable, comparison with dynamic viscoelastic measurements gave rise to possible explanations. In fact there is a strict correlation between P2 and the major changes in the rheological properties occurring during the thermal cycle for both samples. For instance sample A, on heating, shows a peak temperature for P2 at about 58°C which corresponds to the onset of the dramatic increase in elastic modulus (figure 4a and figure 8a). On cooling a similar correspondence between P2 and the end of the major decrease in elastic modulus was also observed. A similar argument applies also to sample C (figure 4b and 8b). Thus it might be reasonable to propose that P2 describes the coming together of the hydrophobic moieties to form a network, in agreement with Sun *et al.* [12] As it was anticipated Sample A shows the presence of the component P1 (figure 9a).

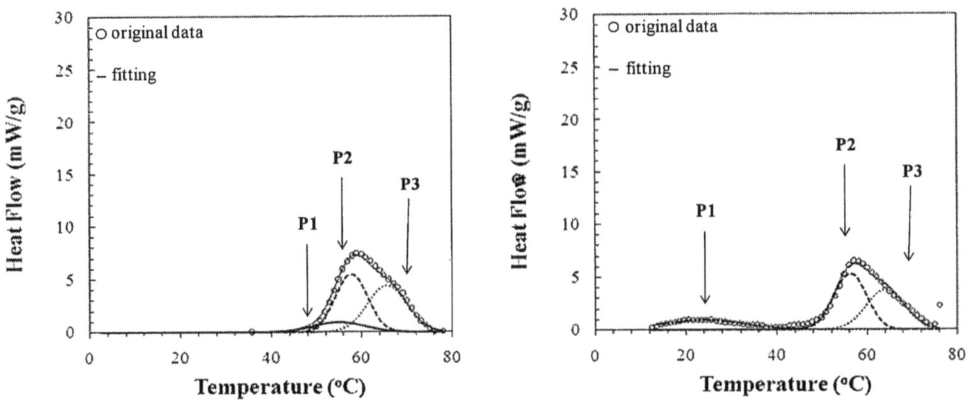

Fig.8 – Fitting for thermal trace of sample A on heating (a) and cooling (b).

In this temperature region however elastic modulus becomes detectable on heating (figure 4a). On cooling P1 is shifted to lower temperature and correlates to the region where the elastic modulus becomes undetectable once again (figure 4b). Sun *et al.* [12] in their study of 3-o-ethyl cellulose also reported three component transitions and the shifting to lower temperature on cooling was also observed. They proposed, based on the Haque et al.[17-18] "bundle" model, that P1 was related to the cellulose structure in solution. As supportive evidence it was shown that on cooling P1 correlates with the gel falling apart. In the present study such evidence was not observed, probably due to the phase separation occurring during the rheological experiment. Indeed, on cooling P1 does correlate with the elastic modulus becoming undetectable again. Comparatively, Sample C does not show any sign of the first component P1 (figure 9). For instance the increased level of propyl seems to suppress P1. The presence of order in the cellulose derivatives, as indicated by the apparent significance of C1, is thought to be due to the possibility of forming hydrogen bonds between P2 and P6 of anhydroglucose units. This was confirmed by wide angle x-ray diffraction in our lab (data not shown). Indeed it is possible that P1 is associated with the existence of residual ordering in the cellulose, weakened by the presence of the two moieties. Increasing level of propyl will reduce the ordering because of the high steric hindrance of the propyl compared to ethyl. Explanation of P3 is rather difficult. All samples show a great interdomain between P2 and P3 indicating that part of P3 is involved in the network formation. Sun *et al.*,[12] suggested that P3 was caused by further molecular rearrangements which stabilised the network. Although in absence of deeper research this cannot be discharged as a hypothesis, it is evident in the present work that the macroscopic phase separation occurs within the temperature range of P3, which therefore might indicate a link between the two.

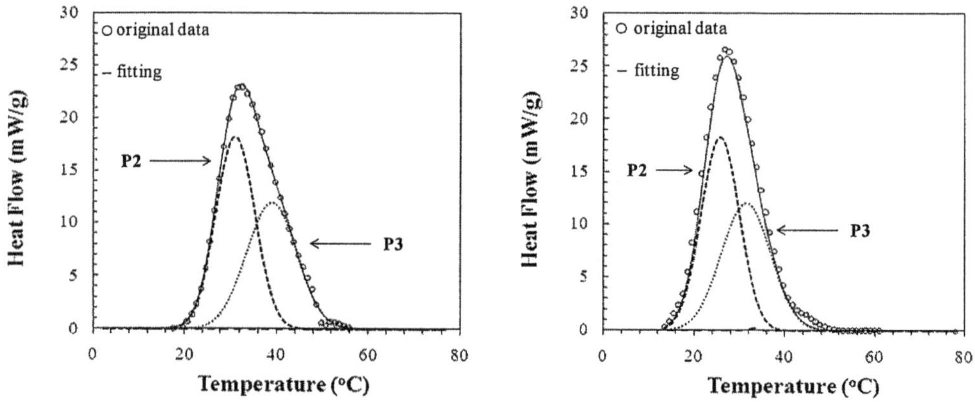

Fig.9 – Fitting for thermal trace of sample C on heating (a) and cooling (b).

5. CONCLUSION

This study showed the interesting thermally induced self-association of aqueous solutions of 3-0-ethyl-propyl cellulose. The aggregation is accompanied by an increase in the elastic modulus and formation of an opaque gel network. Data shows clearly that the ratio of the two alkyl moieties on the cellulose backbone has a great influence on the physical properties of those materials. For instance, the temperature of gelation decreases with increasing level of propyl which was explained in terms of different hydrophobicity among the samples studied. In fact it was assumed that the propyl group, because of its higher molecular size, is more hydrophobic than ethyl group, which gives rise to different hydrophobicity for the different samples studied. An attempt to deconvolute thermal traces highlighted the presence of multiple sequential equilibria characterised by an extensive over-lap of the fitted Gaussian curves which accounts for the complexity of the thermogelation process. However more work is needed to enlarge the knowledge about the physical meaning of those transitions.

Acknowledgment
The research leading to these results has received funding from the European Community's Seventh Framework Program (FP7/2007-2013) under Grant Agreement No. 214015.

Reference

[1] D. Klemm, B. Heublein, H.-P. Fink and, A. Bohn. Angewandte Chemie, Int. Ed. 2005, 44, 3358 – 3393.
[2] V. I. Kovalenko. Russian Chemical Reviews, 2010; 79(3): 231–241.
[3] N. Sarkar, US. Pat. 4,001,211, 1977.
[4] A.R. Rajabi-Siahboomi, R.W. Bowtell, P. Mansfield, A. Henderson, M.C. Davies and C.D. Melia. Journal of Controlled Release, 1994; 31(2): 121–8.
[5] W. Wang, Y.Z. Liu, Y.G. Gu. Colloid and Polymer Science, 2003; 281(11): 1046–54.
[6] T. Sanz, A. Salvador, S.M. Fiszman. Food Hydrocolloids, 2004; 18(2): 227–31.
[7] D. Klemm, B. Philipp, T. Heinze, U. Heinze, W. Wagenknecht. Comprehensive Cellulose Chemistry, 1st ed., Vol. 1 and 2, Wiley-VCH, Weinheim, 1998.
[8] H. A. Kressig. Cellulose-Structure, Accessibility and Reactivity. Gordon and Breach, Amsterdam, 1993.
[9] N. Sarkar. Applied Polymer Science, 1979; 24:1073–87.
[10] N. Sarkar and L. C. Walker. Carbohydrate Polymer, 1995; 27: 177.
[11] J. Guent, Thermoreversible Gelation of Polymers and Biopolymers; Academic Press: London, 1992.
[12] S.M. Sun, T.J. Foster, W. MacNaughtan, J.R. Mitchell, D. Fenn , A. Koschella, T. Heinze. Polymer Science Part B: Polymer Physic, 2009; 47:1743–1752.
[13] A. Koschella, D. Fenn , Fenn D, T. Heinze. Polymer Bulletin, 2006; 57:33–41.
[14] T. Heinze, A. Koschella. Carbohydrate Research, 2008; 7: 343-668.
[15] E. Heymann. Transanction of Faraday Society, 1935; 31: 846.
[16] M. Vigouret, M. Rinaudo, J. Desbrieres, Journal of Chemical Physics, 1996; 93: 858-869.
[17] A. Haque, E.R., Morris. Carbohydrate Polymer,1993; 22: 161–173.
[18] A. Haque, R.K. Richardson, E.R. Morris, M.J. Gidley, D.C. Caswell. Carbohydrate Polymer, 1993; 22: 175–186.
[19] J. Desbrieres, M. Hirrien and M. Rinaudo, Carbohydrate Polymer, 1998; 37: 145-152.
[20] S. C. Joshi and Y. C. Lam, J. Applied Polymer Science, 2006; 101: 1620–1629.
[21] M.Takahashi, M. Shimazaki, J. Yamamoto. J Polymer Science Part B: Polymer Physic. 2001; 39: 91–100.
[22] K. Nishinari, K. E. Hofmann, H. Moritakan, K. Kohyama and N. Nishinari. Macromolecular Chemistry and Physics. 1997; 198: 1217-1226.
[23] M. Hirrien, J. Desbrieres and M. Rinaudo. Carbohydrate Polymers. 1996; 31: 243-252.
[24] C. Tanford. The Hydrophobic Effect: Formation of Micelles and Biological Membranes; Wiley: NewYork, 1980.
[25] P. M. Wiggins. Physica Acta: Statistical and Theoretical Physics, 1997; 238: 113-128.
[26] D. A., Rees. Advanced Carbohydrate Chemistry and Biochemistry, 1969; 24: 267–332.
[27] R.N. Ibbett, K. Philp, D.M. Price. Polymer, 1992; 33(19):4087– 4095.
[28] T. Kato, M. Yokoyama, A. Takahashi. Colloid and Polymer Science. 1978; 256: 15–21.
[29] N. Illy. DiplomaThesis; University of Jena: Jena, Germany, 2006.
[30] J. Armstrong, B. Chowdhry, R. O'Brien, J. A. Beezer, J. Mitchell and S. Leharne. Journal of Physical Chemistry, 1995; 99: 4590-4598.

PRODUCTION OF CORN FIBER GUM UNDER CONDITIONS THAT RETAIN ITS FUNCTIONAL COMPONENTS

M. P. Yadav*, R. A. Moreau and K. B. Hick

Eastern Regional Research Center, Agricultural Research Service, U. S. Department of Agriculture, 600 East Mermaid Lane, Wyndmoor, PA 19038

* Author to whom correspondence should be addressed (e-mail madhav.yadav@ars.usda.gov; telephone 215-836-3783; fax 215-233-6406).

Mention of trade names or commercial products in this publication is solely for the purpose of providing specific information and does not imply recommendation or endorsement by the U.S. Department of Agriculture. USDA is an equal opportunity provider and employer.

1. INTRODUCTION

The most abundant low-valued by-product of industrial corn wet-milling, corn fiber (CF), contains a high percentage of valuable arabinoxylan (AX), which can be used to produce corn fiber gum (CFG) (1, 2). CFG has an arabinoxylan structure, consisting of a β-1, 4 linked D-xylopyranosyl backbone with α-L-arabinofuranosyl substituent attached at positions 2 and/or 3 and most of glucuronic acid at position 2 (3, 4, 5, 6, 7). CFG isolated from corn fiber by an alkaline hydrogen peroxide technology also contains small amounts of non-carbohydrate components that are either covalently linked or otherwise strongly associated with the polysaccharide structure. These components include protein, lipid and also phenolic compounds (8). AX polymers are cross-linked to each other and/or other cell wall polymers through dehydrodiferulate ester bridges (9, 10) and/or dehydrotriferulate ester bridges (11). Oligosaccharide fragments with arabinosyl units esterified at primary hydroxyl groups with ferulic acids obtained by enzymatic (12) and partial acid (13, 10) hydrolysis of corn fiber hemicelluloses further supports the presence of phenolic acids on corn AX. They are believed to be linked by an ester linkage, as they can be easily released from cell walls by alkali treatment. These hydroxycinnamic acids have been suggested to possess many health promoting properties. The ester-linked hydroxycinnamic acids can be released from the cell walls into the colon by the action of bacterial esterases and they may be absorbed or converted into other compounds by colonic bacteria (14). CFG is a very good emulsifier for oil-in-water emulsion system (1) and its emulsifying properties may be due to presence of protein and lipids on it. The aim of the present investigation is to develop CFG isolation conditions which can preserve its functional (protein and lipid) and nutraceutical (phenolic compounds) components.

2. EXPERIMENTAL METHODS

2.1. Standard Proximate Analyses. Moisture, protein (N × 6.25) and ash contents of all CFG samples were determined using "AACC Approved Methods" 44-19, 46-30 and 08-01 respectively (15).
2.2. Isolation of Corn Fiber Gum. CFGs were isolated from de-oiled and de-starched corn fiber by previously reported alkaline hydrogen peroxide technology (2) with some modification. The following three kinds of fibers were used for CFG isolation: (a) wet milling pericarp fiber, WPF (b) wet milling endosperm fiber, WEF and (c) wet milling pericarp and endosperm fiber, WPEF. The isolation was done with 2, 4 and 8 meq sodium hydroxide and calcium hydroxide alkali mixture (equal amount of each) per gram of fiber by heating at 100 °C for 15 min or 1 hr.
2.3. Methanolic KOH Hydrolysis and Extraction of total Lipid with Chloroform/Methanol. The de-oiled and de-starched CFG was treatedwith 1.5M methanolic KOH containing 5% water at 70 °C with stirring for 1 hr and extracted with chloroform and methanol mixture (8).
2.4. Phenolic Acid Analysis. The phenolic acids were analyzed by HPLC containing both UV and evaporative light-scattering detectors (ELSD). P-Coumaric and ferulic acids were quantified by UV absorption at 320 nm by comparing with the peak area of known amount of standard p-coumaric acid and ferulic acid (8).

3. RESULTS AND DISCUSSION

3.1. Isolation of CFGs. From the de-oiled and de-starched CF, CFG was extracted with 2, 4 or 8 meq of alkali mixture (NaOH and Ca $(OH)_2$) per gram fiber at pH 11.5 in the presence of 0.25g H_2O_2/g fiber by boiling the reaction mixture for 15 min or 1 hr. The CFG extraction was done from three fiber sources (WPF, WEF and WPEF) with three alkali concentrations (2, 4 or 8 meq/g fiber) at two different incubation times (15 min. or 1h) to see the effect of alkali concentration and heating time on CFG yield and its functional components retention. The oil present in CF was removed by hexane extraction (16) to avoid its potential oxidation products in CFG. The starch from de-oiled fiber was removed by boiling with heat stable α-amylase to avoid alkali soluble starch contamination in CFG. The yield of CFG from WPF, WEF and WPEF varies from 29.6 to 32.8, 16.0 to 21.2 and 25.6 to 39.2% respectively from de-oiled and de-starched corn fiber (Table 1).

Table 1. Percentage Yields[a] of Corn Fiber Gum made under Different Isolation Conditions

alkali concentration	2 meq/g fiber		4 meq/g fiber		8 meq/g fiber	
	Isolation Time at 100 $^{\circ}$C					
sample	15 min	1 hr	15 min	1 hr	15 min	1 hr
wet milling pericarp fiber	32.8	30.0	36.0	30.0	29.6	31.2
wet milling endosperm fiber	16.0	21.2	17.2	18.8	18.8	18.4
wet milling pericarp and endosperm fiber	26.0	29.2	39.2	32.8	25.6	32.4

[a] Based on the amount of de-oiled and de-starched corn fiber

The yield of CFG from WPF is higher than the endosperm originating fiber (WEF) as previously reported (2). It is clear from **Table 1** that there is no large increase in CFG yield due to increase in alkali concentration or incubation time. Thus it indicates that more that 2 meq/g fiber alkali concentration and longer than 15 min incubation time in the presence of 0.25g H_2O_2/g fiber is not too essential in the extraction medium. It also indicates that most of the AX is linked to the cell wall by ester linkages, which can easily be cleaved by low alkali concentration and shorter heating time.

3.2 Moisture and ash content. Moisture and ash contents of the CFG samples are reported in Table 2 and there are no obvious trends observed with change in extraction conditions.

Table 2. Moisture and Ash Content of Corn Fiber Gum Samples
Wt. percent based on the dry weight of de-oiled and de-starched CFG samples.

alkali concentration	isolation time at 100 °C	moisture	ash
wet milling pericarp fiber			
2 meq/g fiber	15 min	4.77	4.81
	1 hr	5.26	3.05
4 meq/g fiber	15 min	4.0	3.93
	1 hr	5.91	4.55
8 meq/g fiber	15 min	5.42	4.88
	1 hr	5.93	4.44
wet milling endosperm fiber			
2 meq/g fiber	15 min	3.83	5.63
	1 hr	6.56	4.24
4 meq/g fiber	15 min	4.67	5.38
	1 hr	5.78	4.25
8 meq/g fiber	15 min	5.59	5.79
	1 hr	6.75	5.91
wet milling pericarp and endosperm fiber			
2 meq/g fiber	15 min	5.25	3.36
	1 hr	5.44	3.71
4 meq/g fiber	15 min	4.44	3.48
	1 hr	5.98	3.2
8 meq/g fiber	15 min	6.01	4.42
	1 hr	7.33	4.21

3.3. Protein Content in CFGs. The protein content of CFG isolated from three fiber sources with three alkali concentrations (2, 4 or 8 meq/g fiber) at two different incubation times is shown in **Figure 1**. The protein content in CFG isolated from WEF is higher than the CFG isolated from WPF or WPEF supports the previous finding that fiber obtained from corn endosperm contains more protein than the pericarp originating fiber (1). The protein content in CFG isolated with 2 meq alkali mixture/g fiber for 15 min incubation time from all three fiber sources is higher than CFG isolated with higher alkali concentration and longer incubation time. As the alkali concentration is increased, we see a gradual decrease in the protein content in the CFG for all fiber sources. The percent of protein is higher in CFG isolated with 15 min compared to 1 hr incubation time from all fiber sources except CFG extraction with 4 meq alkali mixture/g fiber from WPF in which their yields are almost similar. The importance of protein in the emulsifying activities of naturally occurring emulsifiers, gum Arabic and CFG is reported by many investigators (17, 18, 2, 1, 19). The hydrophobic protein component of these polymers adsorb onto the oil-water interface of emulsions and the hydrophilic carbohydrate groups extend into the aqueous phase to stabilize against droplet aggregation. Thus the CFG extraction conditions which preserve the highest percent of its protein is important for making a good emulsifier.

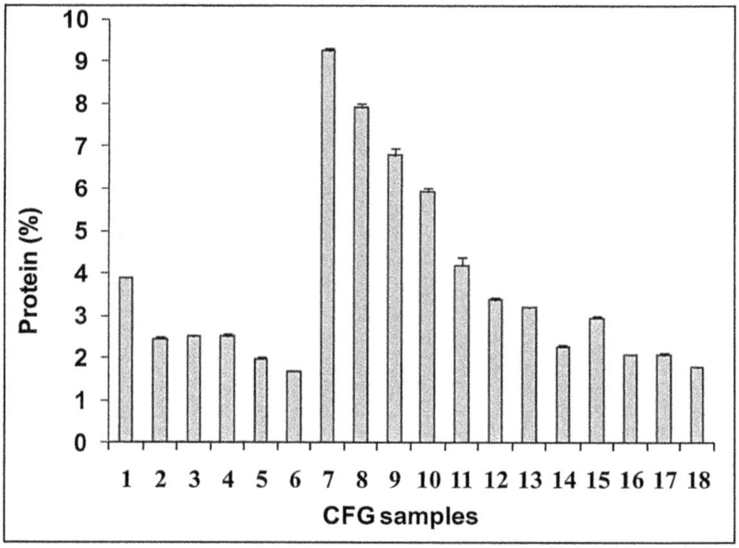

Figure 1. Protein content of corn fiber gum isolated with 2, 4 or 8 meq alkali by heating at 100 ^{0}C with the following alkali concentration and isolation time: (1) CFG (WPF), 2 meq, 15 min; (2) CFG, (WPF) 2 meq, 1hr; (3) CFG, (WPF) 4 meq, 15 min; (4) CFG, (WPF) 4 meq, 1 hr; (5) CFG, (WPF) 8 meq, 15 min; (6) CFG, (WPF) 8 meq, 1 hr; (7) CFG, (WEF) 2 meq, 15 min; (8) CFG, (WEF) 2 meq, 1 hr; (9) CFG, (WEF) 4 meq, 15 min; (10) CFG, (WEF) 4 meq, 1 hr; (11) CFG, (WEF) 8 meq, 15 min; (12) CFG, (WEF) 8 meq, 1 hr; (13) CFG, (WPEF) 2 meq, 15 min; (14) CFG, (WPEF) 2 meq, 1 hr; (15) CFG, (WPEF) 4 meq, 15 min; (16) CFG, (WPEF) 4 meq, 1 hr; (17) CFG, (WPEF) 8 meq, 15 min; and (18) CFG, (WPEF) 8 meq, 1 hr. Data are an average of six trials ± standard deviation.

3.4. Total Lipid content in CFG. The CFG samples were hydrolyzed with a strong solution of alkali in methanol by incubating at 70 °C for 1 hr. The hydrolyzed lipids were extracted with chloroform/methanol, evaporated and their yields were calculated. The lipid content in CFGs isolated from three fiber sources with 2, 4, and 8 meq alkali mixture/g fiber by heating the reaction mixture for 15 min or 1 hr is shown in **Figure 2**. The amount of lipid (mg/g) in CFG isolated from WPF decreases gradually with alkali concentration and its extraction time increases showing that they are linked to arabinoxylan by an alkali susceptible ester linkage. Thus a low alkali concentration (2 meq/g) and short heating time (15 min) is required to preserve lipid on CFG during its isolation from WPF. But the same low alkali concentration and short heating time is not the optimum condition for CFG isolation from WEF and WPEF for maximum preservation of lipid. In the case of WEF, we see the highest lipid content in the CFG which is isolated either with 2 meq alkali/fiber for 1 hr or 4 meq alkali/g fiber for 15 min. But in the case of WPEF, 4 meq alkali and 15 min incubation time gives the CFG with the maximum lipid content in comparison to other alkali concentration and fiber incubation time. Thus it seems that there is not one optimum condition to isolate CFG from different fiber sources preserving most of its lipid. But definitely, alkali concentrations more than 4 meq/g fiber is not desirable, as CFG obtained from all three fiber sources by isolating with 8 meq/g fiber either for 15 min or 1 hr heating has comparatively lower lipid content than the CFG obtained with lower alkali concentration.

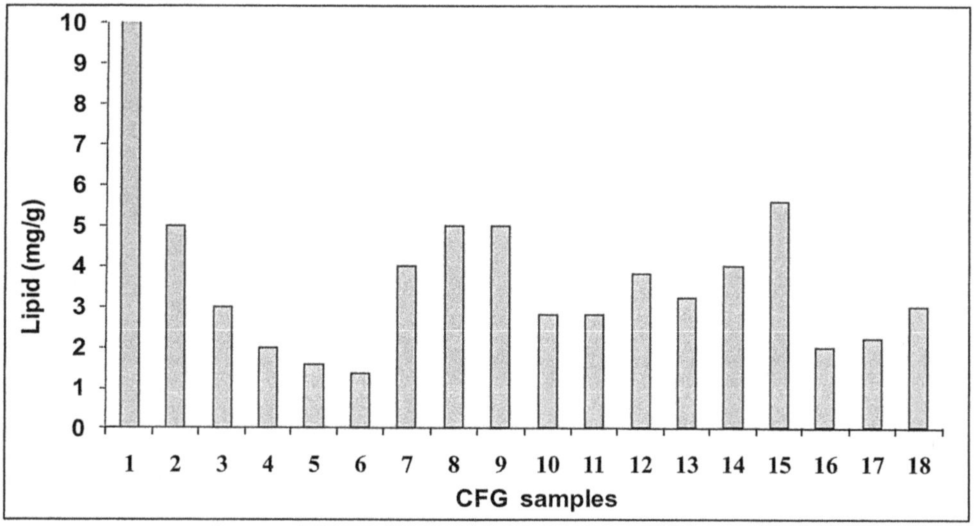

Figure 2. Lipid content of corn fiber gum isolated with 2, 4 or 8 meq alkali by heating at 100 ^{O}C with the following alkali concentration and isolation time: (1) CFG (WPF), 2 meq, 15 min; (2) CFG, (WPF) 2 meq, 1hr; (3) CFG, (WPF) 4 meq, 15 min; (4) CFG, (WPF) 4 meq, 1 hr; (5) CFG, (WPF) 8 meq, 15 min; (6) CFG, (WPF) 8 meq, 1 hr; (7) CFG, (WEF) 2 meq, 15 min; (8) CFG, (WEF) 2 meq, 1 hr; (9) CFG, (WEF) 4 meq, 15 min; (10) CFG, (WEF) 4 meq, 1 hr; (11) CFG, (WEF) 8 meq, 15 min; (12) CFG, (WEF) 8 meq, 1 hr; (13) CFG, (WPEF) 2 meq, 15 min; (14) CFG, (WPEF) 2 meq, 1 hr; (15) CFG, (WPEF) 4 meq, 15 min; (16) CFG, (WPEF) 4 meq, 1 hr; (17) CFG, (WPEF) 8 meq, 15 min; and (18) CFG, (WPEF) 8 meq, 1 hr.

3.5. Phenolic Acids Content in CFG. The alkali-released ferulic and p-coumaric acids of CFG were analyzed by binary HPLC system. HPLC chromatograms of standard p-coumaric (*trans*-p-coumaric) and ferulic (*trans*-ferulic) acids obtained by UV detection were used for identification and quantization. The UV detector peaks in CFG samples were identified by comparing their retention times (T_r) with standard phenolic acids (*trans*-p-coumaric and *trans*-ferulic). The yield of total phenolic acids (ferulic and p-coumaric acids) present in CFGs obtained from three fiber sources with 2, 4, and 8 meq alkali mixture/g fiber by heating the reaction mixture for 15 min and 1 hr are shown in **Figure 3**. The amount of total phenolic acids in CFGs extracted from all three fiber sources by 2 meq alkali mixture/g fiber heating at 100 °C for 15 min is the highest in comparison to CFGs extracted with higher alkali concentration and longer heating time. The amount of phenolic acids in CFGs extracted from WPF, WEF and WPEF with 2 meq alkali mixture/g fiber for 15 min heating is above 600, 275 and 200 μg/g CFG sample respectively. When these fibers are extracted with the same alkali concentration (2 meq alkali mixture/g fiber) but heating for 1 hr, the phenolic acids content in CFGs from WPF and WPEF decreases slightly but the amount of phenolic acids in CFG from WEF decreases dramatically to below 30 μg/g showing a big effect of heating for a longer than 15 min. The amount of total phenolic acids in CFGs extracted with 4 and 8 meq alkali mixture/g fiber by boiling either for 15 min or 1 hr is below 100 μg/g sample and in some cases even below 30 μg/g CFG sample. These results clearly indicate that for preserving the nutraceutical phenolic compounds, CFG needs to be extracted with a low alkali concentration and a short heating time at 100 °C in presence of H_2O_2. These results totally agree with Carvajal-Millan et al. (20), who reported that the weight percent of ferulic in corn arbinoxylan decreased from 1.6 to 0.6, 0.34, 0.22, 0.15 and 0.1 μg/mg as the incubation of corn bran in 0.5 M NaOH increased gradually from 2 hr to 4, 8, 12, 16, and 24 hr respectively. These results also support other investigators' findings that some phenolic acids are esterified to cell wall hemicelluloses in addition to their alkali stable ether linkage (21, 22), which could be susceptible to high alkali concentration and longer treatment time. As previously reported (8), CFG isolated from WPF (pericarp originating fiber) contains a higher amount of phenolic acids than WEF (endosperm originating fiber) agreeing with Seitz, L. M. (23) finding that most of the ferulate ester are present on the inner pericarp portion of the fiber.

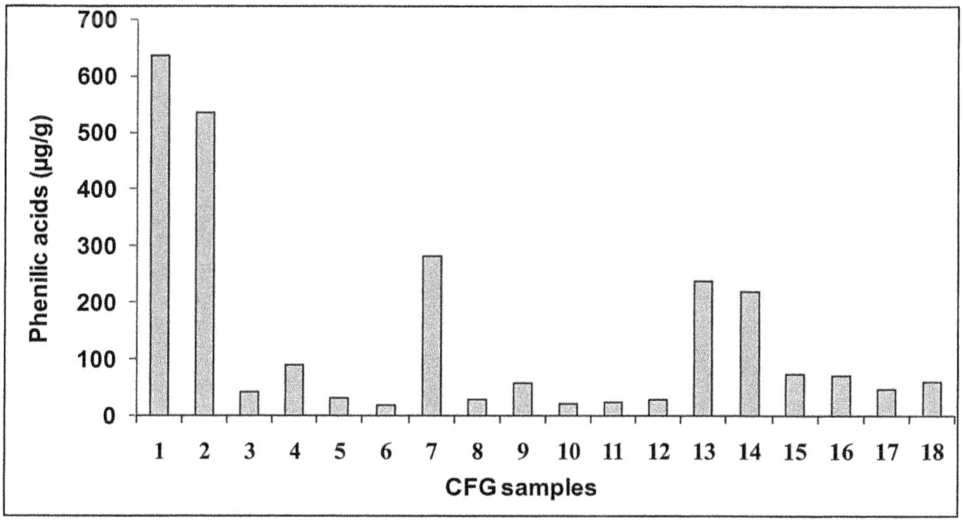

Figure 3. Phenolic acids content of corn fiber gum isolated with 2, 4 or 8 meq alkali by heating at 100 °C with the following alkali concentration and isolation time: (1) CFG (WPF), 2 meq, 15 min; (2) CFG, (WPF) 2 meq, 1hr; (3) CFG, (WPF) 4 meq, 15 min; (4) CFG, (WPF) 4 meq, 1 hr; (5) CFG, (WPF) 8 meq, 15 min; (6) CFG, (WPF) 8 meq, 1 hr; (7) CFG, (WEF) 2 meq, 15 min; (8) CFG, (WEF) 2 meq, 1 hr; (9) CFG, (WEF) 4 meq, 15 min; (10) CFG, (WEF) 4 meq, 1 hr; (11) CFG, (WEF) 8 meq, 15 min; (12) CFG, (WEF) 8 meq, 1 hr; (13) CFG, (WPEF) 2 meq, 15 min; (14) CFG, (WPEF) 2 meq, 1 hr; (15) CFG, (WPEF) 4 meq, 15 min; (16) CFG, (WPEF) 4 meq, 1 hr; (17) CFG, (WPEF) 8 meq, 15 min; and (18) CFG, (WPEF) 8 meq, 1 hr.

4. CONCLUSION

The amount of protein and total phenolic acids in CFGs extracted from all three fiber sources by 2 meq alkali mixture/g fiber heating at 100 °C for 15 min is the highest in comparison to CFGs extracted with higher alkali concentration and longer heating time. The protein content in CFG isolated from WEF is higher than the CFG isolated from WPF or WPEF. The total phenolic acid content in CFG from coarse corn fiber (pericarp fiber) is comparatively higher than CFG from fine corn fiber (endosperm fiber) and the mixture of coarse and fine fiber. The amount of lipid in CFG isolated from WPF with 2 meq alkali mixture/g fiber heating at 100 °C for 15 min is the highest in comparison to CFGs extracted with higher alkali concentration and longer heating time. But CFG extraction from WEF and WPEF requires 4 meq alkali mixture/g fiber to give the highest lipid content.

References

1. M. P. Yadav, D. B. Johnston, A. T. Hotchkiss and K. B. Hicks, *Food Hydrocolloids,* 2007, *21*, 1022.
2. M. P. Yadav, P. Cooke, D. B. Johnston and K. B. Hicks, *Cereal Chemistry, 2010, 87* (2), 89-94.

3. M. P. Yadav, M. L. Fishman; H. K. Chau, D. B. Johnston and K. B. Hicks, *Cereal Chemistry*, 2007, **84** (2), 175.
4. R. Montgomery and F. Smith, *Journal of American Chemical Society*, 1957, ***79***, 695-697.
5. R. Montgomery, F. Smith and H. C. Srivastava, *Journal of American Chemical Society*, 1957, ***79***, 698-700.
6. R. L. Whistler and W. M. Corbett, Journal *of American Chemical Society* 1955, ***77***, 628-6330.
7. R. L. Whistler and J. N. BeMiler, *Journal of American Chemical Society* 1956, **78**, 1163
8. M. P. Yadav, R. A. Moreau and K. B. Hicks, *J. Agric. Food Chem.* 2007, **55**:943.
9. E. Chanliaud, L. Saulnier and J.-F. Thibault, *Journal of Cereal Science*, 1994, ***21***, 195-203.
10. L. Saulnier, J. Vigourous and J.-F. Thibault, *Carbohydrate Research*, 1995, ***272***, 241-253.
11. C. Funk, J. Ralph, H. Steinhart and M. Bunzel, *Phytochemistry*, 2005, ***66***, 363-371.
12. T. Ohta, S. Yamasaki, Y. Egashira and H. Sanada, *J. Agric. Food Chem.*, 1994, ***42***, 653-656.
13. S. Yoshida, I. Kusakabe, N. Matsuo, T. Ono, K. Shimizu, T. Yasui and K. Murakami, *Agric. Biol. Chem.* 1990, **54**:1319-1321.
14. P. A. Karoon, C. B. Faulds, P. Ryden, J. S. Robertson and G. Williamson, *J. Agric. Food Chem.*, 1997, ***45***, 661-667.
15. American Association of Cereal Chemists. *Approved Methods of the AACC.* In *Methods 08-01, 44-19 and 46-30;* 9th ed.; The Association: St. Paul, MN, 1995.
16. R. A. Moreau, M. J. Powell and K. B. Kicks, *J. Agric. Food Chem., 1996, **44***, 2149.
17. R. C. Randall, G. O. Phillips and P. A. Williams, *Food Hydrocolloids* 1988, **2**:131-140.
18. E. Dickinson, *Food Hydrocolloids* 2003, **1**:25-39.
19. M. P. Yadav, P. Nicholas, D. B. Johnston, C. I. Onwulata and K. B. Hicks, *Carbohydrate Polymers*, 2010, **81**, 476-483.
20. E. Carvajal-Millan, A. Rascon-Chu, J. A. Marquez-Escalante, V. Micard, N. P. Leon and A. Gardea, *Carbohydrate Polymers*, 2007, **69**, 280-285.
21. R. C. Sun, B. Xiao and X. F. Sun, *J. Agric. Food Chem. 2001, **49**, 5122-5129.*
22. I. M. Harvey, R. D. Hartly, P. J. Harris and E. H. Curzon, *Carbohydrate Research*, 1986, **148**, 71-85.
23. L. M. Seitz, J. *Agric. Food Chem.* 1989, ***37***, 662.

RELATIONSHIP BETWEEN THE CONFORMATIONAL STRUCTURE AND PROCESSABILITY OF HYDROCOLLOIDS

F. Hajji[1], J.R. Mitchell[1, 2], T.J. Foster[1]

[1] Division of Food Sciences, School of Biosciences, The University of Nottingham, Sutton Bonington Campus, Leicestershire, LE12 5RD, United Kingdom.
[2] Biopolymer Solutions Ltd., Sutton Bonington Campus, Leicestershire, LE12 5RD, United Kingdom.

1 INTRODUCTION

Native starch and cellulose are semi-crystalline biohydrocolloids, by nature, structurally and functionally very weak for a diversified application in the (non-)food industry. Neither starch, nor cellulose can be processed easily by conventional thermoplastic processes, such as extrusion, injection moulding, kneading, casting or compression moulding, because both have to be disordered (i.e., gelatinized, melted or destructured) via high temperature, shearing and a suitable plasticizer [1, 2]. In case of starch, susceptibility to physical processing is determined by the ratio of amylose and amylopectin in the granular structure. High-amylose starch (long chain molecule) requires higher mechanical energy input than amylopectin (short-chain branched molecule) because of the high melt viscosity related to the higher gelatinization temperature and greater molecular entanglements, although the drawback is alleviated by better mechanical properties [3-6]. An intrinsic analogy is given between amylose and cellulose, the latter enunciates its very high melting viscosity due to its highly ordered, chain-extended backbone [7], even though aided by the equatorial conformation of β-1,4-D-glucopyranose units [8].

Hitherto, chemical modification (e.g. derivatization of the hydroxyl groups) was the most versatile pathway to integrate thermoplasticity for creating appropriate processable starch and cellulose based biomaterials. In previous work, however, Tipples (1969) [9] and Sereno et al. (2006) [10] have pointed out the value and effectiveness of physical modification of biohydrocolloids for imparting superior techno-functionalities without resorting to chemistry. In particular, the potential of the low moisture extrusion processing to modify the starch functionalities has been shown [11-13] but has not been considered for native cellulose, mainly due to the high crystalline order, imparted by the strong intermolecular hydrogen bonds.

The question, therefore, arises as to whether crystalline or mechanically induced amorphous cellulose is amenable to thermoplastic processing and the required energy input for processing samples with difference degree of structural order. However, there is the need of more research to find and improve routes towards obtaining easily processable and physically tailored biomaterials.

The present research aims is to quantify the extent of ball milling time on the order to disorder phase transition in biopolymers and estimate the significance of the chemical structure for enabling cellulose thermoplastic processability. Therefore, firstly, the degree of mechanical decrystallization by high energy ball milling and the extent of recrystallization at low water content were examined, and secondly, the ease of thermo-mechanical processing of ball milled cellulose was studied using a twin-screw extrusion process.

2 MATERIALS AND METHODS

2.1 Materials

Waxy maize starch (**WMS**) and high-amylose maize starch (**HAS**), were supplied by an industrial starch supplier. The declared amylose content of the starches was approximately 2 % for WMS and 70 % for HAS. Eucalyptus cellulose (abbreviated by **EC**) was kindly donated by Lenzing AG (Lenzing, Austria). Magnesium nitrate ($Mg(NO_3)_2$) and potassium sulphate (K_2SO_4) were purchased commercially from Fisher Scientific Co. (Loughborough, UK).

2.2 Sample equilibration

Three semi-crystalline polysaccharides were dried in a desiccator over phosphorus pentoxide (P_2O_5) and subsequently were equilibrated isopiestically over saturated salt solutions of $Mg(NO_3)_2$ and K_2SO_4 to known controlled moisture contents, tested by the weight gain or loss. The degree of crystallinity of the original and ball milled samples, which equilibrated to approximately 53 % and 97 % relative humidity for 27 days at ambient temperature, were gained by wide angle x-ray diffraction measurements.

2.3 Wide angle X-ray diffraction

Crystallinity values of the original and treated samples were generated from diffractograms using a Bruker AXS D5005 wide angle X-ray diffractometer (Karlsruhe, Germany), equipped with a Ni-filtered CuKα radiation (λ= 1.5405 Å) operating at a high voltage of 40 kV and a current of 40 mA. Measurements were performed at angular increments of 0.05° 2θ and a counting time of 2.5 s/point in the angular range of $5° < 2\theta < 45°$. The details about the method and the calculation of the degree of crystallinity are given elsewhere [14, 15].

2.4 Scanning electron microscopy

Surface morphology of a few specimens were viewed and photographed with a scanning electron microscope (ESEM XL30 and SEM-505, Philips Co., The Netherlands) at an accelerating voltage of 10 kV. The samples were previously sprinkled onto adhesive carbon tapes, supported by circular silver stubs in which a gold coater was not necessary.

2.5 Processing procedure

2.5.1 Ball milling

A high-energy planetary ball mill (Pulverisette-7, Fritsch GmbH, Germany) was employed for extensive destruction of the samples at a rotating speed of 400 rpm. Two agate bowls (12 ml volume), each charged with five agate balls (10 mm Ø) and 0.2 g (wb) sample. The planetary ball mill "Pulverisette-5" (Fritsch GmbH, Germany) was chosen to produce larger quantities of the amorphicized powder for the extrusion trials. The mechanical treatments were performed at 200 rpm with zirconium oxide bowls (250 ml volume) containing each ten zirconium oxide balls (15 mm Ø) and 8 g (wb) samples. Specimens were milled for varied process time allowing 10 min pause in-between to avoid over-heating.

2.5.2 Extrusion

Control (non-ball milled) and ball milled cellulose was mixed with water as plasticizer (30 % wb) and was processed with a Haake MiniLab II twin-screw extruder (Thermo Electron Corp., Hamburg, Germany). The extrusion trials were conducted using a co-rotating screw configuration with a conical screw design without inherent distributive mixing elements, but being able to continuously re-circulate the melt. The mixture was manually added to the extruder feed port using a funnel in an effort to starve-feed the extruder at a screw speed of 100 rpm. A silver dowel rod with a copper end was used to force the material into the extruder during a process time of 5 min.

3 RESULTS AND DISCUSSION

3.1 Structural degradation during ball milling of biopolymers

The effect of high-energy ball milling on the structural degradation of WMS, HAS and EC are shown in Figures 1 - 3. The diffractograms of untreated WMS and HAS displayed the A-type and B-type polymorph, respectively. The former type of starch polymorphism is represented by the strong reflections at 2θ of about 14.95°, 17.05°, 17.85° and 22.95°, whereas the latter polymorph is described by the peaks at 17.35°, 19.75°, 22.15° and 23.3° (2θ). Natural cellulose is usually biosynthesized into the crystalline structure of cellulose I as indicated for EC through the X-ray diffraction peaks at 15.2°, 22.6° and 34.1° (2θ). The ball milling process caused a loss of macromolecular structural order in starch and cellulose but to a different magnitude. The degree of conversion from a semi-crystalline to amorphous state in WMS and HAS has already started to occur after 60 min ball milling, compared to cellulose that required at least 360 min mechanical treatment. This result affirmed that higher energy is needed to achieve considerable destruction of cellulose, depicted graphically by a "bell-shaped" curve [16]. Furthermore, ball milling is successful in weakening and unlocking the intermolecular hydrogen bonds in starch and cellulose, in the absence of high temperatures and a plasticizer [17], a process known as "cold melting" [18].

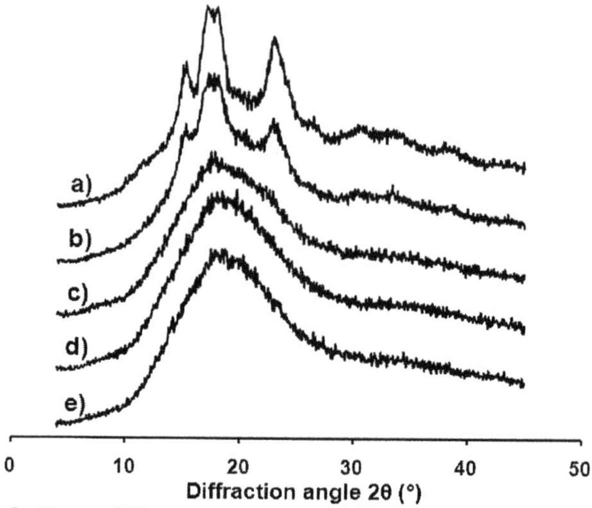

Figure 1 *Wide-angle X-ray diffraction patterns of the WMS at different ball milling time: a) 0 min, b) 5 min, c) 60 min, d) 360 min and e) 480 min.*

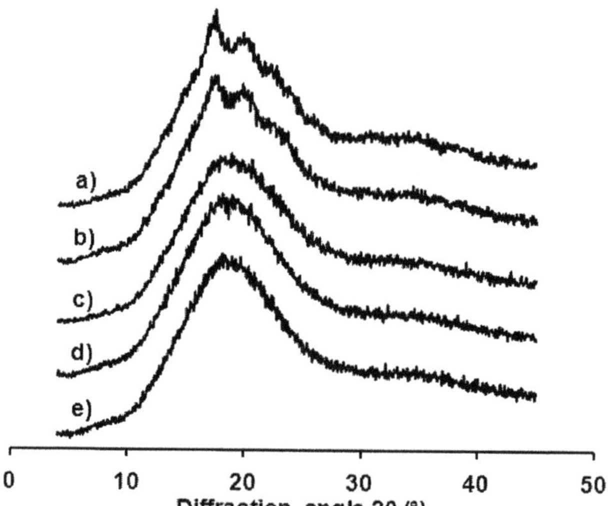

Figure 2 *Wide-angle X-ray diffraction patterns of the HAS at different ball milling time: a) 0 min, b) 5 min, c) 60 min, d) 360 min and e) 480 min.*

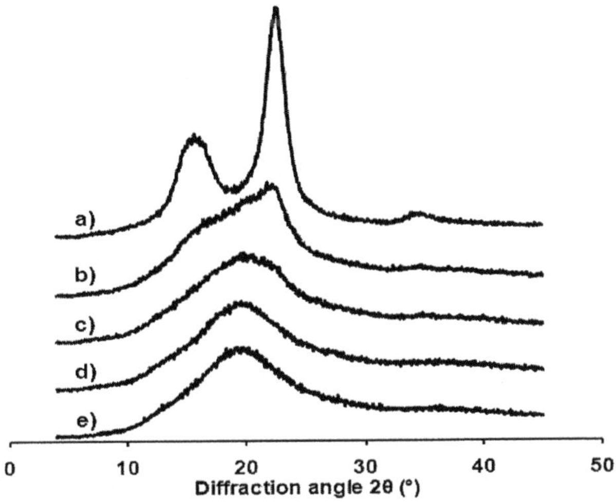

Figure 3 *Wide-angle X-ray diffraction patterns of the EC at different ball milling time: a) 0 min, b) 180 min, c) 360 min, d) 840 min and e) 1080 min.*

The data (Table 1) suggests that WMS (amylopectin-rich starch) was more prone to mechanical stress than HAS (amylose-rich starch) arising from molecular structure (chain length, degree of branching and polydispersity), composition (amylose content) and granular architecture (amorphous: crystalline ratio of amylopectin) [19, 20].

On the other hand EC exhibited a greater stability to mechanical degradation than WMS (Table 1 and 2), probably due to the presence of strong intermolecular hydrogen bonds, which occupies an equatorial position in the D-glucopyranose ring, and thus, can resist higher external stress.

Table 1 Crystallinity values of native and ball milled starch samples

	WMS			**HAS**		
RH (%)	**0**	**53**	**97**	**0**	**53**	**97**
BMT (min)	**Degree of crystallinity (%)**			**Degree of crystallinity (%)**		
0	32.42	33.28	39.07	7.88	11.17	18.34
5	22.46	24.39	33.13	4.37	4.88	13.01
60	2.77	2.89	17.69	2.28	3.12	8.06
360	1.65	2.18	13.97	1.7	1.75	7.46
480	0.69	2.46	15.86	1.28	2.88	8.88

RH denotes the relative humidity, and BMT stands for ball milling time.

Table 2 Crystallinity values of native and ball milled EC

RH (%)	0	53		97	
		Degree of crystallinity (%)			
BMT (min)	C I	C I	C II	C I	C II
0	48.19	48.29	0	48.81	0
360	10.42	12.06	6.64	17.12	9.66
480	4.24	5.73	6.97	6.63	18.24
840	3.05	3.89	7.03	5.09	20.24
1080	2.73	2.99	7.95	4.74	20.45

RH denotes the relative humidity, and BMT stands for ball milling time.

3.2 Crystalline transformation of ball mill-induced amorphous cellulose and starch

The degree of crystallization in amorphous WMS, HAS and EC is displayed in Figure 4 - 6 as response to the extent of mechanical stress by a high-energy ball milling and the relative humidities. The crystallization of ball milled WMS and HAS have experienced after 20 days of equilibration at relative humidities of 53 % and 97 % a return to their original crystalline structure, as evidenced by their characteristic diffraction pattern. The comparison of the diffractograms of WMS (Figure 1 and 4) showed the rearrangement into energetically more favourable and thermodynamically more stable A-type polymorph, whereas HAS (Figure 2 and 5) is converted to B-type structure. The starch diffraction patterns revealed sharper peaks and new shoulders, and thus, an increase in crystallinity and decrease in amorphous regions, respectively (Table 1). However, the level of hydration and the ball milling time determine the rate and degree of recrystallization in which differences between WMS and HAS result from the amylose content and the low-molecular weight fragments of amylopectin, as response of the magnitude of mechanical damage. The event of recrystallization of HAS was found to be fast and reaches completion relative to the WMS, due to the extensively long chain-length of the molecule. Burchard (1962) stated the influence of the degree of polymerization on the rate of crystallization of amylose and formulated an explanation for the observed effects based on the chain-length dependent development of intra/intermolecular hydrogen bonds [21], while Gidley and Bulpin (1987) argued the need of a minimum chain-length of 10 glucose residues required for the formation of double helices (crystallization) [22].

An interesting opportunity arises from the physically modified WMS becoming cold water soluble, concurrent with the capability to form films in low water content environments, as illustrated in Figure 7, by a coherent starch matrix at the macro- and micro-level [23, 24]. Although, the 5 min ball milled WMS does not have film-forming properties based on the stability and intactness of the crystalline structure under this process history that regulated the accessibility of water.

The view on the recrystallization behaviour in ball mill-induced amorphous EC bares two different trends (Figure 6). The short-time ball milled EC (180 min) recrystallized mainly to cellulose I (C I), presumably, because of high amounts of remnants of the original crystal structure in ball milled EC which acts as seeds for producing cellulose I crystals during the 27 days of storage time at 97 % relative humidity, as shown by Paes et al. (2010) [16]. The long-time ball milled samples (360, 840 and 1080 min) were primarily rearranged into the thermodynamically favourable stable form of cellulose II (C II) since the memory of the native polymorph was extensively destroyed by the more prolonged mechanical treatment

(Table 2). The existence of cellulose II is underpinned by the appearance of a slight shoulder at 2θ ~ 12° and a shift of the higher angle peak from a 2θ value of approximately 22.6° to a lower value of 2θ ~ 21.9°. However, the re-organization and the degree of crystallization of amorphicized EC is accelerated by the extent of ball milling and the relative humidities.

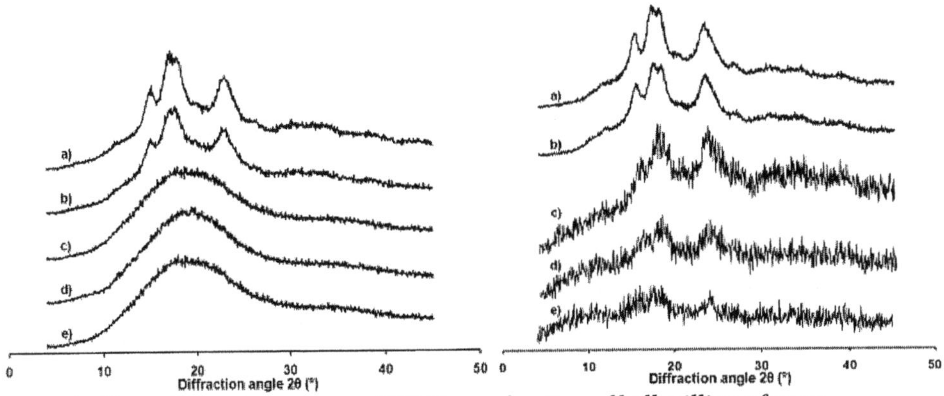

Figure 4 *X-ray diffraction pattern of WMS as a function of ball milling after an equilibration time of 20 days at 25 °C in 53 % (left) and 97 % RH (right). Ball milling times are a) 0 min, b) 5 min, c) 60 min, d) 360 min and e) 480 min.*

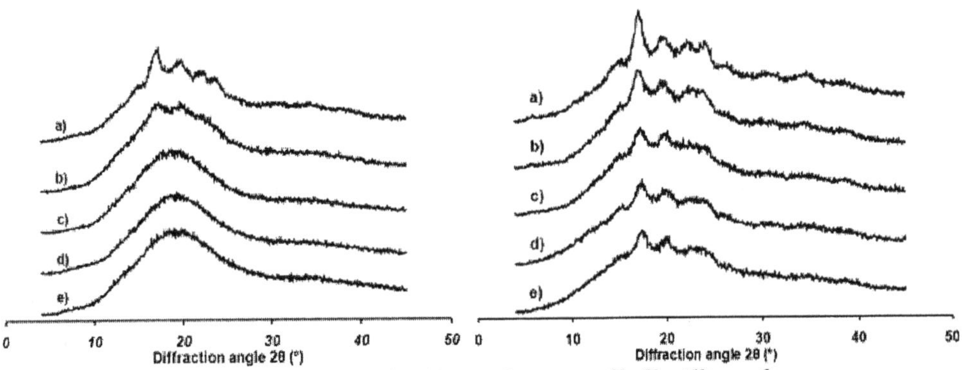

Figure 5 *X-ray diffraction pattern of HAS as a function of ball milling after an equilibration time of 20 days at 25 °C in 53 % (left) and 97 % RH (right). Ball milling times are a) 0 min, b) 5 min, c) 60 min, d) 360 min and e) 480 min.*

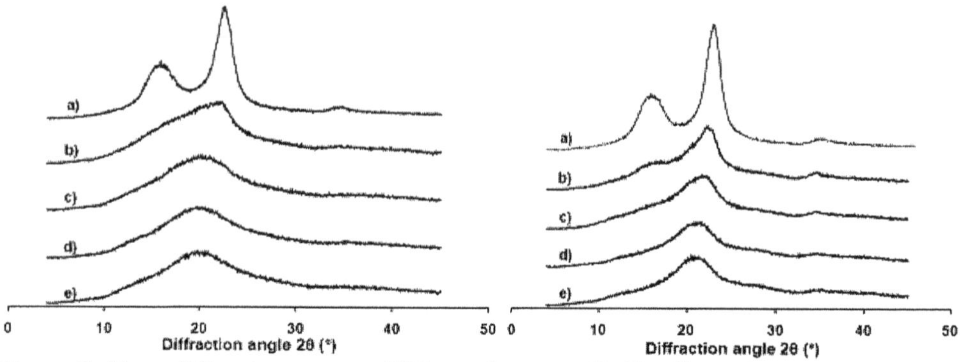

Figure 6 *X-ray diffraction pattern of EC as a function of ball milling after an equilibration time of 20 days at 25 °C in 53 % (left) and 97 % RH (right). Ball milling times are a) 0 min, b) 180 min, c) 360 min, d) 840 min and e) 1080 min.*

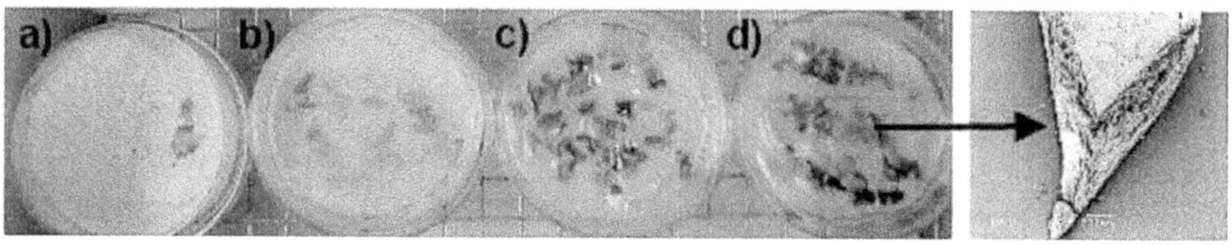

Figure 7 *Film-forming properties of ball milled WMS (a = 5 min, b = 60 min, c = 340 min, d = 480 min, supported by an SEM image, scale bar: 200 μm) equilibrated to 97 % RH at 25 °C.*

3.3 Processability of water-plasticized semi-crystalline and amorphous cellulose

Water plasticized native EC was exposed to high temperatures (100 °C), pressure and shear but was not amenable to thermoplastic process induced transitions as result of extrusion, as confirmed in Figure 8. The X-ray diffractogram of EC shows no change in the crystalline structure, supported by an SEM image illustrating severely compressed cellulose fibers, which, neither were fragmented nor transformed in a coherent biomaterial (Figure 8). The results meet the expectations, due to the extensive intermolecular hydrogen bonding and the high degree of crystallinity prevents cellulose from being processable via normal melt technologies. Although, the process conditions set out correspond to the production of thermoplastic starch, leading to an unsuccessful attempt to destructure cellulose fibres.

However, the cellulose fibers disappeared after utilization of the ball milling procedure representing a technological approach to "cold melt" cellulose and to enhance its machinability. The morphology of ball milled EC showed cellulose fibres, which were extensively degraded and flattened into small, irregular particles with agglomeration tendency. The ESEM micrograph of extruded EC (Figure 9) shows a coherent network with relatively smooth surface, some voids and the absence of fiber pull-out, indicating that amorphous EC might have been plasticized during the extrusion process. Nevertheless, the product yields consisting of homogeneous, opaque and fragile extruded cellulose was considerably low, requiring process optimization solutions to be sought.

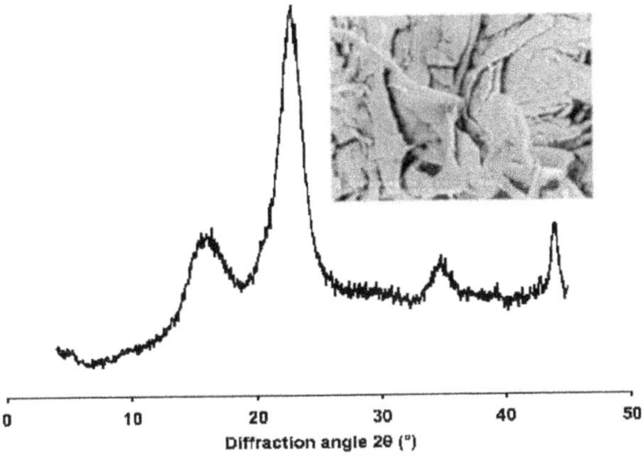

Figure 8 *WAXD pattern and ESEM image of extruded eucalyptus cellulose*

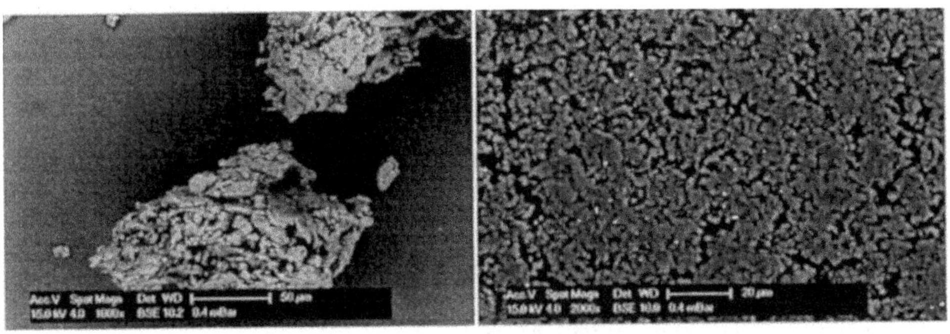

Figure 9 *ESEM images of 6 hours ball milled eucalyptus cellulose (left) and extruded ball-milled eucalyptus cellulose (right).*

4 CONCLUSION

Processability of hydrocolloids is strongly determined by their crystalline conformational structure including short-chain molecules, length of long-chain molecules, the orientation of the glycosidic hydroxyl group in axial or equatorial position and the intra-/intermolecular hydrogen bonds.

Ball milling had a remarkable impact on the structural integrity of starch and cellulose, as proved by the loss of crystallinity in X-ray diffraction pattern. The ball-mill induced order to disorder phase transition in starch occurs faster than for cellulose, whereas cellulose and amylopectin-rich starch is affected more compared to high-amylose starch as a function of ball milling time and magnitude of applied force. The extent of mechanical decrystallization and level of hydration governs the degree of re-crystallization and moreover, the polymorphism of rearranged cellulose.

Finally, twin-screw extrusion has been used to transform and shape amorphous cellulose showing first signs towards susceptibility to thermal-mechanical processing. The preliminary study, therefore, seems to support the hypothesis that “cold melted” cellulose needs less energy input analogue to thermoplastic starch being processable below their thermal degradation temperature. The low yield of extruded cellulose-based material is more a complex function of process history and necessitated further investigation under varying process and operating conditions.

Acknowledgments

The European Community's Seventh Framework Program [FP7/2007-2013] financially supported the present research as part of the STEP ITN project (grant agreement no. 214015). The authors would like to thank Prof. Patrick Navard and Mrs. Edwige Privas at CEMEF for the given opportunity and technical assistance in performing the extrusion trials and scanning electron microscopy.

References

1. J. R. Mitchell, and Areas, J. A. G., in *Food extrusion science and technology*, ed. C. H. J. L. Kokini, and M. V. Karwe, Marcel Dekker., New York, Editon edn., 1992, pp. 345-360.
2. J. Guan and M. A. Hanna, *Bioresource Technol*, 2006, **97**, 1716-1726.
3. R. L. Shogren and B. K. Jasberg, *Journal of Polymers and the Environment*, 1994, **2**, 99-109.
4. M. Thuwall, A. Boldizar and M. Rigdahl, *Carbohydrate Polymers*, 2006, **65**, 441-446.
5. M. Li, P. Liu, W. Zou, L. Yu, F. Xie, H. Pu, H. Liu and L. Chen, *Journal of Food Engineering*, 2011, **106**, 95-101.
6. F. Xie, L. Yu, B. Su, P. Liu, J. Wang, H. Liu and L. Chen, *Journal of Cereal Science*, 2009, **49**, 371-377.
7. R. L. Shogren, *Carbohydrate Polymers*, 1996, **29**, 57-62.
8. B. Lindman, G. Karlström and L. Stigsson, *Journal of Molecular Liquids*, 2010, **156**, 76-81.
9. K. H. Tipples, *Baker's Digest* 1969, **43**, 28-32.
10. N. M. Sereno, S. E. Hill and J. R. Mitchell, *Carbohydrate Research*, 2007, **342**, 1333-1342.
11. R. A. Anderson, H. F. Conway and A. J. Peplinski, *Starch - Stärke*, 1970, **22**, 130-135.
12. C. M. a. P. Feillet, *Cereal Chemistry*, 1975, **52**, 283 - 297.
13. F. Meuser, B. V. Lengerich and F. Köhler, *Starch - Stärke*, 1982, **34**, 366-372.
14. P. H. Hermans and A. Weidinger, *J Polym Sci*, 1949, **4**, 135-144.
15. S. Nara, A. Mori and T. Komiya, *Starch - Stärke*, 1978, **30**, 111-114.
16. S. S. Paes, S. M. Sun, W. MacNaughtan, R. Ibbett, J. Ganster, T. J. Foster and J. R. Mitchell, *Cellulose*, 2010, **17**, 693-709.
17. J. Lelievre, *Starch - Stärke*, 1974, **26**, 85-88.
18. H. Grohn, *J Polym Sci*, 1958, **30**, 551-559.
19. F. Meuser, R. W. Klingler and E. A. Niediek, *Starch - Stärke*, 1978, **30**, 376-384.
20. R. F. Tester, *Food Hydrocolloid*, 1997, **11**, 293-301.
21. V. W. Burchard, *Die Makromolekulare Chemie*, 1963, **59**, 16-27.
22. M. J. Gidley and P. V. Bulpin, *Carbohydrate Research*, 1987, **161**, 291-300.
23. S. Augustat and H. Grohn, *Starch - Stärke*, 1962, **14**, 39-49.
24. W. R. Morrison, R. F. Tester and M. J. Gidley, *Journal of Cereal Science*, 1994, **19**, 209-217.

PHYSICAL MODIFICATION OF XANTHAN GUM

T.J. Foster[1] and J .R. Mitchell[1, 2]

[1] Division of Food Sciences, School of Biosciences, The University of Nottingham, Sutton Bonington Campus, Leicestershire, LE12 5RD, United Kingdom.
[2] Biopolymer Solutions Ltd., Sutton Bonington Campus, Leicestershire, LE12 5RD, United Kingdom.

ABSTRACT

Because of its high chain stiffness xanthan will form an anisotropic liquid crystalline phase at lower concentrations than other food hydrocolloids. The appearance of this phase can result in a **decrease** in viscosity with increasing concentration. A consequence of this is that on concentration of xanthan through phase separation or exclusion, viscosity decreases can be observed. Examples of this unusual behaviour are shown for blends of xanthan with polyelectrolytes and with cold swelling starches. The recovery of viscosity on subsequent dilution has potential in a number of applications including a liquid product for thickening foods for consumers with swallowing difficulties.
A new form of xanthan can be prepared by extrusion. A limited amount of water is injected into the extruder barrel and the product is dried and ground. This form behaves like a polyelectrolyte particle showing excellent dispersibility and a strong salt dependence on the degree of swelling. Under appropriate conditions the viscosity development of dispersions on heating is similar to starch thus xanthan can be produced which combines the desirable dispersibility and heat transfer properties of starch, with the end use advantages of xanthan.

1 INTRODUCTION

Xanthan gum is one of the most popular polysaccharides used in the food industry[1], and is a real success story in the discovery, development and commercialisation of water soluble hydrocolloids. Its discovery and characterisation took place four decades ago, but the research into the molecule is still ongoing. The molecular repeat consists of a cellulosic backbone with a side chain linked through C2 of every other glucose unit which itself consists of an internal mannose, which is occasionally acetylated, a glucuronic acid and a terminal mannose, occasionally pyruvylated. The backbone and side chain interact to form a five fold helix, which promotes a rigid rod conformation and inter-chain interactions, creating weak gel-like properties. Such properties are useful in industrial applications mainly for their shear thinning behaviour, but also high low-shear viscosity, aiding suspension capability for stabilisation of particles such as oil droplets or herb / cocoa particles.

Xanthan is a bacterial exudate gum produced by fermenting the bacteria *Xanthamonas campestris*. An excellent recent review by Palaniraj and Jayaraman[2] not only provides an overview of the parameters required for optimisation / control of xanthan gum production, but also provides a good review of the applications of xanthan in both food and alternative industrial applications. Similar reviews can be found by Sworn[3, 4] Becker and Vorhoelter[5] also review the latest biochemical and genomic information for the xanthan production. Indeed Patel *etal.*[6] shows that through genetic control the viscosity of a given xanthan concentration can be increased due to genetic control of backbone chain length increase. Betlach *etal.*[7] also showed that genetic control of the chemical make up of the xanthan molecule could have sweeping effects on the solution properties. Foster[8] and Morris and Foster[9] using some of these xanthans, showed such changes in molecular architecture could influence the gelation properties of xanthan in admixture with galactomannans or glucomannans. These genetic variations alter the extent of either acetate or pyruvate content, or produce xanthan with either two or one sugar residue side chains (polytetramer or polytrimer, respectively). Alternative ways to change the character of the xanthan molecule can be the use of chemicals. Acidic treatment at high temperatures can depyruvylate the xanthan molecule[10], whereas deacetylation can be achieved by alkali treatments[11, 12, 13]. Recently, Pinto *etal.*[14] has shown that an increase in alkali concentration has a positive influence on viscosity. Mohammed *etal.*[15] also showed that xanthan weak gel rheology can be promoted or inhibited by the levels of divalent cations present. Indeed, trivalent cations are known to gel xanthan[16]. Other chemical modifications by derivatising xanthan with acrylate or maleate groups promote three dimensional hydrogel production[17]. Biochemical changes to the xanthan molecule can also be achieved, with a new endoxanthanase being able to cleave the backbone of xanthan[18]. Physical processes such as shearing for extended processing time in a 2D spouted fluidised bed[19] and sonication[20] can also induce depolymerisation. The later study compared xanthan with guar and pectin, and although xanthan was more stable to sonication, a decrease in viscosity was still seen. Similarly, a decrease in intrinsic viscosity is seen when autoclaving (121°C) xanthan for 30 minutes – 1 hour[21].

A combination of physical and chemical processing can also induce enhanced properties of xanthan. Recently mixtures of xanthan with ionic liquids have produced new hydrogel structures either with time, or solvent exchange with water or salt solutions[22]. The interesting thing about these hydrogels is that they are formed at 9% xanthan concentrations, which far exceeds the concentrations typically employed in the food industry (~1% or lower). Extrusion processes also operate at low moisture contents and therefore high polymer concentrations. Such processing has been used to modify the functionality of xanthan[23] and in admixture with starch[24, 25], producing new thickening options, through to new film formats for the food industry. Wolf[26] has produced an excellent short review of the use of functionalising the properties of other polysaccharides through extrusion processing.

Much information is already acquired for xanthan gum properties in low concentrations[27, 28] however an understanding of the physical properties of such hydrocolloids when the concentration is increased is still an area of debate. There are a number of reports indicating that at high concentrations the xanthan system passes from an isotropic liquid to an anisotropic, nematic liquid crystal structure[29, 30, 31, 32]. Conclusions from these studies indicate that the concentration at which the transition begins to take place is dependant on the ionic environment and M_w of the polymer with lower concentrations required as ionic strength decreases and M_w increases. Departure from Cox-Merz superposition is indicated as an indicator for the transition into the anisotropic phase, described as being broader than theoretically predicted by Flory, passing through anisotropic-in-isotropic and isotropic-in-anisotropic mixtures[30]. The xanthan used in that study was particularly low in molecular weight, with an intrinsic viscosity of 16.3 dl/g, and undergoes the nematic transition between 3-8% xanthan. Carnali[29] (1991) worked with a larger xanthan molecule (M_w = 4mDa) and saw signs of birefringence at concentrations as low as 0.8%, with full birefringence seen at 3%. A consequence of the isotropic-nematic transition is a decrease in viscosity with increasing concentration[30], with the transitions occurring between 2.15-3.11% and 4.52-5.93% for xanthan samples of Mw 480kDa and 154kDa, respectively[31], in similar ionic conditions[29]. As ionic strength increases the transition shifts to higher concentrations[32], indicating electrostatic interactions play a role in determining the concentrations at which these events take place. Attractive forces (molecular interactions) and molecular alignments of rigid rods, are also determining factors in promoting the anisotropic phase[29, 33].

The unifying theme of this paper is the behaviour of xanthan at high concentrations, through physical modification, either through phase concentration when mixed with swelling starch granules or low M_w polyelectrolytes, or through extrusion processing, and considerations from the above discussion are used to attempt to explain the findings.

2 MATERIALS AND METHODS

Materials and methods for some of the work described can be found in Lad etal.[34] (for the starch systems), Boyd etal.[35] (for the polyelectrolyte systems) and Sereno etal.[23] (for the extrusion work).

3 RESULTS AND DISCUSSION

3.1 Phase concentration of Xanthan - Polyelectrolytes

Figure 1 shows the effect of adding a low M_w alginate to a 1% xanthan solution. The work of Boyd etal.[35]shows that as the M_w of alginate increased the birefringent strands became finer. This can be interpreted in terms of nucleation and growth. With increasing alginate M_w the viscosity of the system increases thus reducing the rate of growth of the strands allowing more nuclei to form before the phase concentrations reach equilibrium Such birefringent structures were witnessed for blends of xanthan with other polyelectrolyte polymers such as carboxymethyl cellulose, low methoxy-amidated pectin, λ-carrageenan and polyacrylate, whereas, uncharged methylcellulose and maltodextrin did not show the effect. The insert in figure 1 shows the microstructure of the phase separated material, which departs from the commonly seen spherical droplets of one hydrocolloid in another. We believe that the strand-like phase separation, visualised under cross polarising filters is evidence that the xanthan in this mixture is in the nematic liquid crystalline phase. In figure 2 we see a phase diagram of the xanthan:alginate mixtures and the dashed line indicates that until 1% added alginate phase separation would not be seen, and the initial small decrease in G', in figure 1, witnessed until that point is due to a disruption of xanthan weak-gel behaviour due to volume occupancy of the low Mw alginate chains. At the onset of phase separation the concentration of xanthan in the xanthan-rich phase exceeds 4%, whereby the isotropic to nematic phase transition is expected[29, 30, 31].

With increasing the ionic strength, higher concentrations of xanthan are required to undergo an isotropic-to-nematic phase transition[32]. Figure 3 shows the effect of increasing the ionic strength on the phase separation of xanthan and low Mw alginate. Boyd[35] argues that the xanthan phase transition drives phase separation, which would be similar to the ordering induced phase separation seen by Loren etal.[36]. However, in these systems the xanthan concentration remains at 1%,

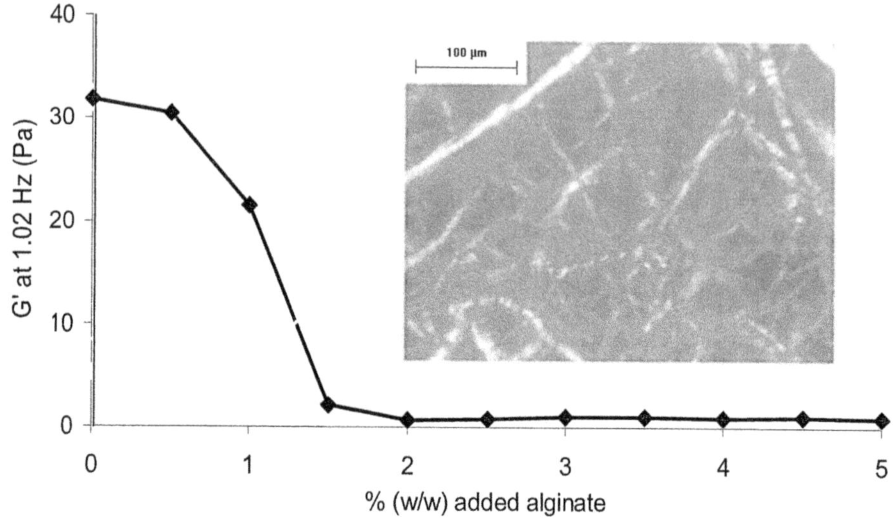

Figure 1 *The decrease in elastic modulus (G') of a 1% xanthan sample with added low Mw alginate, with the insert micrograph indicating a strand-like phase separation.*

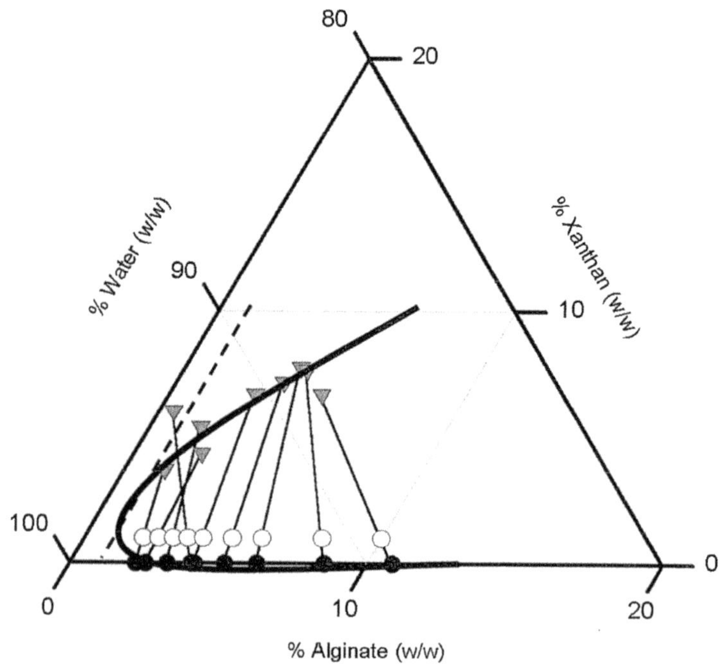

Figure 2 *Phase diagram of xanthan and low Mw alginate. The dashed line shows the concentration of alginate required to induce phase separation.*

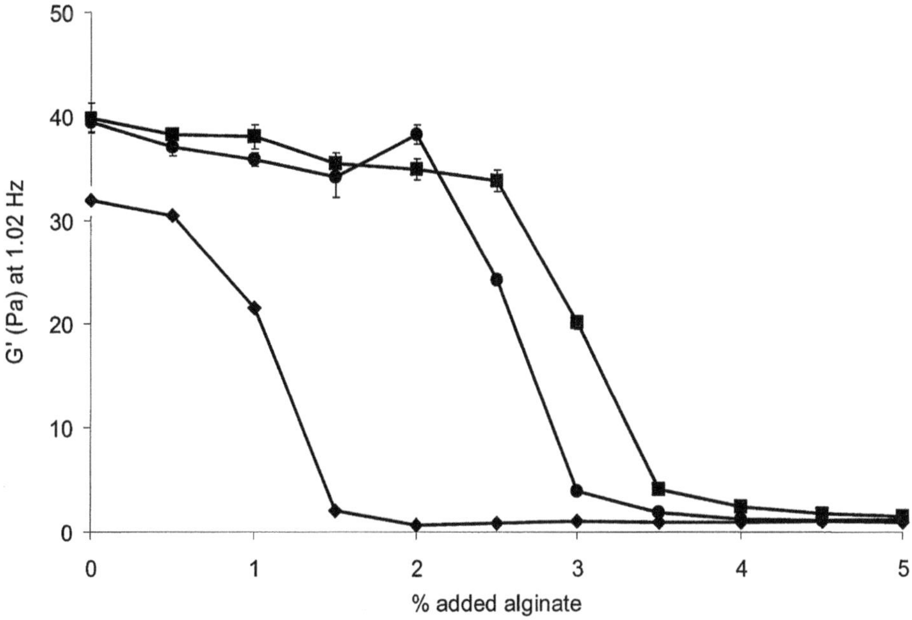

Figure 3 *The decrease in elastic modulus(G') of a 1% xanthan sample with added low Mw alginate, with increasing ionic strength (♦ - No added salt, ● – 0.1M KCl, ■ – 0.2M KCl).*

and it is the increase in alginate concentration that is required to induce phase separation (as indicated by the decrease in G') as ionic strength is increased. Therefore it now seems likely that phase separation leads to phase concentration and isotropic:nematic phase transition, rather than the phase transition inducing phase separation. This is not unexpected, as Kasapis[37] showed that the binodal of the phase diagram shifts to higher concentrations of maltodextrin required to phase separate from gelatin, when the M_w of the maltodextrin decreases. While in the experiment shown in Figure 3 does not involve M_w decrease, it is known that the intrinsic viscosity of both alginate and xanthan is significantly reduced with increasing ionic strength. It therefore seems reasonable that this is the one cause for the requirement of more alginate to induce phase separation at higher ionic strengths. Once the phase concentration of the xanthan is achieved, above a concentration required for the phase transition to be effective, at a given ionic strength, then the nematic liquid crystalline of xanthan is produced. Another possible and complementary interpretation for the ionic strength dependence shown in Figure 3 is reduction in the concentration of the isotropic:nematic transition for xanthan with decreasing ionic strength as previously mentioned.

3.2 Phase concentration of Xanthan – Swelling Starch

Recent work has identified that such phase transitions of xanthan can be induced by mixing xanthan with swellable particulates[34, 38], where exclusion from the swelling particle phase concentrates the xanthan molecules. Figure 4 shows the effect of adding different concentrations of xanthan to 10% cold water swelling starch (CWS). The starch alone swells and produces a dispersion with high dispersed phase volume and subsequent high viscosity (~4500cP). When xanthan is added, initially the viscosity of

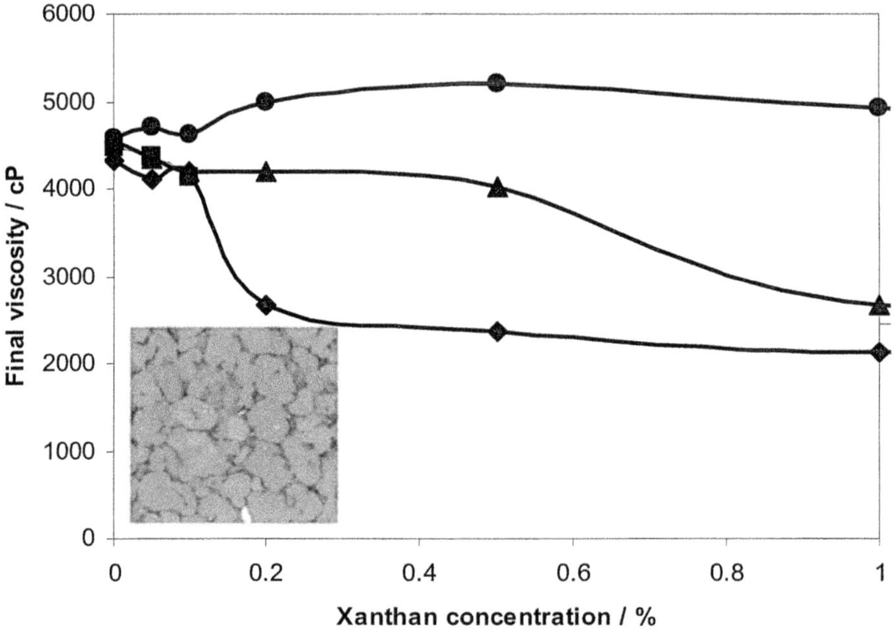

Figure 4 *Decrease in viscosity of a CWS dispersion(insert) upon the addition of increasing amounts of xanthan, with increasing ionic strength (♦ - No added salt, ▲ – 0.1M NaCl, ● – 1M NaCl).*

the mixture remains relatively constant. However, upon further addition the viscosity decreases by half. Following from the discussion above, higher concentrations of xanthan are required for an effective reduction in viscosity when the ionic strength is increased. When the ionic strength is increased to 1M NaCl the viscosity increases (not decreases) upon increasing xanthan concentration because the transition to the anisotropic phase is inhibited. The starch granules appear very swollen at the higher ionic strength, so even though the xanthan phase becomes phase concentrated the concentration is insufficient to induce the xanthan phase transition, which does appear to occur upon addition of 0.2% and 0.6-1% xanthan in water and 0.1M NaCl, respectively. Lad etal.[34] has shown that the viscosity reduction effect is formulation type independent, i.e. starch powder added to a xanthan solution, xanthan powder added to a starch dispersion, or when the two polysaccharides are co-hydrated.

3.3 Exploiting the anisotropic phase in food systems

One might ask why such an understanding of hydrocolloids at high concentrations is of interest. One example which is already on the market is a product (Simply Thick) which thickens foods for consumers with dysphagia or related swallowing difficulties[39]. Unlike the vast majority of thickening products on the market which are powders this is a liquid believed to be 3-4.5% xanthan. The consumer adds the xanthan concentrate to the food product, further diluting the xanthan by (depending on the final viscosity required) a factor of threefold to sevenfold. The anistropic:isotropic transition occurring on dilution results in a much lower viscosity decrease on dilution than would be encountered for an isotropic hydrocolloid solution above the c* concentration.

In the final food product the use of high concentrations *per se* may not be particularly interesting from a cost perspective, unless increased functionality is attained. What is known from a detailed understanding of mixed hydrocolloids is that in such mixtures, if the two hydrocolloids are incompatible, segregative phase separation occurs[40], with consequent phase concentration. Therefore, from what we have seen in the introduction, if such a mixture contains xanthan, and is concentrated above the isotropic:nematic transition concentration, then a very different kind of phase separation, away from the well versed water-in-water spherical emulsion-like microstructures, might be possible. Thus we consider that a major opportunity for the phase concentrated xanthan systems is that upon dilution the viscosity initially increases, due to the re-entry into the isotropic phase (and fully swelling of particulates).

3.4 Extrusion of Xanthan

Typically extrusion of hydrocolloids, in a single, or twin-screw extruder, employs high mechanical and thermal energy, in low water environments, and for starch has led to product developments in thermoplastic extrusion. It has been shown by Farhat etal.[24]and Sereno etal.[23] that such processes can be employed to create new xanthan properties from extruded, dried and ground xanthan. The high concentrations here however, differ from those discussed above. In the barrel of the screw, due to shear and thermal effects, the xanthan molecules melt and align, and upon exit from the die of the screw, and during subsequent cooling a kinetically trapped network of xanthan is produced. The extruded ribbon of material (Figure 5) is then dried and ground to a desired particle size. Upon rehydration these particles swell, but do not dissolve, due to the xanthan network within the particle. Leaching from the particles is minimal, and decreases significantly, along with swollen particle size, as the ionic strength of the rehydrating medium increases. Particles also shrink if the ionic strength is increased after initial rehydration. The network is thought to be due to a reduced number of helices formed during post-extruder cooling, as a result of the high concentration and entanglement. Upon heating and cooling (and re-heating) the full amount of helical structure is regained (from Differential Scanning Calorimetric studies) and the viscosity of the molecular solution is retained upon particle melting. Therefore although high concentrations of xanthan exist within the particles, nematic liquid crystalline phases are not expected, as the polymer is trapped in an amorphous state and the normal rigid rod-like structure are not fully attainable.

It is the swelling nature of the particles that provides the new, intriguing properties. Burey etal.[41] has recently reviewed alternative ways to create, characterise and apply gel particles and identifies dried intermediates as swellable particles, as a result of spray drying a hydrocolloid solution[42, 43]. In such dry particles the hydrocolloid is fully disordered and the particles attain their properties upon the competition for swelling and gelation in the re-hydrating medium. This then relies upon the properties of the medium to provide gelation conditions, whereas the particles produced from the extruder retain integrity irrespective of re-hydrating medium properties. Additionally the extruder particles are produced in limited water environments, without the need for initial dissolution prior to spray drying. Indeed, the extruder particles disperse readily in the re-hydrating medium, without forming clumps or fisheyes, under mild shearing, providing effective cold viscosifying power. Another property highlighted in the work by Sereno is that as a dispersion of the particles is heated, and as the part of the xanthan that is in the helical form undergoes its melting transition, the particles swell further, with a subsequent increase in viscosity.

3.5 Exploiting Extruded Xanthan in Food Systems

This particulate form of xanthan is straightforward to prepare in quantity using a well established food process (extrusion). For this reason, since in involves only the addition of heat and water, it is likely to obtain acceptance as a food ingredient. The use of this process for particle production has the significant advantage over spray drying in that it does not require the addition and subsequent removal of large amounts of water. Typically the water content within the extruder is ~50%. It is also amenable to inclusion of additional materials such as starch[44]. On heating in excess water the xanthan particles swell and disrupt in a rather similar way to starch granules. Thus when viewed against a typical heating and cooling profile in a Rapid Visco Analyser (RVA), the appearance of the melting and setback viscosity is very similar to that of a cook-up starch product (Figure 6), although the concentrations of xanthan required are substantially lower. Thus this form of material in the appropriate food product can give the advantages of the easy dispersibility of starch and viscosity development on heating with some of the end use benefits of xanthan e.g. no retrogradation, high acid and thermal stability, excellent stabilisation of suspensions and nutritional benefits as a low calorie dietary fibre.

4 CONCLUSION

The physical modifications of xanthan by phase concentration and extrusion provide new opportunities for the use of this material. Phase concentration of xanthan into a nematic liquid crystal phase offers a viscosity 'reservoir' which can be tapped by either dilution or increasing salt concentration. Extrusion, however, traps the xanthan in a semi-ordered state. Particles made from such a molecular arrangement can be used can be used to provide rapid cold thickening in products containing very low levels of salt (e.g. fruit juices) without the need for excessive shearing, and in the presence of some salt will disperse and swell upon heating giving rise to starch type viscosity profiles.

Indeed when swellable extruder particles are combined with hydrating xanthan, the xanthan particles can drive the soluble xanthan into a nematic phase[38].

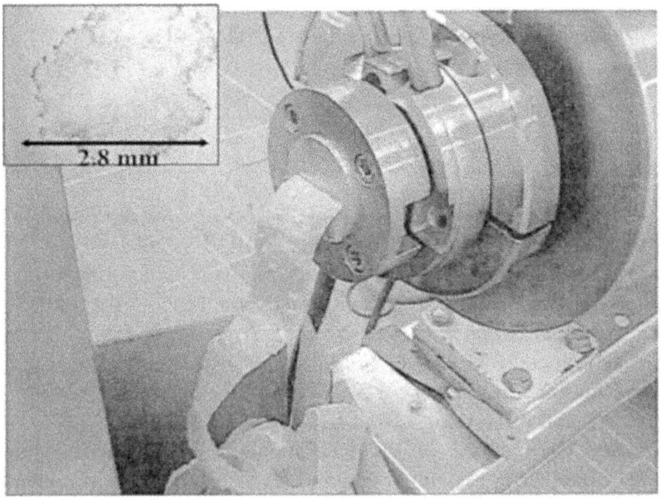

Figure 5 *Ribbon of extruded xanthan being produced. Insert shows a particle swollen in water, after the ribbon had been dried and ground.*

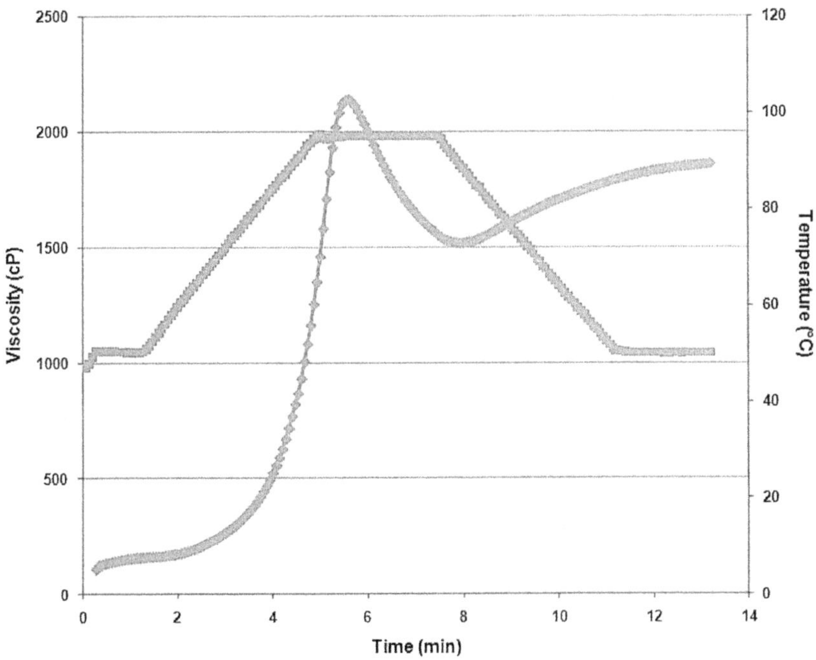

Figure 6 RVA profile of *2% Xanthan extruder particles (♦) dispersed in a solution of 0.4% NaCl and heated and cooled.*

Acknowledgments

The contributions by Nuno Sereno, Mita Lad, Woroud Alesi, Sandra Hill and Bettina Wolf are gratefully acknowledged.

References

1 Seisun D., Market overview, in *Gums and Stabilisers for the Food Industry11*, Eds. PA Williams and GO Phillips, RSC Publications, 3-9, **2002**.

2 Palaniraj A. and Jayaraman V., Production, recovery and applications of xanthan gum by Xanthomonas campestris, *J Food Engineering*, 106(1), 1-12, **2011**.

3 Sworn G., Xanthan Gum, in *Handbook of Hydrocolloids*, Eds. GO Phillips and PA Williams, Woodhead Publishing, **2009**.

4 Sworn G., Xanthan Gum, in *Food Stabilisers, Thickeners and gelling Agents*, Ed. A Imeson, Blackwell Publishing Ltd, UK, 325-342, **2010**.

5 Becker A. and Vorhoelter F-J., Xanthan Biosynthesis by Xanthomonas Bacteria: an Overview of the Current Biochemical and Genomic Data, in *Microbial Production of Biopolymers and Polymer Precursors: Applications and Perspectives* ,Ed. BHA Rehm, Caister Academic Press, UK, 1-11, **2009**.

6 Patel Y., Schneider J. C., Ielpi L. And Ielmini M. V., High viscosity xanthan polymer preparations, US 07439044, **2008.**

7 Betlach M. R., Capage M. A., Doherty D. H., Hassler R. A., Henderson N. M., Vanderslice R. W., Marrelli J. D. and Ward M. B., Genetically Engineered Polymers: Manipulation of Xanthan Biosynthesis, in *Industrial Polysaccharides: Genetic Engineering, Structure/Property Relations and Applications*, Ed M Yalpani, Elsevier Science Publishers, 35-50, **1987**.

8 Foster T. J., Conformation and properties of xanthan variants, PhD Thesis, Cranfield University, **1992**.

9 Morris E. R. And Foster T. J., Role of conformation in synergistic interactions of xanthan, *Carbohydrate Polymers*, 23(2), 133-135, **1994**.

10 Agoub A. A., Smith A. M., Giannouli P., Richardson R. K. And Morris E. R., "Melt-in-the-mouth" gels from mixtures of xanthan and konjac glucomannan under acidic conditions: A rheological and calorimetric study of the mechanism of synergistic gelation, *Carbohydrate Polymers*, 69(4), 713-724, **2007**.

11 Cheetham N. W. H. And Punruckvong A., An HPLC method for the determination of acetyl and pyruvyl groups in polysaccharides, *Carbohydrate Polymers*, 5(6), 399-406, **1985**.

12 Shatwell K. P., Sutherland I. W., Dea I. C. M. and Ross-Murphy S. B., The influence of acetyl and pyruvate substituents on the helix-coil transition behaviour of xanthan, *Carbohydrate Research*, 206(1), 87-103, **1990**,

13 Khouryieh H. A., Herlad T. J., Aramouni F., Bean S. and Alavi S., Influence of deacetylation on the rheological properties of xanthan-guar interactions in dilute aqueous solutions, *J Food Sci*, 72(3), C173-C181, **2007**.

14 Pinto E. P., Furlan L. and Vendruscolo C. T., Chemical Deacetylation Natural Xanthan, *Polimeros-Ciencia E Tecnologia*, 21(1), **2011**.

15 Mohammed Z. H., Haque A., Richardson R. K., and Morris E. R., Promotion and inhibition of xanthan 'weak-gel' rheology by calcium ions, *Carbohydrate Polymers*, 70(1), **2007**.

16 Nolte H., John S., Smidsrod O. and Stokke B. T., Gelation of xanthan with trivalent metal ions, *Carbohydrate Polymers*, 18(4), 243-251, **1992**.

17 Hamcerencu M., Desbrieres J., Popa M., Khoukh A. and Reiss G., New unsaturated derivatives of Xanthan gum: Synthesis and characterization, *Polymer*, 48(7), 1921-1929, **2007**.

18 Li B., Gou J. Q., Chen W. F., Chen X. Y., Chen L.G., Liu Z. W. and Li X. Z., Endoxanthanase, a Novel beta-d-Glucanase Hydrolyzing Backbone Linkage of Intact Xanthan from Newly Isolated Microbacterium sp XT11, *Applied Biochemistry and Biotechnology*, 159(1), 24-32, **2009**.

19 Ciro-Velasquez H. J., Cunha R. L. and Menegalli F. C., Drying of Xanthan Gum Using a Two-Dimensional Spouted Fluidized Bed (2DSFB) with Inert Particles: Performance and Rheological Considerations, *Drying Technology*, 28(3), 389-401, **2010**.

20 Tiwari B. K., Muthukumarappan, K., O'Donnell, C. P., and Cullen P. J., Rheological Properties of Sonicated Guar, Xanthan and Pectin Dispersions, *Int J Food Properties*, 13(2), **2010**.

21 Bindal A., Narsimhan, G., Hem, S. L. and Kulshreshtha, A., Structural changes in xanthan gum solutions during steam sterilization for sterile preparations, *Pharmaceutical Development and Technology*, 12(2), 159-167, **2007**.

22 Izawa H., Kaneko, Y. and Kadokawa, J., Unique gel of xanthan gum with ionic liquid and its conversion into high performance hydrogel, *J Materials Chemistry*, 19(38), 6969-6972, **2009**.

23 Sereno N. M., Hill S. E. and Mitchell J. R., Impact of the extrusion process on xanthan gum behaviour, *Carbohydrate Research*, 342(1), 1333-1342, **2007**.

24 Farhat I. A., Hill S. E., Mitchell J. R., Scharf U., Sereno N. M., and Stolz P., Water-dispersible xanthan gum containing composition, WO2006/064173, **2006**.

25 Flores S. K., Costa D., Yamashita, F., Gerschenson, L. N. and Grossmann, M. V., Mixture design for evaluation of potassium sorbate and xanthan gum effect on properties of tapioca starch films obtained by extrusion, *Materials Science and Engineering C – Materials for Biological Applications*, 30(1), 196-202, **2010**.

26 Wolf B., Polysaccharide functionality through extrusion processing, *Current Opinion in Colloid and Interface Science*, 15(1-2), 50-54, **2010**.

27 Smith I. H., Symes K. C., Lawson C. J. and Morris E.R., Influence of the pyruvaye content of xanthan on macromolecular association in solution, *Int J Biol Macromol*, 3(2), 129-134, **1981**.

28 Ross-Murphy S. B., Morris V. J. And Morris E. R., Molecular viscoelasticity of xanthan polysaccharide, *Faraday Symp of Chem Soc*, 18, 115-129, **1983**.

29 Carnali J. O., A dispersed anisotropic phase as the origin of the weak-gel properties of aqueous xanthan gum, *J Applied Polymer Science*, 43(5), 929-941, **1991**.

30 Oertel R. and Kulicke W. M., Viscoelastic properties of liquid-crystals of aqueous biopolymer solutions, *Rheologica Acta*, 30(2), 140-150, **1991**.

31 Lee H. C. and Brant D. A., Rheology of concentrated isotropic and anisotropic xanthan solutions. 1. A rodlike low molecular weight sample, *Macromolecules*, 35(6), 2212-2222, **2002a**.

32 Lee H. C. and Brant D. A., Rheology of concentrated isotropic and anisotropic xanthan solutions. 2. A semiflexible wormlike intermediate molecular weight sample, *Macromolecules*, 35(6), 2223-2234, **2002b**.

33 Chonakis I. S. and Ramzi M., Isotropic-nematic phase equilibrium and phase separation of kappa-Carrageenan in aqueous salt solution: Experimental and theoretical approaches, *Biomacromolecules*, 3(4), 793-804, **2002**.

34 Lad M. D., Samanci S., Mitchell J. R. and Foster T. J., Viscosity development during competitive hydration of starch and hydrocolloids, in *Gums and Stabilisers for the Food Industry 15*, Eds PA Williams and GO Phillips, RSC Publishing, 126-136, **2010**.

35 Boyd M. J., Hampson F. C., Jolliffe I. G., Dettmar P. W. Mitchell J. R. and Melia C. D., Strand-like phase separation in mixtures of xanthan gum with anionic polyelectrolytes, *Food Hydrocolloids*, 2458-2467, **2009**.

36 Loren N., Hermansson A-M., Williams M. A. K., Lundin L., Foster T. J., Hubbard C. D., Clark A. H., Norton I. T., Bergstrom E. T. and Goodall D. M., Phase separation induced by conformational ordering of gelatin in gelatin/maltodextrin mixtures, *Macromolecules*, 34(2), 289-297, **2001**.

37 Kasapis S., Morris E. R., Norton I. T. and Brown C. R. T., Phase equilibria and gelation of gelatin maltodextrin systems. 3. Phase separation in mixed gels, *Carbohydrate Polymers*, 21(4), 261-268, **1993**.

38 Foster T. J., Mitchell J. R. and Lad M. D., Xanthan gum and swellable particulate containing composition and uses thereof, WO 2010/122332, **2010**.

39 Holahan J. L., Process for preparing concentrate thickener compositions, AU 2008202549, **2008**.

40 Lundin L., Williams M.A.K. and Foster T.J., Phase Separation in Foods, in *Texture in Food Volume1: Semi-solid Foods*, Ed. BM McKenna, Woodhead Publishing, 63-86, **2003**.

41 Burey P., Bandari B. R., Howes T. and Gidley M. J., Hydrocolloid gel particles: Formation, characterization, and application, *Critical Reviews in Food Science and Nutrition*, 48(5), 361-377, **2008**.

42 Gidley M. J. and Hedges N. D., Suspensions of gelled biopolymers, WO95/12988, **1995**.

43 Burey P., Bandari B. R., Howes T. and Gidley M. J., Gel particles from spray-dried disordered polysaccharides, *Carbohydrate Polymers*, 76(2), 206-213, **2009**.

44 Kuhn M., Elsner G. and Graber S., Cooking extrusion of starch with hydrocolloids, *Starch-Starke*, 41(12), 467-471, **1989**.

A COMPARISON OF THE RHEOLOGICAL PROPERTIES OF SUCCINOGLYCAN WITH XANTHAN GUM

G. Sworn and E. Kerdavid
Danisco France SAS, 20 rue Brunel, 75017, Paris, France

1 INTRODUCTION

Xanthan gum and succinoglycan are both microbial exopolysaccharides that are produced commercially by fermentation. They are high molecular weight anionic polymers. The polysaccharide xanthan gum is produced by the microorganism *Xanthomonas campestris*. It consists of a cellulosic backbone of ß-1,4 linked D-glucose units substituted on alternate glucose residues with a trisaccharide side chain. The trisaccharide side chain is composed of two mannose units separated by a glucuronic acid.[1,2] Approximately half the terminal mannose units are linked to a pyruvate group and the non-terminal residue usually carries an acetyl group. The carboxyl groups on the side chains render the gum molecules anionic. X-ray diffraction studies on xanthan gum fibres have identified a right handed, five-fold helix conformation.[3] In this conformation the side chains are aligned with the backbone and stabilise the overall conformation. In solution the side chains wrap around the cellulose-like backbone thereby protecting it. It is thought that this is responsible for the excellent stability of xanthan gum under adverse conditions.

The rheology of xanthan gum can be characterised by the development of very high viscosity at low shear rates and pseudoplastic flow which provide suspension stability to finished products and low viscosity at higher shear rates for ease of filling, pouring and pumping. Xanthan gum is used in a wide range of food and non-food applications.[4]

Succinoglycan is a microbial polysaccharide produced by *Agrobacterium tumefaciens*. It is composed of D-glucose:D-galactose:pyruvate:succinate in the molar ratio of 7:1:1:1.[5] Aqueous solutions of succinoglycan exhibit reversible pseudoplastic flow and viscosity is dependent on salt concentration.[6,7] It exhibits an order-disorder transition as a function of temperature.[6,8]

The primary use of succinoglycan is in non-food applications in very acidic media such as toilet cleaners (pH < 2) where the viscosity is less sensitive and more stable than xanthan gum. Succinoglycan is on the list of permitted additives in food in Japan but not in the rest of the world.

In this study the rheological properties of succinoglycan solutions have been compared with those of xanthan gum in various model systems relevant to food applications.

2 MATERIALS AND METHODS

Succinoglycan (Rheozan lot 4451087743) and xanthan gum (GRINDSTED® Xanthan 80 lot 4450399909) were supplied by Danisco France SAS and were used without further purification.
The hydration profile was determined by measuring the viscosity as a function of stirring time. Gum (1.5g) was dispersed into 2.5% NaCl solution (500 cm^3) using a deflocculating propeller (Rayneri ø 65 mm) at a speed of 650 rpm. Measurements of viscosity (Brookfield LVT, 12 rpm spindle 2) were made at room temperature (22 ± 3°C) after stirring times starting from the introduction of the gum (time = 0). The effect of NaCl on the hydration was determined by dispersing the gum directly into the salt solution and measuring the viscosity at a shear rate of $0.01s^{-1}$ after 30 minutes mixing using a deflocculating propeller (Rayneri ø 65 mm) at 500 rpm.
Solutions at neutral pH were prepared by dispersing the gum in deionised water and mixing at high speed until fully hydrated. NaCl was then added and mixed until dissolved.
Solutions at acidic pH were prepared by dispersing the gum in tap water and mixing until hydrated. 1% NaCl was added and then the pH was adjusted to pH 2.5 with a solution of 12.5% acetic acid.
Rheological measurements were made at 20°C with a controlled stress rheometer (AR-1000, TA Instruments) fitted with a 50 mm diameter, 2° cone. The temperature ramp measurements were made at a heating rate of 2°C min^{-1} with automatic gap adjustment. Viscoelastic measurements were made within the predetermined linear viscoelastic region of the samples.

3 RESULTS AND DISCUSSION

The hydration profile of succinoglycan is compared to xanthan gum in 2.5% NaCl solution in Figure 1. It can be seen that, like xanthan, the succinoglycan hydrates very rapidly. This means that it will have similar tendencies to lump formation as xanthan gum if not well dispersed and, therefore, the same precautions as taken with xanthan gum should be employed when dispersing succinoglycan.
Figure 1 shows that the succinoglycan develops a higher viscosity than the xanthan gum. This could be of benefit for quick suspension of particles such as in brines or dressings. Increasing the sodium chloride concentration above 1% retards the hydration of the succinoglycan and xanthan gum (Figure 2). This is to be expected with anionic polysaccharides. Screening of the charges at high salt levels will reduce the affinity of the gum for water and therefore slow down hydration. However, even at high salt concentrations the succinoglycan gives more viscosity than the xanthan gum.

The flow properties of aqueous solutions of succinoglycan and xanthan gum were measured at varying concentrations from 0.05% to 0.5% in 1% NaCl and in 1% NaCl at pH 2.5. An example of the flow curves in 1% NaCl is shown in Figure 3.

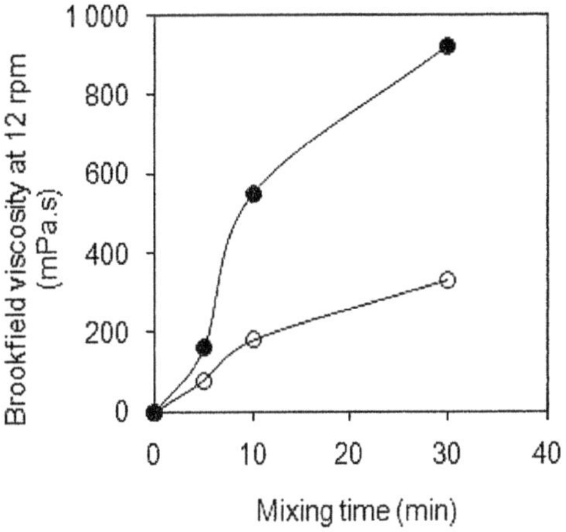

Figure 1. Development of viscosity of succinoglycan (•) and xanthan gum (○) with mixing time in 2.5% NaCl (650 rpm at room temperature).

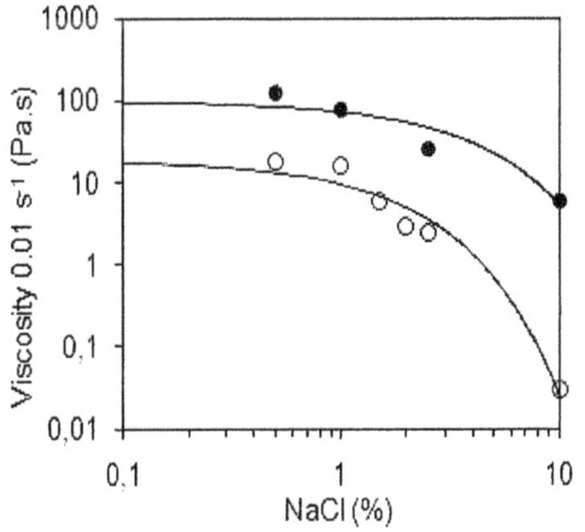

Figure 2. Effect of salt concentration on the hydration of succinoglycan (•) and xanthan gum (○) (Viscosity measured after 30 mins mixing at 500 rpm).

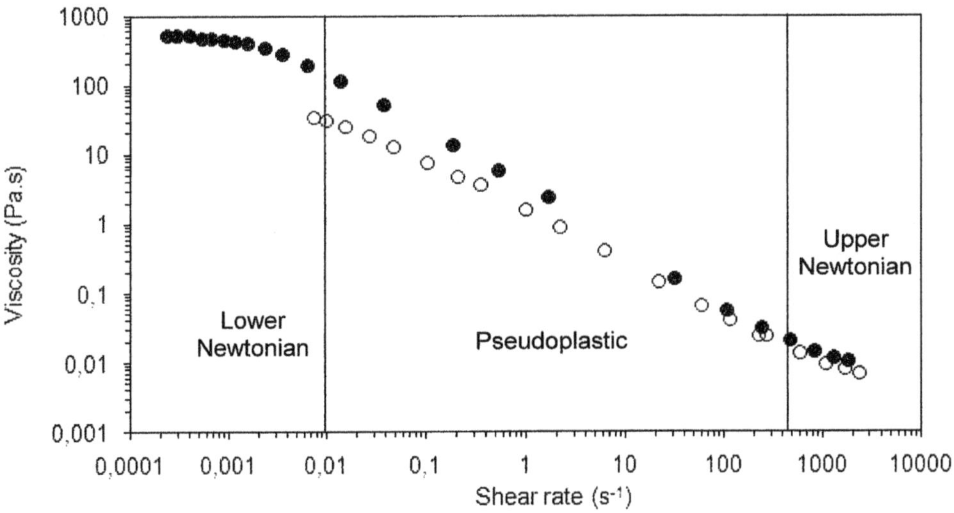

Figure 3. Flow curve of 0.3% succinoglycan (•) and xanthan gum (○) in 1% NaCl.

Both xanthan gum and succinoglycan gum solutions in the concentration range measured tended towards a lower Newtonian plateau at shear rates below 0.01 s^{-1}, exhibited highly pseudoplastic flow behaviour in the shear rate range of 0.01 and 500 s^{-1} and tended towards an upper Newtonian plateau at higher shear rates. The pseudoplastic region could be fitted with a power law model:

$$\eta = K\gamma^n$$

where η is the viscosity, γ is the shear rate and K and n are constants that define the consistency index and degree of pseudoplasticity respectively. The higher the K value the higher the viscosity and the more negative the n value the more pseudoplastic (shear thinning) the flow behaviour. The results from the power law model for solutions in 1% NaCl and 1% NaCl, pH 2.5 are given in Table I.

Table I. K and n values for xanthan gum and succinoglycan solutions at various concentrations in neutral and acidic media (from power law fit of viscosity between 0.01 and 500 s^{-1}, $r^2 > 0.98$ for all fits).

Concentration (%)	Succinoglycan				Xanthan gum			
	K (mPa.s)		n		K (mPa.s)		n	
	1% NaCl	1% NaCl pH 2.5	1% NaCl	1% NaCl pH 2.5	1% NaCl	1% NaCl pH 2.5	1% NaCl	1% NaCl pH 2.5
0.05	128	94.7	-0.58	-0.50	18.5	-	-0.29	-
0.10	427	396	-0.72	-0.65	82.6	7.4	-0.47	-0.40
0.20	1510	1644	-0.79	-0.78	470	421	-0.62	-0.57
0.30	3012	3362	-0.81	-0.81	1433	1298	-0.73	-0.68
0.40	5724	5610	-0.86	-0.84	2682	2600	-0.77	-0.76
0.50	8348	8349	-0.87	-0.86	4706	3685		-0.78

A plot of K and n as a function of gum concentrations in 1% NaCl shows that the solutions increase in viscosity and become more pseudoplastic with increasing concentration and at the same concentration the succinoglycan solution has a higher viscosity and is more pseudoplastic than the xanthan gum solution (Figure 4). The difference in flow behaviour between the succinoglycan and xanthan gum solutions increased with decreasing concentration. For example, at 0.05% the K value of the succinoglycan solution is approximately 7 times that of the xanthan and the n value is double. In contrast, at 0.5% the K value is less than twice that of xanthan and there is only 10% difference in between the n values.

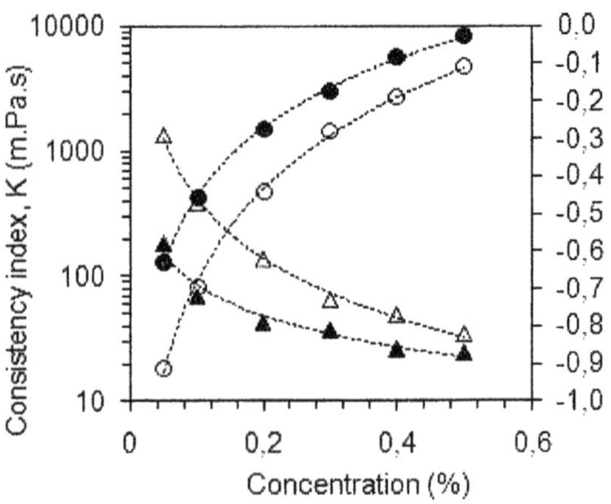

Figure 4. Effect of concentration on the consistency index, K (●, ○) and pseudoplasticity, n (▲, Δ) of solutions of succinoglycan (filled) and xanthan gum (open) in 1% NaCl.

The difference in flow behaviour between the succinoglycan and xanthan gum is even greater at low pH (1% NaCl, pH 2.5). This is because the flow profile of the succinoglycan is virtually identical in neutral and acidic medium whereas the xanthan gum solution has a lower viscosity and is less pseudoplastic in the acidic media. This is seen when comparing the K and n values in Table I and in the sensitivity of the viscosity at low shear rate to pH (Figure 5).

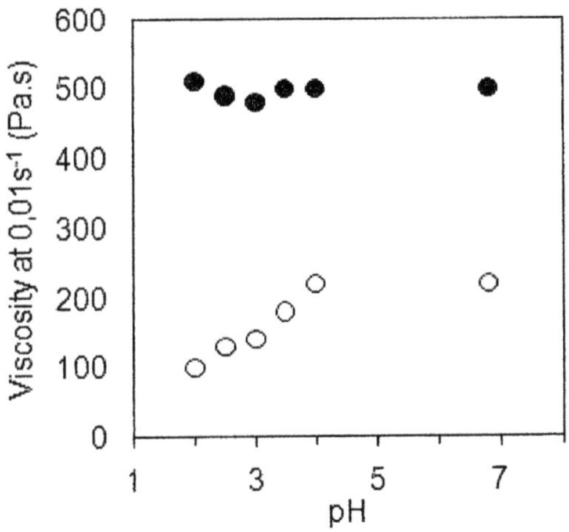

Figure 5. Effect of pH on the viscosity at $0.01s^{-1}$ of 0.5% succinoglycan (•) and xanthan gum (○) in tap water.

Reducing the pH of xanthan gum solutions progressively converts the carboxylate groups from the ionised to the un-ionised form ($COO^- + H^+ = COOH$), with consequent suppression of electrostatic repulsion between xanthan side chains.[9,10] This could reduce the stiffness of the molecule allowing adoption of a more compact molecular shape, resulting in the observed reduction in viscosity.

The insensitivity of succinoglycan flow properties at low pH could be important in products such as salad dressings, vinaigrettes and other acidic foods. For example in a model vinaigrette formulation the concentration of the succinoglycan could be halved compared to a standard xanthan gum whilst still providing the same viscosity at low shear rates (Figure 6). This is an indication that similar product stability could be achieved with half the dosage of succinoglycan compared with xanthan gum. The more pseudoplastic flow behaviour of the succinoglycan also means a lower viscosity at higher shear rates which could lead to lower viscosity when eaten (cleaner mouthfeel).

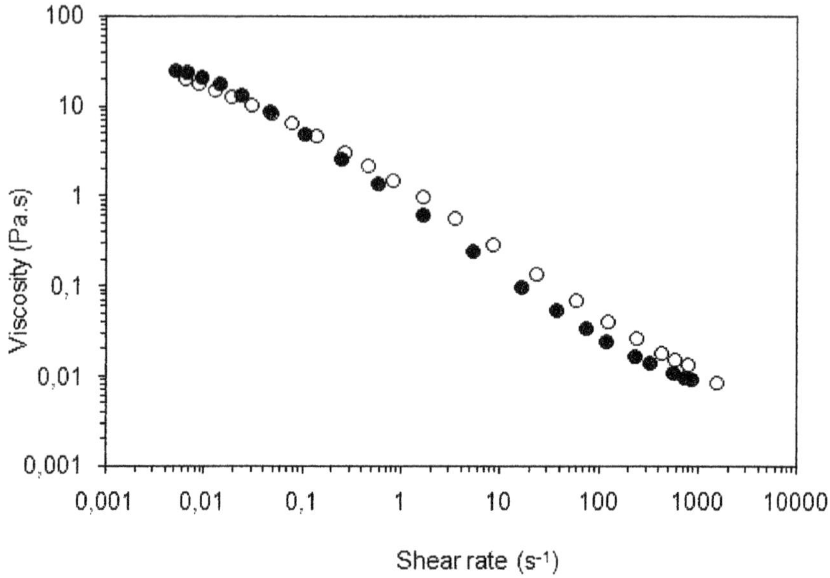

Figure 6. Flow curve of 0.15% succinoglycan (●) and 0.3% xanthan gum (○) in a model vinaigrette (1% NaCl, pH 2.5).

Measurement of the viscoelastic properties of xanthan gum and succinoglycan solutions revealed weak-gel like properties for both solutions (Figure 7). The succinoglycan had stronger gel-like properties than the xanthan gum at equivalent concentrations as evidenced by the higher elastic modulus (G') and lower dependency on frequency of both the elastic (G') and viscous (G") moduli.

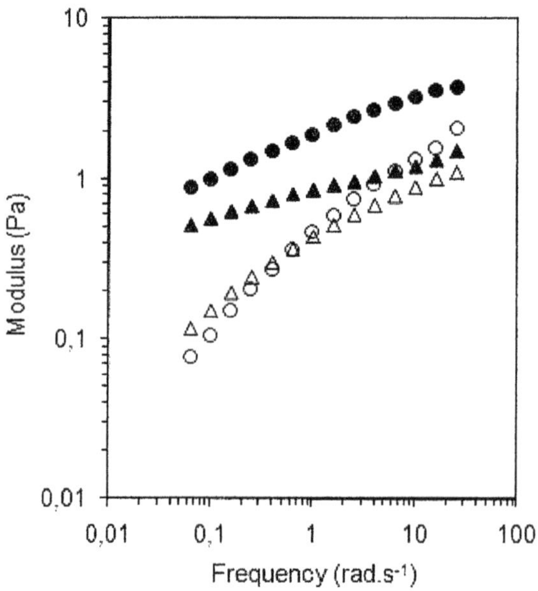

Figure 7. Viscoelastic properties, G' (●,○) and G" (▲, Δ) of 0.2% succinoglycan (filled symbols) and xanthan gum (open symbols) in 1% NaCl.

The change in viscosity of succionglycan and xanthan gum solutions as a function of temperature was measured at 10 s^{-1} in the presence of varying amounts of sodium chloride. An example of the results is given in Figure 8 and it can be seen that both gums undergo a sigmoidal change in viscosity. This has been associated with an order-disorder transition previously observed in xanthan gum and in succinoglycan.[11,6] This transition has also been observed with light scattering techniques.[8]

Figure 8 shows that below the transition temperature the succinoglycan viscosity remains relatively stable compared with the xanthan gum. This could have implications for product stability. For example, if a product is formulated to be stable at 20°C the stability will be reduced at 40°C with xanthan gum but will remain constant with succinoglycan. The succinoglycan can offer a more robust formulation with regard to fluctuations in storage temperature.

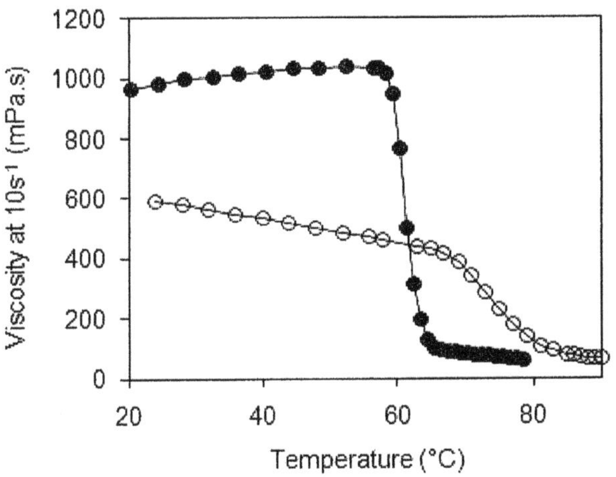

Figure 8. Viscosity changes at 10 s^{-1} during heating of 0.3% solutions of succinoglycan (●) and xanthan gum (○) in 0.1% NaCl.

The change in viscosity of succinoglycan occurs over a much narrower temperature range compared with that of xanthan gum. This was observed at all NaCl concentrations tested.

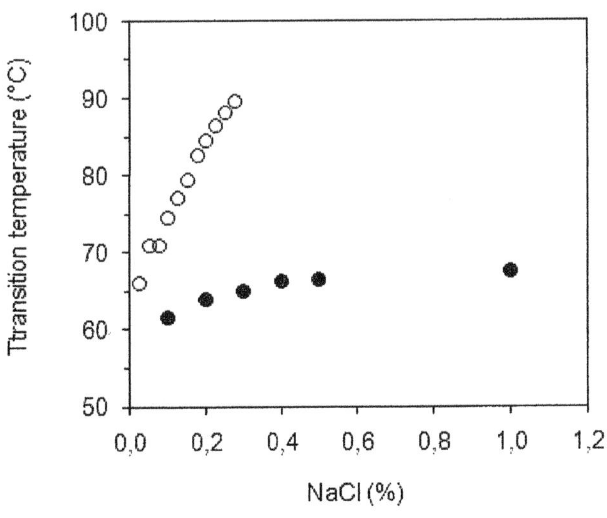

Figure 9. Dependence of the order-disorder transition temperature of 0.3% solutions of succinoglycan (●) and xanthan gum (○) on the concentration of NaCl.

The order-disorder transition temperature was taken as the mid-point of the viscosity change and has been plotted as a function of temperature in Figure 9. This shows that the transition temperature of succinoglycan is much less dependent on NaCl concentration compared with that of xanthan gum. The transition temperature of succinoglycan reached a plateau of 70°C at 1% NaCl whereas the xanthan gum increased to above 90°C at 0.25% NaCl. In many foods that are subject to a thermal treatment this means that the succinoglycan will undergo an order-disorder transition whereas the xanthan gum may not.

4 CONCLUSIONS

Succinoglycan

1. Can viscosify brines (2.5% NaCl) faster than Xanthan gum
2. Is very effective in acidic medium
    - It can bring the same suspension power as standard xanthan at 50% lower dosage.
    - Gives less mouth feel than standard xanthan.
3. Is more stable at higher temperature (<60°C). It can stabilize samples that are stored at 40°C.
4. Solutions have very low viscosity above 65°C
    - Viscosity is partially recovered when cooled.

Succinoglycan has some interesting and unique properties that could be used in a range of food applications. The insensitivity to pH's below 4 enables succinoglycan to provide the same suspension power as xanthan gum at approximately 50% lower dosage. Its greater pseudoplasticity provides less mouthfeel than xanthan gum.
The very low viscosity of succinoglycan above 68°C could be of interest for hot fill foods, or better heat transfer during heat treatment, or for foods that thicken on cooling to ready-to-eat temperatures. In this respect the viscosity development below 68°C could provide a visual indicator of when a food is safe (in terms or temperature) to eat.
Rapid hydration, high viscosity at low shear rates and weak-gel properties make succinoglycan a potentially interesting hydrocolloid for use as a thickening agent in products for the management of dysphagia.
Potential applications of succinoglycan in food

- Dressings: high viscosity at low pH.
- Acidic drinks (pH < 4): high suspension power, low mouth feel.
- Hot drinks or soups: high suspension power below 70°C, low viscosity above 70°C (visual indicator of temperature).
- Thickener for management of dysphagia: viscoelastic rheology, rapid hydration.

References

1 Jansson, P.E., Kenne, L. and Lindberg, B. (1975) Structure of the exocellular polysaccharide from *Xanthomonas campestris*. *Carbohydr. Res.*, **45**, 275 – 282.

2 Melton, L.D., Mindt, L., Rees, D.A. and Sanderson G.R. (1976) Covalent structure of the polysaccharide from *Xanthomonas campestris*: evidence from partial hydrolysis studies. *Carbohydr. Res.*, **46**, 245 – 257.

3 Moorhouse, R., Walkinshaw, M.D. and Arnott, S. (1977) Xanthan gum-molecular conformation and interactions. In Sandford, P.A. and Laskin, A. (Eds) *Extracellular*

microbial polysaccharides, ACS symposium series no. 45, American Chemical Society, Washington, DC, pp 90 – 102.

4 Sworn G. Xanthan gum. (2010) In Imeson (Ed.) *Food Stabilisers, Thickeners and Gelling Agents*. Wiley-Blackwell Publishing Ltd, Oxford, pp 325 - 342.

5 Harada T., Anemura A., Jansson P-E. and Lindberg B. (1979) Comparative studies of polysaccharides elaborated by *rhizobium*, *alcaligenes*, and *agrobacterium Carbohydr. Res.*, **77**, 285 - 288.

6 Gravanis G., Milas M., Rinaudo M. and Clarke-Sturman A.J. (1990) Rheological behaviour of a succinoglycan polysaccharide in dilute and semi-dilute solutions. *Int. J. Biol. Macromol.* **12**, 201 – 206.

7 Simsek S., Mert B., Campanella O.H. and Reuhs B. (2009) Chemical and rheological properties of succinoglycan with distinct structural characteristics. *Carbohydr. Polym.*, **76,** 320 – 324.

8 Kido S., Nakanishi T., Norisuye T, Kaneda I. and Yanaki T. (2001) Ordered conformation of succinoglycan in aqueous sodium chloride. *Biomacromolecules*, **2**, 952 – 957.

9 Agoub, A.A., Smith, A.M., Giannouli, P., Richardson, R.K. and Morris, E.R. (2007) "Melt-in –the-mouth" gels from mixtures of xanthan and konjac glucomannan under acidic conditions: A rheological and calorimetric study of the mechanism of synergistic gelation. *Carbohydr. Polym.*, **69**, 713 – 724.

10 Rinaudo, M. and Moroni, A. (2009) Rheological behaviour of binary and ternary mixtures of polysaccharides in aqueous medium. *Food Hydrocolloids*, **23**, 1720 – 1728.

11 Jeanes, A., Pittsley, J.E. and Senti, F.R. (1961) Polysaccharide B-1459: a new hydrocolloid polyelectrolyte produced from glucose from bacterial fermentation. *J. App. Polym. Sci.*, **5**, 519 – 526.

BULK AND SURFACE RHEOLOGY CHARACTERIZATION OF *STERCULIA APETALA* EXUDATE POLYSACCHARIDE SOLUTIONS

L. M. Pérez-Mosqueda[1], P. Ramírez[1], M. C. Alfaro[1], J. Muñoz[1], F. Rincón[2], G. León de Pinto[2]

[1]Departamento de Ingeniería Química, Universidad de Sevilla, Facultad de Química, 41012 Sevilla (Spain)
[2]Centro de Investigaciones en Química de los Productos Naturales. Universidad del Zulia, Apdo. 526 Maracaibo (Venezuela)

1 INTRODUCTION

Exudate polysaccharides such as gum arabic, gum karaya, gum ghatti and gum tragacanth are tree exudates. These hydrocolloids have many applications in the food and pharmaceutical industries. Gum arabic is the most important and best studied exudate gum, its main application being in the confectionery and soft drink industries[1].

Sterculia apetala exudate gum is obtained from a *Sterculia* genus tree, likewise gum karaya, which can be obtained from *Sterculia urens, Sterculia setigera and Sterculia villosa trees* or the exudate gum obtained from *Sterculia striata.* Chemically, they are partially acetylated branched polysaccharides. They have very high molecular weight (between 9000 and 16000 kDa). Another characteristic of these gums is their high content in rhamnose. However, the sugar composition of each gum is dependent on the botanical source[2,3].

Gum karaya has many applications due to its emulsifying, thickening and stabilising properties, hence it is used as a food additive, in the paper and pharmaceutical industries and for dental applications. Therefore, bearing in mind their wide applications, it is interesting to assess the properties of new exudate gums obtained from other sources, such as *Sterculia apetala*.

Recently, some rheological properties of aqueous gum karaya and *Sterculia striata* dispersions have been reported[4-6]. Above 0.2 % these gum dispersions show shear thinning behaviour and, in addition, exhibit gel-like properties above 2%, the gum karaya gel being stronger than that obtained from *Sterculia striata.* The presence of some electrolyte (0.1 M NaCl) provokes a decrease in the gel strength.

Although gum Arabic and mesquite gum show surface activity, there are no studies regarding the surface adsorption of gum karaya or others gums obtained from *Sterculia* genus trees. Therefore, to the best of our knowledge, this is the first study involving a characterisation of surface activity and surface rheology in addition to the bulk rheology of these gums.

2 MATERIAL AND METHODS

2.1 Materials

The *Sterculia apetala* gum used has been obtained from trees of the same name that grow in some areas of Venezuela. It has a branched structure composed of galactose, arabinose, xylose, rhamnose, β-D-glucuronic acid and its 4-O-methyl derivative[7]. Sodium chloride (99.5 wt% purity) and sodum azide (99 wt%) were purchased from Panreac and used as received.

2.2 Methods

2.2.1 Bulk rheology. The rheological measurements were carried out using a controlled stress rheometer (Mars I, Thermo-Haake) with a serrated plate and plate geometry. The sensor diameter was selected on the basis of the different consistency exhibited by samples. (PP35R has a diameter of 35 mm and was used for samples containing 2% w/v and 3% w/v – and PP60R: 60 mm diameter for the less consistent samples– 0.5% w/v and 1% w/v -). Flow curves were run according to a shear stress stepwise procedure in a shear stress range which depended on the consistency of each sample. Viscosity was calculated in each step after a maximum shear time of 300 s unless steady-state response was previously achieved within 1% tolerance. The mechanical spectra of the samples were performed using a frequency sweep (0.06-2 rad/s) obtained with an oscillatory stress amplitude of 0.1 Pa. The stress amplitude was within the linear viscoelastic range (LVR) according to stress sweeps conducted at 1 Hz. Both oscillatory and steady shear tests were carried out at 25°C. The samples were covered with a thin layer of mineral oil to prevent water evaporation.

2.2.2 Surface characterization. Surface tension measurements were performed with a drop profile analysis tensiometer (CAM200, KSV, Finland). The drop was formed inside a thermostated cuvette at 25°C and controlled using a custom-built control unit consisting of a syringe with a piston that is driven by a stepper motor.

Oscillation perturbations were made using the PD 100 module from KSV. This module consists of a chamber filled with the solution to be measured where there is a piezoelectric device calibrated with the pc sound card which produces harmonic oscillations. The results of oscillating experiments were analysed by the instrument software, which provides values for the elastic (E') and loss (E'') modulus.

Prior to each measurement the surface tension of water, γ_w, was measured and it was checked that its value remained constant between 71-72 mNm^{-1}. Surface pressure values are calculated as: $\Pi = \gamma_w - \gamma$. All measurements were done at 25°C.

3 RESULTS AND DISCUSSION

3.1 Bulk rheology

3.1.1 Flow behaviour. Figure 1 shows the flow behaviour of *Sterculia apetala* aqueous dispersions. Shear stress was plotted against shear rate for gum solutions at concentrations ranging from 0.5% to 3% w/v at 25 °C.

The most diluted dispersion (0.5% w/v) studied showed Newtonian behaviour throughout the whole shear rate range. However, at higher gum concentrations the flow behaviour shifted to shear thinning. This behaviour is characterised by three different regions[8]. First the Newtonian region at low shear rates (the zero-shear viscosity, η_0, is the characteristic parameter defining this region) is observed, then it follows a non-linear region at intermediate shear rates, and eventually a trend to reach the infinite-shear limiting viscosity (η_∞) is found.

The different regions are related with structural changes as a function of the shear rate. This effect is well-known for macromolecules solutions. When a very low stress is applied the Brownian forces are not overcome by shear and macromolecules keep themselves randomly distributed and folded in a network arrangement. When sufficient shear stress is applied the macromolecules tend to orientate in the direction of the applied stress, leading to the unfolding of the original structure. This rearrangement decreases the flow resistance and accordingly its apparent viscosity.

It is worth noting that small concentrations of this gum results in relatively high viscosity values ($\tau/\dot{\gamma}$ = 15 mPas for 0.5% w/v). Yet, *Sterculia striata* gum solutions were reported to show Newtonian behaviour for concentrations below 0.2% w/v at 25°C with a viscosity of just 2.5 mPas, whereas shear thinning behaviour was observed at higher concentrations[5]. The electrolyte concentration (NaCl) used was 0.1 M, while the electrolyte concentration used in the present study was 1M NaCl. It is well-known that high electrolyte concentration provokes a drop of viscosity due to the screening of electrical charges and enhanced flexibility of the macromolecule chain. However, despite the higher NaCl concentration used, the Newtonian viscosity of the *Sterculia apetala* solutions is higher than those of *Sterculia striata,* which support the enhanced thickening properties of *Sterculia apetala* gum compared with other gums of the same or similar nature.

The Carreau model has been used to describe the shear rate dependence of viscosity[9]:

$$\tau = \left[\eta_\infty + \frac{\eta_0 - \eta_\infty}{\left(1 + (\lambda\dot{\gamma})^2\right)^m} \right] \dot{\gamma} \tag{1}$$

where $\dot{\gamma}$ is the shear rate, τ is the shear stress, η_0 is the zero shear viscosity, λ is a time constant, *m* is an dimensionless constant related with the exponent of the power law, *n*, according to $m = (1-n)/2$ and η_∞ is the so-called infinite shear rate viscosity.

The zero-shear viscosity of 1% (w/v) *Sterculia apetala* gum dispersion was 12 times higher than the Newtonian viscosity of the 0.5% (w/v) solution. Greater gum concentrations yielded an exponential increase of the zero-shear viscosity (12371 Pa·s at 2% w/v and 54984 Pa·s at 3% w/v). The infinite-shear viscosity also exhibited an exponential increase with *Sterculia apetala* gum concentration, the most important increment of viscosity taking place between 2% w/v (65 mPa·s) and 3% w/v (500 mPa·s) gum concentration. As expected the lowest characteristic time λ was shown by the most diluted dispersion and the flow index, n, went progressively down as the gum concentration was increased.

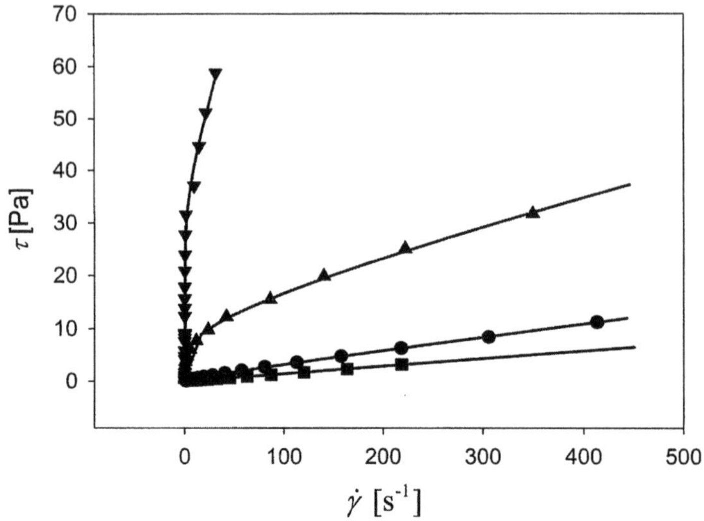

Figure 1. *Shear stress as function of shear rate for Sterculia apetala solutions in 1M of NaCl at 25°C: solid square 0.5% w/v, solid circle 1% w/v, solid triangle 2% w/v, inverse solid triangle 3% w/v. The solid lines are the best fit to Carreau model, except for the solid line of 0.5% w/v solution which is a straight line with a slope of 15 mPas.*

3.1.2 Small amplitude oscillatory shear (SAOS). The viscoelastic behaviour of *Sterculia apetala* dispersions was determined in the (0.06-2 rad/s) frequency range applying a stress amplitude of 0.1 Pa, which is within the linear viscoelastic range. Figure 2 shows the frequency dependence of the complex viscosity for the most concentrated solutions (2% and 3% w/v), which corresponds to the plateau or rubbery relaxation zone. Unfortunately, below 1% w/v a linear viscoelastic response could not be found.

An increase of the *Sterculia apetala* gum concentration from 2% to 3% w/v resulted in a greater complex viscosity value, indicating the occurrence of a higher degree of interactions among the gum macromolecules. The rheological results obtained did not provide direct information to elucidate the microstructure of these exudate food gums. However they point to the formation of transient physical junction zones. The higher the gum concentration the clearer the trend of this exudate gum to form larger and larger networks based on an increasing number of progressively more complex transient junction zones. This is to some extent similar to the evolution with gum concentration of others weak gels, based on polysaccharides such as xanthan, wellan or rhamsan gums[10].

The frequency dependence of the complex viscosity fitted a power-law equation:

$$\eta^* = a\omega^b \tag{2}$$

Parameters a and b were obtained by linear regression fit after plotting the results on a double-logarithmic graph. Parameter b can be also determined from the slope of straight lines plotted in Figure 2. The same value of b (-0.85) was found at both 2% w/v and 3% w/v gum concentrations.

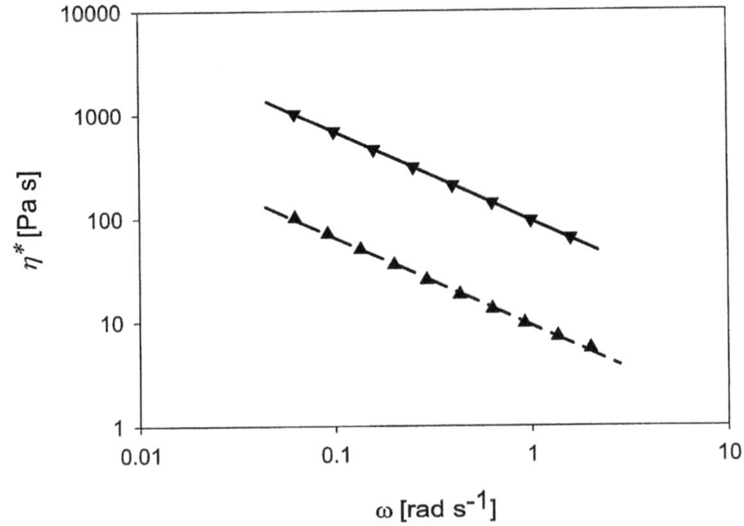

Figure 2. *Complex viscosity as a function of frequency of Sterculia apetala solutions in 1M NaCl at 25°C: solid triangles (2% w/v) and inverse solid triangles (3% /v).*

3.2 Surface characterization

Figure 3 shows the dynamic surface pressure of *Sterculia apetala* gum solutions ranging from 0.1 (w/v) to 3 % (w/v). Data shown are the average of at least three different measurements. The maximum standard deviation was about 1 mNm^{-1}. The error bars are not shown in the plot for the sake of clarity. For the most diluted solution (0.1 % (w/v)) studied, the occurrence of a long lag time was evident, which is a well-known feature of slow polymer adsorption. Figure 3 also illustrates that the time scale of the measurement (12 hours) was not enough to reach the equilibrium surface tension value. The 0.5 % w/v solution showed a dynamic surface tension curve with the three main regions (induction, rapid fall and meso-equilibrium) of the dynamic response.

Unexpectedly, the data of the 1% w/v *Sterculia apetala* gum solution were not significantly different to those of the 0.5% w/v solution. This result was rather surprising since the more concentrated solution should have shown a faster adsorption process. This odd behaviour may be related to the shift from Newtonian to non-Newtonian flow in the (0.5-1) % w/v range. A non-Newtonian material is expected to be more structured than a Newtonian liquid, hence the diffusion of the biopolymer can be affected by the enhanced viscosity of the solution.

For the more concentrated solutions a minimum of Π was observed (Figure 3). This is due to the gel-like behavior of these solutions and it is more evident for the most concentrated (3% w/v) system studied. When the drop is dispensed, it adopts the shape of the syringe, resulting in an anomalous low surface tension value. Then, the drop relaxes and becomes more spherical, but at the same time the surface active gum adsorbs at the interface. These two opposite processes explain the occurrence of a minimum. Eventually, a surface tension equilibrium value is reached.

Figure 4 shows the surface pressure isotherm. For the lowest concentration (0.1%) even a lifetime of 12 hours was too short to achieve the equilibrium interfacial tension value; therefore in this case the equilibrium value was estimated by extrapolation of the Π *vs* $t^{-1/2}$

dependence to infinite time, cf[11]. A first plateau was reached at 0.5 %, with a surface pressure value ca. 16 mNm^{-1}. At higher concentrations the surface pressure increased. This could be due to the occurrence of rearrangements of polymer chains as previously reported for polymeric surfactants[12]. The equilibrium surface pressure value at 0.5% was very similar to that observed for matured Acacia Senegal Gum, and lower than the reported value for the conventional gum[13] at the same concentration.

The viscoelastic properties of the films formed by the gum adsorption were studied by means of the oscillating drop method. To avoid the possible artifacts caused by the more viscous solutions, this study was focused on the 0.5 % w/v solution on account of its Newtonian behaviour. Figure 5 shows the frequency dependence of the elastic modulus, E', and surface viscosity, η, of the 0.5 % w/v solution after it reached the equilibrium surface tension value.

Data shown in Figure 5 were fitted to the following equations (solid lines):

$$E' = E_0 \frac{1+\zeta}{1+2\zeta+2\zeta^2} \tag{3}$$

$$\frac{E''}{\omega} = \eta = \frac{E_0}{\omega} \frac{\zeta}{1+2\zeta+2\zeta^2} \tag{4}$$

where $\zeta = (\omega/\omega_0)^{0.5}$, E_0 is the limiting elasticity and ω_0 is the reciprocal of the relaxation time constant for diffusional process. The values of the fitting parameters were: $E_0 = 41$ mNm^{-1} and $\omega_0 = 0.002$ rad/s. These values confirm the presence of an elastic layer with lower elasticity than that reported for Acacia Senegal Gum[13], but of the same order of magnitude as the elasticity provided by others biopolymers such as chitosan[14].

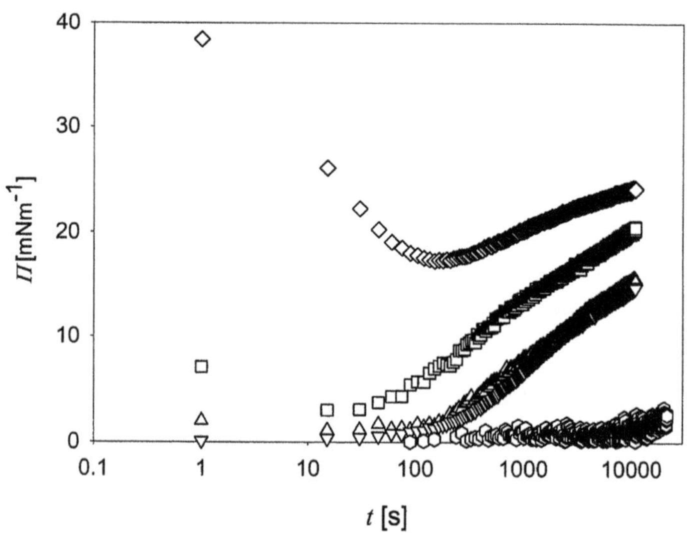

Figure 3. *Dynamic surface pressure of Sterculia apetala solutions in 1 M NaCl at 25°C. hexagon (0.1% w/v); inverse triangle (0.5% w/v); triangle (1% w/v); square (2% w/v) and diamond (3% w/v).*

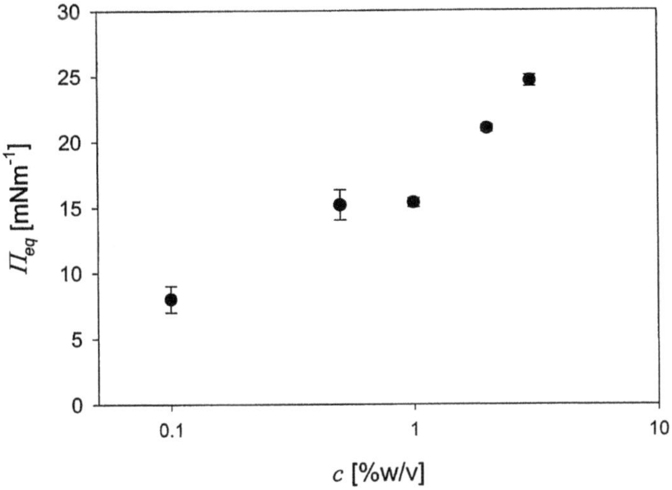

Figure 4. *Surface pressure isotherm of Sterculia apetala solutions in 1 M NaCl at25°C.*

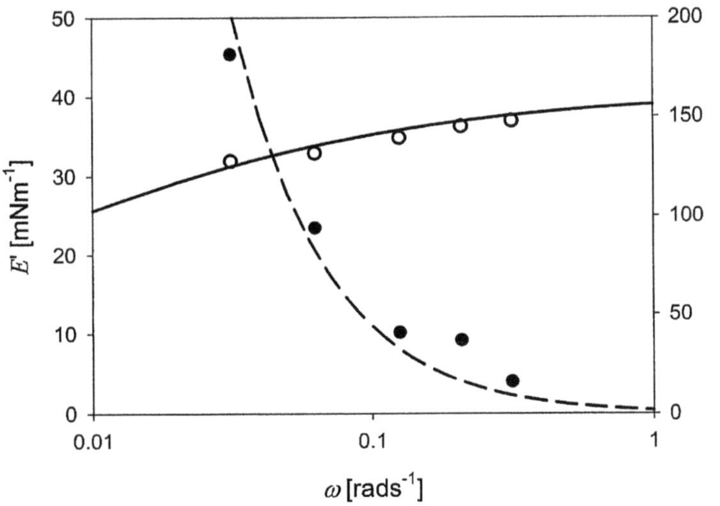

Figure 5. *Dependence of the storage module, E' (open circles) and surface viscosity, η (filled symbols) on the oscillating frequency. The harmonic perturbation was applied after the equilibrium was reached for a 0.5 % Sterculia apetala gum solution in 1 M NaCl at 25°C. The lines are the best fit to equations (3) and (4) as explained in the text.*

4 CONCLUSIONS

The surface activity and the bulk and surface rheology of *Sterculia apetala* solutions containing 1M NaCl have been studied for the first time.

Flow curves demonstrated the occurrence of a transition from Newtonian to non-Newtonian, shear thinning behaviour in the (0.5-1) % w/v range. The viscosity values obtained were not far from those reported for similar exudate gums[5]. However the ionic strength was much higher in the present study, which points to an enhanced thickening performance of *Sterculia apetala gum.* Furthermore, such solutions showed gel-like viscoelastic behaviour at 2% w/v and above, with the elastic component being dominant over the viscous one.

It has been demonstrated that *Sterculia apetala* solutions exhibit surface activity. The surface pressure equilibrium values for 0.5% solutions were quite similar to those reported for gum Arabic[13]. Dilational rheology measurements carried out for the Newtonian fluid solutions supported the formation of a elastic layer with an elasticity similar to that obtained for other macromolecules and polymers[14], although slightly lower than the elasticity reported for gum Arabic[13]. The increase in surface pressure and the significant elasticity of the gum monolayer adsorbed suggest that *Sterculia apetala* exudate gum may play the role of an emulsifier in O/W emulsions, exhibiting in addition some thickening properties, as supported by bulk rheology data.

ACKNOWLEDGMENTS

This study is part of a research project (CTQ2007-66157-PPQ) funded by the Spanish Ministerio de Ciencia e Innovación and by the European Commission (FEDER-Programme). The authors kindly acknowledge the financial support received. L.M. Pérez-Mosqueda also acknowledges the Universidad de Sevilla for its financial support (Beca PIF IV Plan Propio de Investigación).

References

1. P.A. Williams and G.O. Phillips, in *Handbook of Hydrocolloids*, 2nd Edn., ed. G. O. Phillips and P. A. Williams, CRC Press, Cambridge, 2009, ch. 1, p. 2 and ch. 11, p. 270.
2. Y. López-Franco, I. Higuera-Ciapara, F.M. Goycoolea and W. Wang, in *Handbook of Hydrocolloids*, 2nd Edn., ed. G. O. Phillips and P. A. Williams, CRC Press, Cambridge, 2009, ch. 18, p. 496-534.
3. D. Le Cerf, F.Irinei and G. Muller, *Carbohyd. Polym.*, 1990, **13,** 375.
4. D.A. Silva, A.C.F. Brito, R.C.M. de Paula, J.P.A. Feitosa and H.C.B. Paula, *Carbohyd. Polym.*, 2003, **54**, 229.
5. A.C.F. Brito, D.A. Silva, R.C.M. de Paula, and J.P.A. Feitosa, *Polym. Int.*, 2004, **53**, 1025.
6. A.C.F. Brito, M.R. Sierakowski, F. Reicher, J.P.A. Feitosa and R.C.M. de Paula, *Food Hydrocolloids*, 2005, **19**, 861.
7. L. Marvelys, M. Maritza, S. Lilian, L. de Pinto Gladys and H. Julio, *Food Hydrocolloids*, 2006, **20**, 908.
8. J. Ferguson and Z. Kemblowski, *Applied Fluid Rheology*, Elsevier, London, 1991, ch. 2, p. 15.
9. P.J. Carreau, *T. Soc. Rheol.*, 1972, **16**, 99.
10. E.R. Morris, M.G.E. Gothard, M.W.N. Hember, C.E. Manning and G. Robinson, *Carbohyd. Polym.*, 1996, **30**, 165.
11. V.B. Fainerman and R. Miller, *Colloid Surface A*, 1995, **97**, 65.
12. M.G. Muñoz, F. Monroy, F. Ortega, R.G. Rubio and D. Langevin, *Langmuir*, 2000, **16**, 1083.
13. O. Castellini, D. Guibert, S. Al-Assaf, M. Axelos, G. O. Phillips and M. Anton, *Food Hydrocolloids*, 2010, **24**, 193.
14. N. Calero, J. Muñoz, P. Ramírez, A. Guerrero, *Food Hydrocolloids*, 2010, **24**, 659.

EFFECT OF OZONATION ON THE PHYSICOCHEMICAL, THERMAL AND STRUCTURAL PROPERTIES OF WHEAT STARCH SAMPLES

M Gupta[1], BK Tiwari[2], C.S. Brennan[2], PJ Cullen[1] and Colm O'Donnell[3]

[1]School of Food Science and Environmental Health, Dublin Institute of Technology, Dublin 1, Ireland.
[2]Department of Food and Consumer Technology, Manchester Metropolitan University, M14 6HR, UK.
[3]UCD School of Biosystems Engineering, University College Dublin, Belfield, Dublin 4, Ireland.

1. INTRODUCTION

Consumer preference for minimally processed and foods free of chemical preservatives have stimulated demand for novel food processing and preservation systems. Within the food industry ozone is widely employed in various food processing operations. Ozone is a powerful oxidant that reacts with most substances at ambient temperature and residual ozone decomposes to nontoxic products making it environmentally friendly. Apart from fruit and vegetables, ozone is also employed in grain processing and preservation. [1,2]

In addition to food applications native starches as well as modified starches such as oxidized starches have non food applications such as surface sizing agents and coating binders in the paper industry.[3] Starches can be modified, hydrolysed or depolymerised using several treatments such as acid or enzymatic treatments to provide a range of modified starches for various applications. Several oxidizing agents such as sodium hypochlorite, bromine, periodate, permanganate, hydrogen peroxide and ammonium persulfate are employed for starch oxidation. Chan et al.[4] investigated the application of ozone as an oxidizing agent. Generally during oxidation process hydroxyl group of starch granules are oxidized to carbonyl and carboxyl groups as well as cleavage of $\alpha(1\rightarrow4)$ glycosidic linkages.[3,4] Chemical changes occurring during the oxidation process allow modification of starch for improved stability of the paste with reduced viscosity for various food applications. Studies show that oxidized starches have a shorter cooking time, lower viscosity, higher stability, reduced retrogradation and better clarity.[5,6] Application of oxidized starches in food products is reported to improve product stability[7] and to improve adhesiveness that enhances its ability to be used in batters and breadings for fried foods.[8] Chan et al.[9] studied the effects of ozone on physiochemical and functional properties of various starches in dry form including corn, sago and tapioca. They utilized 0.4 to 1.4 mmoles of ozone treatment based on time of exposure with rotation of the starch. The results of Chan et al.[9] showed that starch could be treated with ozone in a dry state to alter functional properties without loss of product and without the use of chemical solvents. The objective of this study was to investigate the effect of ozonation on the physicochemical, thermal and structural properties of wheat starch samples.

2. MATERIALS AND METHODS

2.1 Preparation of samples

Wheat starch was supplied by AllinAll Ingredients, Dublin, Ireland. Starch slurries of 50 g starch and 250 mL of distilled water were prepared for each treatment.

2.2 Ozone treatment

Experiments were carried out in a 500 mL beaker with a glass diffuser (Figure 1). Ozone was generated using an ozone generator (Model OL80, OzoneServices, Vancouver, Canada) and recorded using an ozone gas analyzer (Model OLA-DLS, OzoneServices, Canada). Ozone concentrations of 2.4, 4.0 & 7.8 % (w/w) at a constant gas flow rate of 0.06 $Lmin^{-1}$ were studied. Samples were treated for 0 (control) and 10 min processing times. Wheat starch slurry samples were continuously stirred using a magnetic stirrer. Samples were freeze dried and ground into powder after ozone treatment for further analysis.

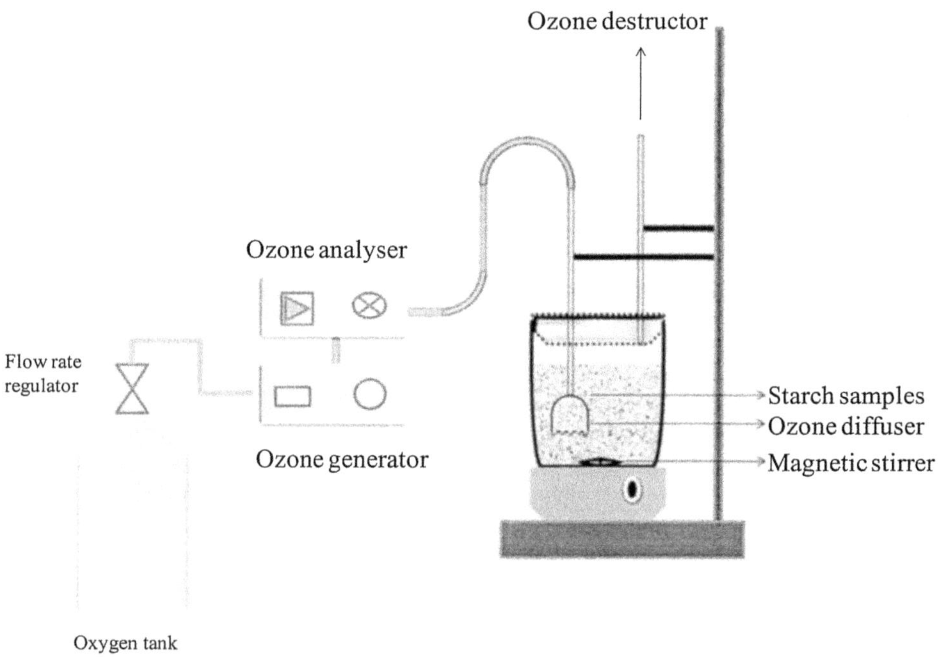

Figure 1. Experimental set up for ozone treatment of starch samples.

2.3 Physicochemical and thermal properties

The amylose and amylopectin content of ozonated starch samples were determined by standard iodine colorimetry method.[10] Light transmittance (%) of the starch pastes made from the starch samples was measured using the method described by Craig et al.[11] A 1% (w/v) aqueous suspension of starch at near neutral pH was heated in a boiling water bath for 30 min with constant stirring. The suspension was cooled for 1 h at 25 °C. The samples were stored for 24 hours at 4 °C and transmittance (%) was measured at 640 nm against a water blank with a UV–Vis spectrophotometer
Thermal analysis was conducted using differential scanning calorimetry (TA Instruments, USA). Ten mg of a sample (dry weight basis) was weighed in a silver pan, and distilled water added to make a suspension of 30% (dry weight basis, w/w). A sealed pan with distilled water was used as a reference. Scans were run at a heating rate of 2°C/min from 25 to 130 °C. The gelatinization onset temperature (T_o), peak temperature (T_P), T_c and enthalpy (ΔH) were recorded.

2.4 Pasting properties

The pasting profile of the starches (8% w/w, dry weight basis) was determined using a Rapid Visco Analyzer (model RVA-4, Newport Scientific Pvt. Ltd., Australia). The samples were equilibrated at 50°C for 1 min and then raised to 95°C in 3.75 min, held for 2.5 min, cooled to 50 °C in 3.75 min, and held for 5 min. The paddle speed was set at 960 rpm for the first 10 s to evenly disperse the starch slurry and reduced to 160 rpm throughout the remainder of the experiment.

2.5 Scanning electron microscopy (SEM)

The granule morphology of ozonated wheat starches was studied using scanning electron microscopy (SEM). Starch samples were mounted on circular aluminium stubs with double sticky tape and then coated with 20 nm of gold and examined and photographed in a Hitachi (5570) scanning electron microscope at an accelerating potential of 5 kV.

3 RESULTS AND DISCUSSIONS

3.1 Physicochemical properties

Table 1 shows the physicochemical properties of control and ozonated wheat starch samples. Amylose content (%) of control (untreated) wheat starch samples were 22.0 % which was found to increase with increase in ozone concentration. A similar increase in amylose content is reported for ozonated rice starch samples.[12] However, Voraputhapron[13] observed no significant change in the amylose content of wheat flour and starch with ozonation at 25 to 100 ppm. The swelling power of ozonated wheat starch samples was found to decrease with an increase in ozone concentration. However this decrease was significant only at highest processing conditions studied (7.8 %w/w for 10 min). Similarly, Chan et al.[9] observed a decrease in the swelling power of ozone oxidized sago and tapioca starch at an ozone treatment time of 10 min. However, they observed an increase in the swelling power of ozone oxidized corn starch. This increase in swelling power was partly due to the degradation of amylose thereby allowing the starch granule to absorb water and swell more easily[9]. Whereas, the observed decrease in swelling power after ozone oxidation may be due to structural disintegration within the starch granule during the process of modification. [9,14] A decrease in the swelling power and simultaneous increase in the solubility of ozonated wheat starch samples was observed for samples processed at 7.8 %w/w for 10 min. A similar trend was observed during oxidation of bean starch using sodium hypochlorite as an oxidizing agent. The increase in solubility following oxidation process might result from depolymerization and structural weakening of the starch granule.[9] Parovuori et al.[15] observed an increase in the solubility of potato starch with an increase in oxidation treatment by hydrogen peroxide.

Table 1 *Physicochemical properties of control and ozonated starch samples processed for 10 min at constant flow rate of 0.06 L.min^{-1}*

	Samples			
Properties	**Control**	**2.4 %w/w**	**4.0 %w/w**	**7.8 %w/w**
Amylose (%)	22.0c	23.1b	24.0ab	24.5a
Swelling power g/g	21.8a	21.7a	21.0a	19.2b
Solubility %	2.7c	3.9b	4.2ab	4.5a
Light transmittance	32.0c	34.2b	36.5a	37.0a
To (°C)	52.8b	53.4a	54.0a	54.8a
Tp (°C)	58.6b	59.2ab	59.0a	59.6a
Tc (°oC)	79.0	79.7	78.1	78.9
ΔHgel (J/g)	10.1c	16.2b	16.4b	18.6a

T_o = onset temperature, T_p = peak temperature, T_c = end set temperature ΔH_{gel} = Enthalpy of gelatinization (dwb, based on starch weight). Values with different alphabets in a row are significantly different ($p \leq 0.05$).

3.2 Thermal properties

Table 1 shows the gelatinization properties of ozonated wheat starch samples using DSC. DSC parameters recorded were onset To, Peak Tp, Final Tc gelatinization temperatures (°C), gelatinization enthalpy ΔHgel (J/g). To, Tp, Tc and ΔHgel for control wheat starch samples were 52.8 °C, 58.6 °C, 79.0 °C and 10.1 J/g respectively. A significant increase in To, Tp and ΔHgel was observed for ozonated starch samples with no significant change in Tc. Similarly, Voraputhaporn[13] observed a higher peak gelatinization temperature (Tp) with a lower enthalpy (ΔHgel) for ozone oxidized wheat starch. The enthalpy of gelatinization reflects the loss of molecular order[16] and gelatinization temperature is considered a parameter of crystallite perfection.[17]

3.3 Pasting properties

The pasting properties of untreated (control) and ozone oxidised wheat starches using the Rapid Visco Analyzer are shown in Table 2. The pasting temperature of oxidized wheat starch was found to decrease with an increase in ozone concentration processed for 10 min. Wheat starch samples processed at 7.8 %w/w for 10 min exhibited significantly lower pasting temperature (74.2 °C) compared to the control (80.6 °C). Similar reductions in pasting temperatures were observed by Chan et al.[9] for ozone oxidized corn and sago starches. The decrease in pasting temperature as a result of oxidation is due to structural weakening and disintegration of starches.[9,14] Ozone oxidized wheat starch samples exhibited an increase in peak viscosity, hot paste viscosity, breakdown viscosity, cold paste viscosity and setback viscosity. A similar increase in peak viscosity, hot paste viscosity, breakdown and cold paste viscosity was observed in ozone oxidized sago starch samples.[9] Voraputhaporn[13] also found that wheat starch treated with ozone had increased peak viscosity compared to native starch. The increase in peak viscosity for ozone oxidized starches could be due to the chemical cross linking as observed in the case of oxidized corn starches.[18] An increase in breakdown viscosity was also observed for ozone oxidized sago starch[9] and sodium hypochlorite (NaOCl) oxidized mucuna bean starch.[19] Higher breakdown viscosity indicates disruption of starch granules. However, An and King[12] observed an increase in peak viscosity of ozonated rice starch samples indicating greater swelling. They also observed an increase in breakdown viscosity compared to untreated rice starch samples resulting in less cooking stability. However, An and King [12] also observed lower final viscosity and total setback viscosity. Slight oxidation of starch by NaOCl is shown to result in greater swelling and peak viscosity due to loss of starch granule integrity.[20] Pasting properties of ozone oxidized wheat starch samples indicate that the ozone oxidized starches also exhibit similar properties as exhibited by other chemical oxidants.

Table 2. *Pasting properties of ozone oxidized (treatment time 10 min) wheat starches*

Samples	PV[a] (cP)	HPV (cP)	BV(cP)	CPV (cP)	SV(cP)	PT (°C)
Control	3249b	2667b	582c	3706b	1039b	80.6a
2.4 %w/w	3687a	3032a	655ab	4280a	1248a	78.9ab
4.0 %w/w	3768a	3061a	707a	4289a	1228a	77.4b
7.8 %w/w	3629a	2995a	634b	4209a	1214a	74.2c

[a]PV: Peak viscosity; HPV: hot paste viscosity; BV: Breakdown viscosity; CPV: cold paste viscosity; SV: Setback viscosity; PT: Pasting Temperature.

3.4 Morphological Properties

Scanning electron microscopic images of untreated and ozonated wheat starches are shown in Figure 2. Images of untreated wheat starch show typical starch granules of mostly spherical shape and some oval shaped granules (Figure 2a). Whereas, ozonated starch samples show some loss of starch granule integrity and subsequent distortion of the granule (Figure 2b-d). Moreover, many small pieces of granules that were probably broken off of large particles. Several investigators reported that oxidized starch showed physical damage and disruption both inside and outside of the starch granule as the oxidizer concentration increased.[7,20]

Figure 2 *Scanning electron micrographs (SEM) of starch samples. 2 a) untreated (control) wheat starch,2 b) ozonated (7.8 %w/w for 10min) wheat starch, c) 10 µm magnification and d) 5 µm magnification of ozonated wheat starch samples.*

Oxidation effects of ozone and depolymerisation of a macromolecule is attributed to the oxidative destruction of β-D-glucoside bonds between units by the electrophilic attack of the C-H bond by ozone. Formation of free radicals due to ozone could also be attributed to degradation of the macromolecules. Ozone oxidation occurs via two pathways, by direct oxidation with ozone molecules and by the generation of free-radical intermediates[22].

4 CONCLUSIONS

Significant reductions in the physiochemical and rheological properties of ozonated starch samples were observed. Results presented indicate that ozonation can be employed for modification of starch physicochemical properties and is a possible alternative to chemical oxidative agents.

5 ACKNOWLEDGMENTS

This work was supported by a Food Institutional Research Measure (FIRM) and the Irish Department of Fisheries, Food and Agriculture.

References

1 Tiwari BK, Brennan CS, Curran T, Gallagher E, Cullen PJ, McDonnell K, O'Donnell, CP (2010). J Cereal Sci 51:248-255.
2 Dubois M, Canadas D, Despres-Pernot AG, Coste C, Pfohl-Leszkowicz A (2008). J Agric Food Chem 56: 1116–1121
3 Sangseethong K, Termvejsayanon N, Sriroth K (2010). Carbohydrate Poly 82:446-453
4 Chan HT, Leh CP, Bhat R, Senan C, Williams PA, Karim AA. (2011). Food Chem 126:1019-1024
5 Kuakpetoon D, Wang Y J (2001). Starch/Starke 53:211-218.
6 Scallet, B. L., and Sowell, E. A. 1967. Production and uses of hypochlorite-oxidized starch, in Starch: Chemistry and Technology. Eds., R.L. Whistler and E. F. Paschall, Academic Press, New York. Pp. 237-251.
7 Boruch, M. L. 1985. Transformation of potato starch during oxidation with hypochlorite. Starch 37 Nr. 3, S. 91-98.
8 Han, J. S. 2002. Food Sci. Biotechnol. 11:231-237.
9 Chan HT, Bhat R, Karim AA (2009). J Agric Food Chem 57:5965-5970.
10 Juliano BO, Perez CM, Blakeney AB, Castillo DT, Kongseree N, Laignelet B, Lapis ET, Murty VVS, Paule CM, Webb BD (1981) Starch/starke 33:157-162.
11 Craig SAS, Maningat CC, Seib PA, Hoseney RC (1989). Cereal Chem 66:173–182.
12 An HJ, King JM 2009. J Food Sci 74:C278-C283.
13 Voraputhaporn W (1996) PhD Thesis, Department of Food Science and Technology The University of New South Wales.
14 Lawal OS (2004) Food Chem 87:205– 218
15 Parovuori P, Hamunen A, Forssell P, Autio K, Poutanen K. (1995) Starch/Starke, 47:19-23.
16 Cooke D, Gidley MJ (1992) Carbohydrate Polymer 227:103-112.
17 Tester RF, Morrison WR (1990) Cereal Chem 67:551-557.
18 Wang Y-J, Wang L, (2003) Carbohydrate Polymers 52:207–217.
19 Adebowale KO, Lawal OS, (2003) J Sci Food Agric 83:1541–1546
20 Kuakpetoon D, Wang YJ 2001. Starch/Starke 53:211-218.
21 Han JS, Ahn SY (2002) J Korean Soc Food Sci Nutr 31:189-195.
22 Staehelin J, Hoigne J (1982). Environ Sci Tech 16:676.

FUNCTIONAL PROPERTIES OF ACETYLATED GLUTENIN AT VARYING pH VALUES

Mahsa Majzoobi*, Elaheh Abedi, Asgar Farahnaky, Mahmoud Aminlari

Department of Food Science and Technology, College of Agriculture, Shiraz University, Shiraz, Iran, Post code: 7144165186; *e-mail: majzoobi@shiazu.ac.ir; Tel: +987116138356; Fax: +987112286110

1 INTRODUCTION

Gluten is a protein mixture remaining after washing of the wheat flour dough with water. It is produced as a by-product in starch producing factories. It is made up of glutenin and gliadin with the ratio of about 1:1. These proteins are the main storage proteins of wheat kernels. They can be separated from gluten or wheat flour based on their difference in solubility. Glutenin is mainly soluble in dilute acid and alkaline solutions, while gliadin is soluble in aqueous ethanol. Each of the separated fractions has unique physicochemical properties and can be used as a source of plant proteins in many food and non-food products.[1]

Hydrated glutenin is both cohesive and elastic and contributes to the elasticity and strength of bread dough and many other bakery products. The amount of glutenin in wheat flour is about 20-40 mg/g. It has varying size range from about 500 to more than 10000 kDa and hence is amongst the largest proteins in nature. Glutenin occurs as aggregates linked together by inter- and intra-disulphide bonds and is divided into two main subunits; high and low molecular weight subunits (HMW and LMW, respectively). About 10% of gluten consists of HMW glutenin subunits. These proteins have great influence on bread dough elasticity and bread volume. Each wheat variety contains three to five HMW-glutenin subunits which can be grouped into two different types; the x- and y- types with molecular weights of 83-88 and 67-74 kDa, respectively. The predominant protein types of glutenin is LMW subunits accounting for about 20% of the total gluten proteins. They are associated with dough resistance and extensibility.[1]

Like other plant proteins, glutenin is fairly low-cost, renewable and biodegradable, however, compared to many other plant proteins, glutenin has distinctive physicochemical properties such as strong elasticity and high water holding capacity. It has been used as biodegradable films[2, 3], however, other applications of glutenin are limited because of its poor water solubility, emulsifying and foaming properties. Nevertheless, the presence of reactive amino groups (i.e. lysine) on this molecule offer a great potential for derivatization. In order to improve the functionality of glutenin, it is possible to modify the structure of glutenin. A common method for this purpose is through chemical modifications. Acylation is a usual method to increase water solubility of the proteins. Acetic, succinic and citraconic anhydride are generally used for protein acylation, resulting

in products with slightly different functional properties. Acetylation of proteins involves the reaction of acetic anhydride with the ε-amino group particularly lysine.
Previous studies have focused on acylation of gluten and its physicochemical properties[4,5], however, little information is available on the functional properties of AC glutenin.
The objectives of this study were to produce acetylated (AC) glutenin and study the solubility, foaming properties, of the AC glutenin at different pH values including 3 (acetic), 6 (nearly neutral) and 9 (alkaline). The AC glutenein can be used as an alternative for some chemical emulsifiers and foaming agents in food products. Meat products (e.g. sausages and burgers), dried soups and bakery products are examples of foods in which the AC glutenin may be used.

2 MATERIALS AND METHODS

Commercial wheat gluten powder was purchased from Fars Glucosin Starch Factory, Marvdasht, Iran. It contained 86.27±0.60% protein, 6.70±0.05% moisture content, 1.68±0.03% lipids, 1.07±0.01 ash (all based on dry weight) and 4.28±0.53% total carbohydrates calculated by subtracting the protein, moisture, fat and ash contents from 100%.
Sunflower oil was purchased from local market. Other chemicals used in this study were obtained from Merck, Darmstadt, Germany. Electrophoresis marker was purchased from Fermentaz, Inc. Canada.

2.1 Separation of glutenin rich fraction from gluten

Gluten powder (100 g) was first dispersed in 500 mL aqueous ethanol 70% (v/v) by means of a mixer (Nasional, Japan), for 2 min at medium speed. The suspension was stirred overnight at room temperature using a magnetic stirrer. Then it was centrifuged (Type GLC-1 Ivan Sorvall Inc., USA) at 1160g for 15 min at 20°C. The pellet (glutenin rich fraction) was collected and was extracted another three times using 100 mL 70% (v/v) aqueous ethanol. The final pellet was lyophilized, then milled using a laboratory mill (Alexander Werck, Model WEL82, Germany) and sieved manually to obtain particle sizes smaller than 40 μm. The samples were packed in polyethylene bags and stored at room temperature for further experiments. Isolated glutenin, called native (N) glutenin, contained 86.31±0.84% protein, 1.30±0.02% fat, 0.94±0.03% ash, 6.00±0.01% moisture (all based on dry weight) and 5.45% total carbohydrates calculated by subtracting the protein, moisture, fat and ash contents from 100%.

2.2 Acetylation of glutenin

Slurry of glutenin in distilled water (1:4, w/w) was made and the pH was adjusted to 8.5 using 1 N NaOH. Acetic anhydride was gradually added to the slurry (0.5 g increment) and the pH was re-adjusted to 8.5 with 2 N NaOH. Then the solution was left stirring by a magnetic stirrer at room temperature overnight. The mixture was centrifuged at 10000g for 10 min and the pellet was recovered, washed with 2-propanol several times and freeze-dried to obtain AC glutenin. The produced AC glutenin had 84.36% protein, 1.2% fat, 1.62% ash, 5.55% moisture (all based on dry weight) and total carbohydrates of 7.27% as calculated by subtracting the protein, moisture, fat and ash content from 100%.

2.3 Determination of the extent of acetylation using 2,4,6-trinitrobenzenesulfonic acid (TNBS) method

The extent of acetylation was estimated by determination of the available lysine content of the glutenin using TNBS method. Acetylation was expressed as percent reduction in available lysine residue.

2.4 Determination of the isoelectric point (IEP)

Suspensions of 0.3% (w/v) glutenin in distilled water were made and the pH of the suspensions was adjusted from 2.0 to 10.0 with 0.2 intervals using either 1.0 mol/L HCl or 1.0 mol/L NaCl. The suspensions were homogenized at 10000 rpm for 30 sec and then centrifuged at 3000 g for 20 min. The protein content of the supernatant was determined using micro-Kjeldahl method (5.7 as the nitrogen to protein conversion factor). The pH in which proteins had the least solubility was selected as the IEP.[10]

2.5 Physicochemical properties of N and AC glutenin at varying pH values

To determine water solubility of the proteins at different pH values, two separate sets of samples were prepared by suspending 25 mg of the N and AC glutenin in 25 mL acetate buffer, separately. These samples were used to obtain pH values of 3 and 6 by addition of 0.5 mol/L HCl. Another set of sample was prepared by suspending of 25 mg of the native or AC glutenin in 25 mL phosphate buffer, separately. The pH of these samples was adjusted to 9 by using 0.5 mol/L NaOH. To determine water absorption, water holding capacity (WHC), emulsifying and foaming activities of the proteins, 0.5 g (instead of 25 mg) N or AC glutenin was suspended in 10 mL (instead of 25 mL) of the buffers as explained before.

2.6 Water solubility

To determine water solubility of the samples, the protein dispersions were mixed for 1 h using a magnetic stirrer at ambient temperature and then they were centrifuged at 7000g for 20 min. The protein content (N×5.7) of the supernatant was determined using micro-Kjeldahl method. The solubility (%) was determined according to using Equation 1.

$$\text{Solubility (\%)} = \frac{\text{Protein content in the supernatant}}{\text{Protein content in the suspension}} \times 100 \qquad (1)$$

2.7 Water absorption

The protein suspensions were stirred using a magnetic stirrer at room temperature for 30 min. Then the samples were left still for 10 min and the weight of separated water was determined. The water absorption was calculated according to the Equation 2.

Water absorption (g H_2O/0.5 g protein) =

$$\frac{\text{Weight of the protein suspension} - \text{Weight of the separated water}}{\text{Weight of proetin in the suspension}} \qquad (2)$$

2.8 Determination of WHC

After adjusting the pH of the protein suspensions, samples were centrifuged at 500g for 25 min at ambient temperature. Then the supernatant was removed and the weight of the pellet was determined. The WHC was calculated according to the Equation 3.

$$\text{WHC (\%)} = \frac{\text{Weight of pellet}}{\text{Weight of protein suspension}} \times 100 \tag{3}$$

2.9 Emulsifying activity and emulsion stability

Five mL of protein solutions were homogenized using a laboratory homogenizer (Digital Ultra-Turrax, IKA, Germany) with 5 mL sunflower oil at 2000 rpm for 5 min and were centrifuged at 1100 g for 5 min. The height of the emulsified layer in a glass measuring cylinder and that of the total contents was measured. The emulsifying activity was calculated according to the Equation 4.

$$\text{Emulsifying activity (\%)} = \frac{\text{Height of emulsified layer in the tube}}{\text{Height of total contents in the tube}} \times 100 \tag{4}$$

To determine the emulsion stability the emulsions were heated at 80 °C for 30 min. They were then centrifuged at 1100g for 5 min and the emulsion stability was calculated using Equation 5.

$$\text{Emulsion stability (\%)} = \frac{\text{Height of emulsified layer after heating}}{\text{Height of emulsified layer before heating}} \times 100 \tag{5}$$

2.10 Foaming capacity and stability

To produce foam, suspensions of N and AC glutenin in buffers were placed in glass measuring cylinders and mixed vigorously using a laboratory homogenizer at 20000 rpm for 2 min at ambient temperature. After that the foam volume of each sample was measured in a glass measuring cylinder.

To determine the foam stability, the volume of the foam was measured every 10 min at ambient temperature until the foam disappeared. Foam stability was expressed as the percentage of the initial foam volume according to the Equation 6.

$$\text{Foam stability (\%)} = \frac{\text{Residual foam volume}}{\text{Initial foam volume}} \times 100 \tag{6}$$

2.11 SDS-PAGE pattern of the N and AC glutenin

To make the samples for electrophoresis, first N and AC glutenin (75 mg each) were dispersed in distilled water and the pH was adjusted to 3, 6 and 9 by addition of a few drops of HCl or NaOH and the final volume was adjusted to 5 mL. The suspensions were mixed well over a vortex mixer at medium speed for 2 min, then centrifuged at 5000 g for 10 min. An aliquot of the supernatant containing 5 mg/mL protein as determined by biuret assay was mixed 1:1 with sample buffer which was then loaded in the electrophoresis gel.

Sodium dodecyl sulfate (SDS) slab gel electrophoresis (Clever, Scientific, Ltd., UK) was conducted using a 10% acrylamide separating gel. Protein samples were added to the loading buffer to give a final concentration of 1 mg/mL protein, 0.1 mol/L Tris-HCl (pH 6.8), 0.4% SDS, 10% glycerol and 0.004% bromophenol blue. The running gel (140×140×1 mm) was a 10% polyacrylamide gel in 1.2 mol/L Tris-HCl (pH 8.8) and 0.3% SDS. The stacking gel contained 3% acrylamide in 0.25 mol/L Tris-HCl (pH 6.8) and 0.2% SDS. The electrode buffer contained 0.025 mol/L Tris-HCl, 0.192M glycine, and 0.15% SDS at pH 8.16. The molecular weight marker was in the range of 20000-200000. Electrophoresis was performed at a constant 15mA current and gels were stained with 0.25% Coomassie Brilliant Blue R-250 in 50% methanol and de-stained with 10% acetic acid/ 7% methanol.

2.12 Statistical analysis

All experiments were conducted in triplicate. The average and standard diviation were calculated. Data were then analyzed using analysis of variance (ANOVA) procedure by means of the statistical software of SPSS 16. The mean comparison was performed using the Multiple Range Duncan's test (α <0.05).

3 RESULTS AND DISCUSSION

The amount of free lysine available to react with TNBS in untreated and acetylated glutenin was used to determine the extent of modification. The results showed that the extent of glutenin acetylation was 53.88 ± 0.34 % as determined by the TNBS method.
Determination of the water solubility of the samples showed that minimal solubility of N and AC glutenin was observed at pH values of 7.8±0.1 and 5.8±0.1, respectively. Most of the proteins have the lowest solubility at IEP due to the minimum repulsion among the constituent amino acids. The balance in positive and negative charges minimizes the electrostatic repulsion resulting in less protein solubility. Therefore, these pH values may be close to the IEP of N and AC glutenins. Upon acetylation the IEP reduced, significantly which can be related to the reduction of NH_2 groups because of the reaction of acetyl groups with lysine.
Acetylation and changes of the pH, had significant effect on the solubility of the samples (Table 1). AC glutenin was significantly more soluble than the N sample at every pH values tested in this study. With increasing the pH from 3 to 6, a decrease in the solubility was observed for AC sample. pH 6 is close to the IEP of AC glutenin and hence it was less soluble at this pH compared to the other pH values. With increasing the pH from 6 to 9, the solubility of AC glutenin increased, significantly. For N glutenin the solubility decreased as the pH of the environment increased and became close to the IEP of the sample.
The results (Table 1) showed that WHC and water absorption of the samples had similar patterns to the water solubility of the samples at varying pH values. Acetylation significantly enhanced water absorption and WHC of the glutenin. Upon acetylation, ammonium cations were replaced by a neutral acetyl group producing less electrostatic repulsion. Acetylation can also promote protein unfolding and producing fewer protein-protein and more protein-water interactions resulting in higher solubility of the AC sample.
Acetylation could significantly increase the emulsifying activity and stability of the glutenin at the pH values tested in this study (Table 1). For AC samples, the emulsifying activity was the lowest at pH 6, while for N glutenin the lowest emulsifying activity was observed at pH 9. Similar trend was observed for emulsion stability of N and AC glutenins at different pH values. These observations can be mostly related to the effect of acetylation

on glutenin solubility, conformation and surface properties. An increase in protein solubility would enhance a rapid migration and adsorption at the water-oil interface.
Unfolding of the proteins during heating, has great influence on the formation of stable molecular layers in the air-water interface which imparts texture, stability and elasticity to the foam. Determination of the foam volume of N and AC glutenin (Table 1) showed that the foam volume of AC glutenin was significantly higher than N sample at every pH values tested in this research. Similar to water solubility of the samples, the minimum foam volume of the N and AC samples were observed at pH 9 and 6, respectively.

Table 1. *Functional properties of native (N) and acetylated (AC) glutenin at different pH values.*

Characteristics	**pH 3**		**pH 6**		**pH 9**	
	N	AC	N	AC	N	AC
Water solubility (%)	35.5^{d}	47.4^{b}	32.7^{d}	41.2^{c}	26.0^{e}	64.1^{a}
Water holding capacity (%)	36.0^{d}	53.0^{b}	32.0^{e}	40.0^{c}	23.0^{f}	80.0^{a}
Water absorption (g H_2O/g protein)	7.1^{b}	7.4^{b}	5.4^{d}	6.6^{c}	3.5^{e}	9.5^{a}
Emulsifying activity (%)	55.2^{d}	72.9^{b}	51.1^{e}	69.4^{c}	48.1^{f}	80.0^{a}
Emulsion stability (%)	59.7^{d}	72.17^{b}	56.0^{e}	65.0^{c}	47.2^{f}	77.8^{a}
Foam volume (mL)	30.7^{c}	38.7^{b}	22.7^{e}	24.2^{d}	7.7^{f}	42.5^{a}

Values are the average of triplicates. Different letters in each row indicate significant statistical difference (α<0.05).

Determination of the foam stability (Figure 1) showed that acetylation increased the foam stability of the sample. Furthermore, the foam stability of the N and AC glutenin was the highest at pH 9 and 6, respectively. These pH values were close to their corresponding IEP. At this point the electrostatic repulsion is minimized and hence cohesive interactions between protein molecules via non-covalent forces increase resulting in more stable foam.
The SDS-PAGE results (Figure 2) showed that the intensity of the protein bands between 85000 and 120000 (related to HMW glutenin subunits) and 40000-60000 (related to LMW glutenin subunits) decreased as a result of acetylation. Moreover, the intensity of protein bands in the range of 20000-40000 (related to the LMW glutenin subunits) increased for the AC glutenin. According to the results, it seems that the amount of large glutenin aggregates reduced upon acetylation. Acetyl groups increase the interaction of glutenin with water molecules and reduce the aggregation of these molecules. Therefore, molecules of lower molecular weights can be formed. Changes of the pH had no considerable effect on the electrophoretic pattern of the samples. Apparently, changes of the pH may only alter the molecular conformation of glutenin without having any significant effect on the molecular weights of these proteins.

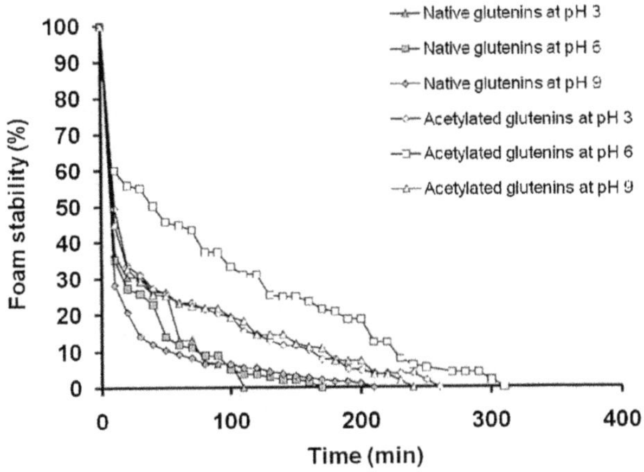

Figure 1. *Foam stability of the native and acetylated glutenins at varying pH values over time.*

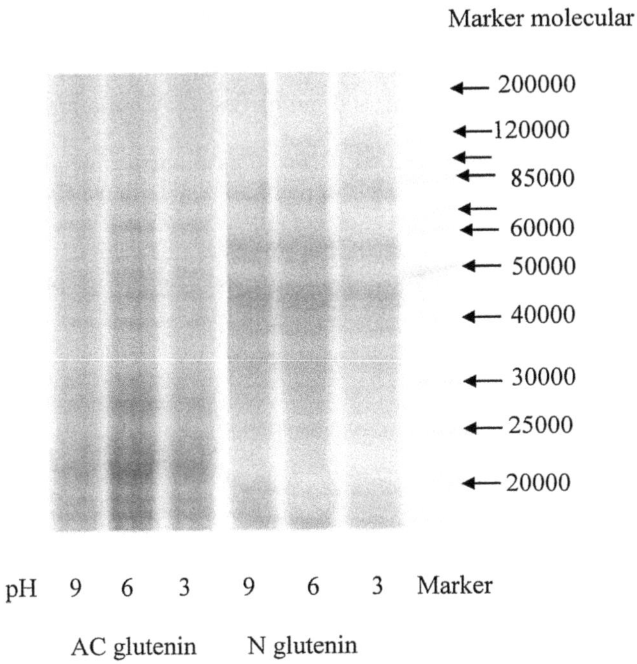

Figure 2. *SDS-PAGE patterns of the soluble part of native (N) and acetylated (AC) glutenins at varying pH values.*

4 CONCLUSION

AC glutenin can be considered as a new source of protein of plant origin which can be used as an emulsifier, foaming agent and water binder particularly at acid and alkaline pH values. Unlike its N glutenin counterpart, AC glutenin showed moderate water solubility and hence can retain its functional properties in aqueous environments. At every pH value tested in this study, WHC, emulsifying and foaming capacity of the AC glutenin were significantly higher than those of N samples. The changes in the functional properties of AC sample are related to the molecular modification caused by acetylation particularly reduction of the glutenin aggregates and formation of more hydrophilic molecules as determined by SDS-PAGE electrophoresis. Based on the electrophoresis results both HMW and LMW glutenin subunits were affected by acetylation, while changes of the pH had no significant effect on the molecular weight of these proteins.

REFERENCES

1- H. Wieser, Chemistry of gluten proteins, *Food Microbiol.*, 2007, **24**, 115.

2- C. Mangavel, J. Barbot, E. Bervas, L. Linossier, M. Feys, J. Guéguen, Y. Popineau, Influence of prolamin composition on mechanical properties of cast gluten films, *J. Cereal Sci., 2002,* **36**, 157.

3- Y. Song, Q. Zheng, Preparation and properties of thermo-molded bioplastics of glutenin-rich fraction, *J. Cereal Sci.*, 2008, **48**, 77.

4- E. A. Żukowska, E. Rudnik, J. Kijenski, Foaming properties of gluten and acetylated gluten studies of the mathematical models to describe liquid drainage, *J. Cereal Sci.*, 2008, **47**, 233.

5- A. H. Saberi, M. Kadivar, J. Keramati, Improvement of functional properties of gluten extracted from two Iranian wheat varieties Sardari and Mahdavi employing chemical and enzymatic modification, *J. Agric. Sci. Technol.*, 2008, **10**, 243.

OPTIMIZATION FOR RAPID SEPARATION OF IMMUNOSTIMULATORY *CHLORELLA* BIOPOLYMERS

P. Lai, Y.-T. Huang, and K.-Y. Jiang

Department of Food and Nutrition, Providence University, Shalu, Taichung, 43301, Taiwan

1 INTRODUCTION

Some species of microalgae *Chlorella* (Chlorophyceae), mainly *C. vulgaris* and *C. pyrenoidosa* (some newly renamed as *C. sorokiniana*)[1], have been regarded as healthy supplements and are popular in many countries. Oral intake of *Chlorella vulgaris* hot-water extract (CVE) at a dosage of 20 g kg^{-1} for 2 wk enhances significantly the resistance of normal and MAIDS mice toward *Listeria* infection potentially via augmenting Th1 signaling pathway and improving phagocyte function of neutrophiles and macrophages[2]. The immunostimulatory activities of CVE are also discovered in casein-immunized DBA/2 mice[3] and BALB/c mice[4] and may be responsible for the lead-expose protection effect on IFN-γ knockout C57BL/6 mice[5]. Besides, six-week supplementation daily with 6.3 g *C. vulgaris* showed favorable effects on antioxidant status in Korean male smokers.[6] *C. pyrenoidosa* hot-water extracts (CPE) are also reported of promising immunostimulatory activities *in vivo* for human[7] and BALC/c mice[8], and have been patented as Respondin™, a proprietary immunomodulator[8,9]. Polysaccharides from CPE are found to activate macrophages via Toll-like receptor 4[10].

The CVE, extracts at 100°C for 20 min, contains 440 g kg^{-1} protein, 400 g kg^{-1} carbohydrate (containing ~14% polysaccharide), and 150 g kg^{-1} nucleic acid[3]. One purified glycoprotein fraction with a molecular weight of 63 k, 670 g kg^{-1} polysaccharide (mainly β-1,6-galactan), and 350 g kg^{-1} protein[3] is reported of some immunostimulatory activity *in vitro*[11]. Generally, large-molecular-weight glycoproteins from CVE (Mw > 10k)[12] and CPE (Mw > 100k)[9] are targeted as the key fraction responsible for imunostimulatory activity. The reported extraction temperatures for soluble *Chlorella* biopolymers differ largely, from room temperature[13], 40 °C[14], 80 °C[9,15], and ~100 °C[16]. However, glycoproteins are naturally easy to denature during intensive extraction and purification, especially involving heating, solvent-caused precipitation and dehydration[17]. Accordingly, this study aimed to investigate an optimal separation condition for rapidly collecting water extracts rich in immunostimulatory glycoproteins from *C. vulgaris*. The combinations of extraction independent variables were set based on the factorial central composite experimental design of response surface methodology. Developing a cost-effective, rapid process involving simple centrifugation unit operation for separating *Chlorella* biopolymers prior to purification is concerned in this study.

2 METHOD AND RESULTS

2.1 Optimal separation condition at room temperature

Spray-dried powder of *Chlorella vulgaris* (gift from Roquette Freres Co., Lestrem, France) was dispersed in distilled water (50 g kg^{-1} *Chlorella* biomass) in a 100-mL, screw-capped glass tube for extraction at room temperature (RT, ~25 °C) for 1 hr. The dispersion was then centrifuged with various combinations of centrifugation force (**F_c** in xg) and time (**t_c** in min) as shown in **Table 1**. Trials 1–13 were set according to the 2^2 factorial central composite experimental design of response surface methodology. Trial 14 (90 °C, 1 hr) was the reference condition to produce the polysaccharide-protein complex proved with promising immunostimulatory activity in dendritic cells from murine bone marrow[18]. Trial 15, extraction during treatment of purified cellulase (C0615, Sigma Co) at 10 U/mL and 50 °C (optimal temperature for cellulase) for 5 hr, was done to breakdown *Chlorella* cell wall materials for improving glycoprotein yield. After centrifugation, the supernatant was properly diluted for protein content analysis using a Qubit® fluorometer with Quant-iT™ protein assay kit (bovine serum albumin as standard) (Invitrogen Co, Oregon, USA.). The rest of supernatant was mixed with 4 volumes of 95% ethanol (EtOH) for 30 min and then centrifuged (8,000 rpm, 15 min) to collect biopolymer precipitate. The EtOH volume added was selected to produce the yield of biopolymer precipitate generally equivalent to that of the large-molecular-weight glycoprotein fraction (Mw >10k), which has been confirmed with immunostimulatory activity *in vivo*[12]. In this study precipitation with EtOH, rather than time-consuming ultrafiltration, was preferably used for rapidly estimating the yield of targeted glycoproteins in extract.

Table 1 indicates that the protein yields of trials 1–13 were in the range of 32.4–37.8 g kg^{-1} on dry basis of *C. vulgaris* powder. These data were mostly higher than those of 4V-EtOH precipitate (ppt, 26.0–48.1 g kg^{-1}). The unusually high yield of trial 5 (1,102 ×g, 16 min) could be ascribed to contamination of swollen *Chlorella* cells in supernatant. By judging the data variation of repeated trials 9-13., the reproducibility in protein yield was apparently greater and more reliable than the yield of 4V-EtOH ppt. Comparatively, extraction at 90 °C for 1 hr (trial 14) increased both protein and 4V-EtOH ppt yields by almost double. Extraction with cellulase treatment at 50 °C for 5 hr (trial 15-CE) showed a noticeably reduced yield in protein.

By statistic analysis with RSREG program of SAS 9.1 software, the protein yield (in % w/w for simplifying illustration) for extraction at RT varied as the function of $2.97 + 8.25\times10^{-5}F_c{}^{**} + 0.0214t_c - 3.06\times10^{-9}F_c{}^{2**} + 2.68\times10^{-7}F_c\cdot t_c - 8.60\times10^{-4}t_c{}^2$, where the significance of F_c ($P < 0.01$) was much higher than that of t_c. This equation accounted for 79% of the data variation ($R^2 = 0.790$). The counter plot for protein yield change with F_c and t_c. (**Figure 1**) produced with SigmaPlot 10 software (SPSS) showed a maximal yield of 3.71% w/w (i.e. 37.1 g kg^{-1}) at 14,119 ×g and 14.6 min. The centrifugation time was therefore fixed at 15 min in the following extraction experiments.

2.2 Optimal extraction condition at elevated temperatures

Chlorella dispersion in water (75 g kg^{-1} *Chlorella*) was prepared in a screw-capped glass tube as mentioned above, followed by extraction in a preheated water bath at a temperature (T in °C) for a certain period (t in min) as listed in **Table 2**. The extract mixture was then centrifuged at a certain force (F_c in ×g) for 15 min to yield supernatant. The supernatant

Table 1 *Separation conditions and biopolymer yields of C. vulgaris according to the 2^2 factorial central composite experimental design*

Trial[a] No	Coded variable X_1, X_2	Practical condition F_c (×g), t_c (min)	Yield of protein (g kg^{-1})[b]	Yield of 4V-EtOH ppt (g kg^{-1})[b]
1	−1, −1	4,000, 8	32.8	29.5
2	+1, −1	18,000, 8	35.9	38.1
3	−1, +1	4,000, 24	32.4	34.9
4	+1, +1	18,000, 24	36.1	29.0
5	−1.414, 0	1,102, 16	32.8	48.1
6	+1.414, 0	20,898, 16	35.7	32.5
7	0, −1.414	11,000, 4.69	36.8	28.7
8	0, +1.414	11,000, 27.31	35.5	26.0
9	0, 0	11,000, 16	35.1	34.6
10	0, 0	11,000, 16	37.8	29.0
11	0, 0	11,000, 16	37.4	31.4
12	0, 0	11,000, 16	35.9	40.8
13	0, 0	11,000, 16	37.8	32.5
14-90	0, 0	11,000, 16	64.7	60.7
15-CE	0, 0	11,000, 16	30.1	32.2

[a] Trials 1–13: extraction at RT forl hr; trial 14, at 90 °C for 1 hr; trial 15-CE, extraction accompanied with cellulase treatment (10 U/mL, 50 °C, 5 hr).
[b] On dry *Chlorella* biomass basis (moisture content = 50 g kg^{-1})

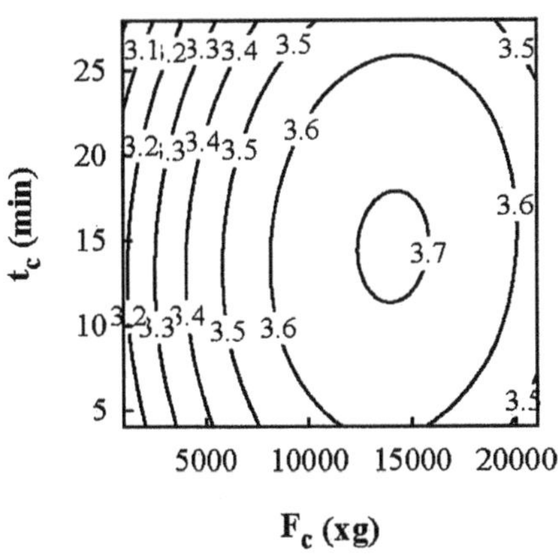

Figure 1 *Counter plots for native protein yield (% w/w) from extracts of C. vulgaris with respect to centrifugation force (F_c) and time (t_c) (extraction at RT for 1 hr).*

was properly diluted for protein content analysis and yielding 4V-EtOH ppt as mentioned above. The (T, t, F_c) combinations were set according to the 2^3 factorial central composite experimental design of response surface methodology and are shown in trials 1-19 of **Table 2**. The resultant yields in protein and 4V-EtOH ppt were in the range of 33.8–67.5 and 49.4–80.3 g kg^{-1}, respectively, on dry *Chlorella* biomass basis. Again, hot-water extraction after cellulase pretreatment (trial 20) showed a reduced yield in either protein or

Table 2 *Extraction conditions and biopolymer yields of C. vulgaris according to the 2^3 factorial central composite experimental design*

Trial No	Coded variable X_1, X_2, X_3	Practical condition T (°C), t (min), F_c(×g)	Yield of protein ($g\ kg^{-1}$)[a]	Yield of 4V-EtOH ppt ($g\ kg^{-1}$)[a]
1	−1, −1, −1	68.1, 35, 5000	35.9	55.3
2	+1, −1, −1	91.9, 35, 5000	48.7	61.5
3	−1, +1, −1	68.1, 105, 5000	36.1	63.0
4	−1, −1, +1	68.1, 35, 15000	35.6	70.4
5	+1, +1, −1	91.9, 105, 5000	48.5	64.1
6	+1, −1, +1	91.9, 35, 15000	53.4	63.7
7	−1, +1, +1	68.1, 105, 15000	33.8	57.5
8	+1, +1, +1	91.9, 105, 15000	53.9	64.2
9	−1.682, 0, 0	60, 70, 10000	34.1	49.4
10	+1.682, 0, 0	100, 70, 10000	65.2	80.3
11	0, −1.682, 0	80, 11.1, 10000	67.5	56.1
12	0, +1.682, 0	80, 128.9, 10000	35.6	58.4
13	0, 0, −1.682	80, 70, 1590	35.5	59.5
14	0, 0, +1.682	80, 70, 18410	36.1	60.3
15	0, 0, 0	80, 70, 10000	40.5	61.4
16	0, 0, 0	80, 70, 10000	41.3	64.4
17	0, 0, 0	80, 70, 10000	39.2	63.1
18	0, 0, 0	80, 70, 10000	41.3	66.9
19	0, 0, 0	80, 70, 10000	41.1	62.4
20-CE[b]	0, 0, 0	80, 70, 10000	36.5	58.8

[a] On dry *Chlorella* biomass basis (moisture content = 50 $g\ kg^{-1}$)
[b] Trial 20-CE, extraction after cellulase pretreatment (10 U/mL, 50 °C, 5 hr).

4V-EtOH ppt, as compared with those without cellulase pretreatment (trials 15–19).

The results of statistic analysis with RSREG program of SAS 9.1 software showed that the protein yield of hot-water extraction varied as the equation of $13.56 - 0.252T - 0.0524t - 9.92\times10^{-6}F_c + 1.83\times10^{-3}T^2 + 5.70\times10^{-5}T{\cdot}t + 2.66\times10^{-4}t^2 + 2.67\times10^{-6}F_c{\cdot}T - 9.29\times10^{-8}F_c{\cdot}t - 9.23\times10^{-9}F_c^2$. This equation accounted for 82.8% of the data variation ($R^2 = 0.828$) with the significance in the order of T** > t > F_c (** means $P < 0.01$). The resultant counter plots illustrate that extraction at a higher T for a shorter t yielded a greater protein yield (**Figure 2A**), where an optimal extraction time could be estimated as: $t = 220 - 2.1{*}T$ (dotted line for guidance) as F_c fixed at 11,000×g. The counter plot of F_c versus t (**Figure 2B**) or F_c versus T (**Figure 2C**) displayed a saddle point as F_c was ~12,000 ×g or ~8,000 ×g, respectively. For overall system, a saddle point was statistically obtained at 61.7°C, 93.2 min, and 7920 ×g. Generally, for spray-dried *C. vulgaris*, the optimal separation condition of maximizing protein yield could be concluded: extraction at 90–100°C for 30–11 min and subsequent centrifugation with 12000×g for 15 min. A satisfactory F_c range was practically 8,000–12,000 ×g (guided by the dotted line in **Figure 2C**).

As for the yield of 4V-EtOH ppt, 57.8% of data variation could be described by the equation of $5.00 - 0.0617T + 0.0176t + 3.28\times10^{-4}F_c + 6.00\times10^{-4}T^2 + 2.49\times10^{-4}T{\cdot}t - 1.50\times10^{-4}t^2 - 1.53\times10^{-6}F_c{\cdot}T - 1.62\times10^{-6}F_c{\cdot}t - 3.61\times10^{-9}F_c^2$ ($R^2 = 0.578$).

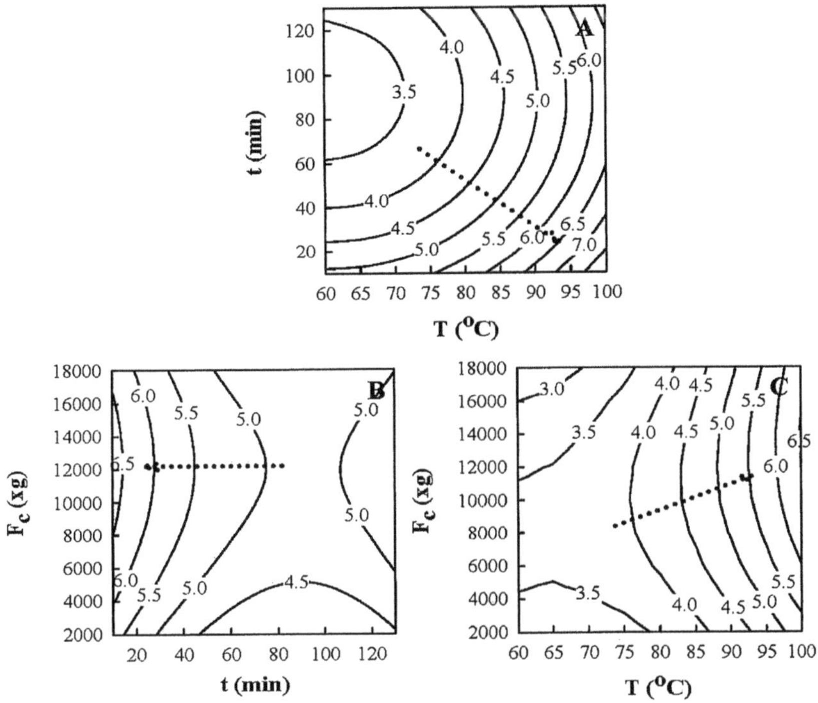

Figure 2 *Counter plots for protein yield (% w/w) of the hot-water extracts from C. vulgaris with respect to various independent variables. A: centrifugation force (F_c) fixed at 11000 ×g; B: extraction temperature (T) fixed at 90°C; and C: extraction time (t) fixed at 60 min.*

2.3 Rapid identification of different chemical species in *C. vulgaris* extracts

Figure 3 and **Table 4** compare the major species in the selected samples including 25WS (the water-soluble extract at 25°C, trials 9–13 of **Table 1**), 90WS (the hot-water extract at 90°C for 1 hr, trial 14-90 of **Table 1**) and its fractions F1 (Mw > 10k) and F2 (Mw < 10k) obtained by ultra-filtration, and 50celWS (water soluble after cellulase treatment at 50°C for 5 hr, trial 15-CE of **Table 1**). Most of the samples, except 90WSF1, revealed a notable absorbance peak at 330 nm (range 280–340 nm, signal I), reflecting the existence of nucleic acids (majority) and proteins[3] that carry carboxyl groups. Most of the samples, except 90WSF2, showed the characteristic peak at ~400 nm (range: 385–415 nm, signal II), mainly attributed to the existence of antioxidant phytocompounds (lutein, chlorophylls and β-carotene)[19], despite they are more hydrophobic than hydrophilic. The 50celWS (**Figure 3A**) showed the highest peak II among the samples. Only 90WS, 90WSF1 and 50celWS, not 25WS and 90WSF2, displayed the signal III peaked at 673 nm, relating to the appearance of chlorophylls[20] or turbidity caused by macromolecular aggregation. The related compounds for the extra peak at 712 nm for 50celWS is under investigation. Absorbance at 260 nm has been used as a simple index for quality control of *Chlorella* growth factor (CGF) content in drinks in industry. Absorbance at ~400 nm or 673 nm may be a simple index for the content of immunostimulatory *Chlorella* glycoproteins in drinks, as suggested by the differences in UV/visible signals of between 90WSF1 and 90WSF2 or 25WS. The above results also imply that *Chlorella* water soluble biopolymers in 90WS and 90WSF1 tended to interact with antioxidant phytocompounds.

For extracts of *C. vulgaris* (**Table 4**), the yield of 90WSF1 containing glycoproteins (containing 702 g kg^{-1} protein) was ~23% of its parent 90WS, agreeing with the results on *C. pyrenoidosa*.[8] The yield of 90WSF1 resembled the yield of 4V-EtOH ppt (trial 14 of **Table 1**). The apparent protein contents of 90WS and 90WSF2 were 524 and 566 g kg^{-1}, respectively, partly contributed by nucleic acids. These compositions are consistent with the findings for CVE extracted at 100°C for 20 min (containing 440 g kg^{-1} protein, 154 g kg^{-1} nucleic acid, and 395 g kg^{-1} carbohydrate)[3].

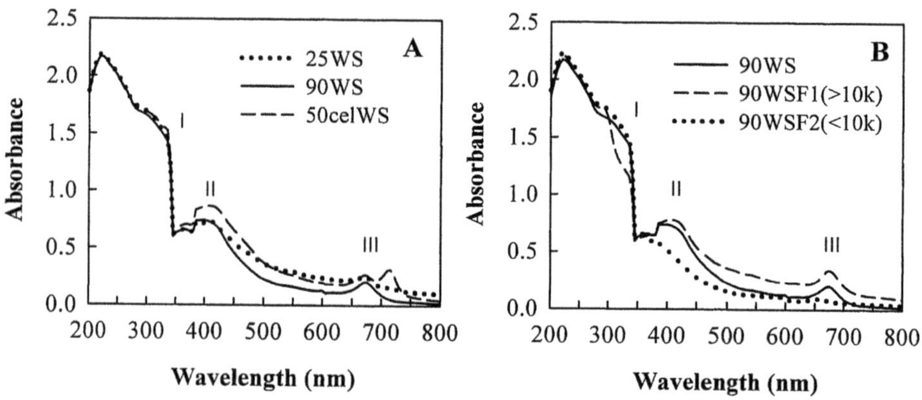

Figure 3 *UV/visible spectra for various water soluble (WS) (A) and fractions of 90WS (B). 25WS, 90WS, and 50celWS = the water soluble extracts at 25 °C for 1 hr, 90 °C for 1 hr, and after cellulase treatment at 50 °C for 5 hr, respectively.*

Table 4 *Yields, total protein contents, and maximal absorbance wavelengths of selected C. vulgaris extracts and fractions*

Sample	Yield of sample[a] (g kg^{-1})	Apparent protein content[b] (g kg^{-1})	λ_{max} (nm) in UV/visible spectrum I	II	III
25WS	–[c]	–	330	385–415	nd[d]
50celWS	–	–	330	385–415	673, 712
90WS	220–230	566±3[f]	330	385–415	673
90WS-F1 (>10k)	50–60	702±1	285sh, 335sh[e]	385–410	673
90WS-F2 (<10k)	~170	524±3[f]	330	nd	nd

[a] On dry *Chlorella* biomass basis.
[b] On dry basis of extract or fraction; determined using a Leco Nitrogen Analyzer (standard: sodium EDTA, N content = 95.8 g kg^{-1}); N factor = 6.25.
[c] Not determined.
[d] Not detectable.
[e] Small shoulder.
[f] Not true total protein content due to containing a significant amount of nucleic acids.

3 CONCLUSION

For spray-dried *C. vulgaris,* extraction at ~25 °C showed a maximal protein yield of ~37 g kg^{-1} under an optimal centrifugation condition of 14,000 ×g for 15 min. But, for extraction at elevated temperatures, a greater maximum in protein yield (>60 g kg^{-1}) could be obtained when extraction at 90–100°C for 30–11 min and subsequent centrifugation with 12000×g for 15 min. Statistic analysis resulted in a function of extraction temperature, time and centrifugation force could accounted for 82.8% of the variation in protein yield (R^2 = 0.828) with the significance in the order of temperature** ($P < 0.01$) > time > centrifugation force. Generally, the yield of *Chlorella* glycoproteins (Mw > 10k) that with reported immunostimulatory activities *in vitro* and *in vivo* was comparable the corresponding yield of 4V-EtOH precipitate. The UV/visible spectra of selected samples suggested that absorbance at ~400 nm or 673 nm may be a simple index for the content of immunostimulatory *Chlorella* glycoproteins in hot-water extract or drink. *Chlorella* water soluble biopolymers tended to interact and coexist with antioxidant phytocompounds. A rapid, simple calibration method for the content of *Chlorella* glycoproteins and their possible interactions with coexisting antioxidant phytocompounds need to be investigated.

Acknowledgement

This work was financially supported by National Science Council, Executive Yuan of Taiwan (NSC 98-2313-B-126-006-MY3). The authors thank Roquette Freres Co., Lestrem, France for providing *Chlorella vulgaris* powder.

References

1 E. Kessler. and V.A.R. Huss, *J. Phycol*, 1992,. **28**, 550.
2 T. Hasegawa, Y. Kimura, K. Hiromatsu, N. Kobayashi, A. Yamada, M. Makino, M. Okuda, T. Sano, K. Nomoto, and Y. Yoshikai, *Immunopharmacology*, 1997, **35**, 273.
3 T. Hasegawa, K. Ito, S. Ueno, S. Kumamoto, Y. Ando, A. Yamada, K. Nomoto, and Y. Yasunobu, *Int. J. Immunopharmacol.*, 1999, **21**, 311.
4 D.C.M. Dantas and M.L.S. Queiroz, *Int. J. Immunopharmacol.*, 1999, **21**, 499.
5 M.L.S. Queiroz, A.P.O. Rodrigues, C. Bincoletto, C.A.V. Figueirêdo, S. Malacrida, *Int. Immunopharmacol.*, 2003, **3**, 889.
6 S.H. Lee, H.J. Kang, H.-J. Lee, M.-H. Kang, Y.K. Park, *Nutrition*, 2010, **26**, 175.
7 S.A. Halperin, B. Smith, C. Nolan, J. Shay, and J. Kralovec, *CMAJ*, 2003, **169**, 111.
8 Kralovec J. A. 2006. AU2001287382B; (2005) US patents 6974576; 6977076; (2003) US patent 6551596.
9 J.A. Kralovec, K.L. Metera, J.R. Kumar, L.V. Watson, G.S. Girouard, Y. Guan, R.I. Carr, C.J. Barrow, and H.S. Ewart, Phytomedicine, 2007, **14**, 57.
10 H.-Y. Hsu, N. Jeyashoke, C.-H. Yeh, Y.-J. Song, K.-F. Hua, and L.K. Chao, *J. Agric. Food Chem.*, 2010, **58**, 927.
11 T. Hasegawa, T. Matsuguchi, K. Noda, K. Tanaka, S. Kumamoto, Y. Shoyama, and Y. Yoshikai, *Int. Immunopharmacol*, 2002, **2**, 579.
12 T. Hasegawa, K. Noda, S. Kumamoto, Y. Ando, A. Yamada, and Y. Yoshikai, *Int. J. Immunopharmacol.*, 2000, **22**, 877.
13 T. Hasegawa, K. Noda, S. Kumamoto, Y. Ando, A. Yamada, and Y. Yoshikai, *Int. J.*

Immunopharmacol., 2000, **22**, 877.
14 J. Sheng, F. Yu, Z. Xin, L. Zhao, X. Zhu, and Q. Hu, *Food Chem.*, 2007, **105**, 533.
15 E.R. Suárez, J.A. Kralovec, M.D. Noseda, H.S. Ewart, C.J. Barrow, M.D. Lumsden, and T.B. Grindley, *Carbohydr. Res.*, 2005, **340**, 1489.
16 F. Yang, Y. Shi, J. Sheng, and Q. Hu, *Eur. Food Res. Technol.*, 2006, **224**, 225.
17 L. Chang and M.J. Pikal, *J. Pharm. Sci.*, 2009, **98**, 2886.
18 N.-S. Yang, W.-T. Chang, J.-G. Lin, K.-Y. Jiang, and Phoency Lai, *Proceeding Handbook of 16th Gums and Stabilisers for the Food Industry Conference*, 2011.
19 L. Gouveia, V. Veloso, A. Reis, H. Fernandes, J. Novas, and J. Empis, *Bioresource Techn.*, 1996, **57**, 157.
20 AOAC International, in *AOAC Official Method of Analysis*, 16th ed, ed. AOAC International, Gaithersbury, MD, 1995, methods 942.04.

Mixed hydrocolloid systems

PHASE BEHAVIOR OF β-GLUCAN/SODIUM CASEINATE MIXTURES VARYING IN MOLECULAR WEIGHT

J.K. Agbenorhevi[1], V. Kontogiorgos[1] and S. Kasapis[2]

[1]Department of Chemical and Biological Sciences, University of Huddersfield, Queensgate, Huddersfield, HD1 3DH, UK
[2]School of Applied Sciences, RMIT University, City Campus, Melbourne, Vic 3001, Australia

1 INTRODUCTION

The phase behavior of protein-polysaccharide mixtures contributes significantly to the stability, structural, rheological and textural characteristics of food products. Protein-polysaccharide interactions may result in phase separation through thermodynamic incompatibility, co-precipitation or complex coacervation. Phase separation is known to occur depending on the molecular characteristics of biopolymers (molecular weight, conformation, charge density) as well as the mixing conditions or parameters affecting them (pH, ionic strength, solvent quality, mixing ratio, relative concentration of both biopolymers, mixing temperature and cooling rate, etc.).[1-9]

The mixed linkage (1→3)(1→4)-β-D-glucans found in cereals (oat, barley, rye and wheat) have received considerable research attention because of their health benefits including cholesterol lowering, modulation of glucose and insulin responses, weight management and improved gastrointestinal function. β-Glucans are water soluble polysaccharides that have been approved as functional bioactive ingredients, hence stimulating new product development activity over the years.[10,11] However, the recommended amount (≥ 0.75 g/serving, 3g/day) for health claims (lowering serum cholesterol levels to reduce the risk of heart disease) presents some challenges in the development of food formulations as β-glucans exhibit thermodynamic incompatibility when mixed with milk proteins.[12-14] Therefore, understanding of the phase behaviour of β-glucans with food proteins will assist development of palatable formulations.

Investigation of phase separated biopolymer systems in the liquid state usually involves construction of phase diagrams of the constituent biopolymers, at the specified conditions of interest, in order to identify their compatibility region. In a previous study, the phase behaviour of high molecular weight β-glucan/whey protein isolate binary mixtures under different solvent conditions was investigated. It was shown that a decrease of pH from 7.0 to 3.0 and sucrose addition enhanced miscibility.[13]

The present study aims to investigate the phase behavior of mixtures of sodium caseinate with β-glucan isolates varying in molecular weight by means of phase diagram construction, rheometry and fluorescence microscopy.

2 METHODS AND RESULTS

2.1 Materials and Sample Preparation

Sodium caseinate, fluorescent brightener-28, rhodamine B, sodium azide (NaN_3) and trifluoroacetic acid (TFA) were purchased from Sigma-Aldrich (Poole, Dorset, UK). Quick Start™ Bradford protein assay kits was purchased from Bio-Rad Laboratories Inc. (Hercules, California, USA). All chemicals used were analytical grade reagents.

The initial *β*-glucan isolate (denoted as OBG) was obtained by aqueous extraction from oat flour (OatWell® 28%, an oat bran with 28% β-glucan content obtained from CreaNutrition (Swedish Oat fibre, Sweden). Three more samples of lower molecular weights (H05, H10 and H15) were produced from the initial isolate by controlled acid hydrolysis (for 0.5, 1.0 and 1.5 hr respectively). Controlled acid hydrolysis caused depolymerization of the initial β-glucan sample, thereby generating products of different molecular weight but with the same structural characteristics: the longer the acid hydrolysis treatment time employed, the lower the molecular weight of the resulting material. The β-glucan content of the isolates produced was determined by the McCleary method using the Megazyme mixed-linkage beta-glucan assay kit purchased from Megazyme International Ltd (Bray, Ireland). The structural features of isolated *β*-glucan samples and molecular weights (MW) were characterized by ^{13}C–NMR spectroscopy, high performance size-exclusion chromatography and dilute solution rheology. The samples with β-glucan content between 78-86 % on a dry weight basis had MW, intrinsic viscosity ([η]) and critical coil overlap concentration (c*) in the range of 142 - 2800 x 10^3 g/mol, 1.7 - 7.2 dl/g and 0.25 - 1.10 g/dl, respectively. The intrinsic viscosity value increases while critical concentration decreases with increasing MW.

Stock dispersions of the biopolymers (14% w/v sodium caseinate, 1.0% w/v OBG, 1.2 % w/v H05, 1.8 % w/v H10, 1.8 % H15) were prepared in 0.1 M Sorensen's phosphate buffer (pH 7.0) with 0.02% w/v sodium azide (NaN_3) as preservative. Sodium caseinate was dispersed with mild heating at 40 °C in the buffer solution whereas β-glucan was dispersed in the pH 7.0 buffer solutions at 90 °C in a sealed vial under continuous stirring for 3 hr. The concentrations of the β-glucan sample solutions differ because of the maximum concentration that could be prepared with each isolate; the higher the molecular weight the lower the maximum experimental concentration achieved. Higher concentrations yielded insoluble β-glucan agglomerates resulting in inhomogeneous dispersions. The dispersions were left to hydrate at 5 °C overnight followed by centrifugation (1500× *g* at 5 °C for 10 min) to remove insoluble particles. Subsequently, the stock dispersions were vortex-mixed at seven different volume ratios (95/5, 80/20, 65/35, 50/50, 40/60, 25/75, 95/5 % v/v (sodium caseinate/β-glucan)) in centrifuge tubes at constant volume (5 mL) yielding binary mixtures of different concentrations.

2.2 Phase Analysis and Phase Diagram Construction

The mixed samples were centrifuged at 3000 × *g* for 30 min at 5 °C in order to accelerate equilibrium phase separation. Centrifugation of the mixtures under these conditions resulted in constant composition of the phases demonstrating that equilibrium was reached.[13] After phase separation, the equilibrium phase was analyzed for the protein and β-glucan content.

The protein content was determined according to the Bradford protein assay.[15] A standard calibration curve was generated with bovine serum albumin (BSA) standards. Absorbance readings were taken at 595 nm using a spectrophotometer (Shimazu UV-VIS 160A).The β-glucan concentration was determined using a modified phenol-sulphuric acid method for the phase-diagram construction.[16] Briefly, 0.2 ml TFA was added in the 5 ml of diluted mixtures and vortexed immediately to precipitate the proteins out of solution. The resulting mixture was left to stand for 40 min at room temperature and immediately centrifuged at 6000 × *g* for 10 min (Beckman, TJ-6). An aliquot of the supernatant was taken and phenol-sulphuric acid method was then carried out to determine the total polysaccharide concentration. All samples were replicated at least twice and the biopolymer concentration in the phases analyzed in triplicates.

After the equilibrium phase analysis for the biopolymer concentration, values obtained were used to construct a phase diagram. The non-linear regression of the experimental data was performed using GraphPad Prism v.5 (GraphPad Software, San Diego, USA) to produce the final binodal curve.

2.3 Viscosity Measurements

The flow behaviour (viscosity) of the mixed systems was determined at 20°C using a Bohlin Gemini 200HR nano rotational rheometer (Malvern Instruments, Malvern, UK) equipped with cone-and-plate geometry (55 mm diameter, cone angle 2°). Flow curves were measured at a range of shear rates in the range 0.01 – 1000 s^{-1}. Data were analyzed with the supporting rheometer software. All rheological measurements were carried out on freshly prepared samples.

2.4 Fluorescence Microscopy

A drop of freshly prepared *β*-glucan/sodium caseinate mixture (at different mixed ratios as previously described) pre-stained with fluorescent brightener-28 (0.01% w/v) and rhodamine B (0.02% w/v) was placed on a slide, covered with a cover slip and placed on the stage of the microscope at room temperature. Microscopy observations were carried out using an Olympus BX-41 epifluorescent microscope equipped with a mercury burner and an Olympus digital camera. Filter cubes with wide band UV excitation/emission filters 350/440 nm and 555/595 nm were used to observe the polysaccharide (fluorescent brightener 28) and the protein phase (rhodamine B), respectively.

2.5 Results and Discussion

2.5.1Phase Diagram. To investigate the concentration levels at which sodium caseinate in mixtures with β-glucans of different molecular weights may exhibit compatibility (co-solubility) or incompatibilty (phase-separation), phase diagrams were constructed (as shown in Figure 1) following the analytical determination of the biopolymer concentrations in the equilibrium phase.

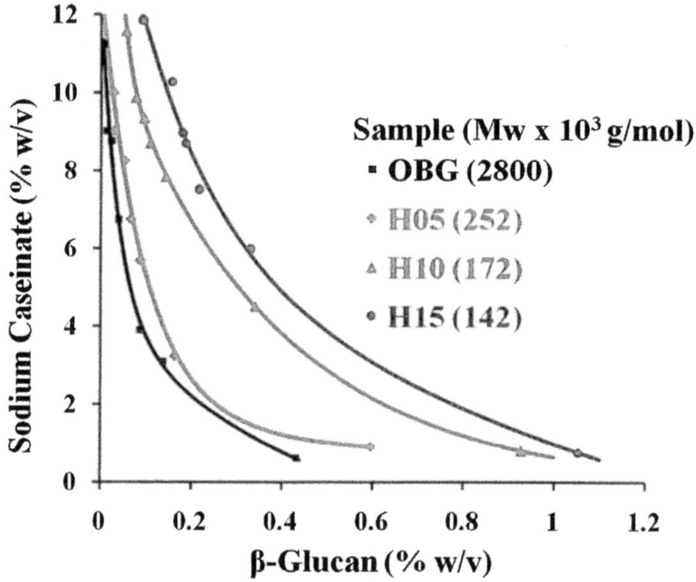

Figure 1 *Phase diagram of binary mixtures of sodium caseinate and β-glucans varying in molecular weight.*

The solid curve represents the binodal, which demarcates the compatible (one phase) from the incompatible (two phase) region. Thus the area below the binodal shows the concentration levels at which the β-glucan/sodium caseinate mixtures are at equilibrium (i.e. stable) whereas the area above the binodal represent the concentration zones where phase separation occurs. Therefore, the results clearly indicate that the compatibility region increases as the molecular weight of *β*-glucan decreases (i.e. OBG < H05 < H10 < H15). This implies that the molecular weight of *β*-glucan samples has a significant effect on miscibility with proteins. Since the experiments were performed at neutral pH, which otherwise would have influence the protein, and *β*-glucans are neutral linear polysaccharides, variation in the thermodynamic compatibility with sodium caseinate as a function of molecular weight can be explained by the excluded volume effect. High molecular weight *β*-glucan yields a large excluded volume, which is inaccessible to sodium caseinate. In contrast, low molecular weight samples (obtained through depolymerisation by controlled acid hydrolysis) have increased molecular flexibility and reduced hydrodynamic volume thereby creating free volume for the protein molecules even at relatively high β-glucan concentrations, resulting in extensive thermodynamic compatibility.

2.5.2 Microscopic observations of the mixed systems. The microstucture of biopolymers in mixtures was probed by means of fluorescence microscopy. The *β*-glucan solutions without sodium caseinate revealed bright patterns scattered homogeneously throughout the sample whereas sodium caseinate solutions in the absence of *β*-glucan result in featureless patterns at all concentrations studied. At various combinations in the mixed system, images taken with two different filter sets revealed a phase-separated system with the *β*-glucan domains surrounded by a continuous protein phase (Figure 2).

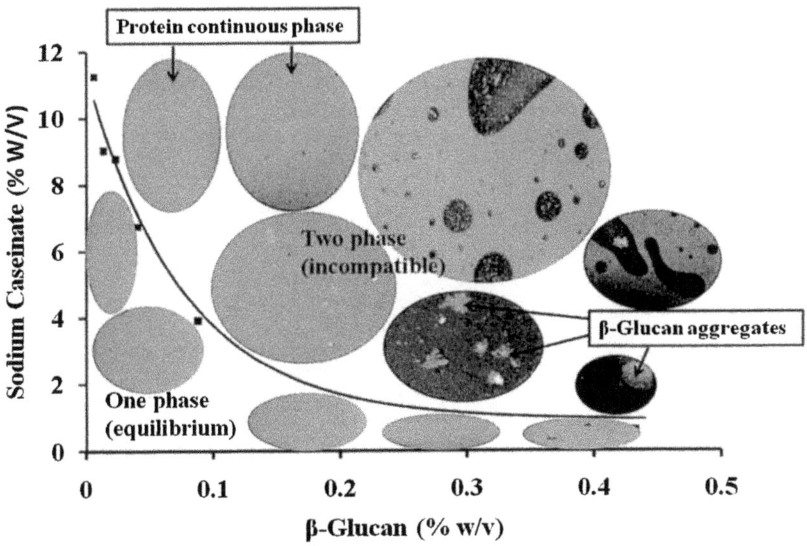

Figure 2 *Typical fluorescence microscopic images of β-glucan/sodium caseinate mixtures at different concentrations of the phase diagram.*

Biopolymer morphology in phases varied depending on the concentration and molecular weight. In the protein continuous phase, *β*-glucans appear as droplets. In the bicontinuous phase, *β*-glucan droplets increase in number especially for the lower molecular weight samples. In the polysaccharide-rich phase, *β*-glucan aggregates become more pronounced in size and irregularity especially for the higher molecular weight materials. Similarly, irregular-shaped morphological elements have been previously observed in high molecular weight *β*-glucan/whey protein mixtures,[12,13] guar gum/sodium caseinate,[16] pea protein/κ-carrageenan,[17] β-lactoglobulin/pectin,[18] or soy protein/κ-carrageenan.[19] In contrast, mixtures of guar/micellar casein,[9,20,21] gelatin/maltodextrin,[22] milk proteins/xanthan[23] or low molecular weight β-glucan/whey protein mixtures[12] exhibited spherical morphology. This indicates that the morphological features of protein-polysaccharide mixtures vary depending on the properties of the solvent, concentration, molecular weight and constituent biopolymers. It was interesting to note that mixed systems at equilibrium were comparatively finer with lesser or no β-glucan aggregates present. At equivalent concentration levels in mixture with sodium caseinate, it was observed that lower MW samples become more dispersed in the protein domain, form small aggregates and achieve homogeneity in the equlibrium phase. This confirms that a reduction in the excluded volume effect of small macromolecules gives rise to improved thermodynamic compatibility.

2.5.3 Flow behaviour of the mixtures. The viscosity of sodium caseinate solutions, increases as the concentration increases, with the flow behavior being mainly Newtonian at all shear rates for the concentrations (0.1 – 14% w/v) studied. For *β*-glucan solutions, viscosity also rises with increasing concentration, and at equivalent concentration levels, viscosity was higher for the higher molecular weight samples. In the sodium caseinate/*β*-glucan mixed systems, however, the viscosity also varied depending on the concentration of the two biopolymers. Figure 3 shows typical flow curves of the *β*-glucan/sodium caseinate mixtures at different mixed ratios (i.e. varying biopolymer concentrations).

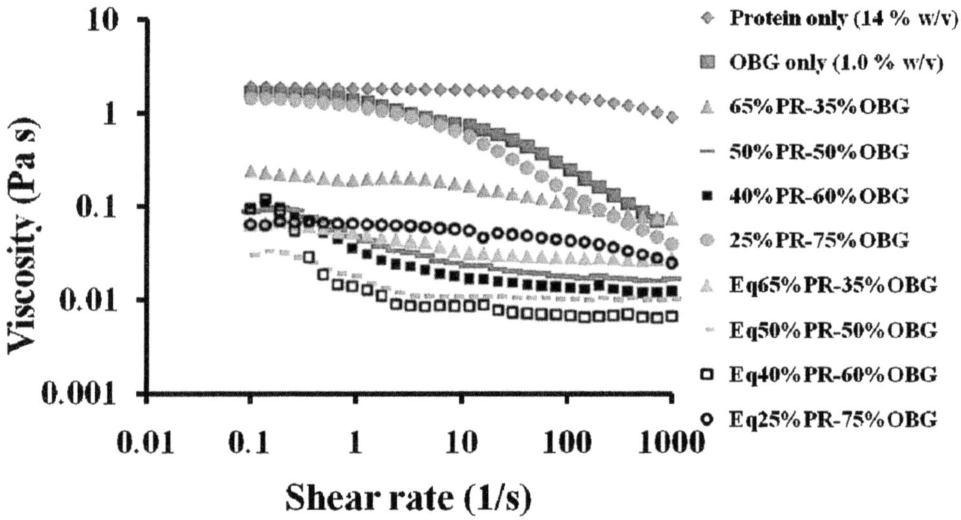

Figure 3 *Typical flow curve of β-glucan/sodium caseinate mixtures at different biopolymer concentrations. The percentage values represent the mixed ratio of the stock solutions (14 % w/v protein (sodium caseinate) and 1.0 % w/v β-glucan. Eq denotes the mixtures at equilibrium phase.*

In the protein-rich phase, the flow behavior was mainly influenced by the sodium caseinate whereas in the polysaccharide-rich phase, the flow behavior was largely determined by β-glucans. At intermediate biopolymer concentrations, a mixed flow pattern was evident as previously reported.[13] Noticeably, the flow pattern remains the same but the viscosity of the binary mixtures at equilibrium (one-phase regime) was lower than that after mixing (two-phase regime) when the system was under kinetic control. This is because centrifugation caused some protein and polysaccharide to precipitate resulting in less total biopolymer concentration in the final one-phase system, hence yielding lower viscosity. As confirmed by the microscopic images (figure 2), mixed systems at equilibrium have finer or lesser β-glucan aggregates present, which influence texture and flow behavior.

3 CONCLUSIONS

In the present investigation, the isothermal phase behaviour at 5 °C of mixtures of sodium caseinate with oat β-glucan isolates varying in molecular weight was investigated by means of phase diagram construction, rheometry and fluorescence microscopy.

Results showed distinct phase behaviour with variable biopolymer concentration in mixtures. Phase diagrams indicated that the compatibility of the *β*-glucan/sodium caseinate system increases as the MW decreases. Images of the mixtures taken at various biopolymer concentrations revealed phase separation, with the *β*-glucan domains being surrounded by a continuous protein phase. Images revealed the presence of β-glucan aggregates whose size depends on MW and concentration. Mixed systems at equilibrium were comparatively finer with lesser or no *β*-glucan aggregates present. Viscosity of the binary mixtures at equilibrium was also lower than that after mixing when the system was under kinetic control.

Results demonstrate that by controlling the structural features of *β*-glucans in mixture with sodium caseinate, intelligent manipulation of rheological properties in these systems can be achieved.

References

1 S. L. Turgeon, M. Beaulieu, C. Schmitt and C. Sanchez, *Curr. Opin. Colloid In.*, 2003, **8**, 401.
2 C. G. de Kruif and R. Tuinier, *Food Hydrocoll.*, 2001, **15**, 555.
3 V. B. Tolstoguzov, *Nahrung*, 2000, **44**, 299.
4 V. B. Tolstoguzov, *Food Hydrocoll.*, 2003, **17**, 1.
5 C Schmitt, C. Sanchez, S. Desobry-Banon and J. Hardy, *Crit. Rev. Food Sci. Nutr.*, 1998, **38**, 689.
6 J.-L. Doublier, C. Garnier, D. Renard and C. Sanchez, *Curr. Opin. Colloid In.*, 2000, **5**, 202.
7 N. Neirnyck, K. Van lent, K. Dewettinck and P. van der Meeren, *Food Hydrocoll.*, 2007, **21**, 862.
8 A. A. Perez, C. R. Carrara, C. C. Sanchez, P. R. J.M. and L. G. Santiago, *Food Chem.*, 2009, **116**, 104.
9 I. T. Norton and W. J. Frith, *Food Hydrocoll.*, 2001, **15**, 543.
10 C. S. Brennan and L. J. Cleary, *J.Cereal Sci.*, 2005, **42**, 1.
11 A. Lazaridou and C. G. Biliaderis, *J.Cereal Sci.*, 2007, **46**, 101.
12 V. Kontogiorgos, S. M. Tosh and P. J. Wood, *Food Biophys.*, 2009, **4**, 240.
13 V. Kontogiorgos, S. M. Tosh and P. J. Wood, *Food Hydrocoll.*, 2009, **23**, 949.
14 A. Lazaridou and C. G. Biliaderis, *Food Hydrocoll.*, 2009, **23**, 886.
15 M. M. Bradford, *Anal Biochem.*, 1976, **72**, 248.

16 J. K. Agbenorhevi and V. Kontogiorgos, *Carbohydr. Polym.*, 2010, **81**, 849.
17 R. M. Musampa, M. M. Alves and J. M. Maia, *Food Hydrocoll.*, 2007, **21**, 92.
18 M. Girard, C. Sanchez, S. I. Laneuville, S. L. Turgeon and S. F. Gauthier, *Colloids Surf. B.*, 2004, **35**, 15.
19 X. Li, Y. Hua, A. Qiu, C. Yang and S. Cui, *Food Hydrocoll.*, 2008, **22**, 845.
20 S. Bourriot, Garnier, C. and J.-L. Doublier, *Food Hydrocoll.*, 1999, **13**, 43.
21 C. Schorsch, A. H. Clark, M. G. Jones and I. T. Norton, *Colloids Surf. B.*, 1999, **12**, 317.
22 S. Kasapis, *Crit. Rev. Food Sci. Nutr.*, 2008, **48**, 341.
23 Y. Hemar, M. Tamehana, P. A. Munro and H. Singh, *Food Hydrocoll.*, 2001, **15**, 565.

STRUCTURE AS RELATED TO RHEOLOGICAL AND TEXTURAL PROPERTIES AND WATER BINDING CAPACITY OF β-LACTOGLOBULIN/XANTHAN GUM ELECTROSTATIC GELS IN ASSOCIATIVE CONDITIONS

Xuan Thang Le and Sylvie L. Turgeon

STELA Dairy Research Center and Institute of Nutraceuticals and Functional Foods, Faculty of Agriculture and Food Science, Université Laval, Pavillon Paul-Comtois, 2425 rue de l'agriculture, Québec, Canada G1V-0A6.

1 INTRODUCTION

Whey protein/polysaccharide mixed gels are often used as a model system for studying semi-solid food products. Mixed gels can be formed from blends containing more than one gelling agent and can be classified into four types: swollen networks, interpenetrating networks, phase-separated networks and coupled networks.[1] The structure of gels generally determines their sensory properties which are mainly related to the large deformation and mechanical properties of gels. A proper understanding of structure – function relationships of mixed gels should enable food scientist to open up new application of these mixed systems.

An interesting gelled system has been discovered using native proteins and polysaccharides at very low concentrations (0.05%). Gelation was induced by *in-situ* acidification, at room temperature, to a pH where both molecules carry net opposite charges. The novelty of this study resides in that no thermal, enzymatic or any other denaturing treatment was applied to the protein or the mixture at any time.[2] Therefore, this gelled system presents potential uses in the food industry to enhance the stability of foods, to protect micronutrients, and to deliver drugs or active molecules for application in the pharmaceutical industry.[2] However, little knowledge about the structure of this gelled system and the relationship with their rheological and macroscopic properties has been reported.

A previous study has reported that the gel network was based on the initial weak network of xanthan gum and proteins aggregated along the xanthan gum chains played the role of cross-linking agent by electrostatic interaction with other xanthan gum chains.[3]

The aim of this study was to investigate the relationships between the rheological and textural properties and water binding capacity of mixed gels with their microstructure.

2 MATERIALS AND METHODS

2.1 Materials

Whey protein isolate (High-Beta, lot # JE 002-8-922, 98.2 % protein, of which 85 % is β-lactoglobulin (βlg), 1.8 % minerals, and 4 % moisture) was obtained from Davisco Foods International Inc. Due to the high content of βlg in this powder, it was assumed that its behaviour was governed by that of βlg[4] and therefore it will be referred to as βlg hereafter.[2]

Xanthan gum (Keltrol RD, lot # 4G3367K, 96.36 % total sugar, 3.02 % protein) was provided by CP. Kelco (Chicago, USA). Glucono-δ-lactone (GDL, lot # 70H0163) was purchased from Sigma (Sigma chemical Co. USA).

2.2 Sample Preparation

Dispersions of βlg and xanthan gum with different total biopolymer concentrations were prepared in filtered deionized water (Modulab Analytical, Continental Water Systems Corporation, San Antonio, TX, USA), left overnight at 4 ^{0}C before preparing the mixtures for analysis. The initial pH of those dispersions was adjusted to 6.60 ± 0.08 using 0.1 N HCl and 0.1 N NaOH. Three βlg/xanthan gum ratios (r) were studied, namely 2:1, 5:1, and 10:1 (r = 2, 5, and 10 respectively), at different total biopolymer concentrations (0.10 - 0.30 %). βlg and xanthan gum dispersions were mixed at the desired r and stirred gently for 30 min. For gel formation, gelation of the mixed systems was induced by the addition of GDL in order to obtain the final pH of gels around 4.4 ± 0.08 for all formulations after 15 h of gelation at 25 ^{0}C.

2.3 Rheology

Small deformation properties of the mixed gels during acidification were determined in dynamic oscillation with a controlled-strain rheometer (ARES-100, Rheometric Scientific, Piscataway, NJ, USA), using couette geometry with a cup (33.93 mm diameter) and bob system (32.05 mm diameter, 33.29 mm length). Immediately after GDL addition, the samples were transferred to the rheometer, and covered with a thin film of castor oil to reduce dehydration during measurement. The development of the storage modulus (G') and the loss modulus (G'') during gelation was recorded during 15 h at 25 ^{0}C, at a frequency of 0.1 Hz and a strain of 0.5 %. At the end of oscillation, a frequency and a strain sweep test was also performed. In this paper, we will present only the results of the frequency sweep tests in the term of rheological properties of the final gels.

2.4 Confocal Laser Scanning Microscopy

Structural features were investigated using Nikon Eclipse TE2000-E Confocal Laser Scanning Microscopy (CLSM) (Tokyo, Japan), equipped with an inverted microscope. The objective lens used was 60X WI, NA 1.20. The excitation wavelength was 543 nm with an emission maximum at 605/675 nm. As proteins do not exhibit intrinsic fluorescence at this wavelength, proteins were stained by adding rhodamine B isothyocyanate (RITC) to protein dispersions under magnetic stirring during 1 h before mixing with xanthan gum dispersions. Mixtures of RITC-βlg/xanthan gum were prepared as described above. After GDL addition, the sample was poured between a 0.5 mm-deep well concavity slide and a cover slip which was hermetically sealed with nail enamel. Micrographs were taken about 24 h after GDL addition. Digital image files were acquired in 512 x 512 pixel resolution. Three images of the final gel of each formulation were taken which were analyzed to determine fractal dimension and pore size. It was checked that labelling did not change the rheological behaviour of the systems (results not shown).

2.5 Image analysis

In order to measure fractal dimension (d_f), lacunarity (A_ε) and pore size of the network in the final CLSM micrographs, an image analysis method, (ImageJ v1.43r; http://rsb.info.nih.gov/ij/), was applied. Final micrographs were transformed into 8-bit gray scale images and then thresholded. The gray level used for thresholding was automatically defined by the ImageJ software based on the isodata algorithm.[5] Finally, those micrographs were transformed into 8-bit binary images.

2.5.1 Fractal Dimension and Lacunarity Measurements.

D_f and A_ε values were measured by the box counting method using the FracLac v2.5-1d plug-in for ImageJ. This method is now widely used in the literature.[6-9] It consists of carrying out an iteration process by placing of several grids of decreasing size over an image, and counting the number of boxes containing pixels (foreground pixels) for each grid size. D_f is obtained as $-D + 1$, D is the slope of the line obtained in a plot of $\log(N_\varepsilon)$ versus $\log(\varepsilon)$.[10] Where N_ε is the number of boxes in the grid at a certain scale containing pixels and ε is the corresponding scale.

A_ε was calculated as the variation in the number of pixels at each grid box placed in the image during a standard box counting, following $A_\varepsilon = (\sigma/\mu)^2$, where A_ε is lacunarity for the grid of size ε. A_ε depends on the square of the pixels coefficient of variation, σ is the standard deviation of the pixel count within all boxes of size ε, and μ is the average number of white pixels per box for the same grid size.

2.5.2 Pore Size Measurement.

For pore size measurement, the 8-bit binary images were watershed in order to automatically separate the pores. The average pore size (μm) was estimated from average area of the pores which was calculated from the count of pores and the total area corresponding to pores.

2.6 Texture Properties

Gel texture parameters were determined by a Texture Analyzer TA TX-2 (Texture Technologies Corporation, NY, USA). The gels formed in the 100 ml-beaker were penetrated with a 12-mm diameter cylinder probe.[11, 12] A force – time curve was obtained at a crosshead speed of 0.50 mm/s for 20 mm displacement. The resulting force-time curves were analyzed using the texture profile analysis (TPA) method in order to determine hardness at fracture.[13] All texture measurements were performed after 24 h of gelation. At that time final pH was stabilized and there were no changes in the rheological properties of gels,[3] and the structure of gels was fully developed.[2]

2.7 Syneresis

The released water was quantified using Millipore Ultrafree-CL centrifugal filter units (Millipore Corporation) with microporous membranes of 0.45 μm pore size (Durapore PVDF Membrane). Mixed gels were formed *in situ* in a 2 ml filter cup with a pinhole at the bottom (the pinhole was obstructed during the gelation process). This filter cup was placed in a filtrate collection tube with cap. After about 24 h of gelation, the pinhole was freed for 4 h, and then the released water was measured. The water loss under gravity will be calculated as:

$$water\ loss\ (\%) = \frac{weight\ of\ water\ released}{weight\ of\ sample}\ x\ 100 \qquad (1)$$

3 RESULTS AND DISCUSSION

3.1 Viscoelasticity of Fully Cured Gels

Figure 1a shows the frequency dependence of the final values of G' and G'' for the gels at total solid concentration 0.30 % and different protein/polysaccharide ratios. Higher values of G' and G'' at smaller ratio, in the range of frequency, may indicate the higher density of gels. Moreover, this figure shows that G' increases slightly with frequency and the G''/G' ratio vary from 0.01 to 0.1 as illustrated in Figure 1b. These two features suggest that the gels are intermediate in properties between the two extremes settled by entanglement networks and true gels, and so are characteristic of weak gels.[14] G'' shows a slight minimum in the experimental frequency range (Figure 1a). Such behavior is related to loss tangents which show a pronounced minimum in Figure 1b and similar to other associating biopolymer systems (e.g., gelatin, guar/borate gels[15] and amylose gel[14]). Those authors reported that minima in tan δ are typical of entanglement networks, and correspond to a period of oscillation which is long compared with the relaxation times of a chain between entanglement points, but short compared to the lifetime of those entanglements.

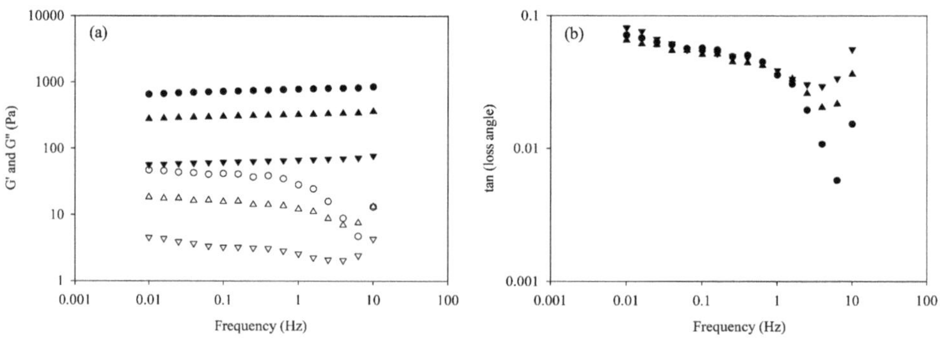

Figure 1 *G' and G" (a) and loss tangent (b) of fully cured gels vs frequency. Total solid concentration: 0.30 %. G', closed symbols; G'', open symbols. r = 2, circles; r= 5 triangle up; r = 10 triangle down.*

Figure 1b shows also the apparent shift, to higher frequency, of the minima in tan δ observed on going from ratio 10 to ratio 2. A possible explanation for this shift has to do with the distribution of elastically active chain lengths as a function of ratio. At lower ratio, the length of the chains between cross-links could be shorter than at higher ratio and contributes to G'' and tan δ down to relatively higher frequencies. As a result, there is a reduced contribution to the relaxation spectrum which in turn shifts the minimum to higher frequency. It can be concluded that in the ratio range from 2 to 10 the gel's strength changed significantly. Such behavior could be attributed to the role of xanthan gum in network organization, higher βlg/xanthan gum ratio led to weak structure formation.[3] This phenomenon will be discussed further on in section 3.2.

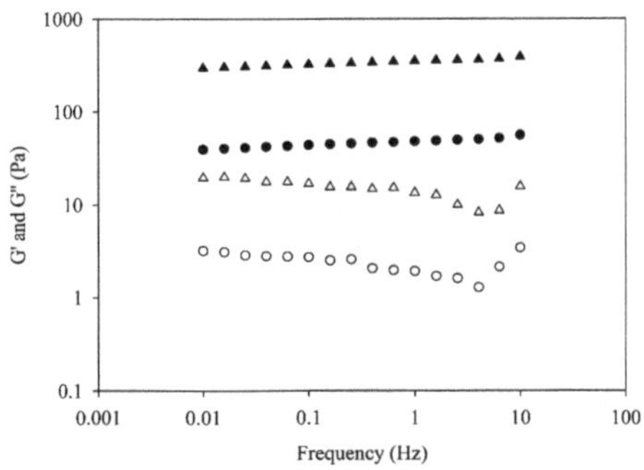

Figure 2 *G' and G" of fully cured gels vs frequency. Ratio 5. G', closed symbols; G'', open symbols. c = 0.15 %, circles; c = 0.30 %, triangle up.*

Figure 2 shows the frequency dependence of the final values of G' and G'' for two tested concentrations at ratio 5. More-concentrated gel yielded a similar profile with much higher values of G' and G''. Minima in G''are also seen in this figure. The shift, to higher frequency at higher concentration, of the minima is also explained by the shorter distance between cross-links which may result in formation of denser networks as seen in Figure 3. The rheological measurements observations are in agreement with the confocal micrographs (Figure 3), in which denser networks were observed at higher concentrations and/or lower ratio (higher xanthan gum contents).

3.2 Microstructure of Fully Cured Gels

Figure 3 presents the final CLSM images of βlg/xanthan gum gels at different ratios and two concentrations. These images show that branched network structures were observed in all formulations. As reported in a previous study[3] that the gel network was based on the initial weak network of xanthan gum and proteins aggregated along xanthan gum chains and played the role of cross-linking agent by electrostatic interaction with other xanthan gum chains. The result qualitatively indicates that more homogeneous structure with smaller pores in more interconnected network was formed at higher total biopolymer concentration and lower ratio. The gel obtained at r = 10 and c = 0.15 % presented the most discontinuous and heterogeneous network with larger strands, indicating larger and coarser electrostatic complexes could be initially formed in the system. This observation coincides with the observation from the small-angle static light scattering measurements.[2] Bigger pores were qualitatively observed in systems with lower total biopolymer concentration and polysaccharide content. Previous results[2] showed that softer gels were formed and the structuration process was slower when microfluidized xanthan gum was used, indicating slower formation of interpolymeric complexes. The formation of interconnection points in the network could be favoured at higher xanthan gum content due to high molecular weight and entanglement of this polysaccharide, which caused less space between these molecules, therefore smaller pores were formed, and stronger gels were obtained; which coincided with the rheological measurements (Figure 1a and b) and image analysis (Table 1). In addition, smaller viscosity in the system with higher ratio causes intrinsic instability. Therefore structural rearrangements can occur much faster, which could lead to locally denser clusters and larger pores, resulting in more turbid gels and a more permeable microstructure.[16]

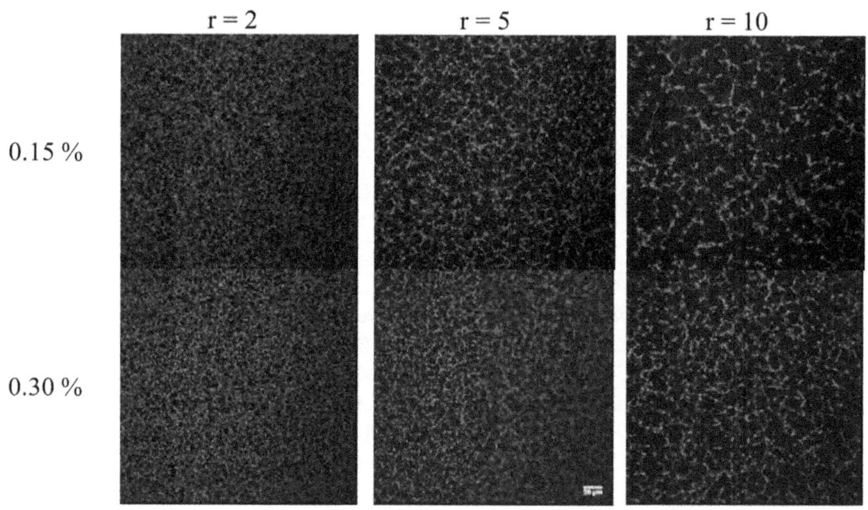

Figure 3 *Final microstructure of βlg/xanthan gum gels at different ratios and concentrations. Scale bar corresponds to 20 μm and is appropriate to all images.*

The gels at higher ratio were more opaque, and easily broken, indicating a shift in structure from fine-stranded to particulate. Pauson, Dejmed and van Vliet[17] reported that transparent protein gels consisted of a molecular homogeneous network, whereas aggregate gels were non-transparent. The availability of proteins in systems with higher ratios favoured also the maximum complexion between proteins and polysaccharides with consequent reduction in junction zones of the gel network.[3] Syneresis also increased with gel whiteness.[13] The change in structure at different conditions may also be explained by competition between the kinetics of phase separation and the gelation processes. For higher ratio, the kinetics of separation are faster than for gelation,[2] favouring the formation of a larger porous structure, resulting in weaker and less elastic gels.[8]

3.3 Fractal dimension, Lacunarity and Pore size

Fractal dimension and pore diameter of βlg/xanthan gum gels were calculated and the results are presented in Table 1. As the ratio increases, d_f values decrease and pore sizes increase indicating the lower density of gels at higher ratio as showed as in Figure 3. In contrast, the total polymer concentration did not significantly influence the d_f values but denser networks (Figure 3), corresponding smaller pore size (Table 1), were formed in systems with higher concentration.

Table 1 *Fractal dimension and average pore diameter of gels obtained at different ratios and total solid concentrations.*

	Total solid concentration (wt %)			
	0.15	0.30	0.15	0.30
Ratio	Fractal dimension		Pore diameter (μm)	
2	2.32±0.06	2.54±0.07	6.59±0.61	4.53±0.02
5	2.29±0.07	2.23±0.01	8.65±0.07	5.49±0.09
10	2.10±0.04	2.08±0.01	16.05±0.38	9.71±0.12

Figure 4 shows the lacunarity distribution of mixed gels at total solid concentration 0.30 % and at different ratios. The lacunarity or heterogeneity of the gels at ratio 2 and 5 increased with the scale until it reaches a maximum at box size of approximately 2.2 and 2.6 μm, respectively, but 3.5 μm for gel of ratio 10. After this maximum point, the heterogeneity reduced until the achievement of steady state values. The larger values of lacunarity reflect the greater heterogeneity of gel network and if the maximum of lacunarity is reached at small scales, the largest variability is mostly explained by small cavities.[7,9] As expected, these results are in agreement with those obtained for the pore size (Table 1).

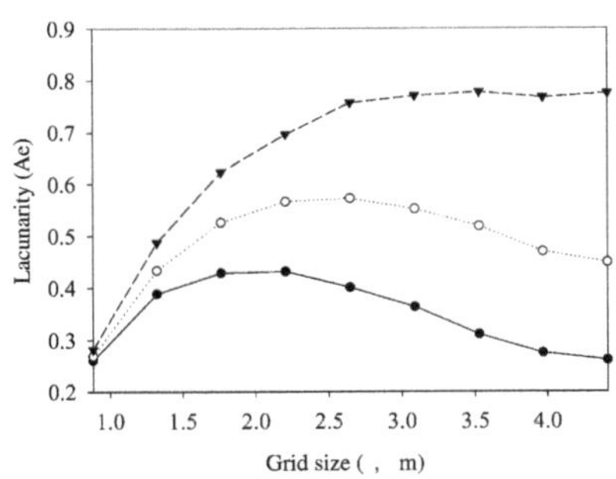

Figure 4 *Lacunarity plot from CLSM binary images of gels obtained at different ratios (R = 2, closed circles; R = 5, open circles; R = 10, triangles down), and total solid concentrations 0.30 %.*

3.4 Textural and Waterbinding properties of Gels

The lowest value of gel hardness was obtained at a ratio of 10 and 0.15 % total solid concentration; the highest hardness corresponded to gels at smaller ratio (r = 2) and higher concentration (0.30 %) (Figure 5a). As mentioned above, this behaviour is related to the density of the network. At high concentration and small ratio, the gel's network is denser due to higher content of xanthan gum in the system. The changes in the gel structure at different ratios and concentrations can be used to explain the observed changes in waterbinding properties of gels (Figure 5b). Generally, syneresis increased with an increased degree of aggregation, resulting in a coarser network with larger pores.[18, 19] Gels with larger pores result in poorer water-holding capacity than a homogeneous network structure.[7] Bigger pores were formed with high ratio, which can be associated with the higher tendency for associative phase separation.[8] Previous work reported that the gelation did not occur at high ratio (r = 20, c = 0.10 %); instead, precipitated electrostatic complexes formed from the beginning.[2] Table 1 highlights that the larger pores obtained with higher ratio and lower concentration, correspond to the greater water loss (Figure 5b). In the mixture of higher concentration or higher polysaccharide content, the initial tenuous network of xanthan gum is denser,[3] so more junction zones were formed (Figure 3), inducing the formation of smaller pores.

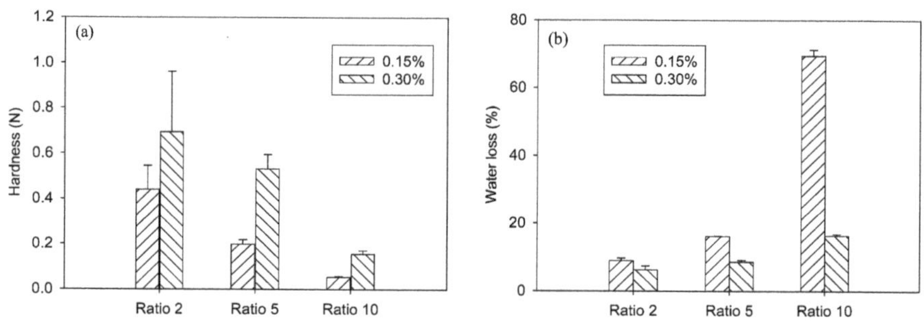

Figure 5 *Hardness (a) and water loss (b) of βlg/xanthan gum gels made at two total solid concentration (0.15%, 0.30%) and different ratios. Error bars represent standard deviation.*

In this work, the microstructural results are in accordance with the rheological and textural measurements. Higher G' values corresponded to a gel with a firmer and more compact structure (Figure 3), leading to lower water loss under gravity (Figure 5b).

4 CONCLUSIONS

It was found that both ratio and total solid concentration are important determinants of microstructure of gels, but the effect of ratio is more pronounced due to the role of xanthan gum in gel organization. These data suggests that the change in rheological, textural properties and water binding capacity of gels is in relation with modification in the microstructure of gels affected by the ratio and the total solid concentration. Correlations between microstructural attributes and macroscopic properties revealed that a decrease in the pore size and the heterogeneity of gaps entailed an increase in textural property, such as hardness, and in the water binding capacity of gels. The amount of water retained in gels was estimated around 611 and 278 g water per g biopolymer for the gels at ratio 2, 10 and at concentrations of 0.15 and 0.30 wt %, respectively. Such a high water binding capacity value is reported for the first time in protein/polysaccharide gelled systems. These experimental findings should enable food industry to develop new functional semi-solid foods with high water and low protein, polysaccharide content which may also include active molecules in their structure. Further studies will focus on the integration of active molecules in gels.

Acknowledgements

ACDI/PCBF is acknowledged for its fellowship for Xuan Thang Le along with Diane Gagnon for her technical assistance. This work is funded by NSERC discovery grant.

References

1. P. Cairns, M. J. Miles, V. J. Morris and G. J. Brownsey, *Carbohydrate Research*, 1987, **160**, 411-423.
2. S. I. Laneuville, S. L. Turgeon, C. Sanchez and P. Paquin, *Langmuir*, 2006, **22**, 7351-7357.
3. X. T. Le and S. L. Turgeon, *submitted*, 2011.
4. F. Weinbreck, R. de Vries, P. Schrooyen and C. G. de Kruif, *Biomacromolecules*, 2003, **4**, 293-303.
5. T. W. Ridler and S. Calvard, *IEEE Transactions on Systems, Man and Cybernetics*, 1978, **vol. SMC-8**, 630-632.
6. E. Dàvila and D. Parés, *Food Hydrocolloids*, 2007, **21**, 147-153.
7. P. Salvador, M. Toldrà, E. Saguer, C. Carretero and D. Parés, *Food Hydrocolloids*, 2009, **23**, 1654-1659.
8. Â. Cavallieri and R. Cunha, *Food Biophysics*, 2009, **4**, 94-105.
9. E. Dàvila, M. Toldrà, E. Saguer, C. Carretero and D. Parés, *LWT - Food Science and Technology*, 2007, **40**, 1321-1329.
10. T. Hagiwara, H. Kumagai and K. Nakamura, *Food Hydrocolloids*, 1998, **12**, 29-36.
11. M. Beaulieu, S. L. Turgeon and J.-L. Doublier, *International Dairy Journal*, 2001, **11**, 961-967.
12. Z. Y. JU and A. Kilara, *Journal of Food Science*, 1998, **63**, 288-292.
13. S. L. Turgeon and M. Beaulieu, *Food Hydrocolloids*, 2001, **15**, 583-591.
14. J.-L. Doublier and L. Choplin, *Carbohydrate Research*, 1989, **193**, 215-226.
15. J. O. Carnali, *Rheologica Acta*, 1992, **31**, 399-412.
16. A. C. Alting, R. J. Hamer, C. G. de Kruif and R. W. Visschers, *Journal of Agricultural and Food Chemistry*, 2003, **51**, 3150-3156.
17. M. Paulsson, P. Dejmek and T. Van Vliet, *J. Dairy Sci.*, 1990, **73**, 45-53.
18. H. Anne-Marie, *Journal of Food Science*, 1982, **47**, 1965-1972.
19. Y. L. Tan, A. Y. H. Singh and Y. Hemar, *Journal of Texture Studies*, 2007, **38**, 404-422.

THE EFFECT OF POLYMER FINE STRUCTURE ON SYNERGISTIC INTERACTIONS OF XANTHAN WITH KONJAC GLUCOMANNAN.

A. Abbaszadeh, T.J. Foster

Division of Food Science, University of Nottingham, Sutton Bonington Campus, LE125RD Loughborough, United Kingdom

Abstract

This work investigates the interaction of xanthan containing different levels of pyruvate and acetate, mixed with konjac glucomannan (KM). The mixture of a xanthan with standard levels of pyruvate with KM shows the previously reported DSC endotherm upon heating. This peak can be seen with almost the same magnitude for a high pyruvate / standard acetate and high pyruvate / low acetate xanthan samples. However, the individual transition of a low pyruvate xanthan sample remains visible at high temperature when combined with KM, with a small endotherm in the region of 40°C, attributable with the synergistic gel. This kind of transition has been reported previously for xanthan with escalating ionic strength. The transition is also detectable for the most stable of the samples (pyruvate-free / standard acetate), which designates that the existence of the terminal mannose is enough to tolerate the low temperature synergistic junction zone. A better understanding of the interaction type has been found by comparing different concentrations of polymer and salt, verified using rheological techniques.

1. INTRODUCTION

Synergistic interactions of mixtures for different types of polysaccharides have the potential for producing novel functionality, like creating a new texture in food or pharmaceutical products, with the advantage of using reduced levels of polysaccharides, and subsequently reduced cost. A requirement is still needed however for more basic goals, such as study the intermolecular interaction between polysaccharides.

A number of mixtures of two polysaccharides presenting synergistic interactions have been identified, like xanthan/galactomannan or konjac glucomannan (KM), carrageenan/ galactomannan or KM, and agarose/carrageenan, cellulose and starch (Cairns et al. 1986; Cairns et al. 1987; Ridout et al. 1998; Fernandes et al. 1991; Fernandes et al. 1992; Williams et al. 1991; Rochas et al. 1990; Cheetham & Mashimba 1988; Shatwell et al. 1990; Turquois et al. 1992; Kohyama et al. 1993; Nishinari 1996; Nishinari & Takahashi 2003; Kim et al. 2006; Goycoolea et al. 1995; Morris & Foster 1994; Morris et al. 1980; Morris et al. 1988). Foster and Morris (Foster & Morris 1994) indicated that two general characteristics of these schemes are that one of the two polysaccharides is made up of a β-(1-4)- linked backbone, and the other polysaccharides undertakes a thermo reversible disorder-order (coil-helix) transition.

One of the well-known polysaccharides in terms of interactions and gelation is xanthan which forms strong gels after mixing with co-synergist such as galactomannan or glucomannan. These types of mixtures has been investigated for more than 30 years, but the mechanism of the gelation or basic questions like whether two polysaccharides create a heterotypic junction zone are still under debate.

Xanthan gum is an extracellular polysaccharide used as a food additive and rheology modifier, however, does not create true gel on its own (Morris et al. 1983). It is produced by fermentation of glucose or sucrose by the *Xanthomonas campestris* bacterium.

The backbone of xanthan is β-(1→4)-D-glucose which is the same as cellulose, however, every alternate glucose reside has a three sugar side chain consisting of two mannose residues with a glucuronic acid residue between them. The internal mannose can be acetylated and the terminal mannose pyruvated. High amounts of pyruvate and low acetate levels would promote low transition temperatures, therefore destabilising the xanthan helix, which affects its interaction with co-synergist.

Konjac glucomannan is a glucomannan obtained from the tuber of *Amorphophallus konjac* comprising 40% by dry weight of the roots and creates a thermally stable gel on addition of alkali (Maekaji et al., 1974). It is also a hemicellulose that is present in large amounts in the wood of conifers and in smaller amounts in the wood of dicotyledons. This gel has long been used in Japanese traditional foods (Nishinari, 1996). KM is mainly a straight-chain polymer, with a small amount of branching. The component sugars are β-(1→4)-linked D-mannose and D-glucose in a ratio of 1.6:1 and the degree of branching is about 8% through β-(1→6)-glycosyl linkages.

The aim in this work is to attain a better understanding of the interaction mechanism, by studying different types of xanthan, having variable amounts of acetate and pyruvate, using different polymer concentrations and different salinity.

2. MATERIALS AND METHODS

A range of xanthan samples, which have variable pyruvate and acetate substitutions, xanthan standard (SA/SP), low pyruvate standard acetate (SA/LP) , high pyruvate standard acetate (SA/HP), high pyruvate low acetate (LA/HP), high pyruvate very low acetate (VLA/HP), very low pyruvate standard acetate (SA/VLP), were kindly provided by Danisco (indicated in Table 1).Glucomannan (Propol RS; High Mw (KM)) was supplied by Shimizu Chemical Corporation .Sodium azide was from Acros and sodium chloride was from Fisher.

Xanthan	Abbreviate	Acetate (%)	Pyruvate (%)
Standard Acetate(SA)/Standard Pyruvate(SP)	SA/SP	6.19	3.85
Standard Acetate(SA)/Low Pyruvate (LP)	SA/LP	6.02	2.20
Standard Acetate(SA)/ Very Low Pyruvate (VLP)	SA/VLP	5.60	0.50
Standard Acetate(SA)/High Pyruvate (HP)	SA/HP	5.10	6.49
Low Acetate(SA)/High Pyruvate (HP)	LA/HP	1.46	6.54
Very Low Acetate(SA)/High Pyruvate (HP)	VLA/HP	0.51	6.98

Table 1 *Acetate and pyruvate level in different xanthan samples*

Xanthan samples and glucomannan samples were prepared by cold dispersion of the polymers individually in distilled water, by shearing using a magnetic flea and then heating at ~90°C for 20 min. When they cooled down to room temperature they were mixed together (ratio of 1:1), and left over night. Micro DSC was carried out (Micro DSC III, Setaram-Caluire, France) using cells made from Hastalloy. By holding roughly 0.8 ml of sample, sealed with O rings on Hastalloy screw tops, the temperature of xanthan helix coil transition and synergic interaction is detected, heating at 1°Cmin^{-1}.Rheology test of the samples were performed using cone and plate geometry (4 cm diameter; 1.581 cone angle) with an Anton Paar Physica MCR-301 rheometer. Samples were loaded at room temperature, coated around the periphery of the sample with light silicone oil to minimise loss of water, and left unperturbed for 30 min before measurements were made. A temperature sweep was measured by heating at 1°C min^{-1} with a 1% frequency and 0.5% strain, ensuring measurements were made in the linear viscoelastic plateau region.

3. RESULTS AND DISCUSSION

3.1 Transition temperature

Xanthan is a helix at low temperatures and upon heating undergoes a change of state from an ordered helical to a disordered coil structure (Holzwarth et al. 1976). Xanthan aqueous solution has a high viscosity, high pseudo-plasticity, significant yield stress, and stability, even at low concentrations over wide pH, temperature and salt content ranges (Whitcomb et al. 1978).

This coil-helix transition is dependent of salt concentration and acetate and pyruvate level typically pyruvate residues can be found on 30-40% of the terminal mannose residues whereas 60-70% of the internal mannose residues may contain acetate groups. The pyruvate and acetate content may vary, depending on the bacterial strain (Cadmus et al., 1976) and fermentation conditions (Sutherland et al. 1981; Tait et al. 1986), genetically modified strains (Hassler & Doherty, 1990) or chemical modification. The relation between the acetate and the pyruvate contents and the midpoint temperature of the transition has been previously reported by different workers (Holzwarth 1976; Cheetham and Mashimba 1988; K Shatwell et al. 1990; Morris 1977) .

In this study, applying different xanthan samples in which the acetate or pyruvate level change while the other one is kept almost in constant level, in contrast with previous works, provides an opportunity for better understanding of their roles on xanthan conformation and interaction.

Figure 1a shows the value of the elastic modulus (*G'*) for a variety of xanthan samples on heating. As can be seen the start of a sharp reduction which is almost coincident with the helix-coil transition temperature as matched with DSC data (Figure 2), confirm the expected role of pyruvate content. By decreasing the pyruvate level in xanthan (SA/LP) the helix-coil transition is stable to higher temperatures, at fixed ionic strength 10mM NaCl; a de-stabilising effect can be achieved at higher pyruvate contents (SA/HP). Accordingly, by changing the pyruvate content when the acetate level of xanthans are kept constant, the transition temperature increases with decreasing pyruvate level (cf. SA/HP < SA/SP < SA/LP).

Additionally, by decreasing the acetate level at LA/HP sample, the transition shifted to lower temperature at almost constant pyruvate level. From figure 1a and 1b it is clear that an increase in ionic strength shifts the helix-coil transition for all xanthans to higher temperatures. What is also apparent is the increase in modulus as pyruvate content is increased. The molecular weight of all of these samples are comparable (results not shown), indicating an increase in weak gel properties of xanthan when carrying the most negative charge. An increased amount of acetate for these high pyruvate samples produces the most effective weak gel at both ionic strengths, implicating a stabilised helix, with charge-screened side chains. A very stable helix, with the lowest pyruvate content produces the lowest moduli in both ionic environments.

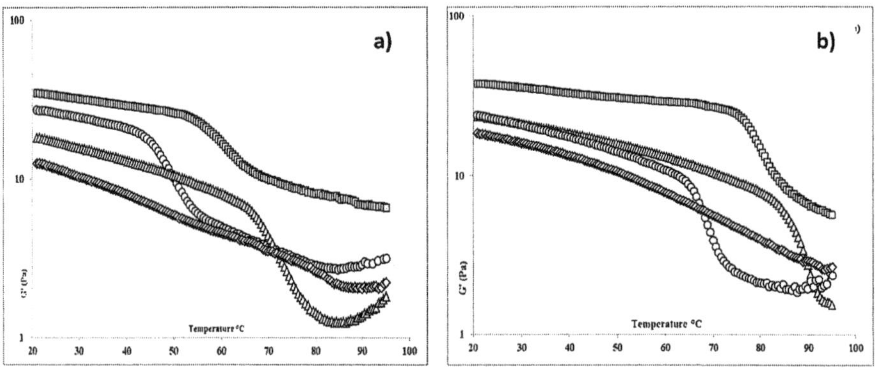

***Figure 1.** Xanthan SA/HP (□) and xanthan LA/HP (○) xanthan SA/SP(Δ), and xanthan SA/LP (◊)on first heating (1 wt %) at **a)**10 mM NaCl, and **b)** 40 mM NaCl.*

The rheometry data can be matched with DSC peaks in (Figure 2 and 3). As can be seen reducing the pyruvate content of SA/ LP promotes the increase of the transition temperature as recorded before and forms a sharper transition. The destabilising effect of pyruvate can be discussed as an effect of increase in internal repulsion between the charged side chains and consequently better interaction (Smith et al., 1981). However, the reason for the stabilizing effect of the acetyl groups is less clear. A polar interaction between the acetyl methyl groups may be responsible or, alternatively, the acetyl substituents could act as hydrogen bond acceptors, and consequently stabilize the molecule (Dentini et al. 1984).

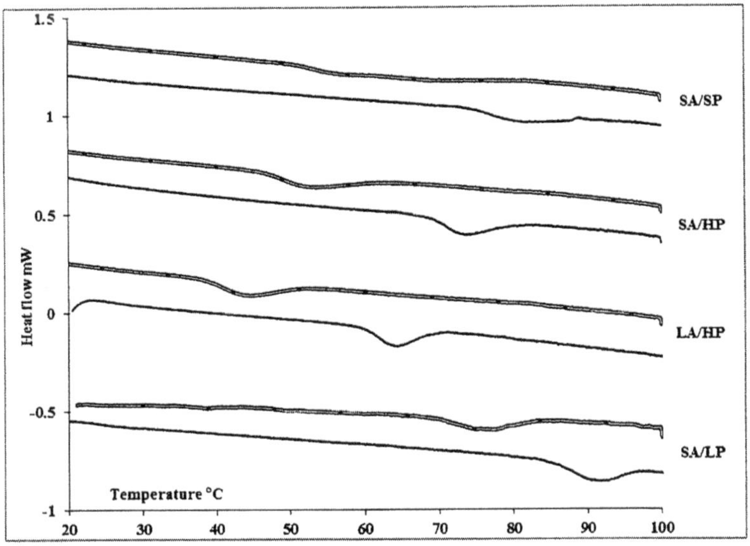

Figure 2. *2nd Heating scans of different xanthan sample at ionic strength of 10mM (=) and 40mM (-) NaCl and a scan rate of 1°Cmin^{-1}.*

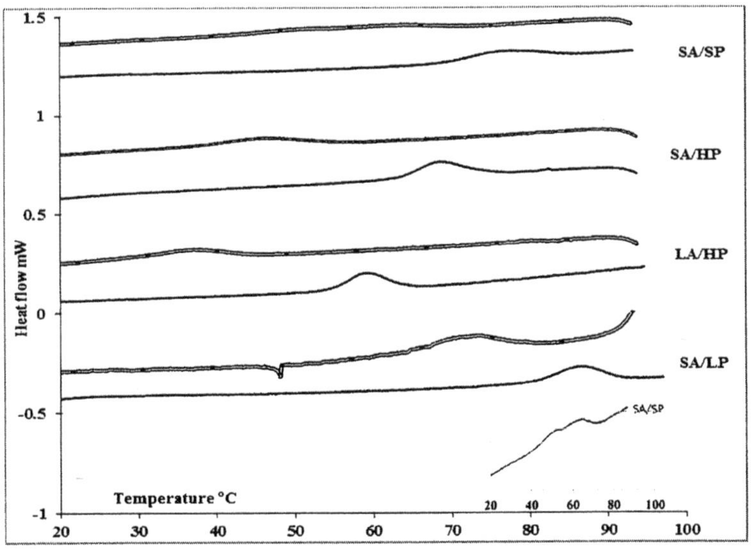

Figure 3. *Cooling scans of different xanthan samples at ionic strengths of 10mM (=) and 40mM (-) NaCl and a scan rate of 1°Cmin^{-1}.*

As discussed previously the degree of substitution by acetyl and pyruvate groups affects the stability of the helix. Holzwarth (Holzwarth & Prestridge, 1977) reported the distribution of acetyl and pyruvic acid groups may affect the range of temperature over which the transition occurs.

However Shatwell (Shatwell et al., 1990) refuted this idea and also the conception of "co-operativity" of transition process i.e. rapid conversion to coil formation when sufficient energy is supplied to disrupt the helix, and reported neither the acetate nor the pyruvate substituents play an important role in determining the breadth of the transition.

It seems reasonable to assume that the (de)stabilising substituents would play a role in such systems, but we have not found this to be true in this study, implicating inter- and intra-molecular distributions being non-uniform as well and molecular weight polydispersity.

Heating and cooling thermographs do not demonstrate large variation in enthalpy that has been seen by Kitamura (Kitamura et al., 1991) and Foster (Foster and Morris 1994) for low pyruvate and polytetramer (xanthan minus the terminal mannose), respectively. The values recorded in this study are consistent with values seen for native xanthans in the past *i.e.* in the range ~4-8 J/g xanthan for low concentration samples. Kitamura (Kitamura et al., 1991) found values increase to 11-13 J/g with the removal of pyruvate and Foster (1994) saw the enthalpy of transition double with a the removal of the terminal mannose and associated pyruvate group. However, neither of these studies compared materials at constant acetate content. It was assumed that acetate had no effect as there was no change in enthalpy upon removal of the acetate group.

Figure 4 represents plots of the reciprocal of the midpoint transition temperature (T_m) versus the natural logarithm of the sodium concentration in the solutions for current work, also comparing the work of Kitamura (1991) and Norton (Norton et al., 1984), according to the van't Hoff equation. As can be seen it is possible to imitate SA/LP 10 mM NaCl, transition temperature at ~75°C by using SA/HP at 45 mM NaCl. By comparing the previous work with our own it is seen that the change in T_m by increasing the salt is almost following same trend in slope. However, the more interesting point is the transition temperature of the same category, for instance high pyruvated samples shows significant diversity at fixed salt level, indicating both differences in total pyruvate and/or acetate and the complexity of comparing across different studies.

3.2 Interaction

Konjac glucomannan shows a limited interaction with stable xanthan helices (Foster and Morris 1994) and is known to create large transitions after mixing with xanthan at low ionic strength (Annable et al., 1994) . The mix of SA/SP-KM shows the previously seen endotherm by Annable (Annable et al., 1994) on DSC. The wide transition endotherm (Figure 5) for SA/SP from 48° C to 80 ° C has changed to a sharper and smaller endotherm around 60 ° C after mixing with KM.

The LA/HP and SP/HP samples, in which the T_m for xanthan alone was at around 43 °C and 52 °C respectively, form a new endothermic peak with almost the same magnitude as the SA/SP sample in combination with KM and the xanthan transition peak disappears (Figure 6). Goycoolea (Goycoolea et al., 1995) studied the stoichiometry of this peak and discovered a 1:1 mix was optimal, which is the ratio used in this study.

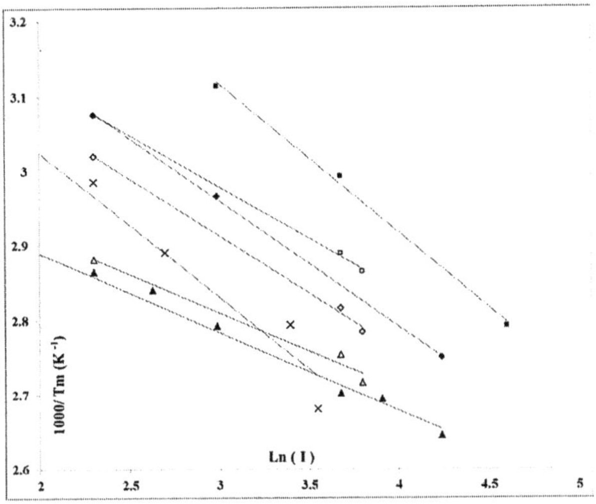

Figure 4. *Plots of $l/T_{1/2}$ vs ln [Na+] for xanthan SA/HP (□), xanthan SA/SP (Δ), xanthan SA/LP (◊)from current work and Kitamura (1991) filled shape with high medium and low pyruvate content respectively, and xanthan in Norton (1984) work with (x) .*

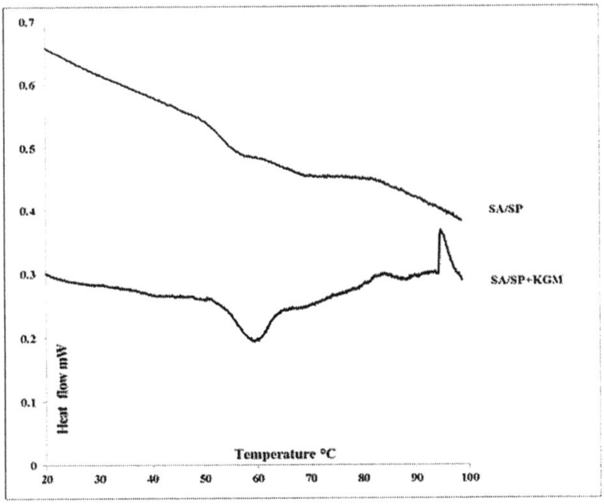

Figure 5. *2st Heating scans of SA/SP xanthan sample at ionic strength of 10 mM NaCl and a scan rate of 1 °C min^{-1} in combination with KM (total 0.5 wt. %)*

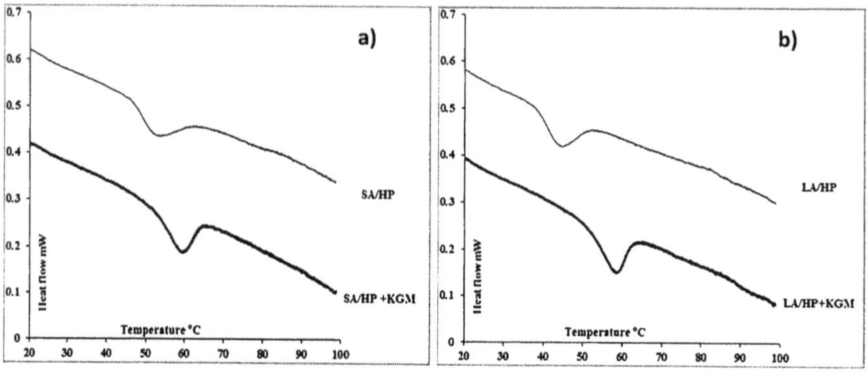

Figure 6. *2nd Heating scans of a) SA/HP b) LA/HP xanthan samples at ionic strength of 10 mM NaCl and a scan rate of 1°Cmin^{-1} in combination with KM (total 0.5 wt. %)*

The SA/LP xanthan shows a different trend (Figure 7), the xanthan transition, at high temperature, remains detectable when combined with KM, however a smaller endotherm in the region of 40 °C represents the synergistic interaction.

This kind of transition has been reported previously by Annable (Annable et al., 1994) with escalating ionic strength (cf. 40mM NaCl instead of 10mM NaCl).

As can be seen in Figure 7, by increasing the salinity to 40mM NaCl the SA/HP sample shows a similar trend and presents a synergistic peak in the region of 40-50 °C, slightly higher than the SA/LP synergistic peak, with a wider endotherm at 75 °C for helix- coil transition.

The synergistic peak is also detectable for the most stable of the samples (SA/VLP – data not shown), which designates that the existence of the terminal mannose is enough to tolerate the low temperature synergistic junction zone. However, Foster (Foster and Morris 1994) did not find a synergistic peak for polytetramer. These current results indicate that although the xanthan helix-coil transition temperatures are matched the destabilisation effect of pyruvate on the helical compactness promotes synergistic gels more stable to temperature.

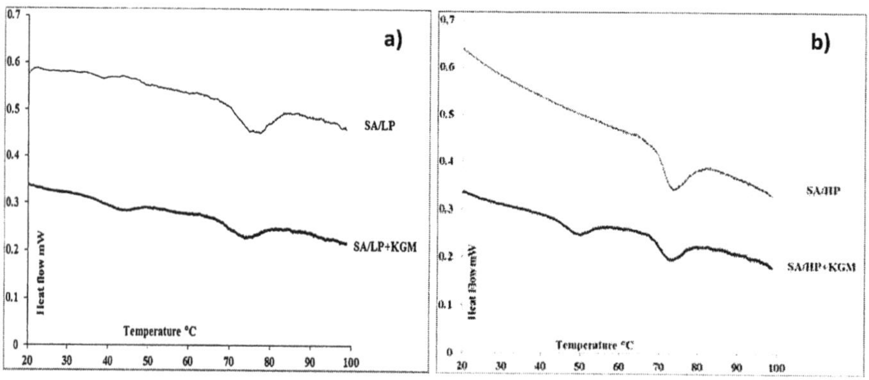

Figure 7. *2nd Heating scans of **a)** SA/LP sample at ionic strength of 10 mM NaCl **b)** SA/HP sample at ionic strength of 40 mM NaCl and a scan rate of 1°Cmin^{-1} in combination with KM (total 0.5 wt. %).*

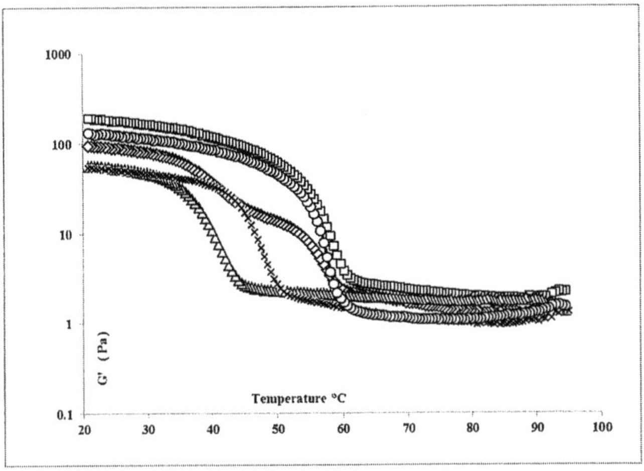

Figure 8. *Xanthan SA/HP (□), xanthan LA/HP (○) xanthan SA/SP (◊), xanthan SA/LP (Δ) at 10 mM NaCl and xanthan SA/HP(x) at 40 mM NaCl in combination with KM (total 0.5 wt. %).*

Figure 8 shows the elastic modulus (G ') in a temperature sweep of different xanthan-KM mixtures on cooling. As can be seen the synergistic endothermic peak on DSC are mirrored by temperatures at which the onset of gel formation is seen. For high pyruvated xanthan sample i.e. SA/HP and LA/HP, G ' starts to increase sharply after 60 °C, independent of acetate level. By increasing the ionic strength to 40mM NaCl the SA/HP sample shows a consequent reduction in T_{gel} to ~50°C, commensurate with the endotherm shown in figure 7b. The T_{gel} for SA/LP at 10 mM NaCl also matched the temperature of the endotherm seen in Figure 9a. The HP synergistic gels show higher final elastic moduli at 20°C, indicating that either there are more junction zones in the gel formed at 60°C, when the xanthan is still disordered, or the homotypic xanthan-xanthan interactions which are enhanced in the HP samples (Figure 1a and 1b) play an important role in stabilising the synergistic gel. Given there is no indication of the individual xanthan transition <60°C in these mixtures and the fact that stoichiometry has been shown for such mixing ratios, would argue against this assumption.

Interestingly the SA/SP xanthan sample, with a broad DSC transition endothermic peak, shows two waves of G' increase upon cooling and gelation, one starting around 60°C similar to the SA/HP and LA/HP samples and another one about 50°C in almost same region of SA/HP at 40 mM NaCl, possibly indicating a different mechanism for its interaction with KM. The insert in figure 3 shows that the broad exotherm of the coil-helix transition traverses the point at 60°C, where gelation is seen for the HP xanthans. Therefore, at the point of the first wave of G' increase (60°C) some of the xanthan molecule is ordered, with ~50% of the transition still to take place. The two waves of G' increase therefore indicates two different modes of xanthan-KM interaction within one mixture.

Acknowledgments

The research leading to these results has received funding from the European Community's Seventh Framework Program (FP7/2007-2013) under Grant Agreement No. 214015" .

We would like to specially thanks Dr. G. Sworn and Danisco Company for useful discussion and kindly providing xanthan samples.

References

Annable, P., Williams, P. A. & Nishinari, K. (1994). Interaction in Xanthan-Glucomannan Mixtures and the Influence of Electrolyte. *Macromolecules*, *27* 4204-4211. doi: 10.1021/ma00093a023.

Cadmus, M. C., Rogovin, S. P., Burton, K., Pittsley, J. E., Knutson, C. a, & Jeanes, A. (1976). Colonial variation in *Xanthomonas campestris* NRRL B-1459 and characterization of the polysaccharide from a variant strain. *Canadian Journal of Microbiology*, *22*(7), 942-8. Retrieved from http://www.ncbi.nlm.nih.gov/pubmed/963616.

Cairns, P., Morris, V.J., Miles, M.J., & Brownsey, G.J. (1986). Comparative studies of the mechanical properties of mixed gels formed by kappa carrageenan and tara gum or carob gum. *Food Hydrocolloids*, *1*(1), 89-93. doi: 10.1016/S0268-005X(86)80010-8.

Cairns, P., Miles, M. J., Morris, V. J., & Brownsey, G. J. (1987). X-Ray fibre-diffraction studies of synergistic, binary polysaccharide gels. *Carbohydrate Research*, *160*, 411-423. doi: 10.1016/0008-6215(87)80327-0.

Cheetham, N. (1992). Proton and carbon-13 NMR studies on xanthan derivatives. *Carbohydrate Polymers*, *17*(2), 127-136. doi: 10.1016/0144-8617(92)90106-Z.

Cheetham, N., & Mashimba, E. (1988). Conformational aspects of xanthan-galactomannan gelation. *Carbohydrate Polymers*, *9*(3), 195-212. doi: 10.1016/0144-8617(88)90025-2.

Dentini, M. (1984). Conformational properties of xanthan derivatives in dilute aqueous solution. *International Journal of Biological Macromolecules*, *6*(2), 93-98. doi: 10.1016/0141-8130(84)90070-9.

Fernandes, P., Goncalves, M., & Doublier, J. (1991). A rheological characterization of kappa-carrageenan/galactomannan mixed gels: A comparison of locust bean gum samples. *Carbohydrate Polymers*, *16*(3), 253-274. doi: 10.1016/0144-8617(91)90112-P.

Fernandes, P., Gonçalves, M. P., & Doublier, J. L. (1992). Effect of galactomannan addition on the thermal behaviour of κ-carrageenan gels. *Carbohydrate Polymers*, *19*(4), 261-269. doi: 10.1016/0144-8617(92)90079-6.

Foster, T. J., & Morris, E. R. (1994). Xanthan polytetramer: conformational stability as a barrier to synergistic interaction. In G. O. Phillips, D. J. Wedlock, & P. A. Williams (Eds.), *Gums and stabilisers for the food industry 7*, (pp. 281-289). Oxford: IRL Press.

Goycoolea, F. M. A, Richardson, R. K., Morris, E. R., & Gidley, M. J. (1995). Stoichiometry and Conformation of Xanthan in Synergistic Gelation with Locust Bean Gum or Konjac Glucomannan: Evidence for Heterotypic Binding. *Macromolecules*, *28*(24), 8308-8320. doi: 10.1021/ma00128a047.

Hassler, R. A., & Doherty, D. H. (1990). Genetic engineering of polysaccharide structure: production of variants of xanthan gum in *Xanthomonas campestris*. *Biotechnology progress*, *6*(3), 182-7. doi: 10.1021/bp00003a003.

Holzwarth, G. (1976). Conformation of the extracellular polysaccharide of Xanthomonas campestris. *Biochemistry*, *15*(19), 4333-9. Retrieved from http://www.ncbi.nlm.nih.gov/pubmed/9135.

Holzwarth, G., & Prestridge, E. B. (1977). Multistranded helix in xanthan polysaccharide. *Science (New York, N.Y.)*, *197*(4305), 757-9. Retrieved from http://www.ncbi.nlm.nih.gov/pubmed/21082815.

Kim, B.-S., Takemasa, M., & Nishinari, K. (2006). Synergistic interaction of xyloglucan and xanthan investigated by rheology, differential scanning calorimetry, and NMR. *Biomacromolecules*, *7*(4), 1223-30. doi: 10.1021/bm050734+.

Kitamura, S., Takeo, K., Kuge, T., & Stokke, B. T. (1991). Thermally induced conformational transition of double-stranded xanthan in aqueous salt solutions. *Biopolymers*, *31*(11), 1243-55. doi: 10.1002/bip.360311102.

Kohyama, K., Iida, H., & Nishinari, K. (1993). A mixed system composed of different molecular weights konjac glucomannan and kappa carrageenan: large deformation and dynamic viscoelastic study. *Food Hydrocolloids*, *7*(3), 213-226. doi: 10.1016/S0268-005X(09)80173-2.

Li, Y., Li, Y., Yao, Y., Liu, B., Chen, M., Song, X., et al. (2009). Two-dimensional scaffold layer formations on a solid surface through xanthan polysaccharide: temperature effect. *Colloids and Surfaces. B, Biointerfaces*, *74*(1), 136-9. doi: 10.1016/j.colsurfb.2009.07.007.

Maekaji, K., & Koch, C. (1974). The Mechanism of Gelation of Konjac Mannan, *Agricultural and Biological Chemistry*, *38*(2),315-321.

Morris, E R, & Foster, T. J. (1994). Role of conformation in synergistic interactions of xanthan. *Carbohydrate Polymers*, *23*, 133-135.

Morris, E R, Rees, D. A., Robinson, G., & Young, G. A. (1980). Competitive inhibition of interchain interactions in polysaccharide systems. *Journal of Molecular Biology*. Retrieved from http://www.ncbi.nlm.nih.gov/pubmed/7411613.

Morris, E. R., Barnes, H. A., & Price, J. (1977). Associations of like and unlike polysaccharides :mechanism and specificity in galactomannans, interacting bacterial polysaccharides and related systems, *Carbohydrate Research*, *57*, 249-272.

Morris, V. J. (1988). X-Ray diffraction of fibres prepared from oriented gels has become a standard method for producing molecular models for the association between like polysaccharides within the junction zones of gels l. Although the preparation of the fibres requires t. *Science*, *176*, 329-334.

Morris, E.R. (1983). Molecular Viscoelasticity of Xanthan Polysaccharide. *Faraday Symposia of the Chemical Society*, 115-129.

Nishinari, K. (1996). Rheological and DSC studies on the interaction between gellan gum and konjac glucomannan. *Carbohydrate Polymers*, *30*(2-3), 193-207. doi: 10.1016/S0144-8617(96)00092-6.

Nishinari, K., & Takahashi, R. (2003). Interaction in polysaccharide solutions and gels. Current Opinion in Colloid and *Interface Science*, *8*, 396-400.

Norton, I. T., Goodall, D.M., & Morris, E.R (1984). Mechanism and dynamics of conformational xanthan polysaccharide ordering in xanthan polysaccharide. *Molecular Biology,* 175,371-394.

Ridout, M. J., Cairns, P, Brownsey, G J, & Morris, V J. (1998). Evidence for intermolecular binding between deacetylated acetan and the glucomannan konjac mannan. *Carbohydrate Research*, *309*(4), 375-9. Retrieved from http://www.ncbi.nlm.nih.gov/pubmed/9764469.

Rochas, C., Taravel, F R, & Turquois, T. (1990). N.M.R. studies of synergistic kappa carrageenan-carob galactomannan gels. *International Journal of Biological Macromolecules*, *12*(6), 353-8. Retrieved from http://www.ncbi.nlm.nih.gov/pubmed/2088491.

Shatwell, K.P. (1990). The influence of acetyl and pyruvate substituents on the helix-coil transition behaviour of xanthan. *Carbohydrate Research, 206*(1), 87-103. doi: 10.1016/0008-6215(90)84009-J.

Shatwell, K. P., Sutherland, I W, & Ross-Murphy, S. B. (1990). Influence of acetyl and pyruvate substituents on the solution properties of xanthan polysaccharide. *International Journal of Biological Macromolecules, 12*(2), 71-8. Retrieved from http://www.ncbi.nlm.nih.gov/pubmed/2078534.

Shatwell, K.P., Sutherland, I., Ross-murphy, S., & Dea, I. (1990). Influence of the acetyl substituent on the interaction of xanthan with plant polysaccharides — III. Xanthan-konjac mannan systems. *Carbohydrate Polymers, 14*(2), 131-147. doi: 10.1016/0144-8617(90)90026-O.

Smith, I., Symes, K., Lawson, C., & Morris, E.R. (1981). Influence of the pyruvate content of xanthan on macromolecular association in solution. *International Journal of Biological Macromolecules, 3*(2), 129-134. doi: 10.1016/0141-8130(81)90078-7.

Sutherland, I W. (1981). Xanthomonas polysaccharides - improved methods for their comparison. *Carbohydrate Polymers, 1*, 107-115.

Tait, M. I., Sutherland, I. W., & Clarke-Sturman, a J. (1986). Effect of Growth Conditions on the Production, Composition and Viscosity of *Xanthomonas campestris* Exopolysaccharide. *Microbiology, 132*(6), 1483-1492. doi: 10.1099/00221287-132-6-1483.

Turquois, T., Rochas, C., & Taravel, F.R. (1992). Rheological studies of synergistic kappa carrageenan-carob galactomannan gels. *Carbohydrate Polymers, 17*(4), 263-268. doi: 10.1016/0144-8617(92)90168-P.

Whitcomb, P. J. (1978). Rheology of Xanthan Gum. *Journal of Rheology, 22*(5), 493. doi: 10.1122/1.549485.

Williams, P. A, Day, D. H., Langdon, M. J., Phillips, G. O., & Nishinari, K. (1991). Synergistic interaction of xanthan gum with glucomannans and galactomannans. *Food Hydrocolloids, 4*(6), 489-493. doi: 10.1016/S0268-005X(09)80199-9.

Hydrocolloid gels

BIOPOLYMER MICROGEL SUSPENSION RHEOLOGY AS A FUNCTION OF PARTICLE MODULUS AND EFFECTIVE PHASE VOLUME

H.M. Shewan[1], J.R. Stokes[1]

[1]Chemical Engineering, The University of Queensland, Brisbane Qld 4072, Australia
Email: h.shewan@uq.edu.au ; Jason.stokes@uq.edu.au

1 INTRODUCTION

Structured soft matter is a class of soft material that is solid-like at rest but flows when a certain amount of force is applied. This rheological behaviour arises from the microstructural arrangement of 'soft' particles in suspension which can be generalised into two forms; a 'soft glass' where particle movement is constrained in a dense disordered structure; or, a 'gel' where particles are arranged in a percolated network structure.[1] Their unique solid and liquid-like properties are widely exploited in such diverse applications as fracturing fluids for oil recovery, paints, inks, personal care products and foods.[2, 3] However, there is increasing need to understand the relationship between the structure and rheology of these systems as manufacturers endeavour to re-engineer products to exclude ingredients that may be detrimental to the environment (e.g. organic solvent-free paints), unsustainable, or harmful to human health and well-being (e.g. low fat foods).

The complexity of most structured soft matter systems makes the relationship between suspension rheology and microstructure difficult to determine. Starch suspensions are a common example of a microgel used as a rheological modifier and fat replacer. Although the rheology of starch suspensions has been widely studied, the fundamental relationship between particle morphology and suspension rheology is not well defined, mainly due to the complexity of both the particles and their suspensions. Starch particle modulus, size and surface structure are interdependent and related to the degree of denaturation. Suspensions become complex due to leaching of amylase and/or amylopectin. Plant cell suspensions such as tomato paste are another common example of structured soft matter. Their behaviour is poorly understood partly due to the difficulty of determining the particle volume fraction.[4, 5] The rheology of foods is generally reported as a function of solids concentration with limited reference to the effect of phase volume, for example Dang et al.[6] discuss the relevance of phase volume to the rheology of their cross-linked waxy maize starch and whey protein isolate suspensions yet present data as a function of dry weight. Steffe[7] presents rheology data for a 53 wt/wt% raw cornstarch water suspension without reference to phase volume. Hence, while it is well known that particle volume fractions are a key parameter governing the rheology of suspensions, most studies on complex food systems only consider the influence of weight fraction. To assist in development of protocols to overcome this deficiency, we note that all these seemingly disparate systems can be considered to contain particles that are 'soft'; which means that the particle volume and/or shape alter in response to its environment or to applied force.

The simplest experimental model for 'soft' particles is spherical microgels. These can be manufactured with controlled size, crosslink density, modulus and surface properties. Microgel suspension rheology is strongly dependent on three parameters: phase volume, particle modulus and interaction potential. Phase volume (Φ) is the volume of particles in suspension (V_P) as a proportion of the total volume of suspension (V_T), ($\Phi = V_P/V_T$). Generally Φ can only be defined under dilute concentrations, referred to as the effective volume fraction (Φ_0). At low Φ_0 microgel suspensions follow hard particle suspension rheology. With increasing phase volume, microgel particles come into close contact and their viscosity diverges from hard particle behaviour with microgel suspensions showing maximum effective packing fractions ($\Phi_{0,m}$) well above 0.64 that is measured for hard spheres.[2, 8] We suggest that the deviation from hard sphere behaviour is a result of soft particle deformation and volume reduction, resulting in Φ_0 being higher than the actual Φ. Without knowing how Φ alters with increasing suspension density, it is difficult to develop quantitative structure-function relationships for the suspension. We aim to elucidate how $\Phi_{0,m}$, and Φ generally, depend on the particle modulus (G_p).

This study utilises suspensions of agarose microgels that are stable at room temperature, do not swell in excess water below 35°C, and are chemically and electrically neutral.[9-11] We chose to use particles that are *ca.* 10µm in size so that long range surface forces and Brownian motion are negligible. The particle modulus can be easily controlled as it is simply a function of agarose concentration, with increased concentration giving a harder particle due to increased cross-link density.[9] It is envisaged that such a model system will allow us to determine the influence of particle elasticity, without any other complicating factors, on the microstructure and rheology of soft particle suspensions.

2 MATERIALS AND METHODS

2.1 Microgel Manufacture

Agarose microgels were manufactured by forming an oil-in-water emulsion and gelling the water droplets to form microgel particles.[5] Agarose solutions were prepared by adding agarose at a concentration of 5, 2.5 or 0.5 wt% to reverse osmosis (RO) treated water containing 0.02% sodium azide and heating to 95°C for 30 minutes while stirring. The agarose was added to sunflower oil containing 1 wt% Admul WOL at 90°C and homogenised using an IKA Ultra Turrax at 80°C and 12000 rpm for 12 minutes to form a water-in-oil emulsion. The emulsion was quenched by placing it in an iced water bath to gel the aqueous phase while stirring slowly to prevent aggregation and coalescence. The spherical microgel particles produced were separated from the oil phase by centrifugation and washing. Microgels were filtered through a 20 µm mesh to remove any large particles; this gave an average particle size of approximately 10 µm for all agarose concentrations. Microgel suspensions were made to known wet weight percentages from 1 to 100% by redispersing the particles in RO water.

2.2 Rheology

The rheological measurements were carried out on the Haake Mars 3 stress controlled rotational rheometer with a Peltier controlled element. Parallel plate geometry (35 mm diameter) was used with titanium plates both smooth and sandblasted to a roughness of 1.8 µm. The gap between the plates was set between 1mm and 50 µm as required. The gap was zeroed at a force of 4 N to account for the squeeze flow of air[12] and gap error calculated using the method of Davies and Stokes.[13] Step shear measurements were used in most cases with the gradient function where an average of 20 data points was taken when the measurement stabilised or at a maximum time of 30 s. Oscillatory measurements were completed for suspensions displaying solid-like behaviour at zero shear. Amplitude sweeps were completed to determine a stress value which was within the linear viscoelastic region (LVR) across the range of frequencies measured. Frequency sweeps were run at a stress within the LVR to determine storage and loss moduli G' and G" (Pa) for the concentrated microgel suspensions. G' describes the solid-like behaviour of the suspension and G" describes the liquid-like behaviour.

The modulus of the bulk gels was determined and used as an approximation for the particle modulus (G_P). Agarose solutions were prepared at 80°C and poured onto the sandblasted rheometer plate surface pre-heated to 80°C. The plates were brought together to a set gap, the sample carefully trimmed and the temperature set point reduced to 25°C to form a gel. The gels were allowed 5 minutes standing time after the temperature set point was reached. An amplitude sweep was run to determine the LVR at high and low frequency followed by a frequency sweep across this range giving G' values of 277, 161 and 7.8 kPa for the 5 wt%, 2.5 wt% and 0.5 wt% agarose bulk gels respectively.

3 RESULTS AND DISCUSSION

3.1 Determining Effective Phase Volume

Relative viscosity ($\eta_r = \eta_0/\eta_s$) of agarose microgels of differing moduli can be presented as a function of dry weight of agarose as shown in Figure 1. This shows that for a single weight fraction of polymer, a suspension of soft particles will exhibit a higher relative viscosity than that of a suspension of harder particles. That is, the hardest particles require a greater dry weight of polymer to achieve the same viscosity increase or thickening effect. To understand this difference we need to determine the specific volume occupied by both the polymer and water included within the microgel. The included water in the particle affects the particle volume, which has a significant influence on suspension viscosity.

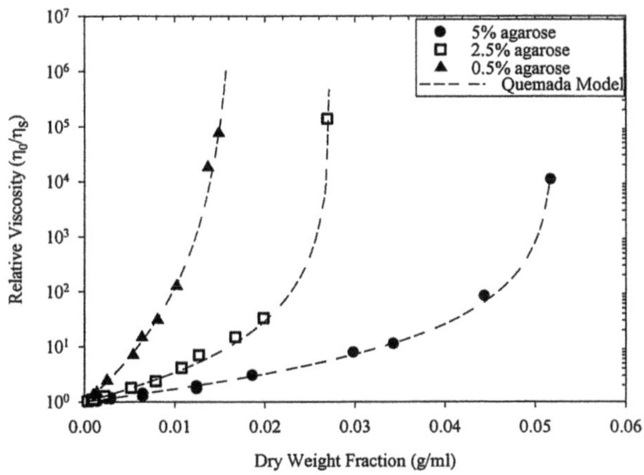

Figure 1 *Microgel suspension relative viscosity against dry weight fraction of microgel suspensions. Measured data is represented with data points. Lines are fitted using the Quemada model, discussed in section 3.2 and shown in equation (4).*

The preferred method widely used to determine volume occupied by particles and polymers in suspension is through the use of Einstein's equation. The effective phase volume (Φ_0) of particles can be determined at low suspension concentration through equation (1), where η_0 = zero shear viscosity and η_s = solvent viscosity.[14] To determine Φ_0, solvent and suspension viscosity is measured at low wet weight concentration. Einstein's theory states that at low concentration (c in g/ml) below that equivalent to $\Phi \sim 0.05$, where particle interaction, deformation and squeezing do not occur, η_0 will increase above that of η_s by an amount proportional to the particle volume; this is known as the intrinsic viscosity $[\eta]$ with units of reciprocal concentration (ml/g). $[\eta]$ is determined by measurement of the reduced viscosity at the zero concentration limit as shown in equation (2) and can be substituted into Einstein's equation to give $\Phi_0 = [\eta]c/2.5$. However, while we found that reasonable values of $[\eta]$ could be determined using this method, there was some experimental uncertainty in their exact value due to the limitations of detecting small viscosity changes in very low phase volume suspensions, particularly when using a low viscosity solvent such as water ; that is, small inaccuracies in solvent or suspension viscosity had a significant effect on the value of $[\eta]$.

An alternative approach is to determine the hydrodynamic volume of the particle by best fit of Batchelor's equation (3), which extends Einstein's equation to higher volume fractions and includes pairwise interactions between particles. Φ_0 is written as a specific volume Φ_0=kc with k (ml/g)[2] , noting that $[\eta]$ =2.5k. The η_r is plotted against Φ_0 for each microgel suspension and Batchelor's model fitted using a least squares method to solve for 'k' (initial value is set at k = 1, the value for hard spheres) to minimise error between the measured data and model fit, up to $\Phi_0 \sim 0.3$. We found that this approach gave more robust values for $[\eta]$ than if equation (2) was used.

$$\eta_0 = \eta_s(1 + 2.5\phi_0) \tag{1}$$

$$[\eta] = \lim_{c \to 0} \eta_{red} \quad \text{where,} \quad \eta_{red} = \frac{\eta_0 - \eta_s}{c\eta_s} \tag{2}$$

$$\eta_0 = \eta_s(1 + 2.5\phi_0 + 5.9\phi_0^2) \tag{3}$$

Figure 2 shows relative viscosity data measured for each microgel suspension and plotted as a function of effective phase volume. Both Einstein's and Batchelor's model are shown, with Φ_0 calculated using 'k' determined by fitting Batchelor's model. All the data follow Einstein's equation up to Φ_0 = 0.05 and Batchelor's equation up to Φ_0 =0.3, and there is no dependence whatsoever on the polymer concentration in the microgel and thus particle modulus.

3.2 Determining Maximum Packing Fraction

The Quemada (equation 4) or Krieger Dougherty (equation 5) models are commonly used to predict the viscosity of suspensions, including microgel suspensions, with volume fraction until the viscosity approaches infinity at some value corresponding to the so-called maximum packing fraction.[2, 15] As commonly performed in the literature, we fit these models for the relative viscosity as a function of Φ_0 as shown in Figure 3. The maximum packing fraction, Φ_{0m} can be determined as a free fitting parameter using least squares error to fit to the models. Figure 3 shows that at phase volumes below ~0.4, microgel suspension viscosity is simply a function of Φ_0 with all three particles following the same curve. The effect of G_P is demonstrated above $\Phi_0 \sim 0.4$, with Φ_{0m} being significantly larger for softer particles.

$$\frac{\eta_0}{\eta_s} = \left(1 - \frac{\phi_0}{\phi_m}\right)^{-2} \tag{4}$$

$$\frac{\eta_0}{\eta_s} = \left(1 - \frac{\phi_0}{\phi_m}\right)^{-[\eta]\phi_m} \tag{5}$$

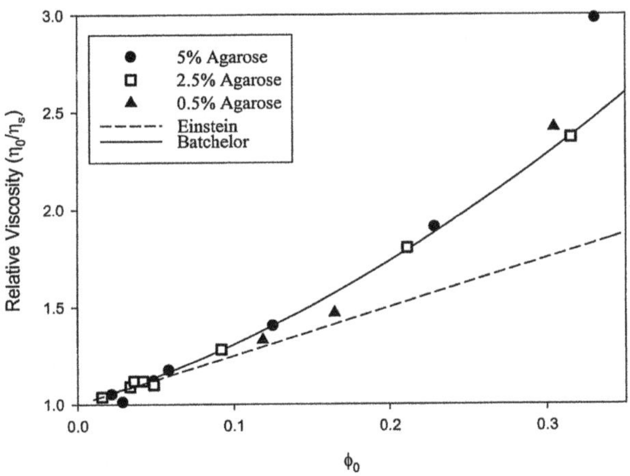

Figure 2 *Microgel suspension relative viscosity against effective phase volume for a 2.5% agarose microgel suspension. Measured data is represented with data points. Lines are fitted using Einstein's (1) and Batchelor's (3) equation.*

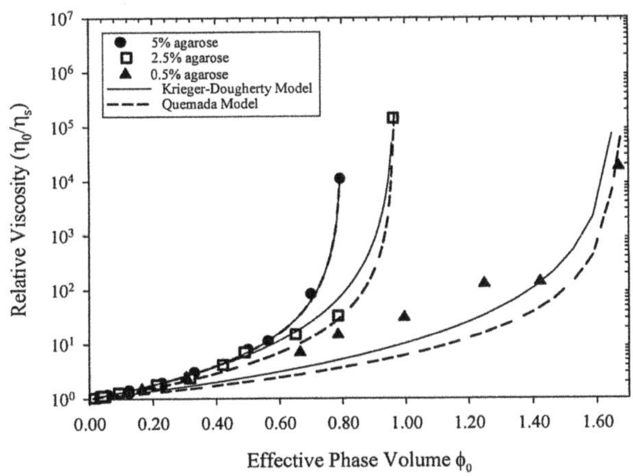

Figure 3 *Microgel suspension relative viscosity against effective phase volume of microgel suspensions. Measured data is represented with data points. Dotted lines are fitted using the Quemada model equation (4), Solid lines are fitted using the Krieger-Dougherty Model (5).*

The problem with free-fitting of Φ_{0m} using a least squares approach is that for accuracy a large number of data points are needed around the value of Φ_{0m}. It is also clear that using this method with either model results in a fit for 0.5% that is not entirely accurate. To investigate these issues, we linearise the Quemada model (see equation 5), so that plotting $1/(\eta_r)^{0.5}$ against Φ_0 should yield a straight line if the model is followed, a method previously used to look at the influence of polydispersity on hard particle suspensions.[16, 17] Φ_{0m} is then simply the intercept when $1/(\eta_r)^{0.5} = 0$. This is utilised in Figure 4 and it is shown that the 3 suspensions follow the same linear relationship for $1/(\eta_r)^{0.5}$ with Φ_0 up to Φ_0~0.4. It also shows that the 5% agarose microgel, which is the hardest particle with a modulus of 277 kPa, fits the linearised Quemada model perfectly for $1/(\eta_r)^{0.5} > 0.01$. This suspension has a Φ_{0m} of 0.80, which is substantially above that known for random close packing of hard spheres ($\Phi_{RCP} = 0.64$) and for hexagonal close packing (Φ_{HCP}) reported at values between 0.71 and 0.74 for both mono and polydisperse emulsions.[17-19] Our high value of Φ_{0m} may be partially accounted for by the degree of polydispersity in the microgel suspension as increasing polydispersity leads to an increase in Φ_{0m}. For example, Luckham and Ukeje[17] found that for sterically stabilised hard sphere suspensions the maximum packing fraction ranged from 0.70 for monodisperse to 0.76 for highly polydisperse suspensions.

While the suspension of 5% agarose followed a linear relationship across the full concentration range, this is not the case for the soft particles. The broken lines in Figure 4 guide the eye to show the phase volume at which the suspension rheology deviates from 5% agarose suspension behaviour and the extent of this deviation. Above Φ_0 ~0.4, the 0.5% and 2.5% agarose microgel suspensions deviate from the straight line fit of the Quemada model. This is not likely to be due to polydispersity effects, since the study previously mentioned by Luckham and Ukeje[17] found linear relationships between $1/(\eta_r)^{0.5}$ and Φ were found for their monodisperse and polydisperse particle suspensions. We suspect that this deviation at Φ_0 ~0.4 is dependent on particle modulus, and may result from particle shrinkage due to water loss from its structure as concentration increases so that the actual $\Phi < \Phi_0$. Supporting this hypothesis is that Φ_{0m} for the softest particles (0.5% agarose) is 1.67, significantly greater than 1 (not possible geometrically for hard spheres) suggesting the particle both deforms and shrinks at high phase volumes. From this we conclude the graphical method shown in figure 4 is a very convenient way in which to determine where Φ_0 is not valid for microgel and other soft particle suspensions, which in our case is Φ_0~0.4, and where Φ is likely to be become a function of G_P.

4 DISCUSSSION

Extra complexity is added in trying to determine a microstructural model for high phase volume suspensions as particles touch and exert forces on each other resulting in their deformation and possible decrease in volume.[20] We propose a new method for determining actual microgel phase volume (Φ) founded on the premise of a linear relationship $1/(\eta_r)^{0.5} = 1 - (\Phi_0/\Phi_m)$ for particles across the full range of moduli assuming no particle overlap. Assuming $\Phi = \Phi_0$ for 5% agarose microgels we use the straight line fit to this data as a basis. We then solve $\Phi = (1/(\eta_r)^{0.5} - 1)\Phi_m$ for the 2.5 and 0.5% suspensions, where $\Phi_m = 0.8$. Using this method the deviation from linear relationship can be accounted for by a change in radius (R), which would correspond to maximum decreases of ca. 6% and 30% for the 2.5% and 0.5% microgels respectively. The source of this deformation and shrinkage could be derived from a combination of processes, including compression force occurring in a close packed microstructure, osmotic deswelling or particle overlap.

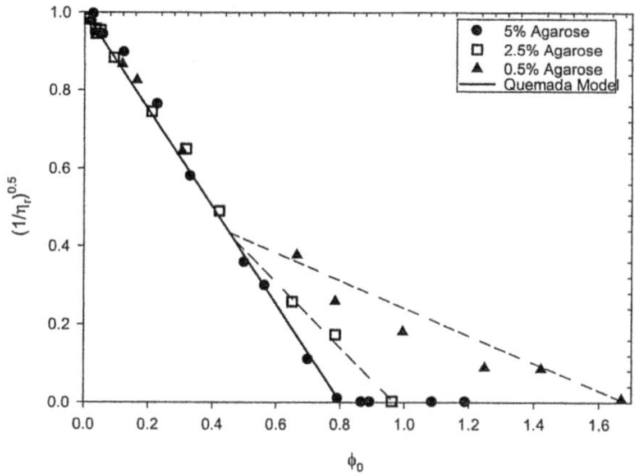

Figure 4 *A linear plot of $1/(\eta_r)^{0.5}$ vs effective phase volume of 3 microgel suspensions. Measured data is represented with points, the solid line represents a fit of the Quemada model to the data with $\Phi = 0.78$, dashed lines are drawn to guide the eye where the data deviates from a straight line fit.*

The surface structure of the agarose microgels is unknown though it is likely to be different for microgels of varying modulus, thus affecting particle interaction and deformation of the surface. TEM images of agar microgel particles reported by Adams et al.[5] show increasing pore size on the microgel surface with decreased agar concentration. A schematic showing two possibilities for microgel surface architecture is shown in Figure 5 on the left ranging from a surface with long agarose strands floating freely to one where the particle has been compressed slightly to form a dense polymer layer on the particle surface. Surface with large pore size or free polymer strands could lead to particle overlap which will affect the relationship between Φ and Φ_0. Exact measurement of the microgel particle surface, particularly the surface roughness and elasticity, would enable quantifiable prediction of the volume of the particles in close contact. We suggest this is likely to contribute to the apparent decrease in $\Phi < \Phi_0$. It is likely that surface interactions are only part of the decrease in effective phase volume and shrinkage of the particle as a whole also occurs.

At phase volumes where multiple particle interactions occur, inter-particle forces are dominated by elastic deformation of contacting particles with the elasticity being related to the particle modulus by $G_P = E/[2(1 + \nu)]$ with E and ν the elastic modulus and Poisson's ratio. Yan et al.[21] used micromanipulation technique to apply a force to single 4% agarose microgel particles and measured the deformation. Comparing their results to available models for linear elastic and viscoelastic deformation, they found the Hertz model, shown in equation (6) , to be applicable to deformation up to 30% with E and ν estimated from bulk gel measurements. They propose two independent relaxation processes occurring as force is applied to the particle, the first being the transport of water from the centre of the microgel into the suspension and the second being related to the viscoelastic nature of the gel network.[21] An applied force (F) of ~5 μN resulted in relative displacement of the plates (Δ) of ~3.5 μm, of the same order of magnitude as that calculated for the 0.5% agarose microgels above. Elastic deformation resulting in water loss could potentially be the sole source of the decrease in Φ. However, osmotic deswelling is another potential reason shrinkage that is considered next.

$$F = \frac{\sqrt{2RE}}{3(1-\nu^2)}\Delta^{3/2} \tag{6}$$

Agarose microgels were chosen as a model system in part because it was reported that they do not shrink with osmotic pressure.[22] However further study has shown that although agarose gels are neutral, they are still able to osmotically deswell when there is sufficient difference in osmotic pressure between the solvent and the gel particle.[23] In agarose gels the rate of deswelling is limited by G_P, as shown by Viallat et al[24] in their study of sucrose induced osmotic deswelling of gel filled lipid vesicles. As the particle shrinks the particle modulus increases[23], although the relationship is not easy to define as preferential water loss from the surface may form a 'skin' on the particle surface as shown in Figure 5.[24]

Both osmostic deswelling and that induced by compressive forces depend on the particle shear modulus (G_P) and elastic properties (E and ν). Many authors [5, 25] assume the modulus of the microgel particle to be the same as that for the bulk gel, which has the potential to be incorrect if the polymer density across the particle is not homogenous or it increases as a result of compression and water loss. It is critical for accurate modelling of suspension behaviour that the single particle properties be accurately measured under a variety of suspension and environmental conditions. A capillary micromechanics approach has been developed by Wyss et al.[26] to determine compressive modulus (K) and the shear modulus (G) of soft particles. Our future work will aim to develop an understanding of micromechanics at the level of individual microgels to investigate the influence of osmotic pressure and compressive forces on their volume reduction, and we will seek to develop scaling relationships to establish whether it is directly related to the change in slope shown in figure 4.

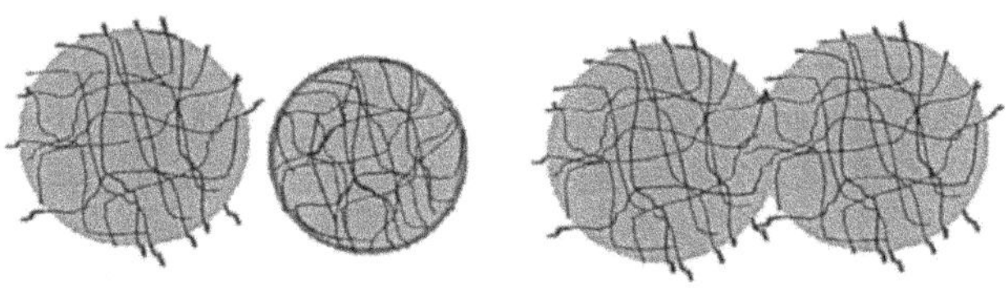

Figure 5 *Schematic showing possible microgel surface architecture from surface with 'loose' polymer strands to a compressed particle with a dense surface 'skin' and two particles with overlapping surface structures.*

5 CONCLUDING REMARKS

Microgel suspension rheology results can be ambiguous when presented as a function of dry weight fraction when the key defining property is phase volume. A more accurate value for Φ_0 can be determined using Batchelor's method than using Einstein's equation as it is applicable up to higher suspension phase volumes of ~ 0.3. Below Φ_0 ~ 0.4 soft microgel suspension rheology is the same as for hard spheres independent of the particle modulus. Above Φ_0 ~ 0.4 particle modulus affects microgel suspension rheology, resulting in an increase in Φ_m of the suspension. Linearisation of the Quemada model is a novel method to show the extent to which soft particle suspension rheology deviates from hard sphere behaviour dependent on particle modulus. The cause of this deviation is due to breakdown of the underlying assumption that $\Phi = \Phi_0$. We suggest that above an effective phase volume of ~0.4 the particle has reached maximum deformation and begins to shrink, that is $\Phi < \Phi_0$. The approach applied here to microgel suspensions may be applicable to other soft particle suspensions and we suggest that use of the linearised Quemada model to determine the change in particle volume is another useful tool for defining the true phase volume – rheology relationship. Future work will focus on developing a greater understanding of the microstructural mechanism for particle deformation and shrinkage to quantitatively understand the relationship between particle morphology and suspension conditions.

References

1. J. R. Stokes and W. J. Frith, *Soft Matter*, 2008, **4**, 1133-1140.
2. J. R. Stokes, in *Microgel Suspensions – Fundamentals and Applications*, eds. Fernandez-Neives A., Wyss H., Mattson J. and Weitz D.A., Wiley-VCH Verlag, Weinheim, 2011.
3. B. H. Tan, R. H. Pelton and K. C. Tam, *Polymer*, 2010, **51**, 3238-3243.
4. L. Day, M. Xu, S. K. Oiseth, L. Lundin and Y. Hemar, *Colloids Surf B Biointerfaces*, 2010, **81**, 461-467.
5. S. Adams, W. J. Frith and J. R. Stokes, *Journal of Rheology*, 2004, **48**, 1195-1213.
6. H. V. Dang, C. Loisel, A. Desrumaux and J. L. Doublier, *Food Hydrocolloids*, 2009, **23**, 1678-1686.
7. J. F. Steffe, *Rheological Methods in Food Process Engineering*, Second edn., Freeman Press, East Lansing, MI 1996.
8. M. S. Wolfe and C. Scopazzi, *Journal of Colloid and Interface Science*, 1989, **133**, 265-277.
9. V. Normand, D. L. Lootens, E. Amici, K. P. Plucknett and P. Aymard, *Biomacromolecules*, 2000, **1**, 730-738.
10. P. Burey, B. R. Bhandari, T. Howes and M. J. Gidley, *Crit. Rev. Food Sci. Nutr.*, 2008, **48**, 361-377.
11. A. Lips, P. M. Hart and A. H. Clark, *Food Hydrocolloids*, 1988, **2**, 141-150.
12. G. A. Davies and J. R. Stokes, *Journal of Rheology*, 2005, **49**, 919-922.
13. G. A. Davies and J. R. Stokes, *Journal of Non-Newtonian Fluid Mechanics*, 2008, **148**, 73-87.
14. B. H. Tan, K. C. Tam, Y. C. Lam and C. B. Tan, *Polymer*, 2004, **45**, 5515-5523.
15. I. M. Krieger and T. J. Dougherty, *Transactions of the Society of Rheology*, 1959, **3**, 137-152.
16. C. Prestidge and T. F. Tadros, *Journal of Colloid and Interface Science*, 1988, **124**, 660-665.
17. P. F. Luckham and M. A. Ukeje, *Journal of Colloid and Interface Science*, 1999, **220**, 347-356.
18. R. G. Larson, *The Structure and Rheology of Complex Fluids*, Oxford University Press, New York, 1999.
19. H. A. Barnes, J. F. Hutton and K. Walters, *An introduction to rheology* Elsevier, Amsterdam, 1989.
20. D. A. Sessoms, I. Bischofberger, L. Cipelletti and V. Trappe, *Philos. Trans. R. Soc. A-Math. Phys. Eng. Sci.*, 2009, **367**, 5013-5032.
21. Y. Yan, Z. B. Zhang, J. R. Stokes, Q. Z. Zhou, G. H. Ma and M. J. Adams, *Powder Technology*, 2009, **192**, 122-130.
22. W. J. Frith, A. Lips and I. T. Norton, *Structure and Dynamics of Materials in the Mesoscopic Domain*, 1999, 207-218.
23. R. T. Bonnecaze and M. Cloitre, in *High Solid Dispersion*, ed. M. Cloitre, Springer-Verlag, Berlin Heidelberg, 1 edn., 2010, pp. 117-161.

24. A. Viallat, J. Dalous and M. Abkarian, *Biophysical Journal*, 2004, **86**, 2179-2187.
25. I. D. Evans and A. Lips, *Journal of the Chemical Society, Faraday Transactions*, 1990, **86**, 3413-3417.
26. H. M. Wyss, T. Franke, E. Mele and D. A. Weitz, *Soft Matter*, 2010, **6**, 4550-4555.

PREPARATION OF STABLE CMC-ACID GEL

M. Takigami[1], N. Nagasawa[2], A. Hiroki[2], M. Taguchi[2] and S. Takigami[3]

[1] TGRI, 2-23, Higashihisakata 1-chome, Kiryu, Gunma 376-0053, Japan.
E-mail : smtakigami623@kuc.biglobe.ne.jp
[2] Radiation Processing Research Group, Japan Atomic Energy Agency, 1233 Watanuki, Takasaki, Gunma 370-1292, Japan.
[3] Center for Material Research by Instrumental Analysis, 1-5-1, Gunma University, Kiryu, Gunma 376-8515, Japan.

1 INTRODUCTION

We found sodium carboxymethyl cellulose (CMC-Na) formed a strong elastic gel when mixed with acid[1]. The mechanism of gel formation was elucidated as follows. When CMC is mixed with acid, replacement of sodium in carboxymethyl group with hydrogen takes place and then aggregation of CMC molecules occurs as a result of reduction of CMC solubility in water. The aggregation of CMC molecules causes hydrogen bond formation between CMC molecules. Any CMC-Na with a degree of substitution (DS) of more than 0.6 can form a gel. Besides, molar mass[2] and concentration of CMC-Na, species and concentration of acid were important factors to prepare gels. Properties of the formed gels depended on the combination of CMC-Na and acid.

Generally speaking, the number of crosslinks effects the mechanical properties of gels[3]: Small number of crosslinks gives a weak gel with high water absorbency and a large number of crosslinks provides a strong gel with less water absorbency. Too many crosslinks result in a brittle gel. Acid species and concentration, gelling time and gelling temperature are factors giving effects on the number of crosslinks.

Here we are going to discuss the factors required to form stable gels which have durability and good mechanical properties.

2 METHODS AND RESULTS

2.1 Preparation of CMC-Acid Gel Using Hydrochloric Acid

CMC Daicel 1380 (Daicel Chemical Industries, Ltd. Japan, viscosity of 1% solution at 25 ℃ =1640 mPa· s, DS= 1.34) , hydrochloric acid (Wako Pure Chemical Industries, Ltd., Japan, reagent grade) and deionized water were used for experiments. CMC and hydrochloric acid aqueous solution were mixed using a planetary centrifugal mixer (ARE-310, Thinky Corp., Japan). The concentration of CMC was 20% and that of acid was changed from 0.25 to 2 M. After mixing, the mixture was moulded to a cylindrical shape with a diameter of 20 mm and height of 10 mm. The mould was put in a plastic bag to avoid weight loss during storage and kept in an oven at 30 ℃ for a given time . After incubating the samples, they were immersed in water to remove soluble material, i.e., uncrosslinked CMC, hydrochloric acid and sodium chloride. Water was exchanged several times until the conductivity of water outside the gel was almost the same with that of the deionized water. Gel was defined as the portion which remained on a 180 μm stainless steel mesh when the wet gel was passed through the mesh. The remaining fraction was collected, weighed (W_{w+g}) and dried at 50 ℃ and reweighed (W_g).

The gel fraction was defined as a weight percentage of the insoluble fraction (W_g) to the initial weight of CMC (W_i) used as a raw material to form gel and was calculated according to the equation.

$$\text{Gel fraction (\%)} = 100 \times W_g / Wi$$

The results are shown in Figure 1. The gelled samples do not dissolve in water. Gels were not formed when 0.25 M hydrochloric acid was used and incubated at 30 ℃. On mixing CMC with 2 M hydrochloric acid, a gel quickly formed. Gel formation depended on acid concentration. With increase of acid concentration, gelling proceeded more rapidly. The time required for gelling of the major part of CMC using 0.5 and 1 M hydrochloric acid were 7 and 1 day, respectively. The appearance and nature of the gels depended on the concentration of hydrochloric acid. The gel was turbid and brittle when 2 M hydrochloric acid was used. It had a similar texture to that of nata de coco (fermented coconut). Other gels were transparent and elastic.

Water absorption by the gel was defined as the weight ratio of absorbed water (g) in the gel to that of the gel (g) as shown in the following equation.

$$\text{Water absorption by gel (g/g)} = (W_{w+g} - W_g) / W_g$$

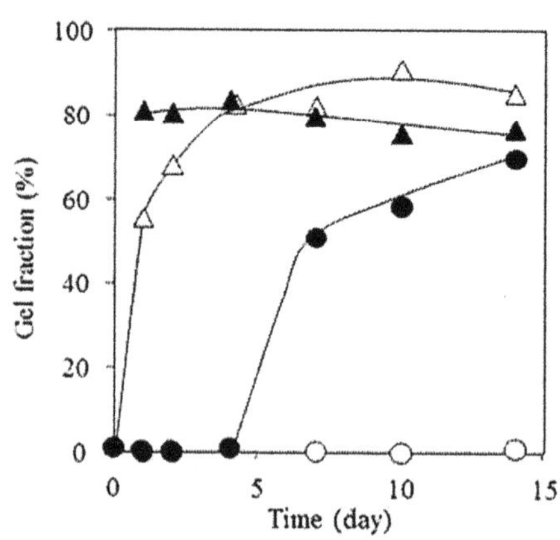

Figure 1 *Gel fraction of CMC- acid gel incubated at 30 ℃. Concentration of hydrochloric acid : 0.25 M (○), 0.5 M(●), 1.0 M(△), and 2.0 M(▲).*

As shown in Figure 2, water absorption by the gel decreased with acid concentration and incubation time. The result infers that crosslinking depends on the incubation time and acid concentration. The number of crosslinks increases with increase of acid concentration and incubation time. The relation of gel fraction and water absorption is shown in Figure 3. With the increase of the gel fraction, water absorption decreased significantly depended on acid concentration. Gels prepared by using 0.5 M hydrochloric acid had more water absorption, i.e. less crosslinking density than gels prepared by using 2 M hydrochloric acid. High crosslinking density causes less water absorption by the gel.

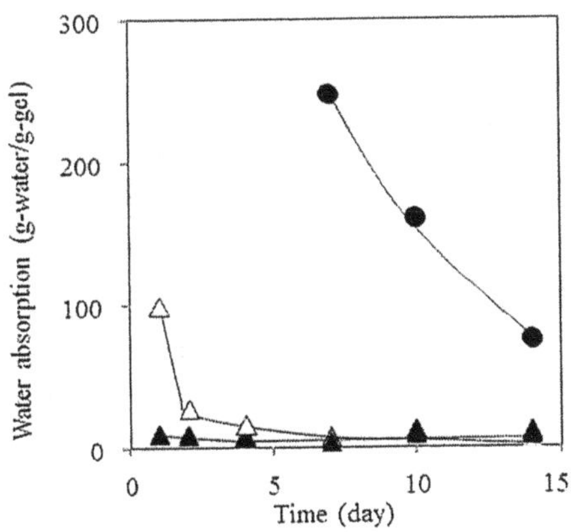

Figure 2 *Water absorption by gel. Captions are the same with those given in Figure 1.*

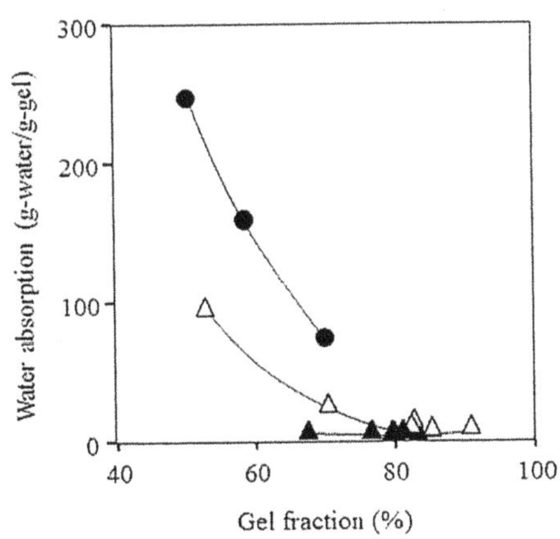

Figure 3 *Relation between gel fraction and water absorption by gel. Captions are the same with those given in Figure 1.*

FT-IR spectra of CMC and gels prepared by mixing CMC and 0.5 M hydrochloric acid and stored at 30 ℃ were measured by ATR using a Magna 560+ Continuμ MII (Nicolet, Thermo Fisher Scientific K. K.). The spectra are very similar regardless of the incubation time. Absorbance around 1600 and 1730 cm^{-1} can be attributed to COONa and COOH. Although the exchange of sodium with hydrogen is a rapid process, crosslinking, namely hydrogen bond formation, proceeds gradually with time.

The physical properties of the moulded samples were measured using a table-top universal tester (EZ-test S, Shimadzu, Japan). Moulded sample containing soluble fractions was compressed with a wedge-shaped jig with a blade length of 20 mm, until the sample was ruptured. Initial compression modulus (1-2N) and force required for rupture was measured and shown in Figures 5 and 6 respectively, after dividing the measured values by the length of the jig blade.

The modulus showed a maximum point for all gels, however, the maximum appeared highly and early in gels prepared with higher concentration of hydrochloric acid. In gels prepared using 2 M acid, the change in modulus was drastic compared to the other gels.

The strength of gels at rupture was completely different for the gels prepared with different concentrations of acid. With 0.25 M hydrochloric acid the gel was not formed regardless of storage time. With 0.5 M hydrochloric acid, the strength increased with time and reached a maximum on the tenth day. With 1 M acid, gel was formed rather quickly and reached a maximum on the fourth day. As mentioned above, gel prepared with 2 M acid was not strong compared to those prepared using 0.5 and 1 M acid and it lost strength rather quickly. In order to apply CMC-acid gel to practical use, it should be stable in strength for a certain period.

The results mentioned above show highly concentrated acid is not suitable to prepare gels with stable physical properties and that it is important to moderate the number of crosslinks to have elastic gels.

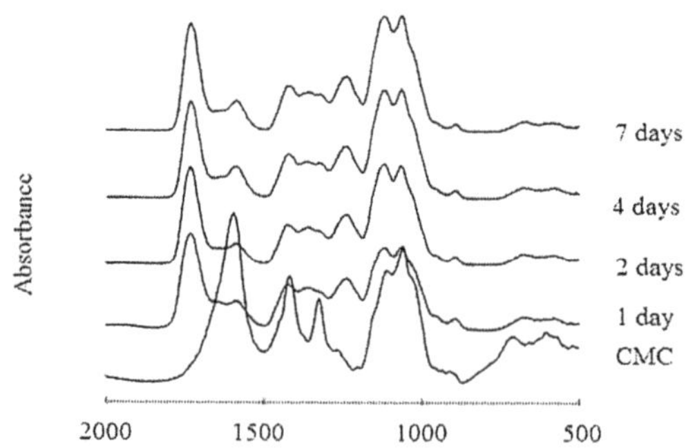

Figure 4 *FT-IR spectra of CMC-Acid gel prepared using 0.5 M hydrochloric acid and incubated at 30 ℃.*

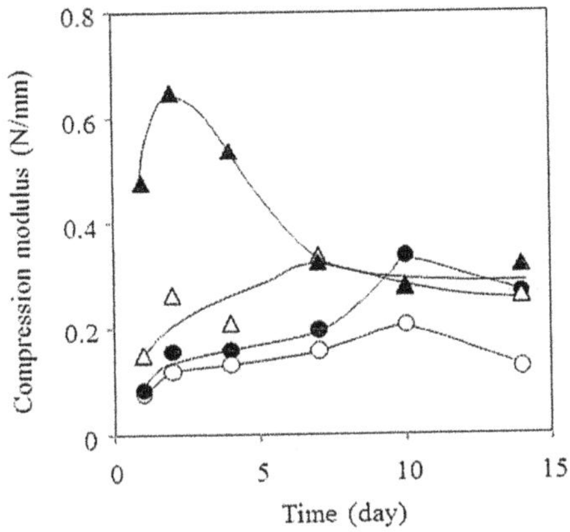

Figure 5 *Initial compression modulus of CMC-acid gel prepared using acid with different concentration Captions are the same with those given in Figure 1.*

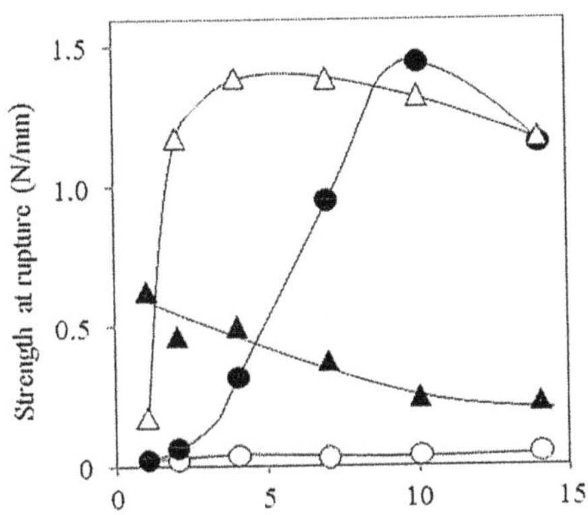

Figure 6 *Strength of CMC-acid gel at rupture. Gel was prepared using acid with different concentration. Captions are the same with those given in Figure 1*

2.2 Preparation of CMC-Acid Gel Using Citric Acid

In order to have some stable CMC-acid gels, the concentration of acid should be as low as possible. However, CMC did not form gels with 0.25 M hydrochloric acid as mentioned above. We have reported that the pH of acid should be less than 1 and 2 for CMC-acid gel formation using inorganic and organic acid, respectively[3]. Instead of hydrochloric acid citric acid was used in this section. CMC with high molar mass forms gels more quickly[2]. Therefore, CMC manufactured by Dai-ichi Kogyo Seiyaku Co., Ltd. (Cellogen HE-1500F, viscosity of 1% solution at 25 ℃ =3000 mPa· s, DS= 1.3) was used for the following experiments. Citric acid was a product of Wako Pure Chemical Industries, Ltd. and was Japanese food additives grade. It was used as received.

CMC was mixed with citric acid aqueous solution with various concentrations and moulded as mentioned in 2.1. The moulded samples were incubated at various temperatures, from 30 to 70 ℃ for 7 days. The results are summarized in Table 1. Gel was not formed using 0.2 M citric acid at 30 to 70 ℃ . Gel was formed when citric acid concentration was more than 0.25 M, however, it took time and required a higher incubation temperature. Gel was not formed at 30 ℃, but did form at 50 ℃ using 0.25 M citric acid. The time required to form the gel decreased with increase of citric acid concentration and temperature except for 70℃. At 70℃, gel was not formed even at higher citric acid concentration, rather it seemed hydrolysis took place.

A concentration of citric acid of 0.25 M, seems to be critical for gel formation. The molar ratio of carboxyl groups in 0.25 M citric acid and sodium existing as a counter ion of carboxymethyl groups is about 1 when CMC and 0.25 M citric acid solution were used in 10 to 90 weight ratio.

The gel fraction is shown in Figure 7 for gels prepared using citric acid with different concentrations and incubated at 50 ℃. In 3 days almost all CMC converted to a gel using citric acid with concentrations higher than 0.25 M.

The physical properties of the moulded samples were measured using a table-top universal tester (EZ-test L, Shimadzu). Samples before removing soluble materials were compressed to 1.6 kN using a disk type jig. Breaking strengths of the samples are shown in Figure 8 after dividing the strengths with cross-sectional areas of the samples before compression. Samples prepared using 0.25 and 0.3 M citric acid were not broken even when compressed to 1.6 kN after incubation at 50 ℃ for 20 days. Samples prepared with 0.5 M citric acid were not broken for 6 days. After 8 days incubation, they were broken. They became brittle little by little. In order to keep the physical properties of samples stable for a long time, samples should be prepared using 0.25 – 0.3 M citric acid.

The FT-IR spectrum of a gel prepared with 0.25 M citric acid is shown in Figure 9 together with the initial CMC-Na and CMC whose carboxyl group is completely converted to COOH. The carboxyl group in the gel prepared with 0.25M citric acid after incubation for 15 days at 50 ℃ was not completely converted to COOH. Some of the groups remained as COONa. Using acid with a lower concentration to form gel does not promote too much hydrogen bond formation. Not all carboxylmethyl groups in CMC-Na turned to COOH, although there are enough carboxyl groups in citric acid. Some of the carboxyl groups in citric acid remain as COOH and some turned to COONa. The ratio of COOH/COONa in the gel seems to be important for stable gel formation.

Table 1 *Gelling of CMC incubated at different temperatures. -: no datum.*

Temperature/℃⁻¹ *Acid Concentration/M⁻¹*	*30*	*50*	*60*	*70*
0.2	No gel	No gel	No gel	No gel
0.25	No gel	2 days	-	-
0.3	No gel	2 days	-	-
0.4	No gel	1 day	1 day	No gel
0.5	4 days	1 day	-	-

The role of temperature during incubation was investigated. Raising temperature accelerates gelling, that is, it promotes hydrogen bond formation. Acid with lower concentration can be used to prepare gels, if the gel is incubated at higher temperature around 50 ℃. Incubation at too high temperature, namely higher than 70 ℃, hydrolyses CMC and gives no gel.

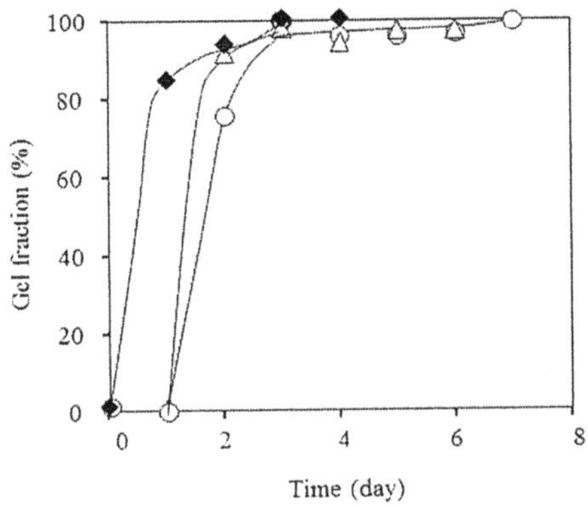

Figure 7 *Gel fraction of CMC-acid gel incubated at50 ℃,. Concentrations of citric acid are 0.25 M (○), 0.3 M (△) and 0.5M(◆).*

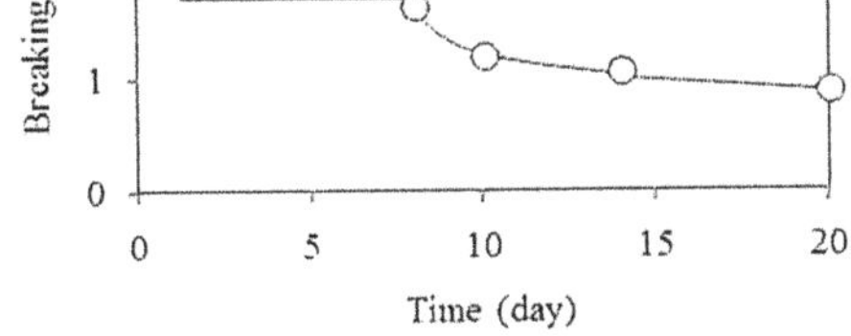

Figure 8 *Breaking strengths of gels prepared with 10% CMC and 0.5 M citric acid and kept at 50 ℃.*

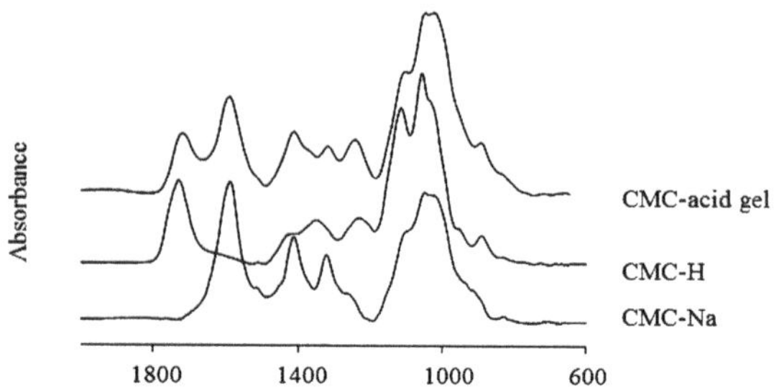

Figure 9 *FT-IR spectra of CMCs in sodium salt (CMC-Na), acid form (CMC-H) and CMC-acid gel prepared using 0.25 M citric acid.*

3 CONCLUSION

Acid concentration and temperature at incubation are important factors to prepare gels with stable physical properties. Decreasing acid concentration and raising temperature at preparation of gels provide stable gels. All conversion from COONa to COOH in CMC gives too many crosslinkings to gel and results in unstable and brittle gel. Some carboxyl groups in CMC remain in the form of COONa in stable gels.

References

1 M. Takigami, H. Amada, N. Nagasawa, T. Yagi, T. Kasahara, S. Takigam and M. Tamada, *Trans. . Mater. Res. Soc. Jpn.* 2007, 32, 713.

2 M. Takigami, A. Hiroki, N. Nagasawa, T. Kasahara, S. Takigami and M. Tamada, *Trans. . Mater. Res. Soc. Jpn.* 2009, 34, 391.

3 M. Takigami, N. Nagasawa, Y. Maehara, S. Takigami and M. Tamada, *Gums and Stabilisers for the Food Indstry 15*., ed. Peter A. Williams and Glyn O. Phillips, RSC Publishing, Cambridge, 2009, p.427.

ON THE LUBRICATING BEHAVIOUR OF KAPPA CARRAGEENAN GELLED NETWORKS

D.A. Garrec and I.T. Norton

Centre for Formulation Engineering, Department of Chemical Engineering, University of Birmingham, Edgbaston, Birmingham B15 2TT, UK

1. INTRODUCTION

The carrageenans are a family of polysaccharides extracted from red-seaweed and are well known for their ability to form thickened liquids and solid gels[1]. The kappa-carrageenan isoform may be used to create aqueous gels in the presence of K^+ ions via a process of molecular ordering. The ordering is understood to occur via a domain model that proceeds via a coil to helix transition, followed by helix side-by-side association[2]. The resulting gels may have a range of properties dependent not only on the carrageenan polymeric structure and molecular weight, concentration, ionic content, and cooling rate, but also on the shear field applied during gelation.

Fluid gels (FGs) are produced by applying shear to a biopolymer solution undergoing gelation. They are known to be a solid suspension of gelled particles dispersed in a non-gelled continuous matrix, and as such they behave as fluids[3]. FGs therefore have very different rheological properties to their counterpart quiescent gels (QGs) that form when the gelation process is allowed to occur quiescently i.e. without an applied shear force[4]. The pronounced distinction between FG and QG material properties is a consequence of their differences in extent of molecular cross-linking. Shear forces restrict the distances that cross-linked networks can span, as well as the length of helices, and thus aggregation regions[5]. During FG production the growth of particles via propagating helix formation, and particle coalescence, is opposed by shear forces causing segregated gel nucleation sites, and particle break-up. When gels are formed quiescently the gelled networks are able to expand over much greater volumes and cross-link the entire system resulting in the production of solid gels. FGs can, however, be formed with inter-particle connections by allowing the molecular ordering process to take place initially under shear, thus forming FG particles, and the latter stage quiescently, to allow disordered polymeric chains at the surface of particles to order with those of neighbouring particles[6].

The nature of FG particles, in terms of internal polymeric network and thus elasticity, has been probed using multi particle tracking microrheology (MPTM)[7]. In their study, Caggioni et al. set tracer particles within FG particles and bulk QGs. Their MPTM data suggested that the tracer particles existed in similar environments, and that the nature of the polymeric networks in FG particles is the same as within QGs. This can be further confirmed by the fact that the storage modulus (G') of bulk FG samples have the same dependence on concentration as do QGs[4].

It is appreciated[8, 9] that the ability of fat-containing foods to lubricate motions made by the tongue and palate during consumption can influence fat-related textural attributes.

Thus solid-surface lubrication by gels is important to their application in reduced fat food formulation. Soft-tribology is a well characterised technique[10] and can be used to correlate friction coefficients with sensorially obtained data[11].

Textural attributes perceived during the consumption of gelled products can potentially be enhanced by understanding the way in which lubrication is influenced by product microstructure, and therefore formulation methods. The entrainment of polymeric networks to confined contacts is poorly understood and so the influence of the extent of cross-linking within gels on lubrication is investigated herein.

2. MATERIALS AND METHODS

2.1 Sample Preparation

Fluid gels were produced using kappa-carrageenan (κC), a gift from Unilever, and KCl, Sigma Aldrich, both of which were used without further purification. κC was first dispersed in agitated de-ionised water containing the required KCl content to produce 0.3% solutions. These were stirred at room temperature for 1 hr before being heated to, and held at, ~ 85 °C for a further 1 hr. The heated κC solutions were then fed into a heated jacketed vessel (to control the pin-stirrer entry temperature) before entering a jacketed pin-stirrer. The pin-stirrer that provided the shear required for FG production was used at 1438 rpm for all tests, whilst cooling from 50 to 10 °C. All samples were collected in vessels cooled below 10 °C and stored at 5 °C for at least 24 hrs - enough time to allow post processing effects[12] to complete - before testing.

2.2 Polarimetry

The degree of rotation of polarised light can be used to monitor the extent of gel ordering[13, 14]. Angles of rotation (α) were measured at 405 nm using an Anton Paar Gyromat automatic Polarimeter in a 5 cm path-length cell. Temperatures were measured using a thermocouple situated within the fluid, but out of the light path, and were controlled by flowing fluid from a recirculating water bath through the cell annulus.

2.3 Rheology

Rheological measurements were made using a TA Instruments AR-G2 stress controlled rheometer. Oscillatory experiments were conducted using a sandblasted bottom plate and sandblasted 40 mm parallel plate at a shear stress of 10 Pa (determined to be within the linear viscoelastic region of all samples) at 1 Hz using a 1 mm gap. Samples were pre-sheared to destroy inter-particle connections[6], 20 mins prior to testing. Rheological data shown herein are the mean of three repeats, all conducted at 10 °C.

2.4 Tribology

A MTM2 (Mini Traction Machine) tribometer (PCS Instruments) was used to measure friction coefficients between rotating surfaces. In these experiments, a ball is lowered to contact a disc, where they are both independently driven to allow fluid to be entrained into the contact and friction to be recorded as a function of entrainment speed (U). The relative sliding to rolling motions can be controlled, and are described by the slide-to-roll ratio

(*SRR*). In a typical Stribeck experiment a normal load, W, is applied by the ball to the disc, and the friction coefficient, μ, ($\mu = F / W$) of the lubricated contact is measured as a function of entrainment speed, U, ($U = U_{ball} - U_{disc} / SRR$) where F is the recorded tangential friction force, and U_{ball} and U_{disc} are the ball and disc speeds, respectively. The resulting Stribeck curves typically show up to three regimes: boundary, mixed and hydrodynamic. Boundary regimes display μ's independent of U, and the applied normal load is fully supported by the asperity contact and adsorbed matter. As U increases, fluid is able to entrain and the load is supported by both asperity contact and the build up of a fluid film. This is known as the mixed regime and μ is seen to decrease with U. At very high U, or lubricant viscosity, the surfaces become separated and the load is fully supported by the build up fluid within the contact; here μ increases with U as the lubricant is sheared. For an idealised schematic of a Stribeck curve, and more information on the technique, the reader is referred to Garrec et al.[15]

In the present study, soft PDMS surfaces (polydimethyl siloxane) elastomer kit (Sylgard 184, Dow Corning) were set in moulds; balls and discs were cleaned by gently rubbing with washing-up liquid, and sonicating in deionised water for 5 mins. The ball and disc were securely attached to the relevant rotating parts, the fluid to be tested was added, and the ball was lowered to the disc. Stribeck experiments were conducted from entrainment speeds spanning 1 to 1000 mm s^{-1}, with W = 2 N and SRR = 50%. These parameters were used as they were determined in a previous study to yield data of high reproducibility and discrimination between samples. Tests were completed by sequentially running U up and down throughout the speed range, until 6 runs took place. Stribeck curves shown are the mean of three tests i.e. 18 runs, all conducted at 10 °C. To monitor frictional changes during gel formation a timed experiment with a temperature ramp was employed. This was conducted at W = 2 N and U = 500 mm s^{-1} (shown to be mixed regime) whilst cooling at a rate of approximately 0.77 °C/min.

3. RESULTS AND DISCUSSION

3.1. Fluid gel formation

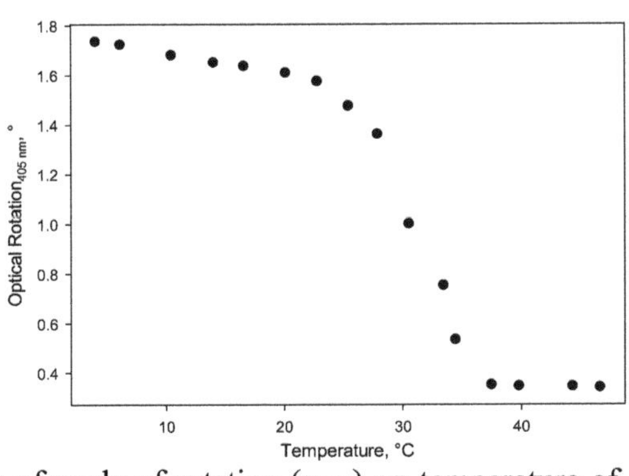

Figure 1. Dependence of angle of rotation (α_{405}) on temperature of quiescently cooled κC 0.5% + KCl 0.3%.

Fig. 1 shows optical rotation with temperature of a κC 0.5% + KCl 0.3% solution being quiescently brought into gelation by cooling. It is known[6] that to produce smooth κC FGs,

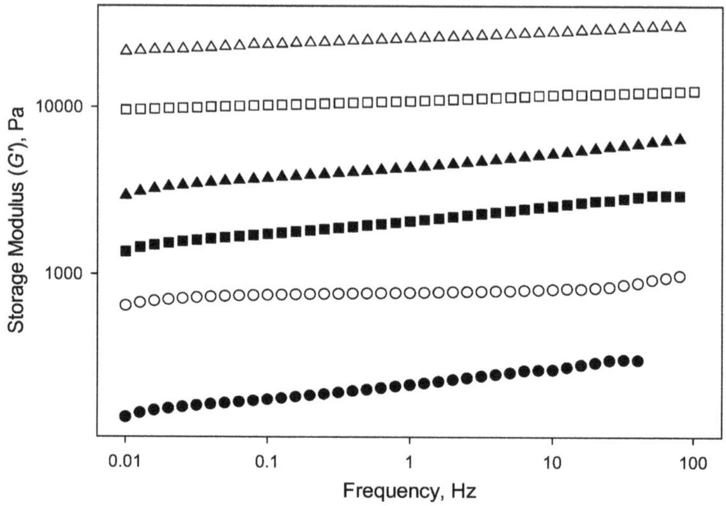

Figure 2. Frequency sweeps of 0.3% KCl FGs (filled symbols) and QGs (empty symbols) of κC concentration 0.5% (circles), 1% (squares) and 2% (triangles). All tests were conducted at 1 Hz and 10 Pa (within the linear viscoelastic region for all samples).

i.e. particles/aggregates smaller that 10 μm, that the relevant shear must be applied whilst the molecular ordering takes place from the onset (~ 37.5 °C, Fig. 1) until at least the beginning of the linear ordered region (~ 20 °C, Fig. 1). Thus the pin-stirrer inlet and exit temperatures are required to be ≥ 37.5, and ≤ 20 °C, respectively, to produce smooth FGs using κC 0.5% + KCl 0.3%. FGs were made using 0.3% KCl and three κC concentrations: 0.5, 1 and 2%. All were produced with pin-stirrer inlet and exit temperatures at 50 and 10 °C, respectively, to ensure smooth FG production. The frequency dependence of the storage modulus for the FGs, and their equivalent QGs (same salt and carrageenan compositions), are shown in Fig. 2. The QGs were formed by quiescently cooling within a rheometer at 3 °C/min, and were subjected to a 15 minute equilibration period prior to testing. Fig. 2 shows that the FGs behave similarly to 'weak-gels', i.e. the storage modulus shows a strong, but linear, frequency dependence, whilst the QGs display the characteristic 'strong-gel' profile, i.e. weak dependence on frequency[16]. As expected, the FGs have lower storage moduli than their equivalent QGs due to restricted cross-linking caused by the applied shear that creates a particulate suspension[7]. Storage moduli also increase with κC concentration ([κC]) as the particle elasticity, and volume fraction, both increase[5].

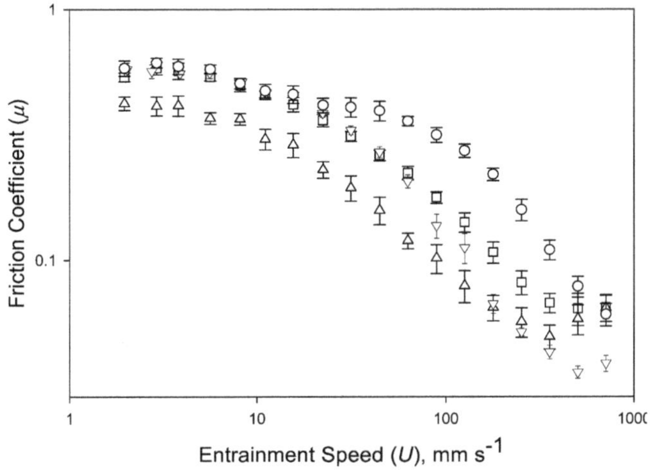

Figure 3. Stribeck curves (friction coefficient vs. entrainment speed) for 0.3% KCl FGs formed with κC concentration 0.5% (○), 1% (□) and 2% (Δ). A Stribeck curve for a diluted FG is also shown: 2% κC FG/water 75:25 (▽).

3.2 Lubrication by κC FGs

The same FGs were also tested tribologically to assess their lubricating behaviour. Fig. 3 shows the Stribeck curves obtained from these samples and from a 2% κC FG/water 75:25 diluted sample. All samples tested display both boundary and mixed regimes which are determined by observing regions at low entrainment speeds where μ is independent of U, and where μ decreases with U, respectively. The 2% κC FGs (both non-diluted and diluted) were viscous enough to separate the ball and disc at speeds over 500 mm s^{-1}, hence the observed hydrodynamic regimes where μ increases with U. Friction coefficients are

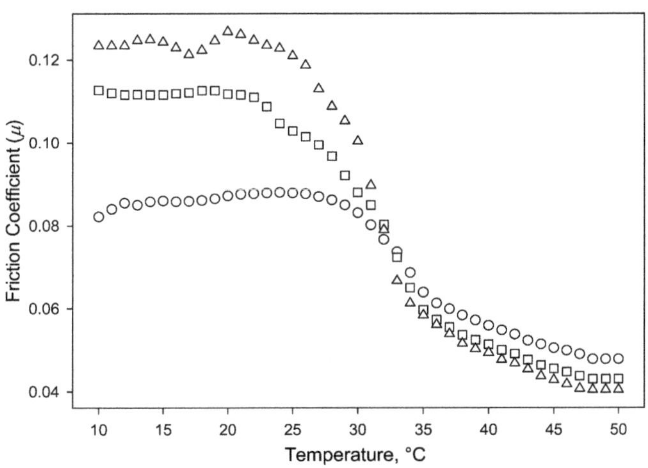

Figure 4. Friction coefficient vs. temperature plots for 0.3% KCl solutions with 0.5% (○), 1% (□) and 2% (Δ) κC. Friction measured at 500 mm s^{-1} (mixed regime).

reduced in boundary and mixed regimes with increases in κC concentration (known to increase particle elasticity and volume fraction[5]) and number of FG particles (compare 2% κC with 2% κC FG/water 75:25). Both of these findings indicate that the particles are entrained into the ball and disc contact at all speeds, and are beneficial to lubricity.

3.3 Impact of molecular assembly on entrainment for lubrication

Testing the lubricating properties of quiescent gels, for the purpose of understanding the influence of applied shear during gelation on lubricity, poses several difficulties. If a QG were to be placed on top of a disc, before lowering the ball, then a solid contact would be tested, and only a boundary regime should be observed. If a QG were to be broken up into small pieces to allow entrainment, then the extent of molecular ordering throughout the sample would be influenced, thus defeating the purpose of the study. To understand the influence of QG network strength on lubricity, and in comparison to FGs, heated solutions of disordered κC coils were placed in the tribometer reservoir, where they were then cooled to allow molecular ordering to take place, whilst measuring the friction coefficient at 500 mm s^{-1} (mixed regime). The resulting gels are 'pseudo quiescent gels' because some shear was provided during the molecular ordering process because the surfaces were rotating to calculate μ's. The results of this test using 0.5, 1 and 2% κC, all with 0.3% KCl, are shown in Fig. 4.

It should be mentioned that a 'volume reducing steel cell' was used to ensure that material was entrained into the contact; however, this created a barrier between the ball-and-disc contact and the tribometer thermocouple, thus the temperatures plotted are not accurately indicative of the sample temperature being tested. At high temperatures (> 32.5 °C), greater κC concentrations give rise to greater solution viscosities and therefore lower μ's (Fig. 4), as expected in the mixed regime. At lower temperatures however (<32.5 °C), the relation between μ and [κC] is reversed i.e. greater concentrations yield greater friction coefficients. Thus gelation results in a reduction of lubrication at the contact, where stronger gels result in greater reductions. Therefore, when molecular ordering is allowed to occur macroscopically i.e. without an applied shear force, the amount of fluid available for entrainment to the contact is restricted as the carrageenan chains build networks surrounding the contact. The observed increase in μ with [κC] of pseudo quiescently cooled gels is the opposite of that observed for FGs, where increasing [κC] was shown to reduce μ in both boundary and mixed regimes. Therefore, allowing the molecular ordering process to occur within a sheared environment forming FGs, where the polymeric molecular ordering is restricted within discrete particles, allows [κC] to be increased to enhance viscosity and storage modulus whilst at the same time, and most crucially, increasing lubricity.

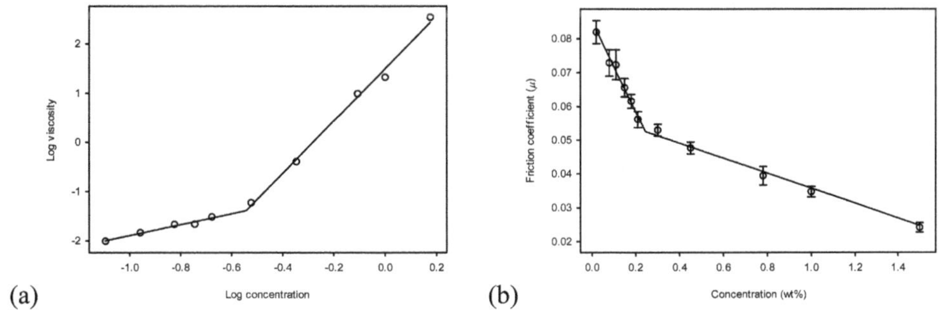

Figure 5. Zero shear viscosity (a) and friction coefficient at 251 mm s^{-1} (b) vs. guar gum concentration.[15] The viscosity plot shows C^* = 0.28 wt%; the friction coefficient plot shows $C_T^* = 0.24$ wt%.

Restrictions of fluid entrainment for lubrication have also been observed by studying the tribology of guar gum (GG) solutions, a non-gelling hydrocolloid. By increasing GG concentration ([GG]), a molecular transition occurs from dilute regions, where GG chains exist as single entities, to entangled networks at high concentrations where GG chain interpenetration occurs. C^* marks the critical concentration where this transition occurs and can be found experimentally by observing a pronounced increase in the concentration dependence of viscosity after C^*. Fig. 5a shows the viscosity/concentration dependence of GG, the intercept between the two power-law fitting lines marks C^* (0.28 wt%). The friction coefficient/concentration dependence of GG (at ~ 250 mm s^{-1}, mixed regime) is shown in Fig. 5b. Here, a tribologically critical concentration (C_T^*) is observed (at 0.24 wt%), where the concentration dependence of μ (gradient) is reduced after C_T^*. Since C_T^* ~ C^*, where C^* marks the onset of chain interpenetration, it seems that the reduction in the dependence of concentration on friction after C_T^* is due to the formation of entangled networks at high concentrations. This was attributed to molecular entanglement restricting fluid entrained to the contact[15], and is analogous to the effect seen in Fig. 4 where entrainment is hindered by molecular cross-linking during the κC sol-gel transition.

4. CONCLUSIONS

Macroscopic biopolymer assembly has been shown to hinder lubrication to a ball-on-disc contact, when compared to their non-assembled counterparts, and is suggested to occur as a consequence of restricted material entrainment. This behaviour was observed when analysing friction coefficients of guar gum solutions throughout a concentration range that spanned from dilute solutions to interpenetrated entangled polymer networks, and during the coil-helix molecular ordering process of κ-carrageenan gelation. Strengthening gels, through increases in carrageenan concentration, where viscosity and elasticity are enhanced, therefore result in reduced lubricity. To overcome this problem, and to gain greater control over material response, fluid gels can be created whereby a shear force is applied to a system undergoing gelation. During this process, the gelled nuclei that form during the onset of spinodal decomposition are segregated by shear forces that restrict the molecular ordering to take place within discrete particles, rather than macroscopically as in the formation of quiescently cooled gels. The fluid gel suspensions were shown to reduce friction coefficients with increasing particle elasticity and volume fraction, therefore allowing enhanced viscosities and storage moduli to be complemented by reduced friction coefficients as the particles are entrained to the contact

Progress is being made in understanding the lubricating ability of hydrocolloid systems relevant to the food industry. The tongue-and-palate contact is one where we tribologically assess products during oral consumption and the determination of mouth-feel attributes, thus we can begin to formulate foods with desired textural characteristics by consideration of the relationship between microstructure and entrainment for lubrication.

Acknowledgements

The authors gratefully acknowledge Cargill for sponsorship and permission to publish this material, the EPSRC for support via the EngD scheme and the Royal Commission for the Exhibition of 1851 for granting David Garrec an Industrial Fellowship. The authors would also like to thank Dr Brian Guthrie (Cargill, Cargill Global Food Research) for useful discussions.

References

1. G. O. Phillips and P. A. Williams, Woodhead Publishing, 2nd edn., 2009.
2. E. R. Morris, D. A. Rees and G. Robinson, *Journal of molecular biology*, 1980, **138**, 349-362.
3. G. Cassin, I. Appelqvist, V. Normand and I. T. Norton, *Colloid & Polymer Science*, 2000, **278**, 777-782.
4. W. J. Frith, X. Garijo, T. J. Foster and I. T. Norton, *Gums and stabilisers for the food industry*, 2002, **11**, 95–103.
5. I. T. Norton, D. A. Jarvis and T. J. Foster, *International Journal of Biological Macromolecules*, 1999, **26**, 255-261.
6. D. A. Garrec and I. T. Norton, *"Influence and control of fluid-gel particle aggregation"*, **In Progress**.
7. M. Caggioni, P. T. Spicer, D. L. Blair, S. E. Lindberg and D. A. Weitz, *Journal of Rheology*, 2007, **51**, 851.
8. R. A. de Wijk and J. F. Prinz, *Food Quality and Preference*, 2005, **16**, 121-129.
9. R. A. de Wijk, L. J. van Gemert, M. E. J. Terpstra and C. L. Wilkinson, *Food Quality and Preference*, 2003, **14**, 305-317.
10. J. H. H. Bongaerts, K. Fourtouni and J. R. Stokes, *Tribology International*, 2007, **40**, 1531-1542.
11. M. E. Malone, I. A. M. Appelqvist and I. T. Norton, *Food Hydrocolloids*, 2003, **17**, 763-773.
12. A. Gabriele, F. Spyropoulos and I. T. Norton, *Food Hydrocolloids*, 2009, **23**, 2054-2061.
13. M. R. C. Rochas, *Biopolymers*, 1980, **19**, 1675-1687.
14. D. M. Goodall and I. T. Norton, *Accounts of Chemical Research*, 1987, **20**, 59-65.
15. D. A. Garrec and I. T. Norton, *Food Hydrocolloids*, 2012, **26**, 389-397.
16. Y. H. Hui, *Handbook of food science, technology, and engineering*, CRC press, 2006.

NEW PARAMETERS FOR THE EXAMINATION OF THE PECTIN GELATION PROCESS

H. Kastner, U. Einhorn-Stoll, B. Senge

Department of Food Technology and Food Chemistry, Technische Universitaet Berlin, Koenigin-Luise-Strasse 22, D-14195 Berlin, Germany

1 INTRODUCTION

Industrial pectins are mainly obtained from by-products of the citrus or apple juice industry by acidic extraction. They are used as gelling or thickening agents in order to improve the texture of food products. Pectins are branched polysaccharides with a backbone of galacturonic acid and neutral sugar side chains, present in the plant cell walls. The galacturonic acid molecules are partly esterified with methanol. Materials with more than 50 % methoxylated carboxyl groups (degree of methylation DM > 50 %) are referred to as high-methoxylated pectins (HMP) and those with less than 50 % are low-methoxylated (LMP). The typical DM for commercial use is about 60 to 77 % for HMP and about 25 to 40 % for LMP.[1] The DM is crucial for the complex pectin gelation process. HMP form gels in the presence of at least 55 % soluble solids (mostly about 65 % sugar) and at low pH (< 3.5).[2,3] Their gelling mechanism is a combination of hydrophobic interactions between methoxyl groups, favored by higher temperature, and hydrogen bonds between undissociated carboxyl groups, dominating at lower temperature.[4] The LMP network formation is less dependent on pH and soluble solids than the HMP gelation, and is promoted by the presence of Ca^{2+}, forming intermolecular ionic junction zones between smooth regions of neighbouring chains (egg-box model).[1,5]
Fundamental knowledge of the material and structuring properties of pectins, especially of their gelling behavior, is of crucial importance for their application in foods as well as for the configuration of technological processes. The gelation temperature of pectins depends not only on their botanical source, manufacturing conditions and molecular and material properties but also on the gel preparation procedure and cooling conditions. There are many experimental studies of the complex pectin gelation process as transition from sol to gel as well as of the "gel point", the temperature at which the material properties change from liquid to solid. Sometimes, however, it can be difficult to define a clear gel point because the pectin gelation process is rather complex.
Mechanical properties of final pectin sugar model gels have been investigated for instance using empirical tests, based on gel breaking, or the °SAG-method.[6,7] They are applied in the pectin industry in order to standardize pectins for commercial use. Nowadays mainly fundamental rheological measurements are applied for gel examination. They allow the

investigation of the structure formation during the gelling process with determination of setting time and temperature (gel point) as well as the final gel properties.

In oscillation measurements, widely used for the examination of the pectin structuring process, the gel point was frequently defined as the cross-over point of storage modulus G′ and loss modulus G′′ ($\tan \delta = G''/G' = 1$). This method was developed initially for chemical gelation.[8] In most food products, however, physical gelation via junction zones occurs and sometimes no clear gel point can be determined for pectin gels, for instance because of pre-gelation.[9,10,11] Therefore, a modified method examined the point of intersection as a function of frequency,[3,8] where the gel point as $\tan \delta = 1$ might be defined only in case it is independent on the frequency.[11] But also this method is not always exact and, therefore, sometimes the described point is named as "apparent gel point".[3] Other indicators of the gel formation were described by a strong decrease of tanδ[12,13] or using the structure development rate $SDR = dG'/dt$ during cooling[14]. Nevertheless, the gel point determination is not completely clear, yet.

The objective of this study, therefore, is to investigate the structuring process of commercial pectin gels using additional new gelling parameters that allow a detailed examination of influencing factors such as botanical origin, preparation conditions, molecular and material properties, gel preparation (pectin concentration, pH, soluble solid, divalent cations) or experimental parameters such as cooling rate on the gelling process.

2 MATERIALS AND METHODS

2.1 Materials

High or low methylated non-standardized commercial citrus pectins were obtained from three pectin companies. For data protection reasons an anonymous pectin declaration 2A, 3B, etc. had to be made. Gelling parameters and the DM of the samples are given in Table 3, further molecular characteristics and the relating determination methods are presented in a parallel paper.[15]

2.2 Gel preparation

The pectin-sucrose-gel composition and preparation was based on the USA-sag method[6,7] (Table 2). In contrast to HMP gel preparation, the LMP gelation requires Ca^{2+} and a different pH (Table 1). All experiments were made at least in duplicate for each pectin gel.

2.2.1 HMP sucrose gel preparation: Dry pectin powder was dissolved in demineralised water. While stirring, the mixture was heated quickly to boiling. The sucrose was added and the HMP-sucrose mass was reduced to a defined value. Afterwards, the pH of the mass was reduced by addition of tartaric acid solution.

2.2.2 LMP sucrose gel preparation: Demineralised water, citric acid solution and sodium citrate solution were mixed in a steel pot and the dry pectin powder was added while stirring. The mixture was heated quickly to boiling, sucrose was added and the mixture was boiled again. The calcium chloride dehydrate solution was added and while stirring the LMP-sucrose mass was reduced to a defined value. The detailed information of the ingredients of the two gels is given in Table 1.

Table 1 *Preparation of gel samples with pectins of different DM*

Ingredients	*HMP gel*	*LMP gel*
Pectin content	2.75 g	6.00 g
Demineralized water	430.00 g	637.50 g
Sucrose	647.25 g	264.00 g
48.8 % tartaric acid solution	7.00 ml	-
54.3 % citric acid solution	-	7.50 ml
6 % sodium citrate solution	-	15.00 ml
2.205 % calcium chloride dehydrate solution	-	37.50 ml
Reduced to	***1015 g***	***900 g***
Total pectin concentration	***0.27 %***	***0.67 %***

2.3 Rheological measurements

Oscillation measurements of pectin-sugar model gels were carried out on a Physica MCR 301 (Anton Paar). The applied geometry was a double gap rotational cylinder CC27/P1 with peltier cylinder temperature system TEZ 150P. After gel preparation, about 15 ml of the sample was transferred immediately into the pre-heated rheometer (105 °C) and cooled to 20 °C (HMP) or 10 °C (LMP) with a cooling rate of 1 K/min. The sample was coated with silicone oil and the cylinder was closed with a special lid in order to avoid evaporation. The dynamic rheological parameters storage modulus G′ and loss modulus G′′ as well as the loss factor $\tan\delta = G''/G'$ of the pectin-sugar model systems were recorded during cooling (temperature sweep) at a frequency of 1 Hz and deformation amplitude of 0.001.

2.4 Ridgelimeter (USA-sag) method

The USA-sag method was implemented by the IFT committee for pectin standardisation. This method is rather empirical but frequently used for routine or quality tests in the pectin industry.

The rest of the hot pectin-sugar solution, prepared as described above, was filled into three special glasses and stored at 25 °C for 24 h before measurement. For an accurate result, the average gel properties of the three gel cones had to be within the limits shown in Table 2.

A Lab850 pH-meter with a special penetration electrode from Schott Instruments was used for the determination of the pH in the gels after Ridgelimeter measurements. Moreover, the soluble solid SS was determined in the gel using a refractometer (Schmidt and Haensch).

Table 2 *Conditions for the different pectin gels*

Ingredients	*HMP gel*	*LMP gel*
pH	2.2 - 2.4	2.8 - 3.2
SS	64.5 - 65.5 %	30 - 32 %
Gel strength	19.5 - 27.0 %sag	

2.4 Examination of structure formation

Three characteristic gelling temperatures were calculated: The (apparent) *gel point temperature (GP)* as cross-over of G′ and G′′, the *initial structuring temperature (IST)* and *critical structuring temperature (CST)* as shown in Figure 1 from the first derivation of the storage modulus (dG′/dt = structure formation velocity) using Origin 8.1 software. The IST is defined as the temperature at which the structure formation velocity was different from 0 for the first time. The CST is the extrapolated temperature for the first strong increase of the structure formation velocity. In contrast to the GP, these two structuring temperatures could be detected for any pectin we have ever investigated. Moreover, from the structure formation velocity curve often different structuring phases could be identified.
The details of the new defined parameters are shown in Figure 1.

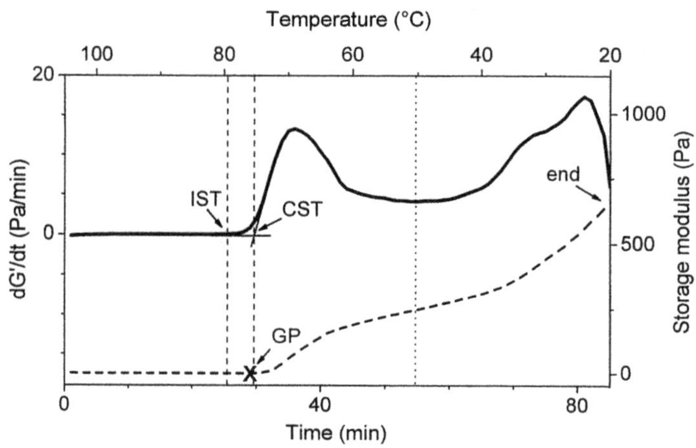

Figure 1 *Evaluation of the structure formation during cooling of an HMP-gel (sample 3A). Full line = dG′/dt; dotted line = storage modulus G′; IST = initial structuring temperature; CST = critical structuring temperature; end = end level at 20 °C; GP = gel point = tan δ = 1.*

3 RESULTS AND DISCUSSION

3.1 Structure formation temperatures

The rheological studies were made using different but constant preparation conditions for HMP and LMP, respectively, (Table 1) and allowed the comparison of the gelling processes and gel properties. Sometimes, the rheological measurements gave no clear setting point as intercept of G′ and G′′, causing uncertainty for identifying the exact gelling temperature. Especially in these cases, the new structuring parameters IST and CST allowed a better description and comparison of the structure formation process.
The structure formation velocity curve dG′/dt characterised not only the gelling kinetics but also different phases in the gelation process by alterations of the structure formation velocity.

Table 3 *Characteristics and gelling properties of the tested pectin samples. DM = degree of methoxylation, GP = gelling point, IST = initial structuring temperature, CST = critical structuring temperature, G'_{end} = final storage modulus, tan δ_{end} = final loss factor.*

Source	*Sample*	*DM*	*Parameters of the gelation process*				
			GP	*IST*	*CST*	G'_{end}	*tan* δ_{end}
		%	*°C*	*°C*	*°C*	*Pa*	
1	**1B**	59.6	93	81	91	1077	0.078
	1C	24.2	51	42	53	91	0.151
2	**2A**	68.9	86	88	85	815	0.068
	2B	55.1	58	62	58	639	0.054
	2C	30.1	43	51	37	81	0.200
3	**3A**	69.8	77	79	76	587	0.061
	3B	57.1	57	60	56	877	0.051
	3C	63.6	67	70	66	788	0.057
	3D	32.8	43	60	31	44	0.166
	3E	30.2	58	66	55	128	0.128
	3F	27.7	-	79	69	293	0.120
	3G	69.0	70	72	69	296	0.090
	3H	56.5	54	54	53	348	0.087

3.2 Gelling process

A higher DM, in general, led to a faster start of the incipient structuring process at higher gelation temperatures (Table 3 and Figure 2a). This is well-known and crucial for the pectin application. In order to investigate the influence of manufacturing conditions on the gelling process, two HMP (2A and 3A) and two LMP (2C and 3E) samples with similar DM but from different companies were compared. Differences in the start of the structuring process and structuring velocity were observed (Figure 2b and 2c, Table 3). As shown in Figure 2b, there was an earlier increase in dG′/dt of the 2A gel, with structure formation temperatures GP, IST and CST about 9 K higher than in the 3A gel. The same was found for the two LMP with a difference > 15 K. This means that the gelling properties of pectins were strongly dependent on their processing conditions and that the DM is one indicator but not the only parameter for evaluating the gelling properties.
An increase or decrease of structure formation velocity dG′/dt can indicate different gelling process phases. It seems that single structuring mechanisms (hydrogen bonds beside hydrophobic interaction for HMP gelation, egg-box junction zones via calcium bridges and hydrogen bonds for LMP) occur in typical temperature ranges of the gelling process. It was found that, even if the IST for two tested HMP or two LMP varied, the typical structuring phases during cooling can behave similarly. Also using the SDR theory[14], two phases of the gelation process have been already described in past publications. The publication of further results in this field with other influencing parameters is in preparation.

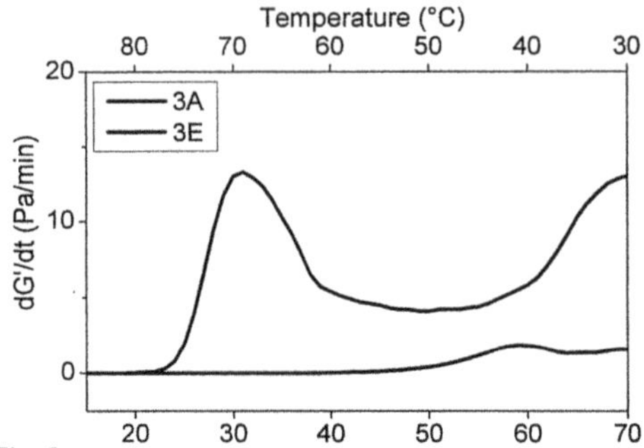

Fig. 2a

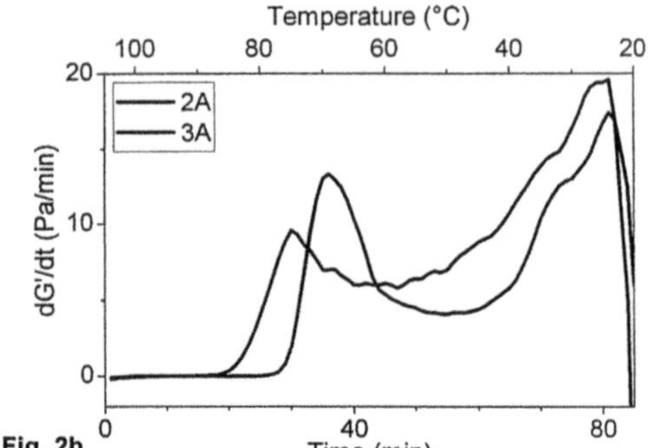

Fig. 2b

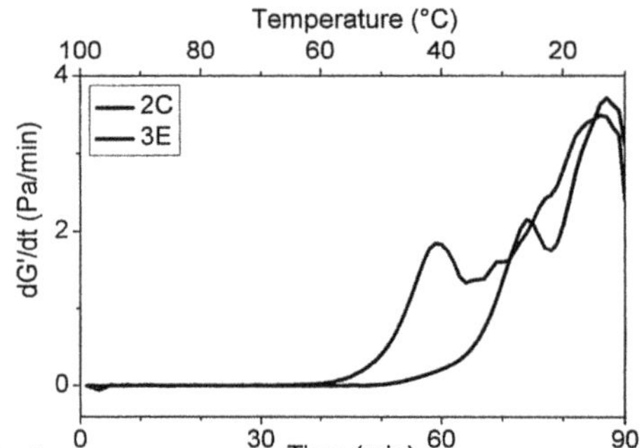

Fig. 2c

Figure 2 *Structure developing velocity dG′/dt during cooling in dependence on degree of methoxylation. a: comparison of a HMP and a LMP from the same company (shown from 85 to 30 °C); b: comparison of two HMP from different companies with similar DM (cooled from 105 to 20 °C); c: comparison of two LMP from different companies with similar DM (cooled from 100 to 10 °C).*

The relationship between CST and GP of the tested gels is shown in Figure 3. From this significant correlation it can be concluded that the newly defined structuring temperatures CST and the nearly CST - parallel IST are a complementary way to evaluate the incipient gelling process, especially for gels without a clear G′-G′′ cross-over. The influence of the manufacturing conditions on the setting behavior of pectins could be sufficiently examined by IST and CST.

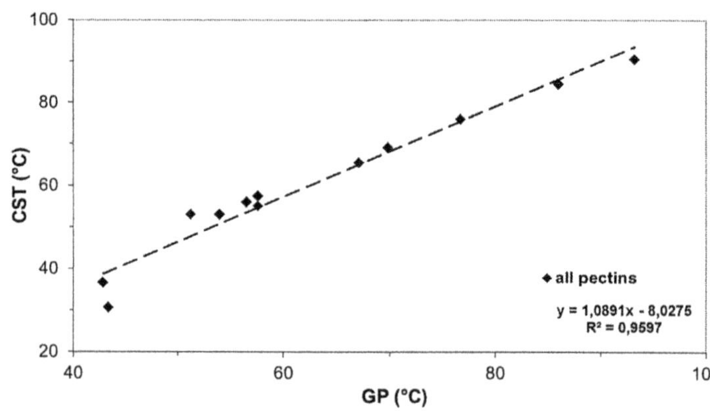

Figure 3 *Correlation of structure formation temperatures GP and CST for all pectin sucrose gels.*

4 CONCLUSIONS

The analysis of the chemical structure of the examined pectins from different companies showed the complexity of the polymer properties and their influence on the pectin gel characteristics, depending on the pectin manufacturing process. The newly defined initial structuring temperature IST and critical structuring temperature CST were, beside the GP, suitable parameters in order to describe the incipient gel structuring process for both types of pectins. Even samples with no clear GP could be evaluated this way. Therefore, IST and CST have proved to be valuable complementary parameters and can help to understand the structure formation processes in more detail. They can be used for optimizing the pectin production process as well as pectin application in food products.

References

1 A.G.J. Voragen, W. Pilnik, J.-F. Thibault, M.A.V. Axelos and C.M.G. Renard, 'Pectins' in *Food Polysaccharides and Their Applications*, eds., A.M. Stephen, New York, Marcel Dekker Inc., 1995, pp. 287-339.
2 B.R. Thakur, R.K. Sing and A.K. Handa, *Crit. Rev. Food Sci. Nutr.*, 1997, **37**, 7.
3 J.A. Lopes da Silva, M.P. Goncalves and M.A. Rao, *Int. J. Biological Macromolecules*, 1995, **17**, 25.
4 D. Oakenfull and A. Scott, *J. Food Sci.*, 1984, **49**, 1093.
5 E.R. Morris, D.A. Powell, M.J. Gidley and D.A. Rees, *J. Molecular Biology*, 1982, **155**, 507.
6 IFT Committee, *Food Technology*, 1959, **8**, 496.
7 R.E. Cox and R.H. Higby, *Food Industry*, 1944, **16**, 441.
8 H.H. Winter and F. Chambon, *Journal of Rheology*, 1986, **30**, 367.
9 D. Lootens, F. Capel, D. Durand, T. Nicolai, P. Boulenguer and V. Langendorff, *Food Hydrocolloids*, 2003, **17**, 237.
10 A. Slavov, C. Garnier, M.-J. Crepeau, S. Durand, J.-F. Thibault and E. Bonnin, *Carbohydrate Polymers*, 2009, **7**, 876.
11 P. Stang-Holst, A.-L. Kjöniksen, H. Bu, S.-A. Sande and B. Nyström, *Polymer Bulletin*, 2006, **56**, 239.
12 A. Dahme, *J. Texture Studies*, 1992, **23**, 1.
13 S. Neidhart, C. Hannak and K. Gierschner, 'Sol-gel transitions of high-esterified pectins and their molecular structure' in *Advances in Pectin and Pectinase Research*, eds., A.G.J. Voragen, H. Schols and R. Visser, Kluwer, 2003, pp. 431-448.
14 J.-T. Fu and M.A. Rao, *Food Hydrocolloids*, 2001, **15**, 93.
15 U. Einhorn-Stoll, H. Kastner and B. Senge, *Gums and stabilisers for the food industry*, 2011, **16**.

COMPARISON OF MOLECULAR PARAMETERS; MATERIAL PROPERTIES AND GELLING BEHAVIOUR OF COMMERCIAL CITRUS PECTINS

U. Einhorn-Stoll, H. Kastner, B. Senge

Department of Food Technology and Food Chemistry, Technische Universitaet Berlin, Koenigin-Luise-Strasse 22, D-14195 Berlin, Germany

1 INTRODUCTION

Pectins are important gelling and thickening agents for the food industry. They can be extracted mainly from citrus fruits and apples but also from other plant materials such as sunflower, sugar beet or other sources. The industrial production has a long tradition and the main steps like extraction or precipitation are well-known.[1] Nevertheless, the parameters and details of the treatments, such as temperatures, pH or drying procedure, can vary considerably between different pectin producing companies and even within one company. Moreover, the pectin sources are biological materials with seasonal and local variations, and it is necessary but not always completely possible for the pectin producers to adapt the technology to the raw material.

Pectin molecules are long and have a mainly galacturonic acid backbone with side chains of neutral sugars. The galacturonic acid molecules are partly methoxylated and the pectins are divided into high-methoxylated (HMP) or low-methoxylated (LMP) with degree of methoxylation (DM) above or below 50 %, respectively.[1,2,3] LMP are mostly made from HMP by chemical demethoxylation procedures at acidic or alkaline conditions during extraction;[1] the resulting pectins are used for different food products.

In previous experiments, pectin modifications such as demethoxylation and amidation were made from HM-pectin of one company in laboratory scale. It was found that the material properties and thermal degradation behaviour of the resulting LMP varied considerably depending on their molecular parameters and the preparation conditions.[4,5] Laboratory preparation and industrial production differ, however, not only with respect to the raw material properties and the processed mass but also in the applied equipment and resulting technological conditions. The question is, whether results of the model pectin investigations can be transferred to industrially produced materials from different companies. Therefore, commercial citrus pectins from three different companies were examined in detail and compared with respect to their molecular and material properties and especially their gelling behaviour. These parameters are relevant for practical pectin application and their interactions and inter-dependencies can give valuable information for pectin producing as well as user companies.

2 MATERIALS AND METHODS

2.1 Materials

All samples were commercial pectins, kindly provided from three pectin companies. For data protection reasons they are named with 1, 2 and 3, and the pectins are labelled with 1A, 1B, etc. The detailed examined parameters are given in Table 1.

2.2 Methods

The molecular parameters galacturonan content GC, degree of methoxylation DM, intrinsic viscosity IV and the material properties, colour and dissolution time as well as thermal degradation properties were determined as described previously.[5,6] The gelling behaviour was tested by oscillation measurements at varying temperature as described in a parallel paper.[7] The particle size of the dry powder was determined using a Horiba particle sizer LA-950 (Retsch, Haan, Germany) and the scanning electron micrographs were made by ZELMI, TU Berlin, using a S-2700 scanning electron microscope, (Hitachi, Tokyo, Japan).

3 RESULTS AND DISCUSSION

3.1 General screening and comparison of high and low methoxylated pectins

3.1.1 Molecular and material properties

Scanning electron micrographs give first impressions of differences between both pectin groups (Figure 1). The HMP particles had a visibly rougher surface and were more porous than the LMP particles. Methoxylation at the industrial scale is often made in an acidic environment[1] where the majority of free carboxylic groups are undissociated. The pectin macromolecules show low electrostatic repulsion and are able to form many strong inter- and intramolecular hydrogen bonds. The result is a compact structure that is trapped during drying in a partly crystalline state and can negatively influence the dissolution properties. The main problem in the hydration and dispersion of pectin powder is to limit the "fish-eye effect". Some of the tested pectins were really difficult to dissolve (Table 1) and confirmed this experience. It took mostly more than 20 min and sometimes even more than 1 h to dissolve 100 mg pectin in 50 ml distilled water, and the smoother surface and a partly crystalline state of LMP particles (in comparison to HMP) additionally delayed the necessary hydration process.
Another general result of the demethoxylation procedure was the smaller ***chain length*** of the pectins. The intrinsic viscosity (IV) of the tested LMP was significantly lower than that of the according HMP (Table 1). The explanation is that any chemical demethoxylation, independent of acidic or alkaline conditions, does not only cleave the ester bonds but also, to a certain extent, the glycosidic linkages in the galacturonic acid backbone.
The third general difference between HMP and LMP was the ***colour***, in particular the b-value (Table 1). LMP of company 1 and 2 hade a significantly higher +b-value (yellow) than the HMP, in case of company 3 this effect, however, was not as clear. It seems that more unsaturated uronides developed during demethoxylation in company 1 and 2 which were able to form brown-coloured reaction products whereas in company 3 this was partly prevented. These differences between the companies were confirmed by the ***galacturonan content*** (GC = purity). Intensive chemical treatment can cause not only formation of browning unsaturated uronides but also removes neutral polysaccharides and impurities

and, thus, increases the GC. The difference in GC between HMP and LMP of company 3 were smaller than in case of company 1 or 2 (Table 1).

The differences in ***particle size and size distribution*** between HMP and LMP (Table 1) depended neither clearly on the DM nor on the pectin company, but probably mainly on the milling equipment and conditions. The particle size should be, however, not completely neglected by the pectin producers because of its influence on dissolution and application as discussed above.

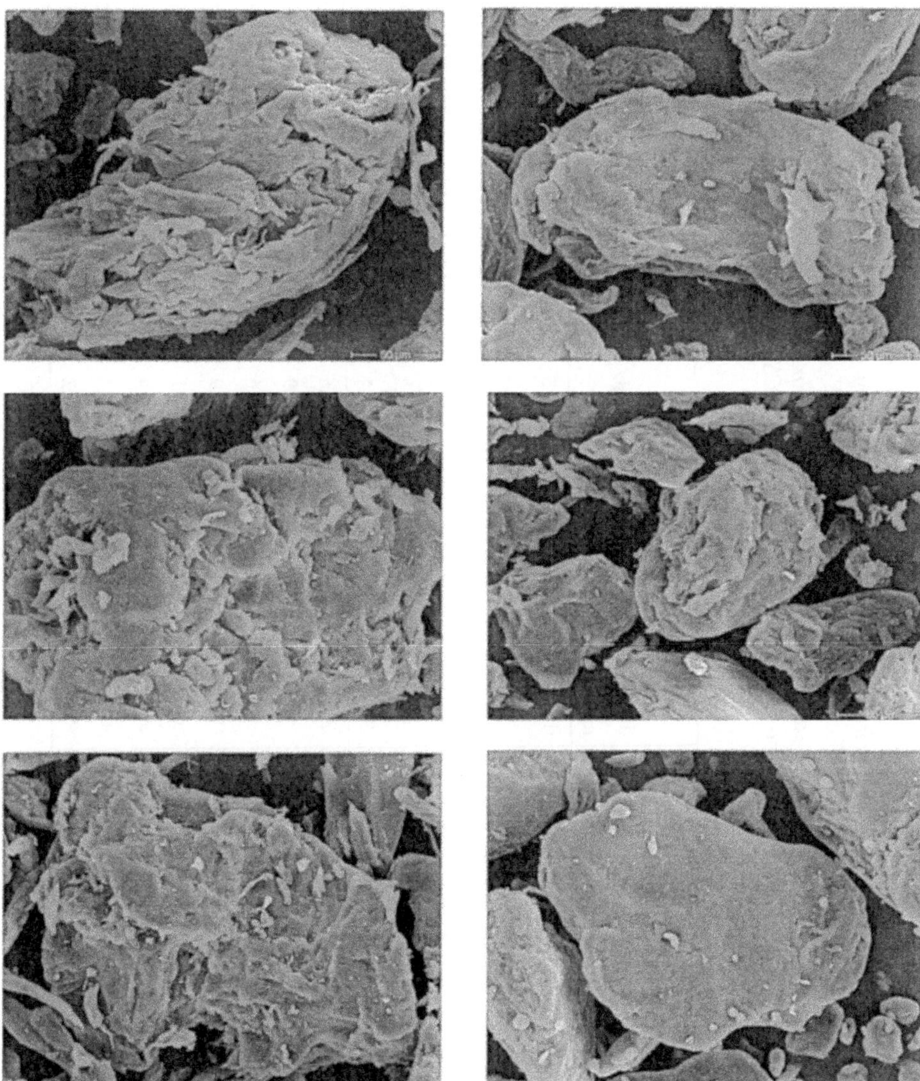

Figure 1 *Scanning electron microscopy of HMP (left side) and LMP (right side) of the three different companies. First line = company 1, second line = company 2, third line = company 3.*

Table 1: *Molecular parameters, material and gelling properties of the tested pectins*
GC=galacturonan content, DM=degree of methoxylation, IV=intrinsic viscosity, E_{500}=absorption at 500 nm, A_o=particle surface, T_{pDSC}=peak temperature in DSC signal, T_{on} and T_{pDTG}= extrapolated onset- and peak temperature in DTG signal, ΔT=peak width, v_{max}=maximum degradation velocity, GP=gelling point, IST=initial structuring temperature, CST=critical structuring temperature, G'_{end}=final storage modulus, $\tan\delta_{end}$= final loss factor.

		molecular parameters			colour			dissolution			particle analysis		thermal degradation					gelation				
source	sample	GC	DM	IV	L	a	b	time	E_{500}	pH	median	A_O	T_{pDSC}	T_{onDTG}	T_{pDTG}	ΔT	v_{max}	GP	IST	CST	G'_{end}	$\tan\delta_{end}$
		%	%	cm^3/g				min			µm		°C	°C	°C	K	%/min	°C	°C	°C	Pa	
1	1A	89.3	60.9	639	88.6	1.1	12.9	1	0.048	4.04	145	607	240.5	220.4	237.8	27.6	19.9					
	1B	85.5	59.6	598	89.7	1.1	10.9	>60	0.045	3.62	178	437	239.4	220.6	236.1	25.8	22.2	93.2	81	90.5	1077	0.078
	1C	93.7	24.2	318	90.0	1.5	16.9	30	0.022	3.52	110	905	238.4	217.3	233.8	28.5	17.3	51.2	42	53	91	0.151
2	2A	81.6	68.9	647	86.4	2.0	12.2	20	0.131	3.42	101	1118	244.0	220.5	240.3	35.6	16.3	86.0	88.0	84.5	815	0.068
	2B	87.7	55.1	492	87.7	1.6	11.4	30	0.166	3.45	96	1154	245.0	220.4	240.7	34.5	16.2	57.6	61.5	57.5	639	0.054
	2C	91.5	30.1	358	81.2	3.4	24.0	30	0.442	3.18	82	1244	246.7	221.0	241.0	33.5	15.8	42.9	51.0	36.5	81	0.200
3	3A	80.9	69.8	554	89.9	1.4	11.0	25	0.056	3.48	101	1114	249.6	231.7	246.1	24.3	22.2	76.7	79.0	76.0	587	0.061
	3B	83.4	57.1	576	88.4	1.7	12.9	>60	0.054	3.56	81	1570	247.9	228.7	243.9	25.7	20.7	56.5	59.5	56.0	877	0.051
	3C	81.5	63.6	608	88.5	1.8	14.0	>60	0.028	3.55	98	1113	249.6	231.1	245.8	24.7	21.4	67.1	70.0	65.5	788	0.057
	3D	84.8	32.8	363	87.7	1.1	13.0	>60	0.022	5.24	118	1127	230.9	213.3	228.4	54.0	17.2	43.4	59.5	30.5	44	0.166
	3E	81.5	30.2	336	84.0	2.7	20.1	>60	0.039	4.91	149	523	234.1	217.0	231.6	50.4	17.8	57.6	66.0	55.0	128	0.128
	3F	78.5	27.7	327	86.9	1.5	15.0	45	0.040	5.24	103	1295	233.9	216.1	231.4	46.9	15.6		78.5	69.0	293	0.120
	3G	82.3	69.0	518	84.0	2.8	15.6	45	0.040	3.64	112	956	248.2	229.6	244.4	25.1	22.0	69.8	72.0	69.0	296	0.090
	3H	87.0	56.5	529	88.2	1.8	12.9	>60	0.065	3.66	89	1129	246.2	226.0	242.3	27.0	20.3	54.0	54.0	53.0	348	0.087

3.1.2 Thermal analysis

A further method for a quick evaluation of pectin is the thermal analysis using a combination of differential scanning calorimetry DSC and thermogravimetry TG. The DSC signals give information on ***transition and degradation enthalpies***. The TG and its first derivation DTG allow insight into ***thermal stability*** by the extrapolated temperatures $T_{on\ DTG}$ and into ***homogeneity*** by the degradation time, measured as peak width ΔT, both from the DTG signals. The values are given in Table 1 and some typical signals in Figure 2.

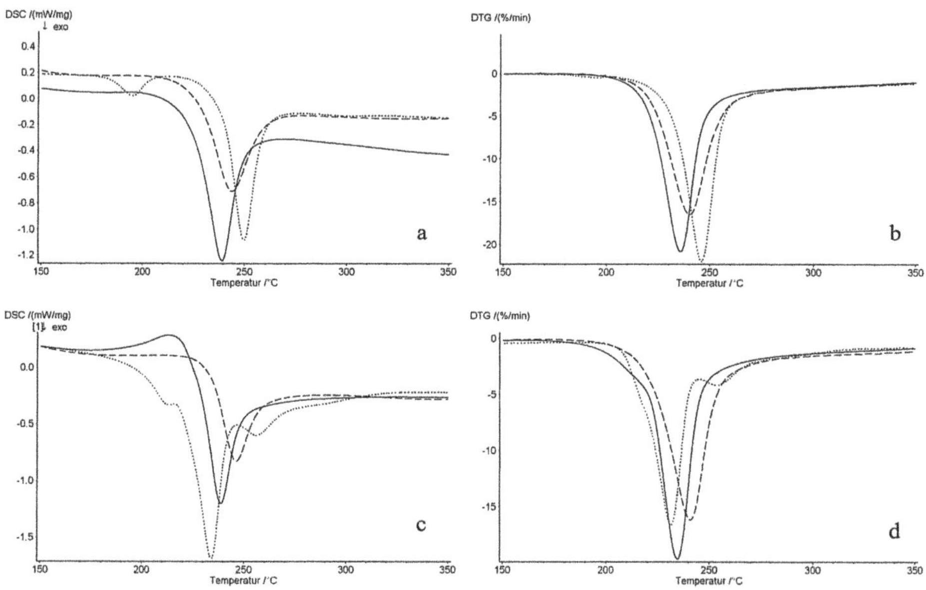

Figure 2 *DSC signal (a,c) and DTG signals (b,d) of HMP (a,b) and LMP (c,d).* *—— = company 1, ---- = company 2, •••• = company 3.*

Comparing HMP and LMP, the main differences are clearly visible: The two HMP of company 1 and 3 were more similar (DTG peak form) and homogeneous (peak width) than the LMP. In the case of company 2 these differences were smaller.

Comparing the DSC signals of the three HMP in detail, the pectins of company 1 and 2 were rather similar and had only one exothermic degradation peak. The HMP of company 3, however, had another smaller exothermic peak before the main fraction degradation and also a small weight loss in the DTG signal; that means another component beside the normal pectin in this sample was degraded. The HMP of company 2 seemed to be the least homogeneous; the peak width in DTG was about 10 K higher than that of the two others. Nevertheless, the DTG signals of the 3 HMP were partly similar.

Comparing the three LMP, the differences were even clearer. LMP of company 1 had an endothermic pre-peak with only small weight loss; it might result mainly from a conformation transition as discussed elsewhere.[5] The LMP of company 3 had the most shoulders and peaks in DSC as well as DTG signals and the broadest peaks, it was the least homogeneous. This confirms the differences in GC as discussed above.

The DSC and DTG signals of the three companies were representative (= found for several single samples). They were a kind of a "fingerprint" and allowed an easy detection of potential differences in the pectin production process which were discussed in the analysis of molecular parameters. Unfortunately, the processing details of the single materials were more or less secret and it was not possible to confirm the assumptions about differences in technology. Nevertheless, it is generally possible to use the thermal analysis for a quick screening in quality control for detection of changes in quality and processing.

3.1.3 Gelling properties

The structuring process parameters and the gel properties of the tested pectins were not completely comparable because of different gel compositions and gelation mechanisms used for HMP and LMP. Nevertheless, ***structuring temperatures*** of HMP were higher than those of LMP with one exception in company 3. The ***final values of G′*** (solid-like properties) were higher for almost all HMP than for the LMP and those of ***tan δ*** (brittleness) were definitely higher for the LMP.

3.2 Comparison of single pectins with similar degree of methoxylation

The degree of methoxylation is often used as a key parameter for pectin application. The following examples show, however, that pectins with comparable molecular parameters (especially nearly identical DM) of different companies and even from one company but different production periods were far from similar in their material properties and gelling behaviour. The examples were the HMP 2A, 3A and 3G with DM about 69 % and the LMP 2C and 3E with DM 30 %, for the detailed values see Table 1.

Beside the general effects of demethoxylation as discussed above, there were several specific differences. The first can be seen by ***thermal analysis*** (see 3.1.2). In particular, the HMP of company 2 was less thermal stable and less homogeneous than the samples from company 3. In case of the LMP, the tendencies were just opposite in thermal stability but the same with respect to homogeneity. These were the first indicators of different processing parameters and resulting properties.

The ***intrinsic viscosities*** were higher for the pectins from company 2, indicating that this company prepared their pectins under conditions (especially pH and T), which caused less cleaving of bonds in the backbone. Another difference was the high ***galacturonan content*** (purity) of the LMP 2C, most of neutral sugars and impurities were removed during demethoxylation. Such an effect was found also for company 1 and it is known from laboratory scale pectin modifications, too, where it was found especially after acid treatments.

The most interesting differences with high practical relevance were found for the ***structuring process*** and the ***gel properties***. In case of HMP, the structuring temperatures varied not only between the two companies for about 10 K but also for about 7 K between 3A and 3G from one company but different production periods (Figure 3). Also the end level of the storage modulus G'_{end} and of the loss factor tan δ_{end} differed not only between the two companies but also within one company. In the case of the LMP, the gelpoint could not be determined for all samples, the structuring temperatures were, however, clearly higher for company 3. The LMP-gel of company 3 was more solid (higher G′) and brittle (lower tanδ) than that of company 2.

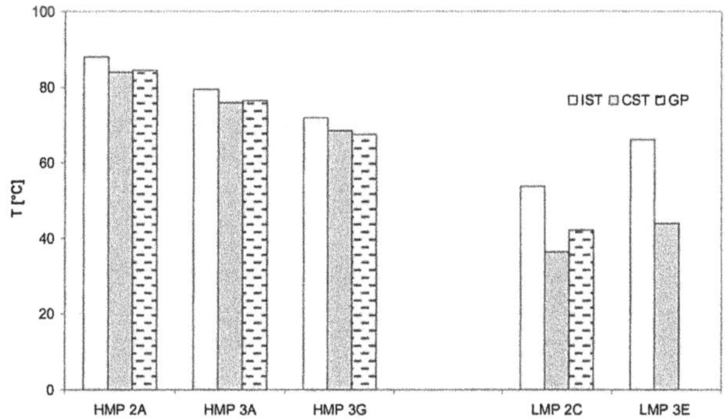

Figure 3 *Structuring temperatures of HMP and LMP.*
IST = initial structuring temperature, CST = critical structuring temperature, GP = gelpoint.

3.3 Statistical analysis

The statistical tests for interdependencies of molecular parameters, material properties and structuring were made by comparing either all pectins together or the groups of HMP and LMP as well as the eight pectins from one company separately (Tab.2). As can be clearly seen, DM was the key parameter for many other properties. It had not only the significant well-known impact on the gelpoint but also on intrinsic viscosity (molecular weight) as discussed above. The interaction of DM and thermal stability was found only for the pectins of one company and confirmed the results of a previous examination of model pectins.[4] Influences of pectin purity (GC) and molecular weight (IV) were found mainly in the LMP-group and are probably indirectly influenced by the DM.

Table 2 *Statistical analysis of the interactions of molecular parameters and material and structuring properties.*
XXX = R > 0.9, XX = R > 0.8. For the single parameters see Table 1.

	all pectins n=32	all HMP n=20	all LMP n=12	one company n=18
DM - IV	xxx			xxx
DM - GP	xx	xxx		xxx
DM - IST		xx		
DM - CST		xxx	xxx	
DM - Tp_{DTG}				xxx
GC - GP	xx			
GC - IST			xx	
GC - Tp_{DTG}			xx	
IV - GP	xx			
IV - IST			xxx	
IV - CST			xxx	
IV - Tp_{DTG}				xxx
GP - IST	xx	xxx		
GP - CST	xxx	xxx		xxx

The correlations of the classical gelpoint and the new structuring parameters IST and CST are of special interest. It should be considered, however, that a gelpoint could be determined not for all LMP, therefore the correlation of gelpoint and CST was significant for all categories except LMP. This supports the application of CST as a complementary measurement to the gel point and is highly important, in particular for gelation processes without clear gel points as sometimes were found in LMP.

4 CONCLUSIONS

The results confirmed the general influence of processing parameters on the pectin quality and application for a collection of commercial pectins from different companies, that was found before for model pectins made from one original sample. These parameters - such as pH, demethoxylation temperature and drying conditions - influenced the molecular weight, colour, purity, thermal stability and homogeneity of the materials. Moreover, they determined the state (amorphous or crystalline) and material properties (surface quality and porosity) of the pectin powder particles, which are crucial for the application properties such as dissolution and gelation.

The degree of methoxylation is a key factor for many tested pectin properties, especially in the gelation process. It is, however, not the only factor what was revealed by a comparison of pectins with similar DM.

The thermal analysis proved to be a very useful and rapid screening method for pectin characterisation. Differences that were found in DSC and TG could be confirmed by the analysis of molecular parameters.

Any pectin producing company should try to promote the favourable material properties, such as rough porous particles and an amorphous state with easily cleavable inter- and intramolecular interactions, in order to produce pectins with excellent application properties.

References

1 C. Rolin, Commercial pectin preparations in *Pectins and their Manipulation,* eds. G.B. Seymour and J.P. Knox, Blackwell Publishing, Oxford, 2002, pp. 222-241.

2 A.G.J. Voragen, W. Pilnik, J.-F. Thibault, M.A.V. Axelos, C.M.G.C. Renard, Pectins in *Food Polysaccharides and Their Applications*, ed. A.M Stephen, M. Dekker, New York, 1995, Chapter 10, pp. 287-339.

3 B.R. Thakur, R.K. Singh, A.K. Handa, *Critical Rev. Food Sci. Nutrition*, 1997, **37**, 47.

4 U. Einhorn-Stoll, H. Kunzek, G. Dongowski, *Food Hydrocolloids*, 2007, **21**, 1101.

5 U. Einhorn-Stoll, H. Kunzek, *Food Hydrocolloids*, 2009, **23**, 40.

6 U. Einhorn-Stoll, T. Salazar, B. Jaafar, H. Kunzek, *Nahrung/Food*, 2001, **45**, 332.

7 H. Kastner, U. Einhorn-Stoll, B. Senge, *Gums and stabilisers for the food industry*, 2011, **16**.

Emulsions

ATOMIC FORCE MICROSCOPY AND FORCE SPECTROSCOPY STUDIES OF THE EMULSIFYING ACTION OF SUGAR BEET PECTIN

V.J. Morris, A. Gromer, A.R. Kirby, R. Penfold and A.P. Gunning

Institute of Food Research, Norwich Research Park, Norwich NR4 7UA, UK

1 INTRODUCTION

Pectin is a complex polysaccharide extracted from non-graminaceous plant cell walls. Pectin extracts from waste cellular material from the citrus fruit and cider industries are sold as gelling agents. Sugar beet pectin (SBP) extracts from sugar beet pulp, after extraction of sugar, are unusual because they do not gel. However, they do show interesting properties as emulsifiers. SBP extracts are also unusual in containing high levels of acetylation, ferulic acid and protein[1-2]. The protein component is difficult to remove by conventional methods and it is believed that the SBP contains stable protein-pectin complexes. Proteins are known for being surface active and for forming irreversibly adsorbed elastic interfacial layers[3]. It has been proposed that the protein component of complexes in the SBP extracts is preferentially adsorbed at oil-water interfaces and that the attached pectin chains form a steric barrier that inhibits coalescence of droplets[4]. This article discusses aspects of the use of atomic force microscopy (AFM) and force spectroscopy, together with conventional surface techniques, to probe[5] the nature of the interfacial structure formed at oil-water interfaces, and the influence of the composition and structure of these interfacial layers on the interaction between oil droplets.

2 METHODS

2.1 Materials

SBP extracts were isolated using methods described elsewhere[6]. The protein content of the sample was ~8.6%. The degree of methylation, determined by Fourier transform infrared spectroscopy, was ~60%. SBP was dissolved in ultrapure water by heating at 40°C and stirring for 2h. For one particular experiment, the SBP was alkali-treated: 0.1M KOH was added to the SBP solution and the mixture was stirred overnight at room temperature. Subsequently, the solution was neutralized with 0.2 M HCl and dialyzed through a 100 kD cutoff membrane. The oil phase (n-tetradecane) was purified using Florisil resins: the oil was kept under mild agitation with the resins for 3h, centrifuged at 12 000 rpm for 30 min in a bench centrifuge before filtering under vacuum and storing under N_2 in the dark.

Sodium polystyrene sulphonate (PSS) with a molecular weight: Mw=32900 Da and polydispersity index: Mw/Mn<1.10 was bought from Polysciences, Inc. (Warrington, PA).

2.2 Interfacial tension (Pendant drop)

Interfacial tension at the oil–water interface was measured using a FTA200 pulsating drop tensiometer (First Ten Angstroms, Portsmouth, VA). An oil droplet was formed at the tip of a Teflon-coated needle (inside diameter 0.94mm and fitted with a syringe with total volume of 100 mL) immersed in the SBP solution. The apparatus is computer controlled and fits the drop profile to the Young–Laplace capillarity equation calculating the drop volume, interfacial tension, and surface area. Adsorption at the oil-water interface was monitored from 0.1% and 0.5% SBP solutions at a constant volume of 10mL.

2.3 Interfacial shear rheology

Interfacial shear measurements were made on an AR2000 controlled-stress rheometer (TA, Crawley, Surrey, UK) operated in the controlled-strain mode using a polished aluminium bicone (diameter 60mm, cone angle 6°). The SBP solution was prepared, and immediately a layer of oil was gently poured on top, with the bicone placed at the oil-water interface. The sample was then subjected to a sinusoidal oscillation (frequency 0.5Hz) at a constant strain (1%) for 90 min. All measurements were made at room temperature.

2.4 AFM Langmuir–Blodgett transfer

Oil-water interfaces were prepared in 250mL glass beakers. A 200mL SBP solution at the required concentration was prepared and poured into the beaker. Two glass pipettes were introduced inside the beaker and maintained in a fixed vertical position at the sides of the beaker. Tetradecane was carefully poured on top of the SBP solution and the sample allowed to incubate for 1h. The glass pipettes were then used to perfuse the subphase with 1L of pure water. Transfer was achieved by dipping a piece of freshly-cleaved mica through the interface at a speed of 8.4mm min^{-1}. A Wilhelmy plate balance was used to monitor changes in interfacial tension during film transfer[5]. The transferred interfacial layers were imaged by contact mode AFM under butanol[5].

2.5 AFM Imaging and force spectroscopy

Images of interfacial films were obtained on an East Coast Scientific (ECS, Ltd., Cambridge, UK) AFM using Nanoprobe (NP) silicon nitride cantilevers with a quoted force constant of 0.38Nm^{-1}. The images were flattened and destriped using the free software Gwyddion (http://gwyddion.net/). Force spectroscopy measurements were made on an Asylum MFP-3D AFM (Asylum Research, Santa Barbara, USA) using Nanoprobe (NP) silicon nitride cantilevers with a quoted force constant of 0.07–0.08Nm^{-1}. Prior to studies on droplets force curves between the cantilever and a clean glass slide were acquired to determine the sensitivity of the optical response to cantilever bending. The cantilever spring constant was measured using the thermal tune function of the MFP-3D software.

2.6 AFM force spectroscopy of droplet-droplet interactions

The preparation of oil droplets and their attachment to cantilevers and surfaces has been described elsewhere[7]. Oil droplets were sprayed onto a clean glass slide to give a fairly uniform coverage of 10–90mm diameter oil droplets[7]. A 100mL drop of water was then deposited onto the slide. Most of the oil drops were displaced on addition of water but a few remained attached to the glass slide. The AFM head was lowered onto the slide to sandwich a layer of water between the head and the glass slide. The cantilever tip was positioned above the centre of one of the droplets on the surface of the glass slide. The cantilever was then driven down into the target droplet, and retracted to detach the droplet from the surface[7]. The droplets attached to the cantilevers were typically of diameters 50–70mm. Once attachment was achieved 100mL of SBP solution was added to the side of the water drop using a pipette. This had to be done after attachment, because otherwise droplet attachment to the cantilever was found to be impossible, as the cantilever could not penetrate into the droplets. The SBP was left to adsorb onto the drops for approximately one hour. When rinsing was required, 100mL of solution were carefully removed from the sample using a pipette before adding 100mL of pure water. This procedure was repeated at least ten times. Finally, the cantilever with attached oil droplet was positioned directly over a second similar sized droplet still attached to the glass slide, and the variation of the interaction force with separation monitored and recorded. Unless stated otherwise force curves were acquired at a constant relative droplet velocity of 100nm s^{-1}.

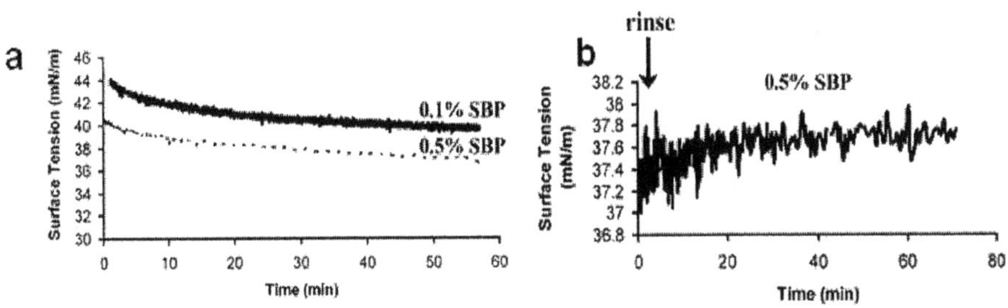

Figure 1 *Interfacial tension data for the adsorption of SBP at a tetradecane-water interface. (a) Data for 0.1% and 0.5% SBP. (b) Effect of rinsing to remove SBP from the bulk phase for 0.5% SBP.*

3 RESULTS AND DISCUSSION

The measurements of interfacial tension using the Pendant drop showed a steady decrease in interfacial tension with time indicating adsorption of the SBP from solution (Figure 1a). No significant change in the interfacial tension was observed after rinsing, suggesting irreversible adsorption of the SBP (Figure 1b). The rheological data showed a steady increase in the elastic component (G') that rapidly exceeded the viscous component (G'') suggesting the formation of an elastic network at the interface (Figures 2). The values observed are similar to those observed previously for protein films[8]. These results are consistent with previous studies on the behaviour of SBP at air-water interfaces[7]. In these studies the protein-pectin complexes were cleaved and the released protein was shown to form elastic networks at the air-water interface. The SBP extracts also produced elastic networks that were considered to arise from adsorption and association of the protein component at the interface. Collectively the data suggests that adsorption of the SBP extracts at the oil-water interface results in an elastic protein-based network structure.

The AFM images of the structures formed at the oil-water interface are shown in Figure 3 for different bulk SBP concentrations. At the lowest SBP concentration (0.1% SBP) the films show holes and visible polymer chains (Figure 3a). The chains are stiff and extended. It has been reported[9] that the adsorbed fraction of SBP at oil-water interfaces has a lower degree of esterification and thus the extension of the pectin chains could result from intra-molecular repulsion between charged galacturonic acid residues or from electrostatic interactions between these charged residues and the protein component of the interfacial film. The holes in the network are consistent with formation of an immobile elastic network: a network would restrict diffusion of the molecules at the interface and closer packing within the film. The holes allow the film thickness to be measured yielding a value of ~0.9nm. At intermediate SBP concentrations (0.5%SBP) the interfacial films are flat, but fewer, smaller holes are present in the network (Figure 3b). The interfacial coverage appears to saturate at a bulk phase concentration of ~2%SBP. The resultant interfacial films are rougher and it is difficult to identify individual pectin chains (Figure 3c).

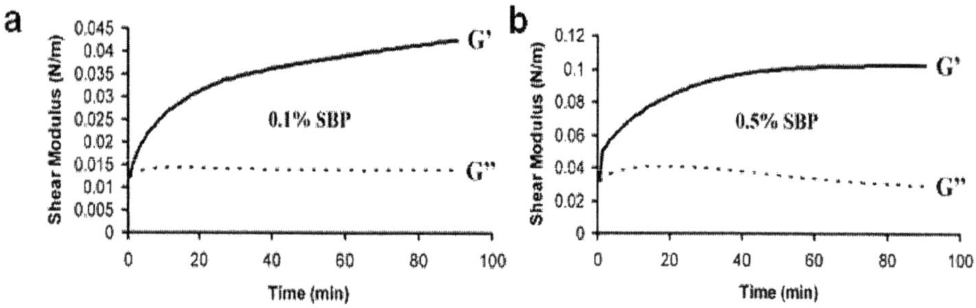

Figure 2 *Changes in interfacial rheology for SBP adsorption at tetradecane-water interfaces. Shear rheology data for (a) 0.1% SBP and (b) 0.5% SBP.*

The roughness of the layer (maximum heights up to 5-10nm) may result from pectin chains which, lacking space during adsorption, protrude away from the interface into the aqueous phase, or due to the adsorption of aggregated structures, known to be present at these concentrations in the bulk phase[10]. The composition and structure of the interface and the nature of the bulk phase was found to influence the nature of the interaction between coated oil droplets. These effects were probed by force spectroscopy.

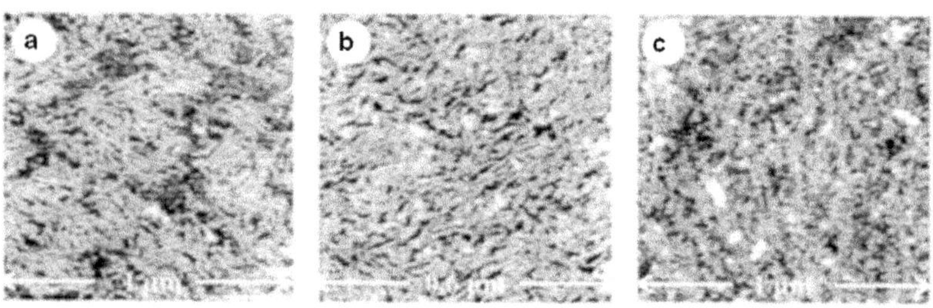

Figure 3 *AFM topographical images of SBP interfacial films transferred from the oil-water interface by the Langmuir-Blodgett method. Bulk SBP concentrations (a) 0.1% SBP, (b) 0.5% and (c) 2% SBP.*

At the highest bulk SBP concentrations studied (2%SBP) the colloidal interactions between the droplets appear to be dominated by a repulsive interaction (Figure 4a). The curves are similar after rinsing to remove SBP from the bulk phase (Figure 4a-b). Comparison with the behaviour of uncoated droplets shows that the repulsive interaction for SBP-coated droplets occurs at larger droplet separations (Figure 4b). Adsorption of SBP will alter the charge on the droplets and thus the nature of electrostatic interactions between droplets. Addition of NaCl to screen the electrostatic interactions did not significantly alter the nature of the interaction[5] suggesting that the repulsive interactions are largely steric in origin. This supports the suggestion by Leroux and co-workers[4] that the associated pectin chains can act as a barrier inhibiting droplet coalescence.

The repulsive interactions between droplets (Figure 4a-b) arise from the unfavourable entropic interactions that result from interpenetration of pectin layers on neighbouring droplets. However, favourable interactions and bridging effects can be introduced by altering the ionic strength and pH of the bulk medium[5]. These effects were observed[5] under conditions (acid pH, high ionic strength solutions of monovalent salts and moderate concentrations of $CaCl_2$) where normal commercial pectin will gel: these results are also consistent with the suggestion[9] that adsorption of SBP favours low acyl pectin chains. Figure 4c shows force spectroscopy curves for bridging induced in the presence of calcium ions: the retract curve shows the extension of the chains linking the droplets, and the detachment force associated with the rupture of the linkage between the chains.

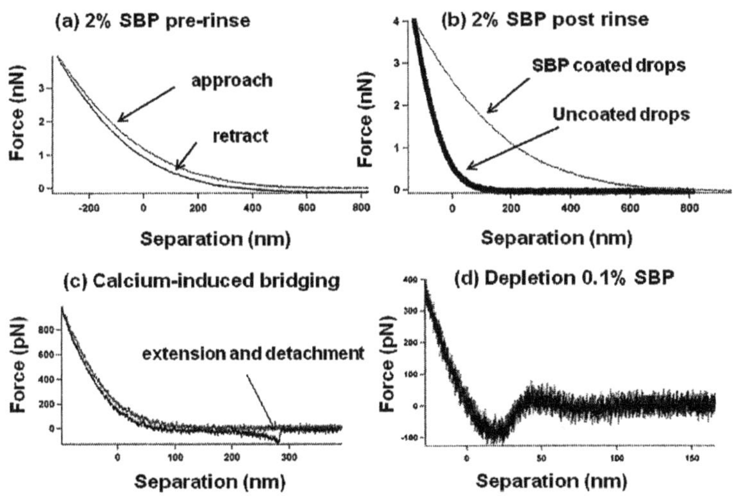

Figure 4 *Force spectroscopy data for interacting SBP-coated oil droplets. (a) Repulsive interaction observed in the presence of 2%SBP in the bulk phase. (b) Comparison of the behaviour of SBP-coated droplets, after removal of the 2%SBP from the bulk phase, with the behaviour of uncoated droplets. (c) Calcium-induced bridging interaction for the SBP-coated droplets. (d) Reversible depletion interaction observed for SBP-coated droplets with 0.1%SBP in the bulk phase.*

The interactions between deformable droplets have been described theoretically by Chan and co-workers[14-15]. As the droplets approach each other they experience a repulsive interaction. When the interaction between the droplets exceeds the forces maintaining droplet shape the droplets start to deform. Deformation requires the drainage of the fluid between the droplets. When the approach and retract speeds are slow compared to the rate of drainage the force-distance curves are reversible (for example Figures 4a & b). When the approach and retract speeds exceed the rate of drainage the retract curves show an adhesive interaction (Figure 5a) and the magnitude of this hydrodynamic adhesion increases with increasing approach speed. These models do not at present include the consideration of depletion effects due to bulk polymers. However, such effects appear in the data for interactions between SBP-coated droplets at low and intermediate SBP bulk concentrations.

At low (0.1%) SBP concentrations the force–distance curves show an attractive well at intermediate separations followed by a small repulsive barrier at longer range (Figure 4d). The approach and retract curves show no discernable hysteresis. These curves resemble those reported for several studies of depletion interactions between solid surfaces[11-13]. The depletion interaction described by Asakura and Oosawa[16] is the low concentration limit of a more general structural mechanism arising from the entropy gain won by solutes in avoiding relatively restricted configurations. Effectively the magnitude of the depletion interaction is determined by the bulk osmotic pressure and the overlap of excluded volume in confined geometries, whilst the range is determined by the solute size and the extent of close packing correlations. Thus depletion interactions are more prevalent between surfaces mediated by solutions of non-adsorbing polymers, but have been shown to also occur for adsorbing polymer[13]. Raising the ionic strength by adding salt to the bulk phase reduced the amplitudes of the attractive well and repulsive barrier (data not shown) indicating that the polyelectrolyte character of the SBP is essential for establishing the structural interaction. Analysis of depletion effects for polyelectrolytes is more complex than that for neutral polymers, but several experimental studies have been reported in the case of solid surfaces[11-13, 17], often using sodium polystyrene sulphonate (PSS) as the depletant. Removal of SBP from the bulk solution causes the attractive well to disappear, yielding repulsive force-distance curves, resembling those for uncoated oil drops in water (Figure 4b). This demonstrates that the attractive interaction is due to a depletion effect resulting from the SBP in the bulk phase, and not due to the SBP coating on the droplets.

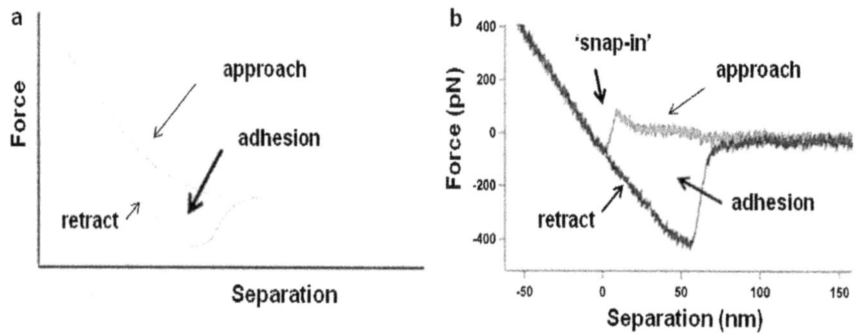

Figure 5 *Interactions between deformable droplets. (a) Schematic diagram showing the appearance of a hydrodynamic attractive interaction. (b) Hysteresis effects observed for SBP-coated droplets in the presence of 0.5% SBP in the bulk phase.*

At higher intermediate concentrations of SBP (0.5%) in the bulk phase the force-distance curves become more complex but reproducible, exhibiting a hysteresis between the approach and retract curves (Figure 5b). The approach curves show a sharp attractive 'snap-in' event suggesting sudden removal of polymeric material from between the droplets, causing the droplets to suddenly jump closer together. The effect is no longer reversible, and the retract curves show a larger attractive well, with a sharp 'jump-out' event suggestive of a sudden breakage of a linkage between the droplets. These effects disappear on removal of the SBP from the bulk phase, suggesting they are depletion effects, and not due to the adsorbed SBP layer on the surface of the droplets.

To investigate these effects model studies were made with the non-adsorbing charged PSS. Similar, but less pronounced hysteresis were observed for semi-dilute 0.5% PSS solutions (overlap concentration of PSS chains is C* ~ 0.6%) in the bulk phase[5]. These effects disappeared when the PSS was removed from the bulk phase, resulting in repulsive interaction slightly less repulsive than that observed for uncoated droplets.
Higher PSS concentrations (3% PSS) revealed the nature of these novel and unexpected effects: as the approach speed was increased the 'snap-in' event was observed to move up the approach curve to closer separations, and then down the retract curve, with the 'snap-in' then occurring as the droplets are moving apart. Furthermore, the energy required to break the adhesive interaction on retraction (the grey area under the curves in Figure 6a) was counter-intuitively found to decrease as the approach speed increased. The decrease in the range and magnitude of the adhesion with increasing approach speed is contrary to the high speed regime observed for interactions between depletant-free droplets, where the attractive hydrodynamic suction increases with increasing approach speed. To the best of our knowledge, this is the first time these unusual hysteresis effects have ever been reported: such effects have never been observed in previous studies on interactions between solid colloidal particles or between non-deformable surfaces.

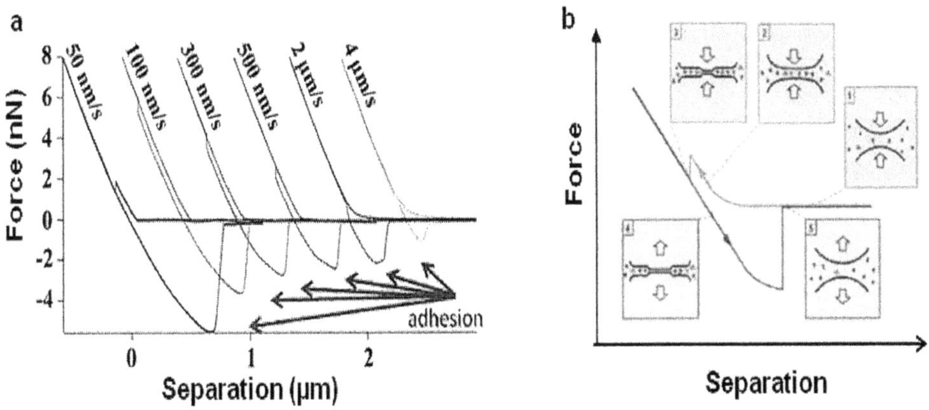

Figure 6 *Depletion interactions observed for semi-dilute bulk solutions of PSS. (a) Effects of approach speed on the interaction for 3%PSS solutions. In order to display the main features the curves at different approach speeds have been shifted along the horizontal axis. (b) Schematic diagram of a suggested explanation for these unexpected effects.*

The origin of these effects has been discussed in more detail elsewhere[5] and figure 6b is a suggested schematic explanation for this behaviour. It is proposed that, as the droplets approach a point is reached at which the droplets start to deform, with the rate of deformation depending on the rate of drainage of fluid between the droplets. The 'snap-in' event is a sudden attraction between the droplets associated with the formation of a 'black spot', due to the complete depletion of polymer from this region between the droplets. It is considered that this black spot will grow with time[18] resulting in a larger adhesive interaction on retraction of the droplets. When the approach speed is increased the rate of formation and growth of the black spot remains the same and hence the snap-in event moves up the approach curve, and eventually down the retract curve. Similarly, restricted growth of the black spot will result in an adhesive interaction on retraction which decreases in magnitude with increasing approach speed. It should be possible to modify the theoretical models of Chan and co-workers[14-15, 19] to include depletion interactions. Adding a depletion contribution, based on the Asakura-Oosawa theory[16], to the Debye-Huckel electrostatic disjoining pressure in the Chan and co-workers model reproduces the behaviour of snap-in event with increasing approach speed, but fails, even qualitatively, to account for the decrease in the size of the adhesive interaction on retraction with increased approach and retract speed. This probably results from a need for a more comprehensive description of the drainage behaviour for the semi-dilute PSS solutions.

4 CONCLUSIONS

This study shows how combined use of AFM and force spectroscopy yield images of interfacial structures and determine the effects of these interfacial structures on a variety of different modes of interaction between oil droplets. The results show that direct studies of the interactions between deformable droplets reveals new and unexpected effects specific to deformable particles, which will be important in understanding droplet interactions in real emulsions, as well as for other examples of soft matter systems.

Acknowledgements: The authors wish to thank Nicola Woodward, Julia Maldonna-Valderrama and Mike Ridout for advice and discussions. This work was supported through the BBSRC grant BB/E002153/1 and the core strategic grant to the Institute.

References

1 P.A. Williams, C. Sayers, C. Viebke, C. Senan, J. Mazoyer and P. Boulenguer, *J. Agric. Food Chem.*, 2005, **53**, 3592.
2 J.-F. Thibault, *Carbohydr. Polym.*, 1988, **8**, 209.
3 V.J. Morris and A.P. Gunning, *Soft Matter*, 2008, **4**, 943.
4 J. Leroux, V. Langendorff, G. Schick, V. Vaishnav and J. Mazoyer, *Food Hydrocoll.*, 2003, **14**, 455.
5 A. Gromer, R. Penfold, A.P. Gunning, A.R. Kirby and V.J. Morris, *Soft Matter*, 2010, **6**, 3957.
6 A.R. Kirby, A.J. MacDougall and V.J. Morris, *Food Biophys.*, 2006, **1**, 51.
7 A.P. Gunning, A.R. Mackie, P.J. Wilde and V.J. Morris, *Langmuir,* 2004, **20**, 116.
8 S. Roth, B.S. Murray and E. Dickinson, *J. Agric. Food Chem.*, 2000, **48**, 1491.
9 C.K. Siew and P.A. Williams, *J. Agric. Food Chem.*, 2008, **56**, 8111.
10 A.G.J. Voragen, W. Pilnik, J.-F. Thibault, M.A.V. Axelos and C.M.G.C. Renard, in Food Polysaccharides and Their Applications, ed. A.M. Stephen, G.O. Phillips and P.A. Williams, CRC Press, Boca Raton, 2nd edn, 2006, pp. 287–340.
11 S. Biggs, J.L. Burns, Y.D. Yan, G.J. Jameson and P. Jenkins, *Langmuir*, 2000, **16**, 9242.
12 A.J. Milling, *J. Phys. Chem.*, 1996, **100**, 8986.
13 A.J. Milling and B. Vincent, *J. Chem. Soc., Faraday Trans.*, 1997, **93**, 3179.
14 S.L. Carnie, D.Y.C. Chan, C.M. Lewis, R. Manica and R.R. Dagastine *Langmuir*, 2005, **21**, 2912.

15 G.B. Webber, S.A. Edwards, G.W. Stevens, F. Grieser, R.R. Dagastine and D.Y.C. Chan, *Soft Matter*, 2008, **4,** 1270.
16 S. Asakura and F. Oosawa, *J. Chem. Phys.,* 1954, **22**. 1254.
17 S. Biggs, R.R. Dagastine and D.C. Prieve, *J. Phys. Chem. B*, 2002, **106**, 11557.
18 G.N. Sethumadhavan, A.D. Nikolov and D.T. Wasan, *J. Colloid & Interface Sci.*, 2001, **240**, 105.
19 D.Y. Chan, R.R. Dagastine and L.R. White, *J. Colloid & Interface Sci.*, 2001, **236**, 141.

ENHANCING PROTEIN-PROTEIN INTERACTIONS AT THE EMULSION INTERFACE TO RESIST DISPLACEMENT BY BIOSURFACTANTS

N.C. Woodward, C. Fernández Fraguas, P.J. Wilde and V.J. Morris

Institute of Food Research, Norwich Research Park, Norwich, NR4 7UA, UK

1. INTRODUCTION

It is well established that the incidence of obesity has increased in recent years and, in view of the rising numbers of children and adolescents who are already obese, and are likely to remain so into adulthood, this trend appears set to continue. Already this rising incidence of obesity is placing an increasing burden on local economies through loss of workdays, and additional health care costs arising from an increased prevalence of obesity-related diseases such as cardiovascular disease, diabetes and cancers. This is despite there being a range of healthy alternatives to foods traditionally high in fat.

Until recently many studies concerning the digestion of fats have focused upon how free fat ingested as part of a meal would be emulsified by the mechanical action of the stomach in the presence of surface active species present in the GI tract and the subsequent digestion by gastric and duodenal lipases. Most work has been carried out to understand the roles that the gastric and pancreatic lipases, colipases, and biosurfactants such as the bile salts, play in the digestion of fats [1,2]. Work has also been carried out to understand the interplay of these components with other moieties such as proteins [3,4] and detergents [5,6] that can inhibit lipolysis.

Many processed high-fat foods contain fat in an emulsified form that is already optimal in size for digestion. The emulsion droplets present in these systems are stabilised by an interfacial layer comprised of proteins or surfactants and until recently, little has been understood as to how this interfacial layer behaves under conditions in the GI-tract. However, in recent years there has been a drive towards understanding how emulsified foods are affected by conditions in the GI-tract particularly with a view to altering the interfacial characteristics of emulsified fats in order to modify their digestion under gastric and/or duodenal digestion [7-10]. It is well understood that following emulsification the adsorption of bile salts is important in clearing the fat substrate of any extraneous material, and conferring a highly negative charge on the interface, thereby improving the 'interfacial quality' of the fat surface [11]. This process is an important step in attracting colipase and subsequently lipase to the interface thereby facilitating lipolysis.

Chu and coworkers [12] demonstrated that incorporating the surface-active galactolipid DGDG, moderates *in vitro* lipolysis: the galactolipid appears to sterically hinder the adsorption of bile salts at the fat surface thereby limiting subsequent adsorption of lipase and colipase. The interest in such an approach is that, if it is possible to slow the rate of lipolysis sufficiently to enable products of digestion to move further along the GI-tract into the ileum, then it is possible that this will result in a physiological response that would promote satiety.

An alternative approach to incorporating molecules into the fat phase for protein-stabilised emulsions is to manipulate the interfacial structure to restrict adsorption of bile salts. Maldonado-Valderrama and co-workers demonstrated that bile salts are incredibly efficient at removing proteins from the air-water [13] and oil-water [14] interface: bile salt adsorption arises due to the growth of bile salt domains within the protein network. There are already several examples of such work in the literature primarily looking at differences between emulsifiers typically used in the food industry [15-17] or by modifying protein networks by enzymatic [18] or heat treatments [19,20]. We have been focusing on whether it is possible to create protein-stabilised emulsions and then enhance interactions between adjacent molecules in the network to strengthen the interfacial film. Previous studies using atomic force microscopy suggest that the ease with which protein networks are broken down during their displacement by surfactants is dependant upon the 'strength' of the protein network. Thus it should be possible to rationally design interfaces to resist complete displacement by bile salts: potentially this would reduce the interfacial area covered by bile salts, thereby reducing the surface concentration of lipase and subsequently limit the rate of lipolysis. However, for such approaches to be effective it is necessary to understand how untreated or modified interfacial structures are affected under digestion conditions.

2. METHODS

2.1 Materials

2.1.1 Proteins & Surfactants. Sodium caseinate samples were supplied by DMV International (Vehgel, Netherlands). Samples of β-lactoglobulin (β-lg) and α-lactalbumin (α-la) were supplied by Sigma Aldrich (Poole, UK). Tween 20 (T20) was obtained as a 10% solution from Pierce (Rockford IL, USA).

2.1.2 Enzymes. Highly purified Pancreatic Porcine lipase (Type VI-S activity ≥ 20,000 U/mg), and Pepsin were obtained from Sigma Aldrich, microbial transglutaminsase ACTIVA-WM (TG) (Ajinomoto, obtained from Vesantti Oy-Helsinki, Finland) was purified from the method described by Lannto and co-workers [21], and the activity determined to be 11,000 nkat using the method of Folk [22].

2.1.3 Bile Salts. sodium taurocholate (NaTC, 95% TLC) and sodium glycodeoxycholate (NaGDC, 97% TLC), were obtained from Sigma Aldrich.

2.1.4 Solutions. All protein, enzyme and bile salt solutions were used as supplied and prepared in 2mM Bis Tris buffer at pH 7. All aqueous solutions were prepared using surface pure water from an Elga Elgastat water purifier, having a conductivity of 18.2 MΩ.

2.1.5 Purification of olive oil. Highly purified olive oil (Sigma Aldrich) was further purified by mixing with Florisil resin (Sigma Aldrich) for 4-6 hours and the mixture centrifuged in a Beckman centrifuge at 13,000 rpm, the supernatant was filtered under vacuum to remove any remaining resin.

2.2 Emulsion preparation

2.2.1 Preparation. Protein solutions were prepared to have a composition of 1.25% w/w and stirred gently for at least one hour. An emulsion premix was prepared from a four: one mixture of 1.25% w/w protein solution: olive oil and pulsed on a vortex mixer for 30s, before passing four-five times through a B3 Emulsiflex homogeniser (Avestin, ON,

Canada) at 10 psi, giving a final emulsion composition of 1% w/w protein and 20% (w/w) olive oil. Samples were left for one hour before carrying out any further treatments and/ or determining the particle size (LS13320 particle sizer – Beckman Coulter, UK).

2.2.2. Heat treatment of β-lactoglobulin emulsions. Emulsions were split into an untreated control sample and a heat-treated sample. The heat-treated sample was placed in an oven and heated from 30 °C to 85 °C at 1°C/min, then held at 85 °C for 40 minutes, before venting the oven and allowing the sample to cool to ambient temperature.

2.2.3 Enzymatic treatment of sodium caseinate emulsions. The emulsion was split into an untreated control and an enzyme-treated sample. The enzyme-treated emulsion was treated with sufficient transglutaminase to give a final composition of 100nkat/g protein and then incubated at 40 °C and shaken at 120 rpm for ~20 hours. The transglutaminase in the enzymatically-treated emulsion was deactivated by placing the sample in a water bath at 90 °C for 5 minutes, followed by immediate immersion in an ice bath for 5 minutes, for consistency the control was treated in the same way in the absence of transglutaminase.

2.3 Protein displacement studies

2.3.1 Sample preparation. To determine how the resistance of treated-protein networks to surfactant displacement compared to their respective controls, emulsion samples were exposed to increasing concentrations of the non-ionic surfactant T20. Samples were prepared to a final composition of 0.5 % (w/v) protein, 10% (%w/v) olive oil, with T20 compositions varying between 0.001 % (w/v) and 5 % (w/v). After mixing, the samples were left for one hour before analysis by either surface loading technique or surface potential measurements.

2.3.2 Surface potential measurements. Prior to analysis samples were diluted in 10mM pH7 phosphate buffer to give a final composition of 0.005% w/v as already determined from a concentration-attenuation sweep. The method has been described previously [18], but in brief samples were placed in a folded capillary cell (DTS1060C – Malvern Instruments, UK), with subsequent analysis carried out on a Malvern Instruments Nano DS. Typically, each sample was analysed three times with each measurement being comprised of 12 sub-runs, with the entire measurement being carried out in triplicate.

2.3.3 Surface loading techniques. Determination of protein coverage at the emulsion interface has been described in detail previously [18]. Emulsions were centrifuged on a bench-top centrifuge at 13,000 rcf for 10 minutes, followed by recovery of the subnatant, and this process was repeated for the subnatant twice more. Protein concentration was measured using the *ortho*-phthaldialdeyde (OPA) assay where 100μl protein solution is added to 3ml of OPA reagent inverted and left for 5 minutes before measuring its UV absorbance at 340nm, correction for scattering from residual emulsion was carried out as described previously [18].

2.4 AFM Imaging

Interfacial protein films were prepared using a 150 ml glass jacketed dish connected to a waterbath. The surface pressure (π) was monitored by means of the Wilhelmy plate method using either a roughened glass or platinum plate of known dimensions. Protein was spread at the bare air-water interface to give a surface pressure of ~10mN/m. Typically, after stabilisation for 1 hr the heat treated β-lg films were prepared by increasing the temperature of the subphase to ~70 ° C and held at this temperature for 40 minutes. The total time for the heat-treatment (including heating and cooling was typically ~3 hours), so control films were left for a comparable period, as aging can alter film strength, before beginning displacement with surfactant. The surface pressure is increased by sequential additions of T20 from a starting concentration of 0.5μM. As the surface pressure is increased the interface was sampled by carrying out successive Langmuir-Blodgett dips onto freshly-cleaved mica. Once dry the interfacial structures transferred onto the mica substrates are imaged under redistilled n-butanol, using an East Coast Scientific AFM (ECS Ltd, Cambridge, UK) in dc (contact mode), using a SiNi short-legged soft contact mode lever, having a force constant of 0.27 N/m (Budget Sensors, Sofia, Bulgaria).

2.5 *In vitro* digestions

In vitro digestions were carried out using a Titrando 836 titration device, equipped with a pH meter and Dosino 800 dosing unit (Metrohm Instruments), which allows *in situ* titration of the digestion mixture as the lipolysis progresses. 0.5ml of a 20% w/w protein-stabilised emulsion was added to ~27.5ml buffer (2mM Bis Tris, 150mM NaCl buffer, pH 7.0) at 37 °C in a jacketed vessel. Samples were exposed to simulated gastric conditions (pH 2.5, pepsin (20:1, g protein: g pepsin) for one hour, before the addition of 10 mM $CaCl_2$ and addition of NaOH to increase the pH to 7.0. The addition of duodenal reagents were made in the following order; 0.97 mM bile salt mixture (NaTC:NaGDC, molar ratio 47.3:52.7) 2 nM porcine pancreatic lipase and a ~five-fold excess of porcine pancreatic colipase. Lipolysis was monitored for 2 hours in pHSTAT mode: where the fatty acids evolved during lipolysis are neutralised by addition of 0.01 M NaOH. The volume of NaOH required to maintain neutral pH reflected the extent of lipolysis under different treatment regimes. The mixture is stirred vigorously to ensure that the emulsion is well dispersed throughout the titration.

3. RESULTS & DISCUSSION

Understanding how protein networks are affected by the presence of enzymes and biosurfactants in the gastrointestinal tract is essential to enable the rational design of emulsion interfaces to promote satiety. Atomic force microscopy studies have demonstrated that the ease with which proteins are displaced from interfaces by surfactants appears to be related to the mechanical strength of the protein film. The displacement occurs through an orogenic mechanism in which surfactant domains expand within a protein network with the expansion opposed by the local elasticity of the network. Protein displacement requires the failure of the network at a characteristic surface pressure [23,24]. The process is generic and similar displacement mechanisms apply for air-water and oil-water interfaces. Thus studies at air-water interfaces, which are easier to measure experimentally, can be used to evaluate the effects of strengthening protein networks. Typically proteins that form strong interactions with one another such as the whey proteins, tend to be less easily displaced than protein networks composed of random coil proteins [23]. The displacement of commercial sodium caseinate is determined by the displacement of the random coil α- and β-caseins during the early stages of displacement, with the later stages of displacement dominated by the presence of a minor (less than 15% w/w) component κ-casein [25]. Figure 1 demonstrates the difference in the levels of protein coverage observed for a β-lg film displaced from an air-water interface by Tween 20 at 27mN/m and the behaviour of a commercial whey protein film at an equivalent surface pressure. Whey protein isolate is composed primarily of β-lg and the differences are attributed to differences in preparative conditions of the isolate. It has been shown previously that the displacement of β-lg by bile salts from an air-water interface is also complete at a surface pressure of ~27mN/m [13]. Thus there is an interest in determining if enhancing the resistance of protein networks to displacement by surfactants, would give us the potential to create systems able to resist displacement by biosurfactants such as bile salts: providing a potential route by which to moderate lipolysis.

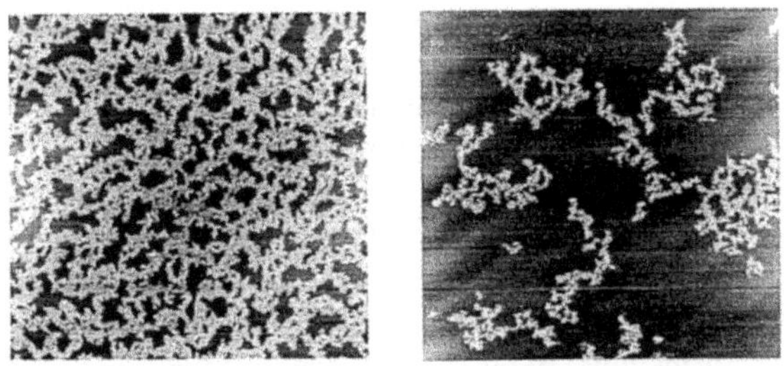

Figure 1. AFM images obtained for spread whey protein isolate (left) and β-lactoglobulin (right) films at an air-water interface, at a π ~27mN/m during their displacement with the non-ionic surfactant Tween 20. The grey areas represent protein, the black areas represent the area occupied by Tween 20.

Protein-protein interactions have been enhanced in this study in one of two ways; enzymatic crosslinking or heat treatment. Figure 2 shows the protein loading and surface potential measurements made on sodium caseinate emulsions prepared in the presence or absence of transglutaminase (TG). The resistance to surfactant displacement has been monitored by following displacement with Tween 20 (T20).

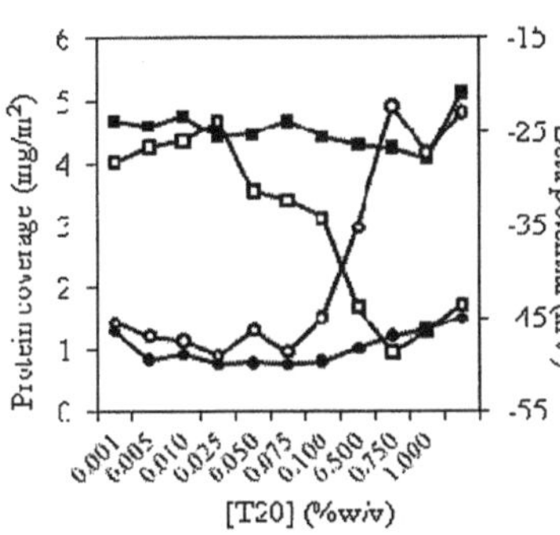

Figure 2. Protein loading measurements for 1% sodium caseinate stabilised emulsions +TG (■) and –TG (□). Surface potential data obtained +TG (●) and -TG (○).

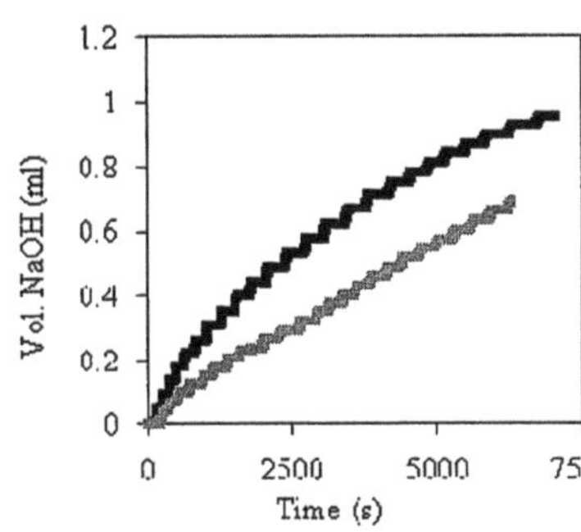

Figure 3. *In vitro* duodenal lipolysis of 1% sodium caseinate stabilised emulsions; +TG (■), and -TG (■), following a gastric conditioning step in the presence of pepsin.

Surface-potential measurements demonstrate that the protein-stabilised emulsion has a negative surface potential (~-46mV). As the concentration of the non-ionic surfactant T20 in the continuous phase of the emulsion increases it displaces the negatively charged protein and the surface potential tends to zero. At the same time protein-loading measurements demonstrate a concomitant reduction in surface protein coverage. Conversely, the transglutaminase-treated emulsion shows very little decrease in protein coverage and this is reflected by little change in the surface potential of the emulsion. Previous studies on a similar system demonstrated that, under *in vitro* duodenal conditions, enzymatically-crosslinked protein networks resulted in the production of less fatty acid during digestion when compared to the untreated control, suggesting that the enzymatically-treated emulsion was also more resistant to lipolysis [18]. Preliminary *in vitro* duodenal digestion following a gastric step in the presence of pepsin (Figure 3), indicate that the enzymatically treated protein networks are also more resistant to lipolysis than their respective control emulsion under the conditions investigated.

Further studies have been carried out on the displacement of heat-treated (β-lg)-stabilised emulsions. Surface potential measurements (Figure 4a) have demonstrated that the heat treatment itself causes a slight decrease in the surface potential of the emulsion, suggesting the surface properties of the protein at the emulsion interface have been altered. Surface potential measurements obtained for emulsions that have been exposed to increasing concentrations of the surfactant Tween 20, demonstrates that the heat-treated β-lg network is less susceptible to displacement by surfactant, than the untreated control sample (Figure 4a). Preliminary AFM images obtained for protein networks transferred from the air-water interface have shown that the surface area occupied by the heat-treated β-lg network is still almost complete at surface pressures greater than those by which the unheated β-lg network is completely displaced (Figure 4b).

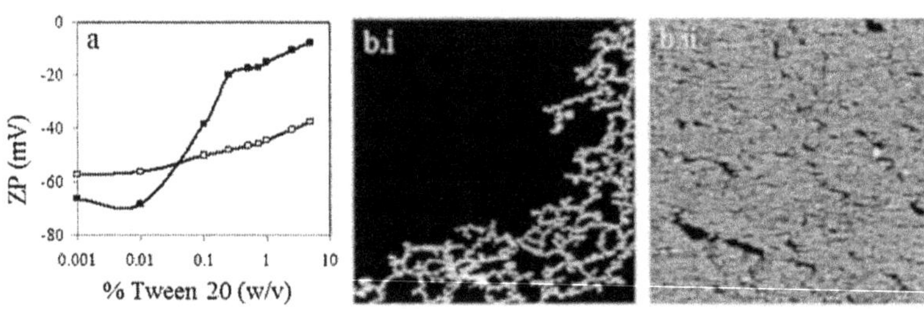

Figure 4. Effect of heat treatment on the displacement of β-lactoglobulin (β-lg) networks by Tween 20 (T20) as demonstrated by; (a) surface potential measurements for β-lg stabilised emulsions; unheated (■) and heated (□); (b) AFM images obtained for β-lg:T20 networks substrates sampled at the air-water interface; i. unheated, π -26mN/m; ii. heated, π - 29.5mN/m.

Since completing this work it has been demonstrated that heat-treated β-lg absorbed at the oil-water interface is more susceptible to pepsinolysis than untreated networks [19]. It was considered of interest to follow the digestion of a heat-treated β-lg network following a gastric step in the absence of pepsin, as it has been suggested that emulsion-based foods may be poorly mixed in the stomach, so some emulsion may pass through to the duodenum relatively untouched by acid conditions or pepsinolysis [26].

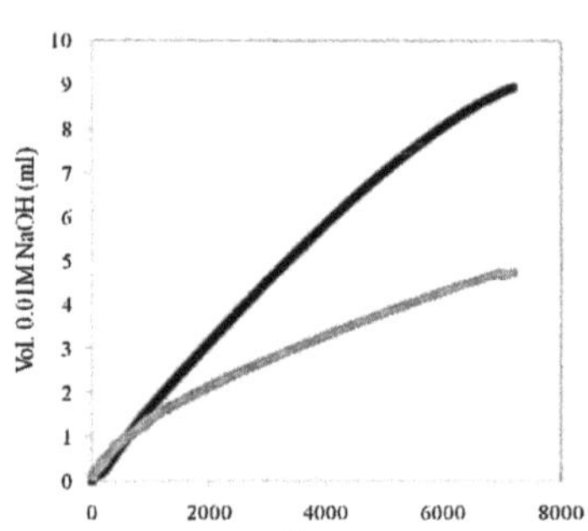

Figure 5. *In vitro* duodenal lipolysis following a gastric-conditioning step in the absence of pepsin, of a 1% β-lactoglobulin-stabilised emulsions; heat-treated at 85°C for 40 minutes (■) and untreated (▒),.

The results from these lipolysis studies shown in figure 5, suggest that the heat treatment may also result in an increased susceptibility of (β-lg)-stabilised emulsions to lipolysis. As yet, we do not know if this increased susceptibility has arisen from the heat-treatment making the β-lg more susceptible to displacement by bile salts or/ and enzymatic degradation.

In addition to observing that heat-treated β-lg is less resistant to pepsinolysis when adsorbed to the oil-water interface, Nik and co-workers [19] have also demonstrated that heat-treated α-la is more resistant to pepsinolysis when adsorbed to the oil-water interface than when free in solution. It is well established that α-la and β-lg interact with one another via disulphide and hydrophobic interactions during exposure to elevated temperatures. We have been interested in using heat treatment to modify α-la: β-lg interactions in emulsion systems with an increasing proportion of α-la. Surface potential data shown in Figure 6 demonstrate the effect of heat treatment on the resistance of α-la: β-lg stabilised mixtures following exposure to T20. Data obtained for the mixtures having α-la:β-lg ratios of 1:1 and 3:1 demonstrate that the heat-treatment increases the resistance of the protein network to displacement by surfactant. Conversely, the heat treatment had little effect on the displacement of 100% (α-la)-stabilised emulsions (data not shown).

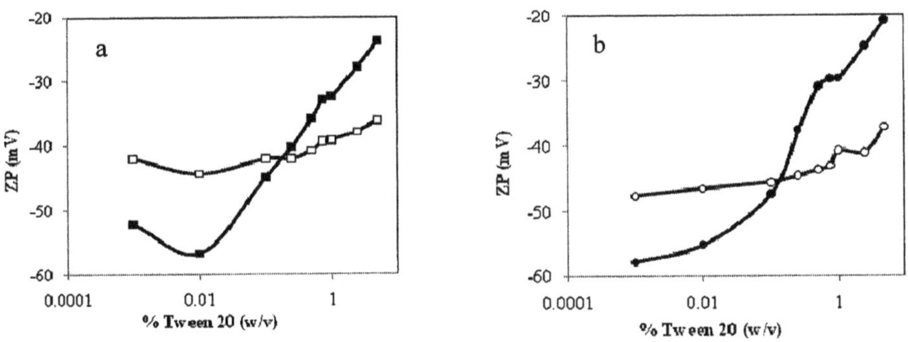

Figure 6. Surface potential data obtained for emulsions stabilised by α-lactalbumin: β-lactoglobulin (a) 3:1 ratio (□ heated; ■ unheated) and (b) 1:1 ratio (O heated, ● unheated), in the presence of Tween 20.

4. CONCLUSIONS

Treatment of protein-stabilised emulsions to enhance the resistance to displacement of protein networks by the non-ionic surfactant has been successful using both the enzymatic treatment of sodium caseinate networks and the heat-treatment of β-lg networks. Subsequent duodenal lipolysis measurements following exposure to a gastric step, demonstrated that the enzymatic treatment of sodium caseinate with transglutaminase does appear to reduce the extent of lipolysis of this emulsion in comparison to the untreated analogous emulsion. Conversely, despite the resistance of heat-treated β-lg networks to displacement by T20 and even following a gastric step in the absence of pepsin the extent of lipolysis was greater for this system than the control. Initial studies have demonstrated that emulsions stabilised by heat-treated (α-la)-(β-lg) mixtures are more resistant to displacement by T20. Whilst using a non-ionic surfactant to probe the resistance of protein networks is useful for assessing resistance to displacement by surfactant, it is clearly important to assess the effect of modifications using *in vitro* methodologies.

ACKNOWLEDGEMENTS

The authors would like to thank the BBSRC for funding this research through the responsive mode grant (BB/F020996/1) and the competitive and strategic grant to the Institute, and also the Fundación Española Para la Ciencia y Tecnología (FECYT) and the Spanish Ministry of Education for the award made to Dr Cristina Fernández Fraguas.

REFERENCES

1. M.E. Lowe, *Annu. Rev. Nutr.,* 1997, **17**, 141
2. P. Reis, K. Holmberg, H. Watzke, M.E. Leser and R. Miller, *Adv. Coll. Interface Sci.*, 2009, **147-148**, 237
3. Y. Gargouri, R. Julien, G.Pieroni, R. Verger and L. Sarda, *J. Lipid Res.*, 1984, **25**, 1214
4. Y. Gargouri, G. Pieroni, C. Rivière, A. Sugihara, L. Sarda and R. Verger, *J. Biol. Chem.*, 1985, **260**, 2268
5. Y. Gargouri, R. Julien, A.G. Bois, R. Verger and L. Sarda, *J. Lipid Res.*, 1983, **24**, 1336
6. P. M. Reis, T. W. Raab, J.Y. Chuat, M.E. Leser, R. Miller, H.J. Watzke, and K. Holmberg, *Food Biophysics,* 2008, **3**, 370
7. E. Dickinson, *Soft Matter*, 2008, **4**, 932
8. M. Golding and T.J. Wooster, *Curr. Opin. Colloid Interface Sci.,* 2010, **15**, 90
9. D.J. McClements and Y. Li, *Adv. Colloid Interface Sci.*, 2010, **159**, 213
10. V.J. Morris, *Trends in Biotechnology*, 2011, **29**, 509
11. M. Wickham, M. Garrood, J. Leney, P.D.G. Wilson and A. Fillery-Travis, *J. Lipid Res.,* 1998, **39**, 623
12. B.-S. Chu, G. T. Rich, M. J. Ridout, R. M. Faulks, M. S. J. Wickham and P. J. Wilde *Langmuir,* 2009, **25**, 9352
13. J. Maldonado-Valderrama, N.C. Woodward, A.P. Gunning, M.J. Ridout, F.A. Husband, A.R. Mackie, V.J. Morris and P.J. Wilde, *Langmuir*, 2008, **24**, 6759
14. Maldonado-Valderrama, A.P. Gunning, M.J. Ridout, N.C. Woodward, P.J. Wilde, and V.J. Morris, in *Gums and Stabilisers for the Food Industry* 15, (Eds) P.A. Williams and G. O. Phillips, RSC publishing, Cambridge, 2010, p.384
15. S. Mun, E.A. Decker and D.J. McClements, *Food Res. Int.*, 2007, **40**, 770
16. S.J. Hur, E.A. Decker and D.J. McClements, *Food Chem.,* 2009, **114**, 253
17. A.M. Nik, A.J. Wright and M. Corredig, *Colloids and Surf. B:*, 2011, **83**, 321
18. N.C. Woodward, A.P. Gunning, P.J. Wilde, B.-S. Chu and V.J. Morris, in *Gums and Stabilisers for the Food Industry* 15, (Eds) P.A. Williams and G. O. Phillips, RSC publishing, Cambridge, 2010, p.367
19. A.M. Nik, A.J. Wright and M. Corredig, *JCIS*, 2010, **344**, 372
20. S. Sandra, E.A. Decker and D.J. McClements, *J. Agric. Food Chem.*, 2008, **56**, 7488
21. R. Lannto, E. Puolanne, N. Kalkkinen, J. Buchert and K. Autio, *J. Agric. Food Chem.*, 2005, **53**, 9231
22. J.E. Folk in *Metabolism of amino acids and Amines* (Eds.) H. Tabor and C.W. Tabor, Academic Press, New York, 1970, p.889
23. A.R. Mackie, A.P. Gunning, P.J. Wilde and V.J. Morris, *JCIS,* 1999, **210**, 157
24. V.J. Morris and A.P. Gunning, *Soft Matter*, 2008, **4**, 943
25. N.C. Woodward, A.P. Gunning, A.R. Mackie, P.J. Wilde and V.J. Morris, *Langmuir,* 2009, **25**, 6739
26. L. Marciani, M.S.J. Wickham, D. Bush, R. Faulks, J. Wright, A.J. Fillery-Travis, R.C. Spiller and P.A. Gowland, *British Journal of Nutrition*, 2006, **95**, 331

EFFECT OF pH ON O/W EMULSIONS FORMULATED WITH POTATO PROTEINS AND CHITOSAN

N Calero, J Muñoz and A Guerrero

Departamento de Ingeniería Química. Facultad de Química. Universidad de Sevilla. c/P. García González, 1, E41012-Sevilla (Spain).

1. INTRODUCTION

Emulsions formulated with healthy ingredients are becoming increasingly popular in the developed world, due to the consumer concern on diseases related to fatty diets.
Oil in water emulsions consist of oil droplets dispersed in an aqueous phase. They are intrinsically unstable systems and tend to separate into the aqueous and oil phases by a number of mechanisms[1]. However, emulsions may be kinetically stabilised for a period of time by using two kinds of substances, emulsifiers and thickening agents, which may inhibit or retard the destabilisation mechanisms.
Emulsifiers are surface-active molecules that absorb at the surface of droplets forming a protective membrane that prevents droplets aggregation. The most commonly used emulsifiers are amphiphilic proteins, polysaccharides, low molecular weight surfactants, and phospholipids. The second class of compounds, thickening agent, is often used in order to retard destabilisation and to control texture of emulsions[2].
Polysaccharides induce a viscosity increase of the emulsion and also promote a network of oil-droplets[3]. Also, some polysaccharides show surface activity and therefore they act as excellent emulsifiers[4]. Chitosan is a natural amino-polysaccharide, which offers numerous physiological properties[5,6]. Thus, the use of chitosan in emulsions is not only desirable to produce healthy food products, but also because of its role as stabiliser and on account of its surface activity, which contributes to the emulsifying capacity when formulated with proteins[7].
The electrical charge on the droplets is the result of the adsorption of emulsifier molecules that contain ionized groups[8]. The magnitude and sign of the electrical charge at the droplet interface largely depends on the type and concentration of surface-active molecules present at the interface, as well as the pH and ionic composition of the aqueous phase.
This contribution deals with the influence of the pH of a vegetable protein suspension used as emulsifier on O/W emulsions with high-oleic sunflower oil stabilised with chitosan.

2 METHODS

2.1. Materials

Potato Protein Isolate (PPI) (c.a. 80 wt%) was supplied by Protastar (Reus, Barcelona, Spain). The chemical composition of potato protein isolate was as follows, 80± 2 wt % protein; 3.1± 0.4 wt % lipids; 5.9± 0.6 wt % carbohydrates; 0.8± 0.1 wt % ashes; 10± 2 wt % moisture. The isoelectric point of PPI was ca. 4.

Chitosan (CH) of medium molecular weight (MW 190,000-310,000 Da, 75-85% deacetylated), anhydrous sodium acetate and glacial acetic acid were used as supplied by Sigma Aldrich (St. Louis, USA).

2.2. Preparation of Oil-in-Water (O/W) Emulsions

1st Step.
A potato protein dispersion was prepared by dispersing 5.0 wt% potato protein powder in Milli-Q water. This protein dispersion was stirred for 10 min and was then adjusted to pH 2, 3, 6, 8, 10 and 11.5 with 1 M NaOH or 1M HCl. A concentrated emulsion was prepared by mixing 50 wt% oil and 50 wt% protein suspension with a high-shear mixer (Ultraturrax T-50) (pre-emulsion 1)
2nd step.
The second processing step consisted of passing the pre-emulsion 1 through a high-pressure valve homogenizer (Emulsi-Flex C5, Avestin) three times at 1000 bar (pre-emulsion 2).
3rd step.
Final emulsions were obtained after mixing with a chitosan solution (2.5 wt% chitosan, 0.2 M acetic acid and 0.05 M sodium acetate, pH=3.9), using again the high-shear mixer for homogenisation. Thus, final emulsions containing 40 wt% high-oleic sunflower oil, 40 wt% aqueous potato protein dispersion (2 wt% potato protein) at different pH and 20 wt% chitosan solution (0.5 wt% chitosan in the final emulsion) were produced.

Table 1. Evolution of pH during processing.

Protein suspension	2.00	3.00	6.00	8.00	10.0	11.5
Pre-emulsions	3.54	3.87	6.17	7.77	9.55	10.65
Emulsions	4.46	4.65	4.88	5.00	5.15	5.55

Addition of chitosan buffered solution modified the pH of final emulsion to give values within the pH range used for food emulsion.

2.3. Rheological characterisation

The rheological characterisation involved stress and frequency sweeps in oscillatory shear experiments and flow curves at steady shear tests. Both tests were conducted with a Haake MARS controlled-stress (CS) rheometer (Thermo-Scientific), equipped with a serrated surface plate-and-plate geometry (60 mm diameter).
All the samples had the same recent past thermo-mechanical history. Thus, the samples were kept in a bath at 25°C for at least half an hour prior starting any rheological test. Then they were allowed to equilibrate at the measuring position of the sensor system for 10 minutes. All the rheological tests were repeated in duplicate for each system, the results showed corresponding to average values. The maximum deviations of experimental data from average values were always lower than 10% for both G' and G'' and 5% for flow curves.

2.4. Particles size distribution

The mean particle diameter of the protein dispersion was measured using a laser diffraction instrument (Malvern Mastersizer X).

2.5. ζ.potential measurements

The z-potential of oil droplets in the emulsions was determined using a (Nano Zetasizer, Malvern Instruments). All emulsions were equally diluted with water prior to the analysis to avoid multiple scattering effects.

2.6. Microscopic observation

The microstructure of emulsions was observed using a confocal laser-scanning microscope (CLSM) (Leica TCS SP2, Leica Microsystems CMS, Mannheim, Germany) using the 100 X objective lens.

A proper amount of emulsion was placed in a test tube and subsequently Nile red solution (oil staining dye) (0.01% in ethanol) was added and mixed thoroughly. The mixture was dropped on a microscope slide, which was covered with a cover slip and observed under the microscope.

3 RESULTS

Preliminary oscillatory torque sweep tests at 1 Hz were carried out in order to estimate the maximum amplitude value of the sinusoidal shear stress function, which guarantees a linear viscoelastic behaviour.

Figure 1 shows the variation of loss tangent (tan δ) versus angular frequency (ω) plots for final emulsions at different pH values, which were prepared with potato protein suspensions whose pH varied between 2 and 11.5. On the whole the loss tangent (tan δ) decreased with emulsion pH at all frequencies. This behaviour is due to the fact that an increase of pH provoked an increase of the elastic response of emulsions. Figure 1 also shows that the frequency dependence of the loss tangent was weaker and weaker as pH was increased, indicating a progressively slower relaxation mechanism. This is a clear indication of a stronger microstructure of emulsions with increasing pH. Analysing the mechanical spectra (data not shown), it was observed that at the lowest pH, the emulsion exhibited fluid-like viscoelastic behaviour as demonstrated by the fact that G'' values lay above those of G'. As emulsion pH was increased, a transition from fluid-like to a gel-like behaviour occurred. Thus, at pH 4.65, the elastic component was not significantly different to the viscous one throughout the frequency range covered. This has been related to the onset of a gel-like structure for polymeric systems. This gel-like structure was increasingly reinforced in the 4.88-5.55 pH range as demonstrated by the increasing extension of the plateau or rubbery relaxation zone.

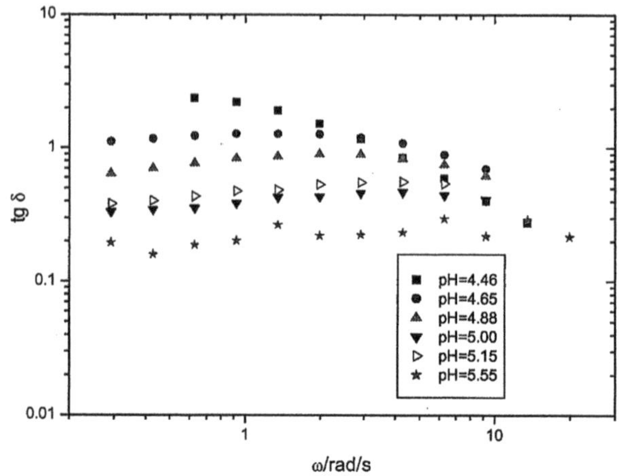

Figure 1. *Frequency dependence of the loss tangent as a function of emulsion pH. Stress amplitude: 0.1 Pa. T = 25 °C.*

Figure 2 shows the influence of emulsion pH on the shear rate dependence of shear stress. An increase of pH provoked a clear rise of shear stress and, therefore, of viscosity regardless the shear rate. This demonstrates that the effect of pH on emulsion structure keeps the same sequence as that observed under linear viscoelasticity, despite the effect of shear rate. All emulsions exhibited shear thinning behaviour over the whole shear rate range studied, with no evidence of a tendency to reach a Newtonian region at low shear rate or towards the infinite-shear rate viscosity.

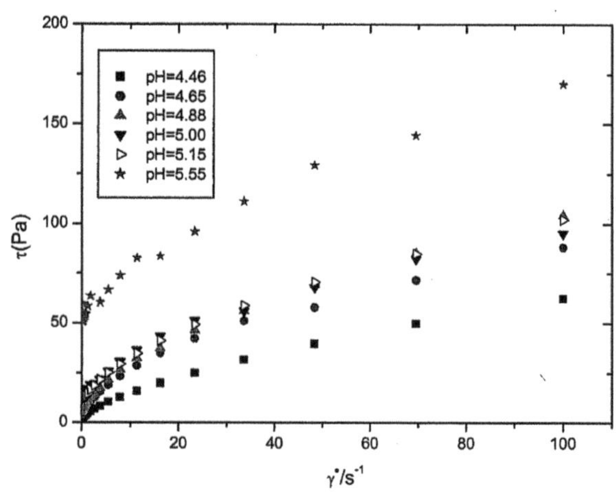

Figure 2. *Steady shear flow curves as a function of emulsion pH at 25 °C.*

The influence of potato protein suspension on ξ-potential of both pre-emulsion and emulsion droplets is shown in Figure 3. pH plays an important role in the ionization of the functional groups of the biopolymers and therefore it determines the formation of chitosan-potato protein complex. It is widely known that proteins carry a net positive charge at a pH below their isolectric point and on the contrary, a net negative charge above their isoelectric point. Droplets charge in pre-emulsions was only conditioned by the charge of potato protein at a given pH. Thus, the surface charge of particles in protein suspension was positive at pH 2 and 3 since the pH was below the potato protein isolectric point (4) (see table 1). On the contrary, the surface charge was negative above the potato protein isoelectric point (from pH 6 to pH 11.5). Measurement of ξ-potential of the final emulsions confirms that at this protein/chitosan ratio there was enough chitosan present to balance the negative charge on the protein. Thus, as the pH of the protein suspension was increased from 2 to 11.5, ξ-potential gradually became less positive. Given that both chitosan and potato protein exhibit opposite charges at pH above 4, it may be expected that positively charged chitosan macromolecules form a complex with the negatively charged surface of potato protein. The protein-polysaccharide protective layer formed around oil droplets must be responsible for the increasing emulsion stability with pH.

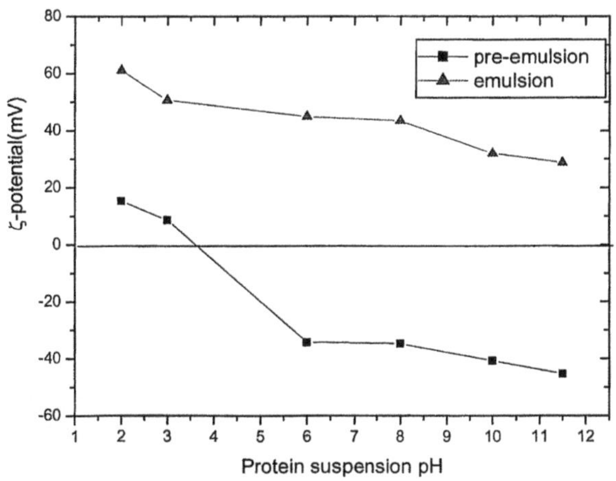

Figure 3. *Effect of protein suspension pH on the ζ-potential of pre-emulsions (potato protein isolate, high oleic sunflower oil) and final emulsions (potato protein isolate, high oleic sunflower oil and chitosan)*

Figure 4 shows the droplet-size distribution expressed as volume distribution for all emulsions studied. Distribution was a non-gaussian curve and indicated that at least two populations were present. The first peak corresponds to the population of droplets and the second one may be ascribed to protein aggregates. However, it is important to emphasize that both size ranges may partially overlap and distort the results. Therefore, just a qualitative analysis can be provided. The minor population was centred around 2 μm and the second population was centred between 10 and 20 μm, which pointed out the different solubility of protein as a function of pH. Interestingly, the volume of the second population decreased as the emulsion pH increased, revealing an enhanced protein solubility, which could be checked by visual observation.

Figure 5a and 5b show optical and confocal micrographs, respectively, for the emulsion with highest pH (5.55), which showed the best stability results. Confocal microscopy provides a 3D measurement of the dispersed phase morphology within the bulk of the sample, which revealed a polydisperse emulsion. Most droplet sizes were in the (2 -10) μm range, although smaller droplets were also observed.

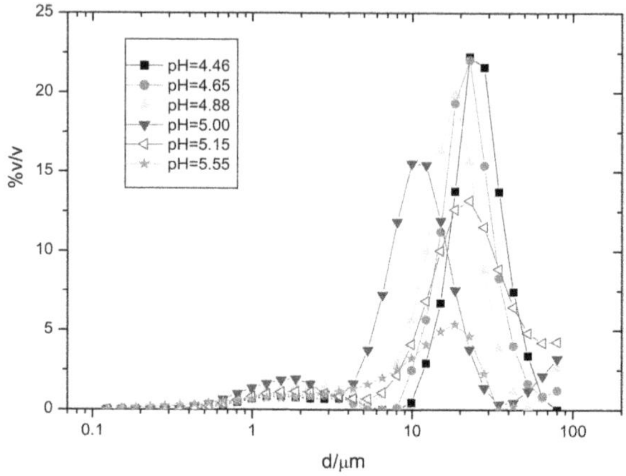

Figure 4. *Effect of emulsion pH on the particle size distribution of final emulsions (potato protein isolate, high oleic sunflower oil and chitosan)*

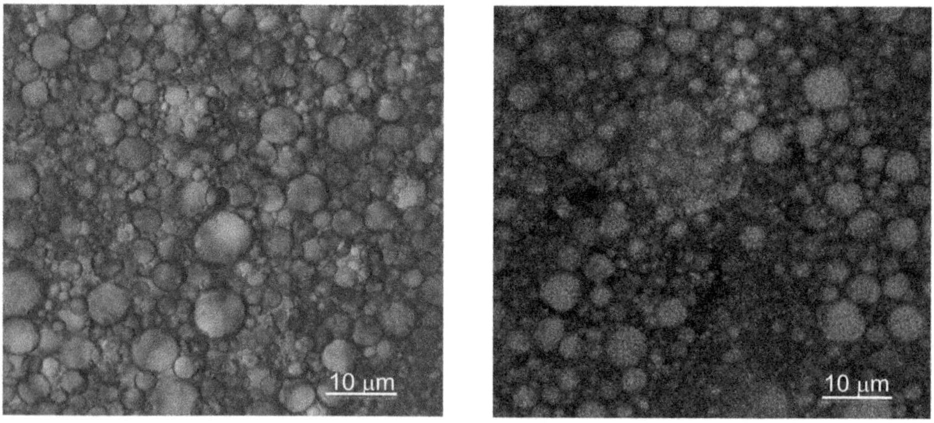

Figure 4. *Optical and CSLM micrographs of the potato protein-based emulsion stabilised with chitosan at pH 5.55.*

4 CONCLUSIONS

Potato protein-based high-oleic sunflower oil in water emulsions stabilised with chitosan exhibited viscoelastic properties which shift from fluid-like to gel-like as pH was increased. The increasing elastic response of emulsions with pH above 4.0 has been ascribed to the occurrence of a progressively stronger microstructure of emulsions, which must be based on the formation of thicker and thicker protective layers of potato protein-chitosan complex around oil droplets. The progressive formation of this microstructure turned out to be also efficient to withstand the effect of steady shear flow. Yet, the greater viscosities were found at the highest pH studied throughout the whole shear rate range studied.

ξ-potential and droplet-size distribution measurements supported the emulsion stability information, which could be visually assessed. Thus, the bimodal droplet-size distribution of these emulsions revealed the occurrence of a peak corresponding to oil droplets centred around 2 μm and another one centred between 10 and 20 μm, which may correspond to protein aggregates. Interestingly, the volume of the second population decreased as the emulsion pH was increased, indicating a higher protein solubility, which resulted in enhanced emulsion stability.

ACKNOWLEDGMENTS

The results presented are part of a research project (reference: P08-AGR-03974) supported by Consejería de Innovación, Ciencia y Tecnología (Junta de Andalucía, Spain). The authors kindly acknowledge their financial support.

REFERENCES

1. McClements, D. J. (2005). Food emulsions. Principles, practices, and techniques (2nd ed.). Boca Raton: CRC Press.
2. McClements, D. J. and Weiss J. (2005) Lipid emulsion. In F. Shahidi (Ed.), Bailey's Industrial Oil and Fat Products (6th Ed.) John Wiley & Sons, Inc.
3. Dickinson, E. (2009) Hydrocolloids as emulsifiers and emulsion stabilizers. Food hydrocolloids, 23 (6) 1473–1482.
4. Dickinson, E. (2003). Hydrocolloids at interfaces and the influence on the properties of dispersed systems. Food Hydrocolloids 17, 25–39.
5. Borderías, A. J., Sánchez-Alonso, I., & Pérez-Mateos, M. (2005). New applications of fibres in foods: addition to fishery products. Trends in Food Science and Technology, 16(9), 458-465.
6. Muzzarelli, R. A. A. (2009). Chitin and chitosan hydrogels. In G. O. Philips, & P. A. Williams (Eds.), Handbook of hydrocolloids (2nd ed.). (pp. 849-876) Boca Ratón: CRC Press.
7. Calero, N., Muñoz, J., Ramírez, P., Guerrero, A. (2010) Flow behaviour, linear viscoelasticity and surface properties of chitosan aqueous solutions. Food Hydrocolloids 24, 659-666.
8. D. J. McClements, Food Emulsions: Principles, Practice and Techniques, CRC Press, Boca Raton, Florida, 1999.

Fibres and films

ELECTROSPINNING OF FISH GELATIN FROM BLACK TILAPIA (*Oreochromis mossambicus*) SKINS

M.H. Norziah[1], H.L. Farm, A.B. Khairulnizam[1] and K. Sudesh[2]

[1]Food Technology Division, School of Industrial Technology,
Universiti Sains Malaysia, 11800, Penang, Malaysia. E-mail: norziah@usm.my
[2] School of Biological Sciences, Universiti Sains Malaysia, 11800 Penang, Malaysia.

1 INTRODUCTION

Electrospinning (e-spinning), a combination of two techniques (electrospray and spinning) is a novel and efficient process where polymer liquid is fabricated into ultra-fine fibres with diameters in the nanometer to sub-micrometer range[1]. A high electric field is applied to the droplet of a fluid which may be a melt or solution coming out from the tip of a die, which acts as one of the electrodes. This leads to droplet deformation and finally to the ejection of a charged jet from the tip of the cone accelerating towards the counter electrode leading to the formation of continuous fibres[2]. Recently, electrospun nanofibre matrices have gained a lot of attention and are being researched. Electrospinning has been regarded as an effective way to fabricate ultrafine fibres from a variety of synthetic or natural polymers. Due to their high surface to volume ratio, high porosity and good inter-pore connectivity, potential uses of e-spun fibres includes immobilized enzyme in nanofibrous membrane[3,4], tissue engineering scaffolds[5], platform for drug delivery[6] and carrier for DNA[7]. According to Huang et al.[8], synthetic materials such as polylactide (PLA), 3-phosphoglyceric acid (PGA), polylactic-co-glycolic acid (PLGA), and polycaprolactone (PCL) were common biopolymers used in the study of electrospinning. Recently, several studies were conducted on the electrospinning from natural biopolymers such as silk fibrion[9], collagen[10], chitosan[11] and fish gelatin[1].

As a naturally occurring biopolymer, gelatin is one of the most abundant natural biomaterials which can be obtained from skins and bones of animal and marine sources. The unique functional and physical properties of gelatin have made it a popular biopolymer used in food, pharmaceutical and cosmetic applications and has been used for several biomedical applications due to its biodegradability and biocompatibility. Commercial gelatins are mostly derived from pig skin, cow skins and bones. However, within the past few decades, gelatin from fish sources has been highlighted as an alternative to mammalian gelatin. Recently, the farming and production of tilapia, through aquaculture industry in Malaysia has increased steadily and become an important source of freshwater fish supply within the last few years. The ensuing production of fillet and tilapia based food products means more bones and skins would be discarded as waste[12]. Thus fish gelatin production could increase to serve as a potential alternative to mammalian sources for gelatin. Recently, many research attempts have been reported to form gelatin nanofibers by electrospinning using polar organic solvents such as 2,2,2trifluorothanol (TFE), hexafluoro isopropanol (HFIP) and water in which organic acid such as formic acid, acetic acid were usually added to improve the spinnability of the solutions (Chang et al.[13]; Yin et al.[14]; Zhang et al.)[15]. Although gelatin has been successfully electrospun into ultrafine fibers, the solubility of the electro spun gelatin fibers is a drawback since the fibres are hydrophilic in nature, have poor mechanical properties and the fibrous structure may be destroyed irreversibly when being exposed to a high humidity environment. These properties have limited the application of nanofibres in many fields in the past. Thus it is necessary to improve the water resistance of these nanofibers.

In this work, fish gelatin extracted from skins of black tilapia fish (*Oreochromis mossambicus*) was used for the fabrication of nanofibre mats using the electrospinning method. The objective is to evaluate the properties of electrospun fish gelatin, including the morphologies and the mechanical properties of the nanofibres as well as crosslinking treatment with glutaraldehyde.

2 MATERIALS AND METHODS

2.1 Materials

Black tilapia fish (*Oreochemis mossambicus*) weighing about 700g each were kindly provided fresh by a local aquaculture farm in Perak. Upon arrival at the laboratory, the fish were filleted, scales removed and skins were collected and kept below -20 °C until further use. All the chemicals used were of analytical grade.

2.2 Extraction of Tilapia Fish Skin Gelatin

Gelatin was extracted from Tilapia fish skins following the method optimised in our preliminary work[16]. Tilapia skin was thawed, cut into pieces (approximately 3 x 3 cm) and rinsed with excess tap water to remove superfluous material. The cleaned tilapia skin was soaked in 0.4 % (w/v) sodium hydroxide, followed by 2 % (v/v) sulphuric acid and 1 % (w/v) citric acid at a skin/solution ratio of 1:7 (w/v) for 40 min each, respectively. The skin was rinsed with distilled water after each soaking. The final extraction was carried out in distilled water at a skin/water ratio of 1:2 (w/v) at 45 °C for 8 hrs. The extract was then filtered through cheese cloth and dried in a hot air oven in a plastic tray at 50 °C for 18 hrs. The dried gelatin was ground, kept in an air-tight bottle and stored in a desiccator for further analysis.

2.3 Characterization of Extracted Tilapia Fish Gelatin

Characterization of the extracted Tilapia fish skin gelatin was carried out for moisture and ash content, bloom gel strength, viscosity and pH value. The yield of fish gelatin obtained was calculated as:

$$\text{Percentage of yield (\%)} = \text{(dry weight gelatin)/(wet weight fish skin)} \times 100$$

The moisture content of gelatin was determined on drying the samples (5 g) at 105 °C using IR-30 Moisture Analyzer (Denver Instrument Company, Ltd. England). The ash content of gelatin was determined according to the AOAC method: 923.03[17]. The pH value of 1.0 % (w/v) gelatin solution was measured using a pH meter (Mettler Toledo FE20/EL20, Switzerland). The gel strength of gelatin gel was determined according to the

Gelatin Manufacturers Institute of America[18] testing standard method. Gelatin solution (6.67 %, w/v) was prepared in a standard glass bloom jar (Schott Duran, Germany) and equilibrated for 16-18 hrs at 10 °C and was tested using a Texture Analyzer (TA.XT2, Stable Micro System, Surrey, England) with a load cell of 5 kg and test speed 1 mm/s. The gel strength was determined as maximum force (in g) when the 1.27 cm diameter cylindrical probe proceeded to penetrate into the gelatin gel to a depth of 4 mm. The viscosity of gelatin solution (6.67 %, w/v) was also determined using a Vibro Viscometer SV-10, (A & D Company, Limited, Japan) at 60 °C.

2.4 Electrospinning of Extracted Fish Gelatin Solutions

Initial experiments were conducted to evaluate the suitability of acetic acid and formic acid as solvents to dissolve extracted fish gelatin for the electron spinning process. Acetic acid was found to be the more suitable solvent and thus was used in this study. In the initial study, the concentrations of both acetic acid and fish gelatin were varied in order to prepare gelatin solutions that are suitable to produce gelatin nanofibre mats. All gelatin solutions was prepared by dissolving extracted fish gelatin powder in acetic acid solution under constant stirring with a magnetic stirrer for 12 hrs at 40 °C in a covered beaker to avoid evaporation, prior to the electrospinning process. Initially, fish gelatin (25 %, w/v) was dissolved in different concentrations of acetic acid, ranging from 10 to 60 % (v/v) to select the most suitable concentration of acetic acid for the electrospinning process. 30% acetic acid solutions produced smooth fibers and was used for further studies. This was then followed by a series of experiments in which several concentrations of fish gelatin were prepared (10 to 40 %, w/v) in 30% (v/v) acetic acid to investigate the effects of gelatin concentration on the electrospinnability of the solutions. Prior to e-spinning, these solutions were characterized for their viscosity and morphology by SEM.

The electrospinning process was performed using the device, Esprayer Multi Models ES-2000S series (Fuence CO., Japan). Gelatin solution was placed in a 5 ml glass syringe (outer diameter 1.6 mm) and a blunt tip stainless steel hypodermic needle (diameter 100-200 μm) was used as the nozzle. A fixed electrical potential of 30 kV was applied across a finite distance between the tip of the needle and the collecting plate of 15 cm. The feed rate was set at 3μL/min whereas the collection time was fixed at 45 min. For the nanofibre mat sample used for mechanical testing, selected gelatin solutions were electrospun continuously for 3 hrs and the plastic template with rectangular hole (dimension: 50 mm x 40 mm) was used as the collecting device. Nanofibre mat samples were removed from collecting device using forceps. The obtained nanofibre mat was stored in a desiccator, until further treatment and analysis.

2.5 Crosslinking of Gelatin Fibres

The nanofibres containing fish gelatin are water soluble and thus water can destroy the nanofibrous structure. By placing the nanofibres into a desiccator filled with glutaraldehyde vapour, the fish gelatin nanofibres can be crosslinked appropriately. The nanofibre mat sample was clamped between a pair of supporting frames with adhesive tape and kept in a sealed chamber saturated with the vapour of glutaraldehyde (20 ml) for 3 days. The sample was then left in the fume hood for 2 hrs followed by heating in the oven at 100 °C for 1 hr to evaporate the GTA vapour.

2.6 Characterization of Electrospun Gelatin Products

The properties of crosslinked and non-crosslinked nanofibre mats were evaluated and compared. The morphology of non-crosslinked and crosslinked electrospun products (fibres and/or beads) were evaluated by a scanning electron microscope (Leo Supra 50VP Field Emission SEM) equipped with Oxford INCA 400 energy dispersive x-ray microanalysis system. Dried nanofibre samples were glued on double-sided tape mounted on a specimen holder and coated with 100-200 Å thickness of gold. For each sample, 4 dimensions were used to view the surface of the nanofibre mat i.e. 2000x and 5000x. The fibre thickness was measured from average values of 100 points of SEM images obtained using SIS Image Processing software.

The colour measurements (L*, a* and b* values) of the nanofibres were conducted using a colorimeter (Minolta Spectrophotometer CM-3500d, Minolta Co. Ltd., Japan). Water solubility was also performed by immersing the nanofibre mat samples in distilled water at 37 °C. Tensile strength and elongation at break nanofibre strips were measured using TA-XT2 Texture Analyzer (Stable Micro System, Survey, UK) equipped with the tensile grip following the method suggested by Huang et al.[8] with initial grip separation of 30 mm, trigger force of 5.0 g and crosshead speed of 0.2 mms^{-1}.

2.7 Statistical Analysis

All measurements were conducted in triplicate. Data were subjected to Independent-Sample T-Test. Analysis was performed using SPSS V 14.0 software (SPSS Institute Inc., Cary, NC, USA).

3 RESULTS AND DISCUSSION

3.1 Properties of Extracted Tilapia Skin Gelatin

The gel strength, pH value, moisture and ash content of extracted tilapia skin gelatin was 255 ± 1.5 g, 4.3 ± 0.02, 4.2 ± 0.1 % and 0.5 ± 0.01 %, respectively. The solution viscosity of gelatin tested was 7.8 ± 0.2 mPa.s. The fish gelatin yield obtained was 20.6 ± 0.02 %. The yield was very much higher than those reported for Tilapia spp. by Grossman and Bergman [19] and Jamilah and Harvinder[12] which were less than 10 %. Grossman and Bergman[19] have also reported a bloom strength of 263 g for the tilapia spp. Jamilah and Harvinder[12] have reported for black and red tilapia: bloom strength (181 g and 128 g); viscosity (7.12 mPa.s and 3.20 mPa.s); and pH values (3.9 and 3.1), respectively. The difference in pH value of the gelatin may be due to the different type and strength of acids employed during the extraction procedures.

3.2 Electrospinning of Fish Gelatin Solutions

In a set of preliminary studies, solvents i.e. glacial acetic acid and formic acid of varying concentrations were used to prepare extracted fish gelatin solutions and tested for their ability to produce electrospun gelatin nanofibers. Solvents influence the surface tension and viscosity of the solution and affect the morphology of fibres. It was found that formic acid was not suitable to give electrospun gelatin fibres (data not shown). With glacial acetic acid, fish gelatin solutions at 25% (w/v) in low acetic acid concentrations of 10 and 20 % (v/v) did not produce electro spun gelatin fibres due to the low solution viscosity, however fish gelatin in 30 % (v/v) acetic acid, gave continuous gelatin fibre jets during the electrospinning process and a smooth homogeneous nanofibre mat was successfully deposited. Increasing the acetic acid concentration to 40% (v/v), fibre strands were observed at the needle tip during the electrospinning process. The high concentration of acetic acid increased the volatility of the electrospun solution and fast evaporation of solution causing the fibres to solidify rapidly before reaching the collecting plate, and thus resulting in fibre strands oscillating producing fibre balls at the syringe needle tip instead. Further increase in acetic acid concentration (50 and 60 % (v/v)), resulted in high solution viscosities and poor aspiration rates. When gelatin dissolves in an acid solution, some of the free amino groups of some acid residues bind with dissociated protons (H^+) causing the chains to carry positive charges. This causes an expansion of the coil-like structure of the gelatin chains, hence the observed increase in the viscosity of the solutions[1]. Thus, 30 % (v/v) concentration of acetic acid was chosen as the concentration of the solvent for further experiments.

The effects of varying concentrations of extracted fish gelatin solutions from 5 to 40 % (w/v) in 30 % (v/v) acetic acid to produce good nanofibres during the electrospinning process were observed as given in Table 1. The viscosity of electrospinning solution increased with increasing concentration of gelatin. This could be attributed to the increase in the molecular chain entanglement with increasing the solution concentration[20]. At lower concentration of gelatin, 5 % (w/v), dripping was observed during electrospinning due to its low viscosity. At such low viscosity, the viscoelastic force (a result of the low degree of chain entanglements) in a given jet segment was not large enough to counter the higher Coulombic force. This caused the break-up of the charged jet into smaller jets. Smaller jets were later rounded up to form droplets as a result of the low surface tension[20]. At 10 % (w/v) gelatin concentration, viscosity of the solution was 17.4 mPa.s, whereas the surface tension was relatively high. Therefore, the solution jet at the end of the tip could not maintain its own shape due to the high surface tension and thus formed small drops among the fibres instead of nanofibre strands.

Table 1 *Characteristics of electro spun gelatin fibres of different gelatin concentration 30% (v/v) acetic acid*

Gelatin concentration, % (w/v)	Viscosity of the gelatin solution (mPa.s)	Average diameter of gelatin fibres (μm)	Diameter range (μ) of gelatin fibres
5	7.8	-	-
10	17.4	-	-
15	32.1	-	-
20	60.7	0.11 ± 0.04	0.04 – 0.26
25	87.5	0.18 ± 0.09	0.07 – 0.75
30	220	0.22 ± 0.07	0.09 – 0.49
35	520	0.38 ± 0.12	0.23 – 0.93
40	890	-	-

3.3 Morphology and structure property

From SEM images (Figure 1a), no fibre formation was found and only spherical beads and droplets were observed. A similar finding was also reported by Fong et al.[21] and explained that this phenomenon was the effect of the viscosity and surface tension. Low viscosity induced electrospraying instead of electrospinning and thus caused the solution to be sprayed instead of forming nanofibres. At 15 % (w/v) concentration of gelatin, the solution still could not form nanofibres and produced a "bead on a string" structure as shown in Figure 1b. Increasing the solution viscosity with increase in gelatin concentration favored the formation of smooth fibres. The beads were bigger when the viscosity of the solution was increased. As a result, the average distance between beads was longer, the fibre diameter was larger and finally the shape of the beads changed from spherical to spindle-like. Thus, less bead formation and thicker diameter nanofibres were observed in the samples with higher concentration of gelatin as was also observed by Fong et al.[21].

Thus, the nanofibre formation was facilitated when the concentration of fish gelatin increased from 20 % to 35 % (w/v) (Figure 1c-f). At higher viscosity of the solution, the charged jet did not break up into small droplets. This was due to the increased chain entanglements (and hence an increase in the viscoelastic force) which were sufficient to prevent the break-up of the charged jet.

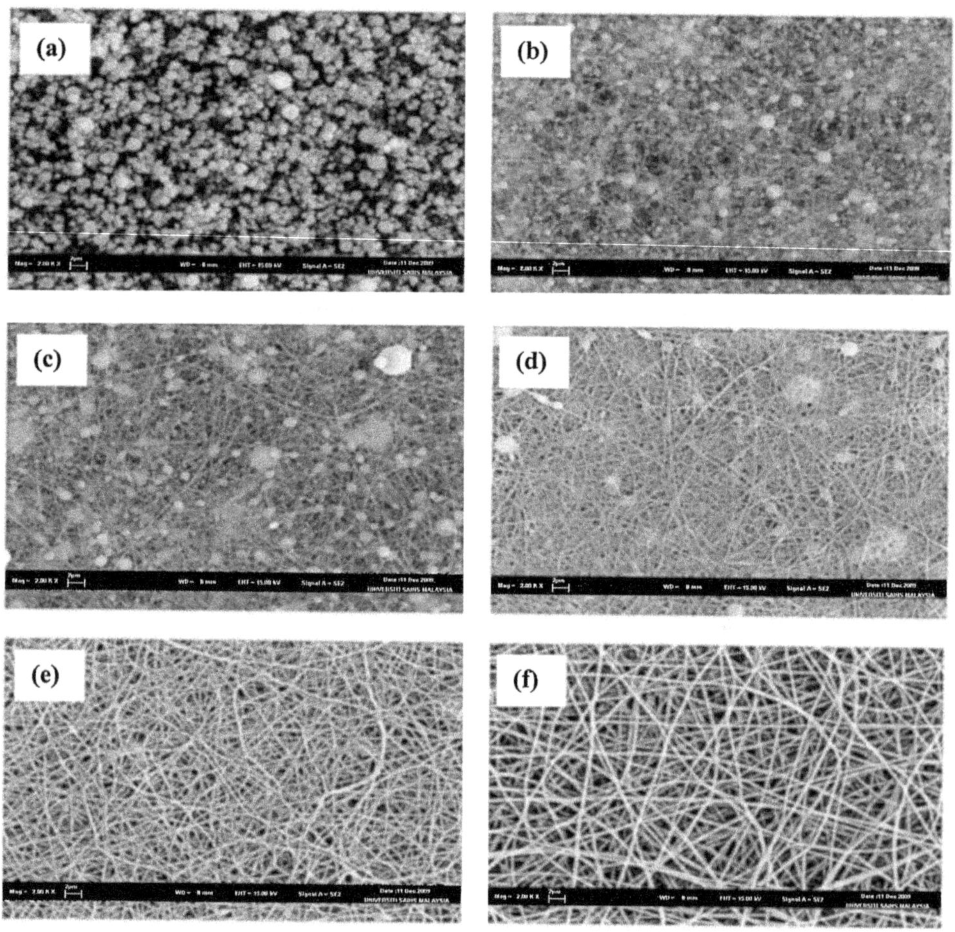

Figure 1 *SEM images (2000x) of gelatin nanofibre mat with gelatin concentration of (a) 10% w/v (b) 15% w/v (c) 20% w/v (d) 25% w/v (e) 30% w/v and (f) 35% w/v.*

This allowed the Coulombic stress to further elongate the charged jet during its flight to the grounded target, which ultimately thinned down the diameter of the charged jet[22]. However, too high concentration of gelatin, 40 % (w/v) inhibited the electrospinning process as the suction clogged the syringe needle and made aspiration of the solution with the syringe too difficult.

The average diameter of gelatin nanofibre increased with increasing the concentration of the solutions (from 0.11 to 0.38 μm) (Table 1). Since nanofibre formation results from evaporation or solidification of polymer fluid jets, the diameter of gelatin nanofibre depends primarily on the jet size as well as on the polymer contents in the jets. It has been recognized that during the traveling of a solution jet from the nozzle onto the metal collector, the primary jet may or may not be split into multiple jets, resulting in different fibre diameters[23]. Solution viscosity is the most significant parameter influencing the fibre diameter if no splitting was involved. A higher viscosity resulted in a larger fibre diameter[24]. From Figure 1f, it was observed that bead-free gelatin nanofibres were formed without any defects. This suggested that the concentration of 35 % (w/v) gelatin in 30 % (v/v) acetic acid was a suitable solution for fabricating gelatin nanofibre by electrospinning method.

3.4 Effects of Crosslinking Treatment on Fish Gelatin Nanofibres

Crosslinking is a common method used to improve mechanical and physical properties of a film. Gelatin nanofibre mat produced with 35% (w/v) gelatin in 30% (v/v) acetic acid was further subjected to treatment with glutaraldehyde as the crosslinking agent because of its high efficiency in stabilizing collagen materials[25]. Figure 2 shows the SEM images of non-crosslinked and crosslinked gelatin nanofibre. After crosslinking treatment, the fibres at junctions were fused together and forming coalesced joints. This phenomenon happened because the moisture-rich glutaraldehyde vapour partially dissolved the fibre segments at points where the fibres came into contact. Thus, the diameter of the resultant fibres increased and the porosity of the nanofibre mat decreased. Crosslinking involved the reaction of free amino groups of lysine or hydroxylysine amino acid residues of the polypeptide chains with the aldehyde groups of glutaraldehyde. The physical properties of the gelatin nanofibres changed slightly after treatment with glutaraldehyde vapour. Slight shrinkage from the original dimensions was observed, with evidence of fibres fusing to one another at contact points (Figure 2). The gelatin nanofibre became visibly darker and turned to a yellowish colour after crosslinking treatment. The crosslinked gelatin nanofibre had lower L^* value (86.9) than the non-crosslinked gelatin nanofibre L^* (94.0) which means that crosslinked nanofibre was darker (Table 2). Zhang et al.[15] reported that the colour change of crosslinked nanofibre was caused by the formation of aldimine linkages (–CH-N–) between the free amino groups of lysine or hydroxylysine amino acid residues of the protein and the aldehyde groups of glutaraldehyde. One of the major weaknesses of the original gelatin nanofibre mat was its high sensitivity to water. Hence, treatment with glutaraldehyde vapour had changed the hydrophilic properties of gelatin nanofibre. The non-crosslinked gelatin nanofibre dissolved immediately when in contact with water. However, the crosslinked sample did not dissolve in the warm water at 37 °C after 12 hrs of soaking. Aldehyde groups of glutaraldehyde have reacted with free amino groups of lysine or hydroxylysine amino acid residues of the polypeptide chains, thus reduced the free hydroxyl groups in the gelatin nanofibre.

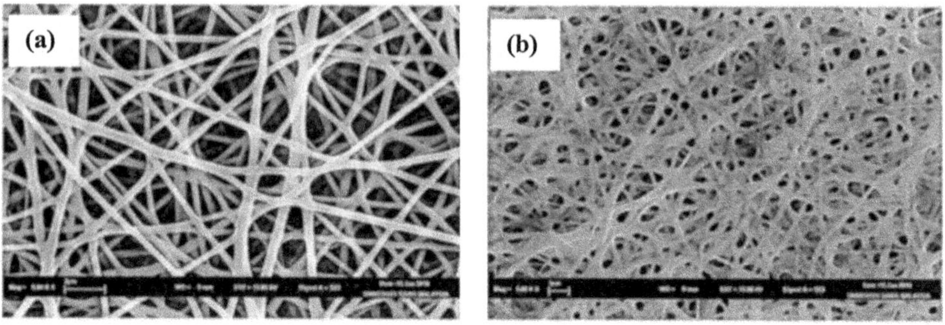

Figure 2 *SEM images (5000x) of gelatin nanofibre mat (a) non-crosslinked and (b) crosslinked with glutaraldehyde vapour.*

Hence, the solubility of the nanofibre mat in water decreased. The mechanical properties of gelatin nanofibre also were affected by crosslinking treatment. The tensile strength of non-crosslinked gelatin nanofibre increased from 1.22 mPa to 5.04 mPa after treatment with glutaraldehyde vapour. However, the elongation of crosslinked nanofibre decreased (5.22 %) compared to the non-crosslinked nanofibre (7.61 %). Thus, the crosslinked nanofibre had a stiffer texture compared to non-crosslinked sample. The inter- and intra- molecular covalent bonds and the bonding between the junctions gave structural integrity to the nanofibre mat[26]. Thus, the glutaraldehyde treatment seemed successful in improving the mechanical properties of gelatin nanofibres. However, the bonding formed between the junctions has limited the mobility of the polypeptide chains. Therefore, the gelatin nanofibre mat was stiffer but on other hand the extension ability of the nanofibre was reduced.

Table 2 *Properties of non-crosslinked and crosslinked gelatin nanofibres.*

Parameters	Gelatin nanofibre mat: Non-crosslinked	Gelatin nanofibre mat: Crosslinked
Colour $L*$	94.0 ± 0.04^{a}	86.9 ± 0.06^{b}
$a*$	-0.7 ± 0.03^{a}	-0.6 ± 0.01^{a}
$b*$	-1.3 ± 0.04^{a}	7.1 ± 0.03^{b}
Water solubility test	Dissolved immediately when came in contact with water	Did not dissolve after being immersed in warm water (37°C) for 12 hrs.
Tensile strength (mPa)	1.224 ± 0.5^{a}	5.047 ± 1.0^{b}
Elongation (%)	7.611 ± 2.1^{a}	5.218 ± 1.2^{b}

Values are means ± standard deviation. Values in the same row with different superscripts are significantly different ($p<0.05$).

4 CONCLUSION

Electrospinning is a versatile and relatively easy technique to produce nanofibres. The quality and characteristics of the final nanofibres are determined by the temperature, viscosity, elasticity, conductivity, and surface tension of the solution, strength of the electric field, distance between the needle tip and collector, and humidity. Solvents influence the surface tension and viscosity of the solution and affect the morphology of fibres. From this study, fish gelatin was successfully extracted from the skin of black tilapia (*Oreochromis mossambicus*) and characterized for its gel strength, viscosity and pH value. The concentration of solvent (acetic acid) and the concentration of biopolymer solution (fish gelatin) were important parameters that affected the electrospinning process. The crosslinking treatment using glutaraldehyde (GTA) enhanced the properties of the nanofibre such as tensile strength and lowered water solubility but caused the nanofibre mat to shrink and turn to a yellowish colour.

Acknowledgement

The authors wish to express their thanks to Trapia (Malaysia) Sdn. Bhd. for supplying the black tilapia fish. This work was funded by RU grant from Universiti Sains Malaysia.

References

1 P. Songchotikunpan, T. Jirarat, and S. Pitt, *Int. J. of Biological Macromolecules*, 2007, **42**, 247-255.
2 S. Agarwal, J.H. Wendorff and A. Greiner, *Polymer*, 2008, **49**, 5603-5621.
3 L.L. Wu, X.Y. Yuan and J. Sheng, *J. of Membrane Science*, 2005, **250**, 167-173.
4 Y. Wang and Y.L. Hsieh, *J. of Membrane Science*, 2008, **309** (1-2), 73-81.
5 D. Liang, B.S. Hsiao and B. Chu, *Advanced Drug Delivery Reviews*, 2007, **59** (14), 1392-1412.
6 J. Zeng, X. Xu, X. Chen, Q. Liang, X. Bian, L.Yang and X. Jing, *J. of Controlled Release,* 2003, **92** (3), 227- 231.
7 Y.K. Luu, K. Kim, B.S. Hsiao, B. Chu and M. Hadjiargyrou, *J. of Controlled Release,* 2003, **89**, 341-353.
8 Z.M. Huang, Y.Z. Zhang, S. Ramakrishna and C.T. Lim, *Polymer*, 2004, **45**, 5361-5368.
9 C. Chen, C. Cao, X. Ma., Y. Tang and H. Zhu, *Polymer*, 2006, **47**, 6322-6327.
10 S. Scott, A. Micheal, J. McClure, G. Koyal, W. Patricia, L. Gary and Bowlin, *Advanced Drug Delivery Reviews*, 2009, **61**(12), 1007-1019.
11 X.K. Geng, O. Hyeong and J. Jang, *Biomaterials*, 2005, **26** (27), 5427-5432.
12 B. Jamilah and K.G. Harvinder, *Food Chemistry,* 2002, **77,** 81-84.
13 S.K. Chang, H. B. Doo, D.G. Kyung, H. L. Ki, C.U. In and H.P. Young, *Polymer*, 2005, **46**, 5094–5102.
14 G.B. Yin, Y.Z. Zhang, W.W Bao, J.L. Wu, B. De and Z.H. Dong, *J. of Applied Polymer Science,* 2009, **111(3),** 1471–1477.
15 Y.Z. Zhang, J. Venugopal, Z.M. Huang, C.T. Lim and S. Ramakrishna, *Polymer*, 2006, **47**(8), 2911–2917.
16 M.H Norziah and A.B. Khairulnizam, unpublished results.
17 AOAC. *Official methods of analysis of AOAC International 17th ed.*, 2005.

18 GMIA (Gelatin Manufacturers Institute of America). (2006). Standard methods for the testing of edible gelatin. In: *Official procedure of the Gelatin Manufacturers Institute of America, Inc.* Revised September 2006 [on line].<http://www.gelatin-gmia.com> Accessed 10.06.09.

19 S. Grossman and M. Bergman, *United States Patent* ,1992, No. 5,093,474.

20 C. Mit-uppatham, M. Nithitanakul and P. Supaphol, *Macromolecular Chemistry and Physics*, 2004, **205**, 2327-2338.

21 H. Fong, I. Chun, and D.H. Reneker, *Polymer*, 1998, **40**, 4585-4592.

22 C.J. Buchko, L. Chen, Y. Shen and D.C. Martin, *Polymer*, 1999, **40**, 7397.

23 M.M. Bergshoef and G.J. Vancso, *Advanced Material*, 1999, **11** (16), 1362-1365.

24 P.K. Baumgarten, *J. of Colloid and Interface Science*, 1971, **36** (1), 71-79.

25 E. Khor, *Biomaterials*, 1997, **18**, 95-105.

26 K.H. Lee, H.Y. Kim, Y.J. Ryu, K.W. Kim, and S.W. Choi, *J. Polymer Science, Part B: Polymer Physic*, 2003, **41**, 1256-1262.

ENCAPSULATION BY ELECTROSPINNING OF LIVE BACTERIA USED IN THE FOOD INDUSTRY

R. Hans Tromp, Carolien Vink and Ann C. Stijnman

NIZO food research, Kernhemseweg 2, 6718 ZB Ede, The Netherlands

ABSTRACT

Electrospinning was tested as a fast and energetically favourable way to dry and encapsulate live bacteria. Bacterial suspensions with 15 % pullulan as a carrier polymer were electrospun and tested with respect to their survival. The results were compared to those of freeze drying maltodextrin based systems of the same bacteria. It turned out that the survival rates after electrospinning and freeze drying are comparable, but different for different species of bacteria. Activity of freeze dried and electrospun yoghurt bacteria was tested and compared by making yoghurt with dried and untreated cultures. A weak reduction in activity was observed, while the quality, taste and flavour of the final yoghurt appeared unaffected.

1 INTRODUCTION

Electrospinning is a method to prepare long, thin fibres from polymer solutions. The spinning solution may contain particles, emulsion droplets or microorganisms, which after spinning end up in an encapsulated state in the fibres. In electrospinning, droplets of viscous polymer solution are exposed to a strong electric field, which stretches the droplet to a thin liquid jet. The solvent evaporates from the jet, which as a consequence changes into a macroscopically long, but microscopically thin polymer fibre[1,2].

The technique of electrospinning is fairly well established for manufacturing filters and high surface area materials for medical purposes[3,4]. Because the upscaling of electrospinning is still in development, high added value applications are needed for economic feasibility. The low yield (typically kg per hour at most) is probably part of the reason electrospinning has hardly been explored with respect to its applicability in the food industry. However, work reported here may suggest that in some specific cases electrospinning of food grade polymers shows some promising results. These may contribute either to improvements and a higher sustainability of existing processes in the food industry, or to food ingredients in a novel form, e.g. long, thin vitrified threads (See Figure 1)

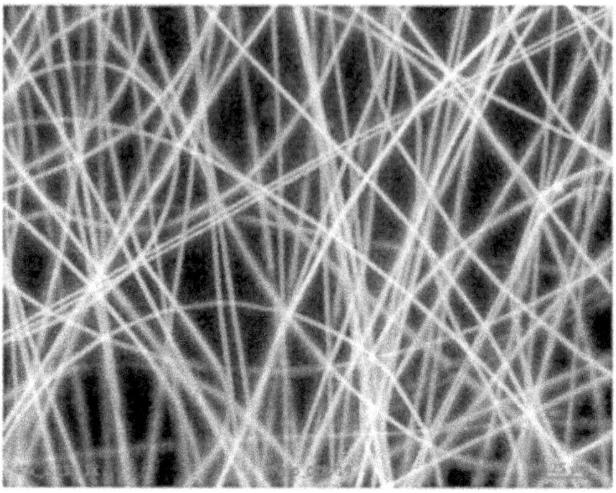

Figure 1. Dextran (473 kDa) spun from a 40% aqueous solution. The diameter of the fibres is between 1 and 2 μm.

In this paper some proofs of principles are described of the application of electrospinning for encapsulation and protection of live bacteria. In this case, electrospinning turns out to be a fast and mild way of drying suspensions of bacteria with a relatively limited effect on viability. The survival of bacteria dried and encapsulated by electrospinning and freeze drying was compared. Electrospinning may be an alternative to freeze drying with respect to drying bacterial suspensions. They both result in a better survival than spray drying, which involves high temperatures. However, in comparison to freeze drying, electrospinning of small quantities of bacterial suspension is two orders of magnitude faster, and much less energy is used.

The method of drying may not only affect the viability of the bacteria, but also the functionality, e.g. as a fermenting agent. Therefore, a practical test was carried out, in which a comparison was made between three yoghurts. The first of these was prepared using a bacterial yoghurt culture dried by electrospinning, the second by using freeze dried culture and the third using an untreated culture.

2 ELECTROSPINNING

Figure 2 shows the basic set up of electrospinning. A syringe containing a polymer solution is fitted with a metal needle, which is kept at a high electric potential, typically between 10 and 20 kV, relative to a metal collector screen positioned at approximately 15 cm from the tip of the needle. The syringe is in a syringe pump which is set at (typically) 1 ml/h. In the absence of an electric potential the polymer solution flows dropwise from the needle exit. When the potential is turned up, the droplets are extended to form a so-called Taylor cone, at the tip of which the accumulation of electric charge cancels the interface tension of the drop. This cancellation results in a jet of charged solution, pulled towards the collector screen. Because the jet is charged and liquid, it is unstable against shape perturbations such as bending. As a consequence, within millimetres from the needle tip, the jet turns into a chaotically moving spiralling coil, which forms a cone-like volume, with the cone base on the collector plate. During this process the jet is strongly extended and obtains a high surface-to-volume ratio. The large surface causes the solvent to evaporate, resulting in a desolvated and often vitrified polymer fibre reaching the collector screen. Figure 3 shows an example of pullulan spun on a collector screen, in this case a metal grid. A thorough discussion of the current status of the electrospinning technology can be found in ref. 2.

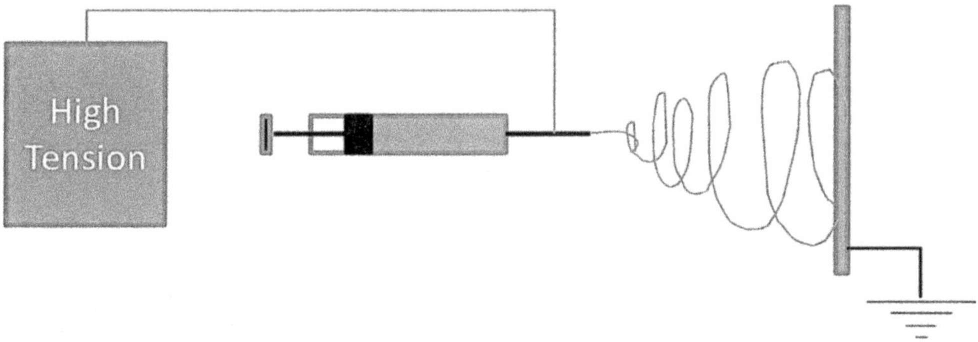

Figure 2. Schematic overview of an electrospinning setup.

Figure 3. A grid as collector screen, with electrospun polysaccharide.

3 EXPERIMENTAL

Lactococcus lactis and *Lactobacillus johnsoni* were compared with respect to their survival after encapsulation in a carbohydrate matrix by freeze drying and encapsulation in a polysaccharide fibre by electrospinning. In the present case electrospinning is used as a mild method of drying. These two bacterial strains were chosen because they differ strongly in survival after spray drying, which is currently the most common way to dry bacterial suspensions, and which usually leads to strong losses in viability. *Lactococcus lactis* is considered relatively resistant to spray drying, whereas the viability of *Lactobacillus johnsoni* is found to suffer from spray drying.
For a solution to be suitable for electrospinning, it has to meet certain rheological conditions. It should have a concentration of typically 15-25 % (w/w) polymer, which corresponds to 10-20 times the so called 'overlap' concentration, above which polymer coils overlap each other and are entangled. Another necessary property of the solution is a weak tendency to thin on shearing, typically above 1000 s^{-1} (Figure 4). It turns out that pullulan and dextran are polysaccharides that easily meet the conditions for electrospinning. Therefore, the polymer used was pullulan (15%, 200 kDa, kindly provided by Hayashibara, Okayama, Japan).
In the case of freeze drying, it is well known that the type of low molar mass carbohydrate present in the suspension is important for the viability. Therefore, in order to test if this is also the case for electrospinning, was carried out without low molar mass carbohydrate, and with 10% (w/w) glucose, lactose and trehalose.

For encapsulation of the bacteria, pullulan and low molar mass carbohydrate was dissolved in bacterial suspension (from which the growth medium was removed by repeated centrifugation and washing), to obtain a 15% solution of pullulan and a 10% solution of low molar mass carbohydrate. Before and after encapsulation the bacterial systems were plated and colonies counted. The increase in bacterial concentration due to drying (from approx. 75% to approx. 5/10% water) was taken into account.

Freeze drying of bacterial suspensions is commonly carried out with maltodextrin as a protectant and carrier material. However, maltodextrin can only be electrospun at concentrations above 40% (w/w), which is not feasible for a bacterial suspension. Therefore, in order to get a practical comparison between drying by electrospinning and by freeze drying, pullulan-based systems for electrospinning were used, and maltodextrin-based systems for freeze drying.

Yoghurt was prepared from reconstituted milk powder, manufactured at NIZO by a low temperature spray drying process, which avoids denaturation or lactolysation. Two identical solutions of 15% pullulan in bacterial culture suspension were made. One was dried by electrospinning, the other by freeze drying. The milk was inoculated with a standard yoghurt culture (0.02% of a mixture of *Streptococcus thermophilus* and *Lactobacillus bulgaricus*). The dried suspensions of yoghurt cultures were added in dosages that were equal to that of the untreated culture, assuming that the survival was 100%.Acidification of the milk was monitored at 32°C until the pH had reached 4.5.

4 RESULTS AND DISCUSSION

Figure 5 shows a comparison of bacterial survival after drying by electrospinning and freeze drying. The fibres and the freeze dried material were stored at 7°C and at a relative humidity corresponding to a water activity (a_w) of 0.4. It turns out that the survival of the two strains after drying with the two methods is comparable, with *Lactococcis lactis* surviving better the drying by electrospinning, and *Lactobacillus johnsoni* by freeze drying. However, the survival appears to depend strongly on the type of low molar mass carbohydrate present. The same strong effect of the type low molar mass carbohydrate is concluded from Figure 6, which shows the survival after a much longer storage time of 4 weeks at 7°C. It turned out that trehalose improves most strongly the survival after electrospinning, while lactose and glucose have similar smaller effects. For both bacterial strains electrospinning without any low molar mass carbohydrate gives a much lower survival. The electrospun fibres were also stored at two different values of relative humidity, corresponding to a_w values of 0.2 and 0.4. It turned out that the reduction of a_w from 0.4 to 0.2 is very significant. This strong effect of relative humidity may be caused by the high surface-to-volume ratio of the fibres (typically 8-10 m^2/g), which tends to make them slightly hygroscopic.

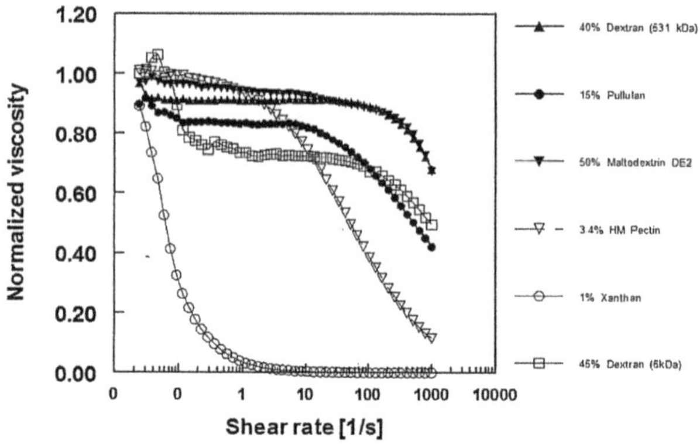

Figure 4. Normalized shear viscosity of some polysaccharide solutions. Solutions suitable for electrospinning have filled symbols.

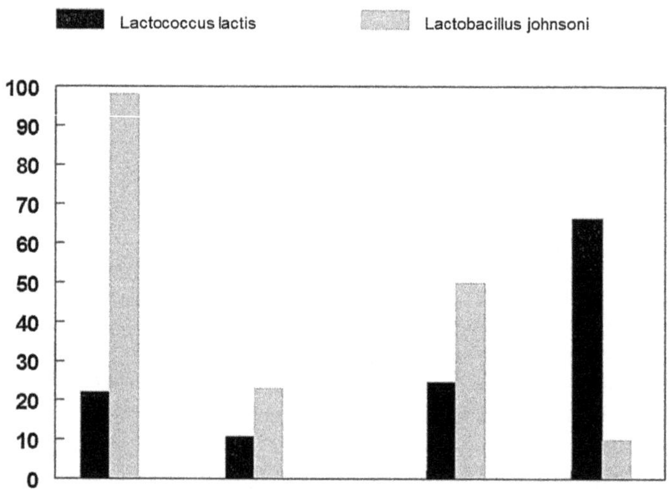

Figure 5. Viability after 1 week at 7°C and a_w 0.4 of bacterial suspension dried by electrospinning and freeze drying.

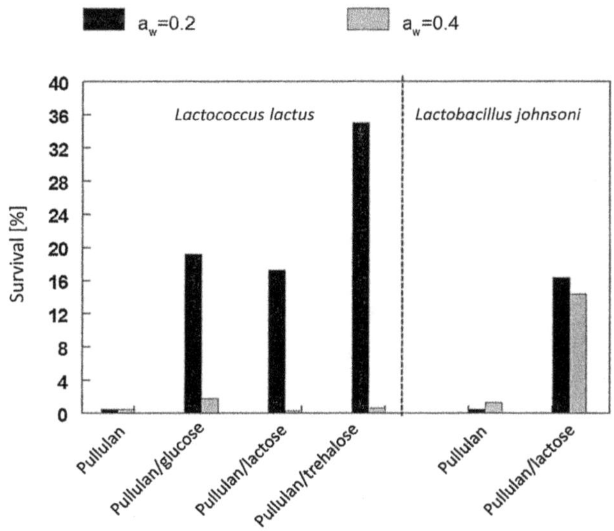

Figure 6. Viability after 4 weeks at 7°C of bacterial suspensions dried by electrospinning, stored at two values of relative humidity.

Yoghurt was made using untreated and dried yoghurt culture, in order to test if a significant survival of electrospun bacteria also means the functionality is largely intact. The survival was tested by plating the finished yoghurt. The functionality was tested by measuring the rate of acidification of inoculated milk. The taste, flavour and structure of the yoghurt were also superficially tested. The acidification is shown in Figure 7. It turns out that the rate of acidification, which is a measure of the activity of the bacteria, is about 10% slower for the freeze dried culture and the culture dried by electrospinning, compared to the untreated culture. The two methods of drying weaken the activity to the same extent. The survival of the two constituent species is listed in Table 1. In spite of the fact that both methods of drying lead to the same weak reduction in activity, the survival turns out to be better after freeze drying than after electrospinning. In particular survival of *Lactobacillus bulgaricus* is reduced significantly by electrospinning (two orders of magnitude). Apparently, other factors than just the numerical survival affect the activity of the mixture of cultures.

A superficial comparison between the taste, flavour and structure of the yoghurts made with dried and untreated cultures did not show significant differences.

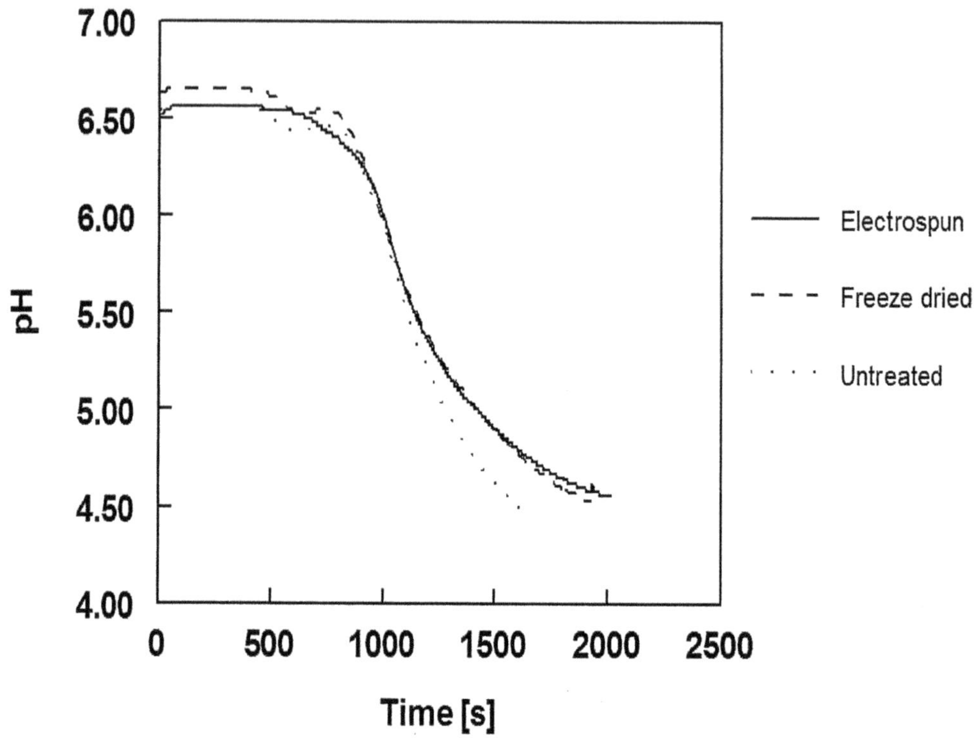

Figure 7. Progress of acidification during yoghurt preparation at 32°C for two methods of drying of the culture, and for the untreated culture.

5. CONCLUDING REMARKS

From the results of the exploratory experiments reported here, it can be concluded that electrospinning of bacterial cultures offers an alternative method, besides spray drying and freeze drying, to obtain bacteria included in a dry material. The material in this case consists of fibres with a regular, cylindrical shape with a typical thickness between 0.5 and 3 μm. Considering this small diameter, the bacteria are therefore probably lined up along the fibres. The exact arrangement of the bacteria inside the fibres has not yet been investigated.

The survival after electrospinning and freeze drying is comparable. However, results with the pairs *Lactococcus lactis/Lactobacillus johnsoni* and *Streptococcus thermophilus /Lactobacillus bulgaricus* show that the effects of the two methods of drying differ significantly for different species. Adding low molecular mass carbohydrate is necessary for a survival of more than a few percent in the case of electrospinning. The type of low molar mass carbohydrate is important, with trehalose appearing to be the most promising candidate.

Considering the strong effects of type and presence of these additives concluded from these preliminary experiments, it is expected that further work to optimize the type and quantity of low molar mass carbohydrate will improve the performance of electrospinning.

The preparation of yoghurt from electrospun yoghurt culture shows that it is possible to maintain both quantity and quality of the bacterial culture as a food ingredient. However, the quality is somewhat reduced, to about the same extent as in the case of freeze dried culture. This again suggests electrospinning as an alternative to freeze drying for drying and immobilizing bacteria.

Table 1. Occurrence of bacteria from yoghurt culture in yoghurt.

Treatment	*Streptococcus thermophilus* [CFU/ml yoghurt]	*Lactobacillus bulgaricus* [CFU/ml yoghurt]
Electrospinning	$6.6 \cdot 10^{8}$	$2 \cdot 10^{5}$
Freeze drying	$1.6 \cdot 10^{9}$	$4.1 \cdot 10^{7}$
No treatment	$5.4 \cdot 10^{8}$	$4.9 \cdot 10^{7}$

References

[1] Baumgarten, B.K. (1971), Electrostatic spinning of acrylic microfibers. J. Coll. Interface Sci 36, 71-79

[2] Reneker, D. H., Yarin, A. L., Zussman, E. and Xu, H. (2007). Electrospinning of nanofibers from polymer solutions and melts. In Advances in applied mechanics: Vol. 41 (pp. 43-195).

[3] Patanaik, A., Jacobs, V., & Anandjiwala, R. D. (2010). Performance evaluation of electrospun nanofibrous membrane. Journal of Membrane Science, 352, 136-142.

[4] Agarwal, S., Wendorff, J. H., & Greiner, A. (2009). Progress in the field of electrospinning for tissue engineering applications. Advanced Materials, 21, 3343-3351.

[5] Kriegel, C., Arrechi, A., Kit, K., McClements, D. J., & Weiss, J. (2008). Fabrication, functionalization, and application of electrospun biopolymer nanofibers. Critical Reviews in Food Science and Nutrition, 48, 775-7.

EFFECTS OF SUCROSE PALMITATE ON PHYSICAL AND MECHANICAL PROPERTIES OF SAGO STARCH-GELATIN EDIBLE FILMS

M.H. Norziah[1], J.L. Wong

[1]Food Technology Division, School of Industrial Technology,
Universiti Sains Malaysia, 11800, Penang, Malaysia. E-mail: norziah@usm.my

1 INTRODUCTION

Edible and biodegradable films are becoming an important area of food research with enormous commercial potential which can be used to extend the shelf-life and improve the quality of almost any food system by serving as mass transfer barriers to moisture, oxygen, carbon dioxide, lipid, flavour and aroma between food components and the surrounding atmosphere[1]. Edible films can be formulated from protein, lipids, several polysaccharides or the combination of these components[2]. Starch is one of the most commonly used raw materials for edible film preparation since it is a renewable source, inexpensive, widely available and relatively easy to handle[3]. Gelatin has a potential to form biopolymer film or coating component because of its unique colloidal properties, which include thickening, stabilizing, film forming and emulsion stabilizing. Several studies on the preparation of edible film from gelatin have been reported by Jongjareonrak et al.[4], Avena-Bustillos et al.[5] and Sobral et al.[6] Films made from polysaccharides or proteins usually show good mechanical properties, but are sensitive to moisture. On the other hand, films made from lipids show good water vapor barrier properties, but are opaque, only slightly flexible and brittle. The addition of lipid and hydrocolloid in starch-based film forming solution can serve as a good barrier to water vapour permeability and gas permeability (O_2 and CO_2)[7,8]. Addition of plasticizers such as such as sorbitol, glycerol and propylene glycol into the film forming solution is necessary to improve mechanical properties of protein or polysaccharide-based edible films[9]. However, all these plasticizers present hydrophilic characteristics, which contribute to the increase of water permeability and susceptibility of the matrix to environmental humidity. The inclusion of hydrophobic components provide an alternative to reduce this behavior by closing the micro-voids in the film, leading to a decrease in water uptake. However, water insoluble plasticizers can cause phase separation leading to flexibility losses or yet to the formation of discontinuity zones during film drying[10]. Hydrophobic plasticizers are much used in the pharmaceutical industry for the production of biodegradable materials, but are little used in filmogenic formulations. Incorporating lipid materials into edible films or coatings provides a way to retard the water vapour permeability of the film. Stearic acid was reported as the most effective fatty acids in decreasing water vapour permeability and carbon dioxide permeability of cellulose-based film[11]. Emulsifiers are extensively used in the food industry to improve texture and stability, among other features. The sucrose mono- and diesters of long-chain fatty acids form a class of food-grade emulsifiers with varying hydrophilic - lipophilic balance (HLB) values. When sucrose esters served as part of the composition of gelatin films, they decreased the films' water-vapor permeability and tensile strength[4]. Jongjareonrak et al[4] also reported that the chain lengths of the fatty acids and their sucrose esters affected the properties of the films differently, depending upon the gelatin source. Sucrose esters also improved the moisture-barrier properties of emulsified edible-film structures containing arabinoxylans and hydrogenated palm kernel oil[12].

However, to the best of our knowledge, there has been no research on starch/gelatin films concerning the effect of fatty acid sucrose ester on the film properties. The objective of this study was to develop films based on starch and gelatin and plasticized with 25, 30 and 35% sorbitol, and substituted by different amounts of palmitic acid sucrose ester (sucrose palmitate). The effects of incorporation of sucrose palmitate on the physical and mechanical properties of the biopolymer film were investigated.

2 MATERIALS AND METHODS

2.1 Materials

The following materials were used to produce the biopolymer films: sago starch (*Metroxylan sagu*) (purchased from Sim Company Sdn. Bhd. Penang, Malaysia); fish gelatin (extracted from the skins of black tilapia according to the method described by Norziah and Khairulnizam[13]; sorbitol (Neosorb P20/60, Rhodia Inc, USA), sucrose palmitate or palmitic acid sucrose ester (Mitsubishi-Kagaku Foods Corporation, Japan). All the chemicals used were of analytical grade.

2.2 Preparation of sago starch/gelatin edible films

The film forming solutions (FFS) were prepared from a mixture of sago starch and fish gelatin (3:1) based on total weight basis (5 g) in 200 ml distilled water as described by Hassan and Norziah[14] From previous work[14], the protein content in this starch:gelatin film was 16.3%. Sorbitol, as the plasticizer was incorporated into the film forming solutions at various concentrations (25 %, 30 % and 35%). For each FFS with the different sorbitol concentration levels, sucrose palmitate was added at sorbitol to sucrose palmitate (S:SP) ratios of 100:0, 90:10, 80:20 and 70:30 to produce a total of 13 formulation of FFS including the control.

In the preparation of films, sago starch (db) was initially dissolved in distilled water and heated with magnetic stirring in a water bath at 85 °C for 30 min until completely gelatinized. Fish gelatin was dissolved separately in distilled water at 60 °C for 30 min until a clear solution was obtained. Fish gelatin solution was then added to gelatinized sago starch at 60 °C and stirring was continued for 30 min, followed by addition of sorbitol and sucrose palmitate with constant stirring for another 30 min. The mixture was then cooled to room temperature before casting. The solution mixture (85 g) was cast onto polyacrylic plates (16 cm x 16 cm x 3 mm) and dried in a ventilated oven at 40 °C for 24 hrs. The dry films obtained were peeled off and stored in a desiccator containing saturated sodium bromide (NaBr) solution with 56% (RH) at 30 °C before determination of their properties. Control films were prepared in the same way without the addition of sorbitol and sucrose palmitate. Each film formulation was prepared in triplicate.

2.3 Visual aspect,

Visual analysis on the films was performed, aiming to consider only homogeneous (uniform color) and reasonably flexible (absence of rupture or brittle zones) films.

2.4 Film thickness measurement and light transmission

Film thickness was measured using a manual micrometer (Mitutoyo Co., Tokyo, Japan) with an accuracy of 0.001 mm. Six different positions of the samples were measured and average thickness was calculated. The light transmission of the film sample was measured at wavelength 600 nm using UV spectrophotometer (Shimadzu, Japan) following the method described by Fang et al.[15]. The transparency of the film sample was calculated as:

$$\text{Transparency} = -\log T_{600}/x$$

where T_{600} is the transmittance at 600 nm and x is the film thickness (mm).

2.5 Tensile Strength and Elongation tests

Tensile strength and elongation at break of film sample was measured using TA-TX2 texture analyzer (Stable Micro System, Surrey, U.K.) with a load cell of 30 kg and crosshead speed of 60 mm/min following ASTM method D-882[16]. The film strips (14×2 cm) of each formulation were clamped between tensile grips and stretched at 0.2 mm/s with an initial distance between the grips of 100 mm.

2.6 Water vapour permeability

Water vapour permeability is defined as the amount of water vapour transferred through the film per unit of time, area, vapour pressure difference and thickness. Water vapour permeability (WVP) of film sample was measured following ASTM method E96-80[17]. The test film sample was sealed onto a glass permeation cell (4.5 cm in diameter and 3.2 height) containing silica gel (0 % RH) with 1.5 cm deep headspace. The cells were then placed in a desiccator maintained at 30 °C with controlled relative humidity (56 % RH) using a saturated sodium bromide solution. The cells were weighed at 24 h intervals over a 7-day period. The cells were recorded to the nearest 0.0001 g and plotted as a function of time. The slope of each line was calculated by linear regression ($r^2 \geq 0.99$). The measured WVP of the films was determined as follows:

$$\text{WVP} = (\text{WVTR}\ L)/\Delta P$$

where WVTR is the water vapor transmission rate (g $m^{-2}h^{-1}$) through a film, calculated from the slope of the straight line divided by the exposed film area (m^2), L is the mean film thickness (mm), and ΔP is the partial water vapor pressure difference (Pa) across the two sides of the film. For each type of film, WVP measurements were replicated three times for each batch of films.

2.7 Water sorption isotherms

Sorption isotherms of film samples were determined according to the method described by Chang[18] with slight modifications. The edible films were cut into small pieces (3×5 mm), and placed in a desiccator maintained at 30 °C with (P_2O_5) for 8 days. The samples were then weighed to the nearest 0.0001 g into pre-weighed bottles and equilibrated versus saturated salt solutions (lithium chloride, potassium acetate, magnesium chloride, potassium carbonate, strontium chloride, sodium chloride and potassium chloride) with known relative humidity (RH) in air tight containers at 30 °C and the equilibrium was assumed to be achieved when changes in weight did not exceed 0.1% for 3 consecutive measurements.

2.8 Water solubility

Water solubility of the films was evaluated according to the method by Romero-Bestida et al.[19]. Film sample (2 × 3 cm) of each formulation was stored in a desiccator containing silica gel (0% RH) for 7 days. Film sample was weighed, placed into a beaker with 80 mL deionized water and maintained under constant agitation for 1 hr at room temperature (approximately 25 °C). The remaining pieces of film after soaking were filtered through filter paper (Whatman no.1), followed by oven drying at 60 °C to constant weight. The percentage of total soluble matter (% solubility) was calculated as:

$$\% \text{ Solubility} = (\text{initial dry weight} - \text{final dry weight})/\text{initial dry weight} \times 100$$

2.9 Scanning electron microscopy (SEM)

Films were cut into small strips and pretreated with 2% (v/v) glutaraldehyde, 0.1M sodium cacodylate buffer and post fixed in 0.4% (w/v) osmium tetroxide prior to SEM analysis following procedures described in Hassan and Norziah [14]. The dried samples were mounted on aluminum stubs and then coated with a layer of gold by "Baizer SCD SPUTTER coater", allowing surface visualization using an accelerated voltage of 5 kV. Films were observed using Scanning Electron Microscopy, SEM (Leo Supra 50vp, Oberkochen, Germany).

2.10 Statistical analysis

SPSS 12.0 was used in this study to analyze the data. One-way analysis of variance was carried out using Duncan's test with confidence level as $p<0.05$.

3 RESULTS AND DISCUSSION

3.1 Film characterization, visual aspects, thickness and light transmission

The thickness of the sago starch/gelatin films with sorbitol and sucrose palmitate (PASE) added varied from 0.063 ± 0.001 to 0.100 ± 0.001 mm with 35% sorbitol films and S:SP ratio of 70:30SP) having highest thickness. The variation is possibly caused by the increase in concentrations of the hydrophobic emulsifier especially in films containing the highest plasticizer level in the formulations, an effect observed previously[20, 21]. Films of sago starch/fish gelatin prepared were transparent and homogeneous. All films were found to be flexible and easily removed from the acrylic plate. Transparency is the physical property that allows the light to pass through a material and normally transparent and colourless films are used to preserve the appearance of food[22]. The decrease in film transparency (or greater transparency index value) of films was observed with addition of sucrose palmitate at increasing levels of sorbitol substitution (Figure 1). The dispersion of sucrose palmitate caused a higher barrier property to light transmission compared to the control film. A reduction in light transmission of films with sucrose palmitate have also been reported by Jongjareonrak et al.[4]. The droplet of dispersed sucrose palmitate in the gelatin film might prevent the light transmission. Increasing amounts of sucrose palmitate added into the film caused lower light transmission and thus lower film transparency.

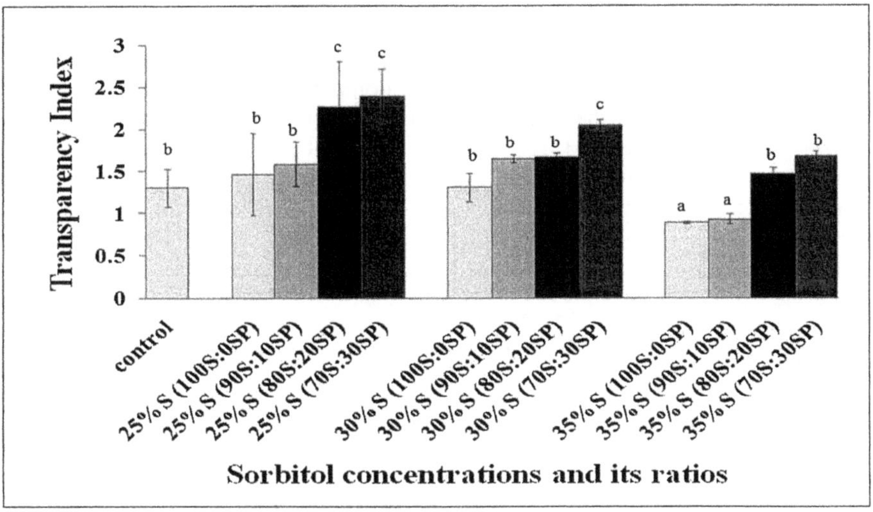

Figure 1 *Effect of sorbitol (S) and sucrose palmitate (SP) on transparency index of sago starch-gelatin films. Mean values ± SD (n=4).*

3.2 Mechanical properties

Tensile strength of film is defined as maximum force that a film can sustain[23]. Figure 2a shows that in all film formulations plasticized at sorbitol concentration levels (25%, 30% and 35%), increasing amounts of sucrose palmitate (SP) substitution, resulted in an increase in tensile strength (TS). However the increase in TS was more pronounced in films with lower sorbitol concentration (25 %) compared to films plasticized with 35 % sorbitol. Thus films with added SP increased TS compared to control films (0% sorbitol substitution or films with S:SP of 100:0). Jongjareonrak et al.[4] also reported that tensile strength of films from fish skin gelatin gradually increased with increasing addition of fatty acid sucrose ester (FASE). This was possibly due to the nature of FASE, which is an emulsifier that consists of both hydrophilic and hydrophobic properties. Incorporation of sucrose palmitate into film forming solution may induce intermolecular interaction between protein and sucrose palmitate possibly via the hydrophilic head of palmitic acid sucrose ester (PASE). As a result, the structural integrity between gelatin and sucrose ester molecules was obtained in the resultant film. Higher substitution of palmitic acid sucrose palmitate (PASE) might enhance the formation of hydrogen bonds or other inter-chain interactions between gelatin and PASE which could result in a stronger film matrix as shown by the increase in film TS. The addition of PASE to the starch/gelatin films also increased the elastic modulus of films (data not shown) but decreased the elongation at break for all formulations studied (Figure 2b).

Thus at higher sorbitol concentration levels (35%), tensile strength decreased as a result of increase in hydrogen bonding. Incorporation of plasticizer into the polymer matrix also caused a competition for hydrogen bonding between polymer-polymer and polymer-plasticizer and thus direct interaction between polymer chains are reduced due to hydrogen bond formation with plasticizer[24]. Compatibility between plasticizer and polymer is of major significance for effective plasticization

Elongation at break (EAB) of film is defined as maximum change in length of the film before breaking[23]. Higher elongation indicates the films are more flexible when subjected to tension and mechanical stress. EAB values were much higher in all films plasticized with sorbitol with no sucrose palmitate added, compared to the control film (no sorbitol added). In general, EAB of the film decreased with increasing substitution of PASE at all plasticizer concentration levels (Figure 1b). Laohakunjit and Noomhorm[25] reported that as the film structures are softening by addition of plasticizer, the tensile strength decreased whereas the elongation at break increased.

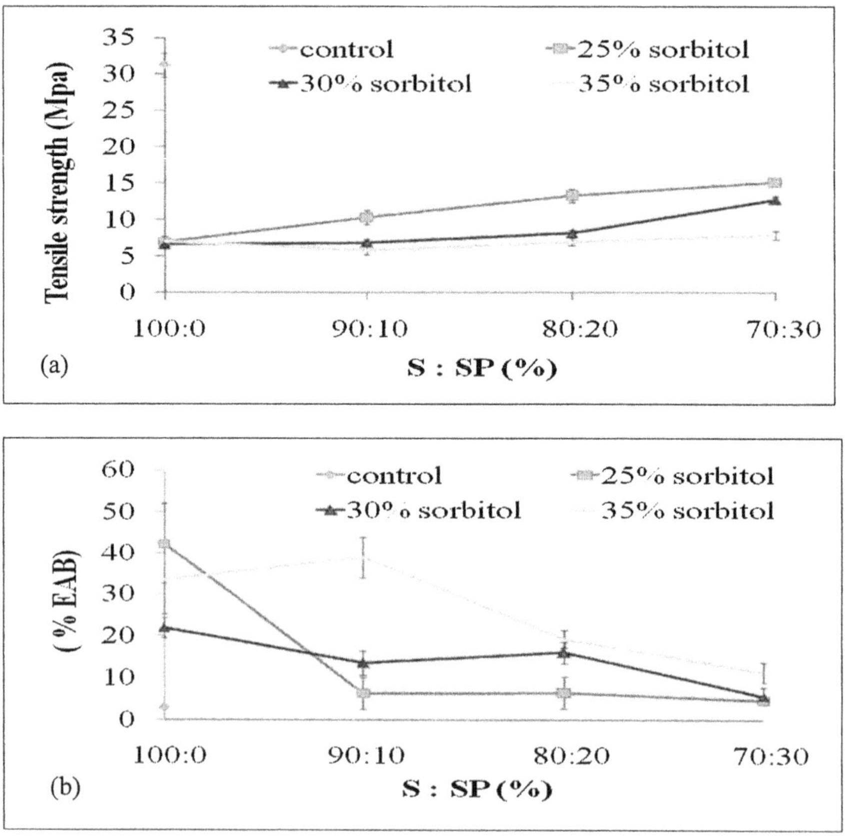

Figure 2 *Effect of sorbitol (S) and sucrose palmitate (P) on (a) tensile strength and (b) elongation at break of sago starch-gelatin films. Mean values ± SD (n=5).*

3.3 Water vapour permeability

The results show that the control starch/gelatin film (unplasticized) had the highest WVP value compared to the plasticized films (Figure 3). This result is in contradiction to most reported results in the literature, in which WVP of plasticized films increased as plasticizer content in film increased [8]. Talja et al.[26] and Garcia et al.[7] have reported that high WVP of films without plasticizer could be caused by micro-cracks in the films (as observed in SEM micrographs in Figures 6 a-e), resulting in higher WVP values. Results from SEM studies on these films (Figure 6 a-e) showed microcracks which supported the findings of WVP results. From Figure 3, it was observed in films plasticized at lower sorbitol concentration (25%), higher PASE substitution caused an increase in WVP within this type of film. The increase in WVP could be due to the incompatibility of sucrose palmitate and the polymer chains at that sorbitol concentration. As a consequence, a discontinuous polymer matrix due to the incompatibility of sucrose palmitate may cause cracks within the films. However, a reverse effect was obtained in 30 and 35% sorbitol plasticized films. The effects of higher sucrose palmitate substitutions, seemed to decrease the WVP slightly. This suggested that the effects of hydrophobicity of PASE could be more pronounced and contributed to reduced WVP of the films as explained by Ayranci and Tunc[11]. Sucrose palmitate is an emulsifier which contains both hydrophilic and hydrophobic properties, and can easily be dispersed and can interact with polymer chains (in this case both starch and protein) via the formation of hydrogen bonds with the hydrophilic head group of the molecule, while exposing its hydrophobic tails to the surrounding. Yang and Poulson[27] also reported that hydrogen bonds formed between plasticizer and polymer will replace the interactions between the polymers in the films, resulting in reduced available polar (OH) groups to interact with water, thus resulting in lower WVP. Cheng et al.[28] had also highlighted those available polar groups in the system will highly influence the WVP of the films.

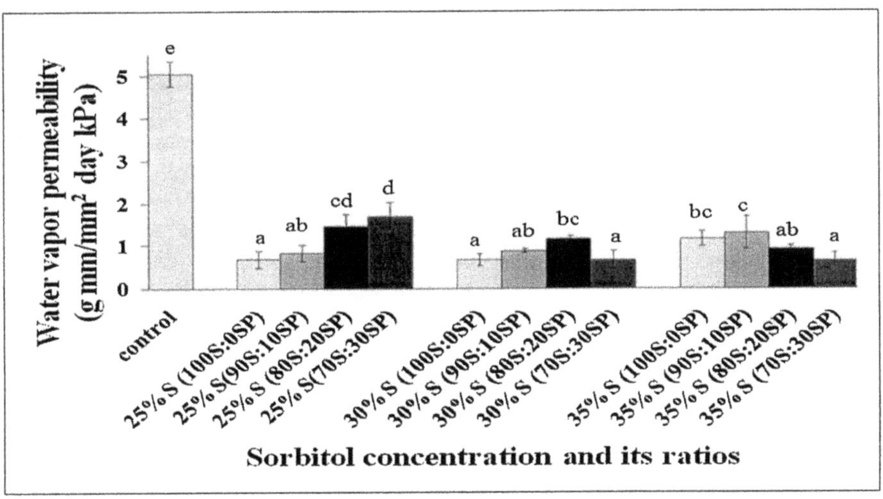

Figure 3 *Effect of sorbitol (S) and sucrose palmitate (SP) on water vapour permeability of sago starch-gelatin films. Mean values ± SD (n=3).*

3.4 Water sorption isotherms

Water sorption isotherms are important for predicting the stability and quality changes during packaging and storage of food[29]. Figure 4 shows that water content of all films increased with increasing RH. Films plasticized with sorbitol showed a typical sigmoidal shape, whereas control films (without plasticizer) showed a slightly different curve and are less sigmoidal in shape. The control film had a higher water content at RH below 60% (at lower water activity) compared to the other films with plasticizers. This results coincides with the WVP values obtained as shown in Figure 4.

3.5 Water solubility

Solubility of edible films indicates their integrity in an aqueous environment and higher solubility would indicate lower water resistance. Film solubility is an important factor that determines biodegradability of films when used as packaging. Figure 5 shows that the water solubility of starch-gelatin based films was higher in plasticized films at all sorbitol concentration levels without addition of PASE compared to the control film. It can be concluded that unplasticized film had greater polymer-polymer intermolecular forces than plasticized films, thus lower solubility. These results were in agreement with Laohakunjit and Noomhorn[25] who also reported that plasticized rice starch films have higher water solubility compared to unplasticized films. Much lower % solubility was observed in 35% sorbitol film at all amounts of PASE substitution compared to the 25 and 35% sorbitol films and this concurs with results obtained for WVP.

3.6 Scanning electron microscopy (SEM)

Scanning electron microscopy (SEM) was conducted to observe the morphology of the films. Figure 6a-e shows SEM micrographs of sago starch-gelatin films. The control unplasticized film (Fig 6a) shows a highly irregular surface with channels or holes and higher porosity which was reflected in higher WVP results compared to plasticized films.
The SEM micrographs show that there were differences in the morphologies of films with different sorbitol concentration levels without the presence of sucrose esters PASE.
The film with addition of sorbitol showed a relatively smooth, more compact and homogeneous surface. At higher sorbitol concentrations without the presence of PASE, the films exhibited larger diameters of channels and were less compact. This resulted in a greater the rate of water diffusion through the polymer and thus increase in the WVP. Similar observations were reported by Andreucutti et al.[30] where lower quantities of plasticizer produced a more compact internal structure and less discontinuous zones, whereas higher amounts of plasticizer resulted in higher heterogeneity and a less compact surface. The addition of sucrose palmitate also caused surface irregularity of films.

4 CONCLUSION

The effects of incorporation of sucrose palmitate on the physical and mechanical properties of the biopolymer films were investigated. Incorporation of sucrose palmitate into starch-gelatin based film resulted in an increase of tensile strength and water solubility and improves water vapour permeability. The surface of sago starch-gelatin based film with sorbitol and sucrose palmitate was homogeneous, smooth and more compact compared to unplasticized film.

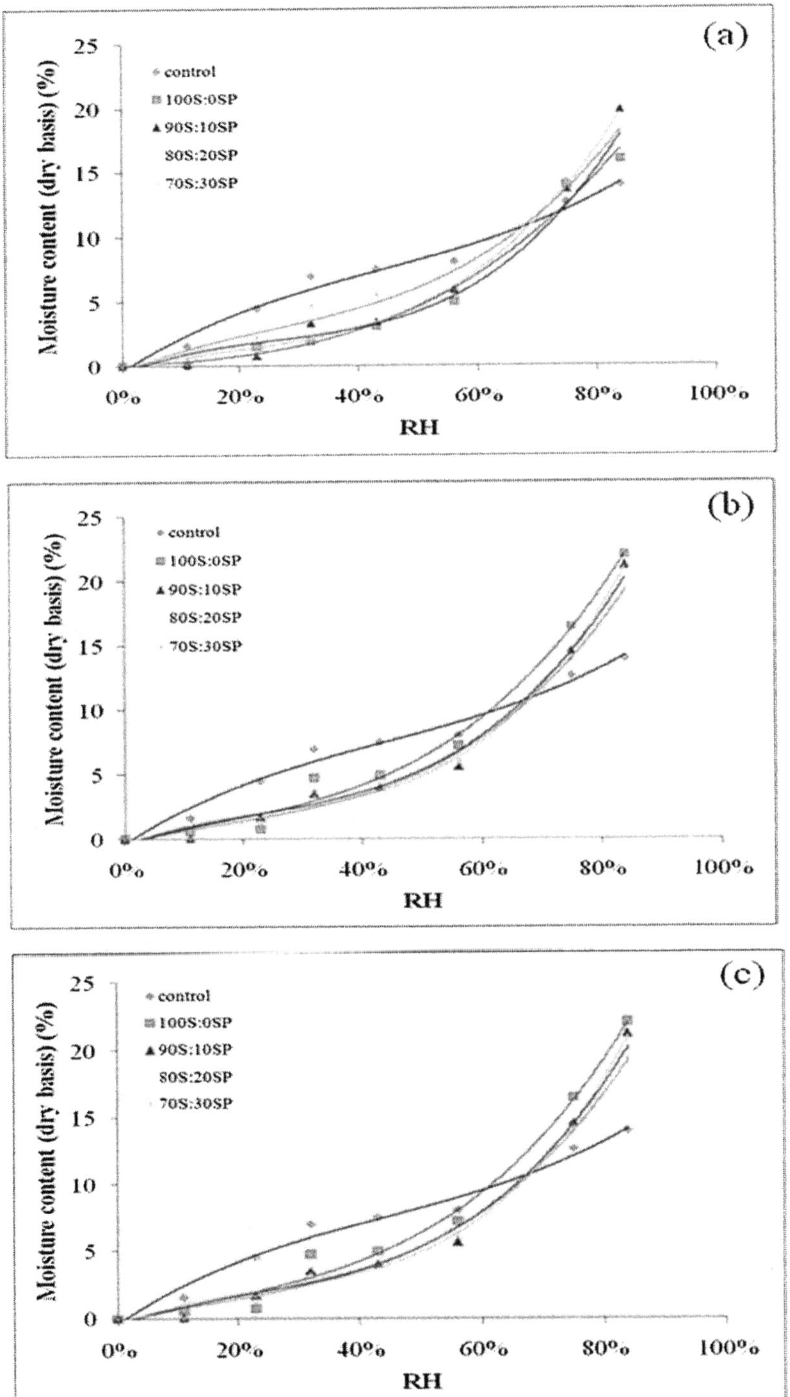

Figure 4 *Moisture sorption isotherm of sago starch-gelatin films incorporated with (a) 25 % (b) 30 % and (c) 35 % sorbitol (S) and sucrose palmitate (SP).*

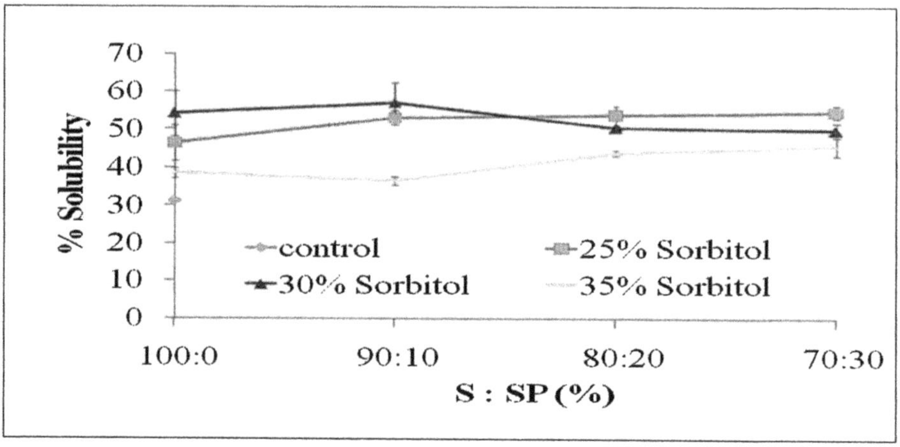

Figure 5 *Effect of sorbitol (S) and sucrose palmitate (SP) on % solubility of sago starch-gelatin films. Mean values ± SD (n=3).*

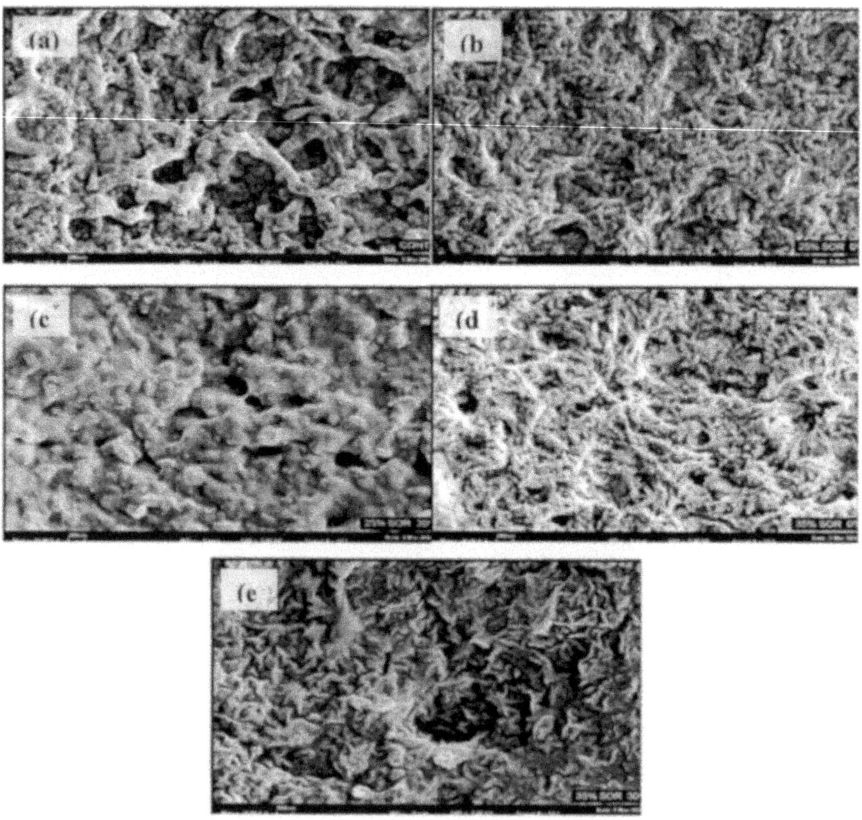

Figure 6 *SEM images (500 x) of sago starch-gelatin films (a) control, (b) 25 % sorbitol (100S: 0 SP), (c) 25 % sorbitol (70S: 30 SP), (d) 35 % sorbitol (100S: 0 SP) and (e) 35 % sorbitol (70S: 30 SP). S is sorbitol; SP is sucrose palmitate*

Acknowledgement

The authors wish to express their thanks to Universiti Sains Malaysia for the RU grant funding to support this work.

References

1. L.H. Cheng, A.A. Karim, M.H. Norziah, and C.C. Seow, *Food Research International*, 2002, **35**, 829-836.
2. N. Cao, Y. Fu and J. He, *Food Hydrocolloids,* 2007, **21**(7), 1153-1162.
3. D. Lourdin, G. Della Valle and P. Cooba, *Carbohydrate Polymers*, 1995, **27**, 261-270.
4. A. Jongareonrak, S. Benjakul, W. Visessanguan, T. Prodpran and M. Tanaka, *Food Hydrocolloids*, 2006, **20**(4), 492-501
5. R.J. Avena-Bustillos, C.W. Olsen, D.A. Olson, B. Chiou, E. Yee, P.J. Bechtel and T.H. McHugh, *J. of Food Science*, 2006, **71**, E202-E207.
6. P.J. Sobral, F.C. Menegalli, M.D. Hubinger and M.A. Roques, *Food Hydrocolloids*, 2001, **15**, 423-432.
7. M.A. Garcia, M.N. Martino and N.E. Zaritzky, *J. of Food Science*, 2000, **65**, 941-947.
8. I. Arvanitoyannis, E. Psomiadou and A. Nakayama, *Carbohydrate Polymers*, 1996, **31**, 179-192.
9. F. Debeaufort and A. Voilley, *J. Agricultural Food Chemistry,* 1997, **45**, 685-689.
10. M.G.A. Vieira, M. A, da Silva, L.O. dos Santos and M.M. Beppu, *European Polymer Journal*, 2011, **47**, 254-263.
11. E. Ayranci and S. Tunc, *Food Chemistry*, 2001, **72**(2), 231-236.
12. D.P. The, F. Debeaufort, C. Peroval, D. Despre, J.L. Courthaudon and A. Voilley, *J. Agric. Food Chem.* 2002, **50**, 2423–2428.
13. M.H. Norziah and A.B Khairulnizam (unpublished data).
14. A. A. Al-Hassan and M.H. Norziah, Food Hydrocolloids, 2011, http://dx.doi.org/10.1016/j.foodhyd. 2011.04.015.
15. Y. Fang, M.A. Tung, I.J. Britt, S. Yada and D.G. Dalgleish, *J. of Food Science*, 2002, **67**, 188-193.
16. ASTM. *Standard test methods for tensile properties of thin plastic sheeting*, 1981a, 380-388.
17. ASTM. *Standard test methods for water vapour transmission of materials*, 1981b, 87-95.
18. Y.P. Chang, P.B. Cheah and C.C. Seow, *J. of Food Science,* 2000, **65**(3), 445-451.
19. C.A. Romero-Bestida, L.A. Bello-Perez, M.A. Garcia, M.N. Martino, J. Solorza-Feria and N.E. Zaritzky, *Carbohydrate Polymers,* 2005, **60**, 235-244.
20. L.C. Bertan, P.S. Tanada-Palmu, A.C. Siani, and C.R.F. Grosso, *Food Hydrocolloids*, 2005, **19**, 73-82.
21. M. Thomazine, R.A. Carvalho and P.J.A. Sobral, *Journal of Food Science*, 2005, **70**, 172–176.
22. J.Q. Gallo, F. Debeaufort, F. Callegarin, and A. Voilley, *J. of Membrane Science*, **180**, 37-46.
23. T.H. McHugh and J.M. Krochta, *J. of Agricultural and Food Chemistry*, 1994, **42**(4), 841-845.
24. K.N. Turhan, Z.O. Erdohan Sancak and D. De Wit, *J. of Food Process Engineering*, 2007, **30**, 485-500.
25. N. Laohakunjit and A. Noomhorm, *Starch*, 2004, **56**, 348-356.

26. R.A. Talja, H. Helen, Y.S. Roos and K. Jouppila, *Carbohydrate Polymers*, 2007, **67**, 288-295.
27. L. Yang and A.T. Paulson, A.T. *Food Research International*, 2000, **33**, 571-578.
28. L.H. Cheng, A.A. Karim, M.H. Norziah and C.C. Seow, *Food Research International,* 2002, **35**, 829-836.
29. P.C. Srinivasa, M.N. Ramesh and R.N. Tharanathan, *Food Hydrocolloids*, 2007, **18**, 1113-1122.
30. C. Andreucutti, R.A. Carvalho and C.R.F. Grosso, *Food Research International*, 2009, **42**, 1113-1121.

EFFECT OF PALM OIL INCORPORATION ON THE PROPERTIES OF BIODEGRADABLE COBIA (*RACHYCENTRON CANADUM*) SKIN GELATIN FILMS

Amiza M. A.[1]* and Wong, Y.P.[1]

[a]*Department of Food Science, Faculty of Agrotechnology and Food Science, Universiti Malaysia Terengganu. 21030 Kuala Terengganu. Terengganu. Malaysia. *E-mail: ama@umt.edu.my*

1 INTRODUCTION

Recently, proteins have been extensively tested to develop biodegradable films as they offer good mechanical and barrier properties (Cuq et al., 1998). Among all proteins, gelatin has attracted the attention for the development of biodegradable edible films due to its abundance (Bigi et al., 2002). Gelatin is generally produced from the mammalians, deriving from skin and bone of cattle and pigs. However, the outbreak of bovine spongiform encephalopathy (BSE) as well as religious concerns has led to intensive research to identify and develop alternatives to mammal-derived gelatin (Karim and Bhat, 2009). In this regard, fish gelatin has been studied intensively in recent years and have been shown to exhibit excellent film-forming properties (Avena-Bustillos et al. 2006). Cobia (*Rachycentron canadum*) is a warm water marine finfish species with emerging global potential for mariculture, most notably in Taiwan, mainland China, and Vietnam (Holt et. al., 2007). Cobia skin accounts for approximately 6% of the total fish weight and this skin has been described as a good raw material for the production of gelatin (Chow, 2004). Thermal extraction of gelatin from cobia skin with the yield of about 18.9% has been reported (Amiza and Siti Aishah, 2011).

Possible applications of the cobia skin gelatin include biodegradable film. However, gelatin-based films are highly water and moisture absorbant, owing to the hydrophilicity of the amino acids in the protein molecules (Prodpran et al., 2007). In addition, hydrophilic plasticizers such as polyols and mono-, di- and oligosaccharides, necessarily incorporated into the gelatin films to improve the film flexibility, decrease the water vapor barrier properties of the films (Gennadios et al., 1996). To improve water vapor barrier properties of the protein films, hydrophobic substances such as fats, oils, and waxes could be added into the gelatin films (Prodpran et al., 2007; Jongjareonrak et al, 2006b). Generally fish gelatin films are highly sensitive to moisture and exhibit poor water vapor barrier properties (Guilbert et al., 1996). Moreover, Avena-Bustillos et al. (2006) reported that the water barrier properties of warm water fish gelatin films were significantly poorer than cold water fish gelatin. Furthermore, the pure gelatin films are normally brittle (Karnnet et al., 2005). Until now, no study has reported the effects of introducing palm oil into cobia skin gelatin film. In this study, the effects of incorporating palm oil on cobia skin gelatin film properties were examined.

2 MATERIALS AND RESULTS

2.1 Materials and Methods

Marine cultured whole cobia (*Rachycentron canadum*) was obtained from Langkawi Island, Kedah, Malaysia. Vesawit, a commercial brand of palm oil was purchased from a local store. The fish were filleted and the skin was manually removed, descaled, and cut into small pieces (1.0 × 1.0 cm). Excessive water was drained by hand squeezing in cheesecloth. The cleaned and cut fish skins was then packed into polyethylene and kept frozen at -40 °C until required for gelatin extraction. Gelatin was extracted from cobia skin according to Kasankala et al. (2007) with slight modification. Gelatin films were prepared according to method mentioned by Pérez-Mateos et. al (2009) with slight modification. To prepare the film-forming solution, first, gelatin powder (4% w/v) was dissolved in distilled water by heating in a water bath at 60°C for 2 h. Then, 0.75% glycerol and 0.75% sorbitol (w/v) were added into solution as plasticizers. Seven formulations of gelatin films containing different amounts of palm oil (0%, 7.5%, 15%, 25%, 30%, 40% and 50% (w/w) of gelatin powder) were prepared. All dispersions were homogenized at 14000 rpm for 8 min using a homogenizer and allowed to stand for 2 min at room temperature (30 ± 2°C) before subjecting to film casting. The films were cast onto 21 × 28 cm Teflon- tray and dried in a ventilated oven at 45°C for 18 h. Dried films were peeled manually from the casting moulds and conditioned in a desiccator with a saturated solution of magnesium nitrate at 50 ± 5% relative humidity and 23 ± 2 °C for 48 h period prior to testing.

The prepared films were analyzed for their thickness, tensile strength (TS) and percentage elongation at break (EAB) using a texture analyser, water vapor permeability, light transmission against ultraviolet and visible light (200-800 nm), color, solubility, biodegradability and sample surface and cross section using scanning electron microscopy.

2.2 Results and Discussion

Cobia skin gelatin based-film had slight but significant increase in thickness as the levels of palm oil added increased ($p<0.05$) as shown in Figure 1. Only two films were not significantly different from other films i.e. 15% and 40% palm oil incorporated. Similar trend was observed in the study done by Pérez-Mateos et al. (2009) on the addition of sunflower oil of different concentration to cod gelatin films. Prodpran et al. (2007) also observed an increment in thickness on adding palm oil to chitosan added round scad protein based-film. In another study on porcine skin gelatin with corn oil, a linear increase in thickness was also observed with increasing corn oil addition (Wang et al., 2009). Pérez-Mateos et al. (2009) suggested that the increment on film thickness was due to conformational changes in the film matrix at high added oil concentrations. The films made with gelatin alone or with low oil concentrations were relatively denser.

Tensile strength (TS) of the cobia skin gelatin films decrease significantly ($p<0.05$) with increase in oil addition as shown in Figure 2. However, there was a slight increase ($p<0.05$) in TS at 7.5% palm oil addition. The TS of the gelatin film decreased with subsequent increasing concentration of palm oil addition. A similar trend was observed when triacetin was incorporated into catfish gelatin (Jiang et al., 2010). When sunflower oil was incorporated into cod skin gelatin, a decreased puncture force was also observed with increase in oil incorporation (Pérez-Mateos et al., 2009). Jongjareonrak et al. (2006b) also found a decrease in tensile strength in a fish gelatin film with the addition of fatty acids, though tensile strength increased with the concentration of sucrose fatty acid esters. Comparing the TS of fish gelatin with mammalian gelatin, Bertan et al. (2005) reported a decrease in the tensile strength made from bovine hide gelatin with the addition of stearic acid, palmitic acid and elemi. The lower TS may be due to the ability of lipid substances to act as plasticizers which reduces the polymer-polymer interactions responsible for film strength. However, the slight increase in TS when 7.50% and 15% palm oil was incorporated into the gelatin film could be due to the polymers which were arranged in a way that could strengthen the film matrix as suggested by Prodpran et al. (2007). Moreover, both glycerol and sorbitol are polar plasticizers which could distribute uniformly throughout the film matrix. Since oil was added at low concentration and was not dissolved in the film forming solution thoroughly, the protein molecules could interact with each other to a greater extent.

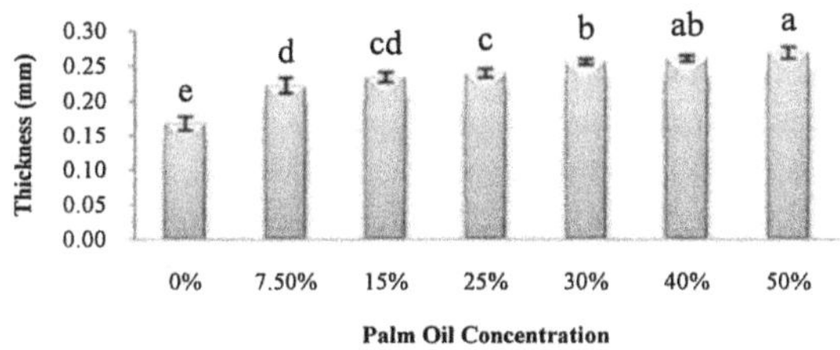

Figure 1: *Thickness of cobia skin gelatin films prepared with different amount of palm oil*

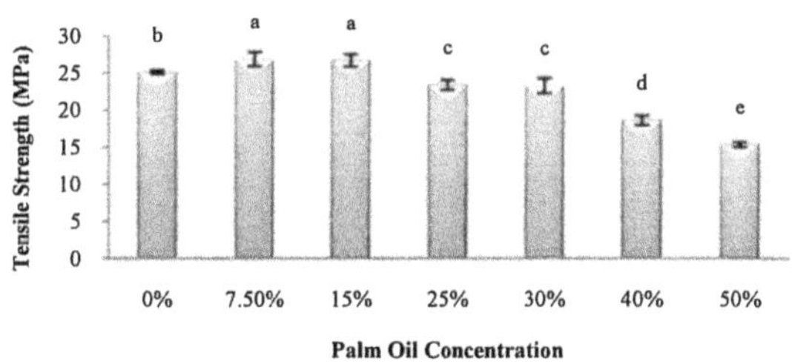

Figure 2: *Tensile strength of cobia skin gelatin films prepared with different concentration of palm oil*

In this study the tensile strength of the cobia skin gelatin films without added oil was higher than the values reported for catfish gelatin films (17.3 MPa) (Jiang et al., 2010) and for bigeye snapper skin gelatin (15.4 MPa) (Jongjareonrak et al., 2006b), both prepared using comparable amounts of plasticizer. However, higher tensile strength was reported for brownstripe red snapper skin (33.5 MPa) (Jongjareonrak et al., 2006b) and bovine hide gelatin (115 MPa) (Bertan et al., 2005). These differences could be explained by the origin of the gelatin (Pérez-Mateos et al., 2009). Gómez-Guillén et al. (2002) discussed the rheological characteristics (viscoelasticity and gel strength) and chemical/structural properties (amino acid composition, molecular weight distribution, and triple helix formation) of gelatins of different species as factors responsible for differences in behavior.

Figure 3 shows the elongation at break percentage (EAB %) of cobia skin gelatin films prepared using various concentration of palm oil. Elongation at break (EAB) of cobia skin gelatin films increased when the concentration of palm oil increased up to 15% and then decreased afterwards. However, the difference in EAB is not siginifacntly different for 25% and 40% palm oil concentration incorporation. The same trend was observed by Shellammer and Krochta (1997) whereby they showed an increase in elongation of milk whey protein films with increased lipid concentration caused by their plasticizing effect. Lim et al. (1998) described the action of plasticizers as a "lubricant" that allows the polymer molecules to more readily slide pass one another and thereby increases the extensibility of polymer. In contrast, Péroval et al. (2002) argued that some lipids were unable to form a continuous, cohesive matrix which results in decreased elongation. Such phenomenon may explain the decrease in EAB when more than 30% palm oil concentration was incorporated into the cobia skin gelatin film.

In this study, EAB of the cobia skin gelatin films without added oil, was much higher than the values reported for films made from catfish gelatin (68%) (Jiang et al., 2010), bigeye snapper skin gelatin (24%) and brownstripe red snapper skin gelatin (39%) ((Jongjareonrak et al., 2006b) prepared using comparable amounts of plasticizer. Moreover, when comparing with mammalian gelatin, Bertan et al. (2005) reported a much lower value for bovine hide gelatin (≈7%). Chiou et al. (2008) reported that these differences might be related to the different levels of renaturation developed during the film forming process. They explained that for mammalian gelatin, film drying occurred at temperatures below their gelation temperature. This allowed the gelatin chains to partially revert back to a triple helical structure in solution. The chains then became locked into this structure in solution. In contrast, film drying for cold fish gelatin occurred at a temperature above their gelation temperature. Since cobia fish is a warm water fish, the high elongation value may be due to its skin gelatin exhibiting attributes between mammalian and cold fish gelatin.

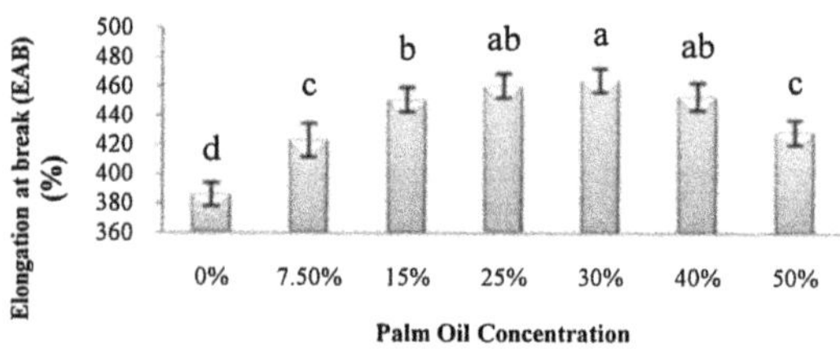

Figure 3: *EAB of cobia skin gelatin films prepared with different concentration of palm oil*

The effect of palm oil concentration on the water vapor permeability (WVP) of the cobia skin gelatin films is presented in Figure 4. WVP increased steadily up to palm oil concentrations of 30% and then decreased afterwards. Adding lipids can increase the hydrophobicity of the film matrix thus improving the WVP, depending on the kind and amount of lipid (Bertan et al., 2005). A tendency of the WVP to decrease has also been reported in fish gelatin films containing fatty acids and sucrose fatty acid esters (Jongjareonrak et al., 2006b). Gontard et al. (1992) observed that when lipid substances were used in composite film formulations, large molecules are incapable of associating with the protein chain and could cause breakage of the matrix structure, resulting in a loss of water barrier properties. Therefore, the increased WVP of film incorporated with 40% and 50% palm oil may due to the occurrence of such phenomenon. Jiang et al. (2010) also reported a similar trend when 50% triacetin was incorporated in catfish gelatin film, the WVP reduced, however, when triacetin contents increased to 100% and 150% levels, the WVP of the film increased. They suggested that when hydrophobic substances were added at a moderate amount, the substance may pose certain barriers to the water vapor but excessive amounts do not necessarily reduce the moisture transfer in a composite film. Shellhammer and Krochta (2007) also reported that permeability and mechanical properties of film were affected by lipid type and amount.

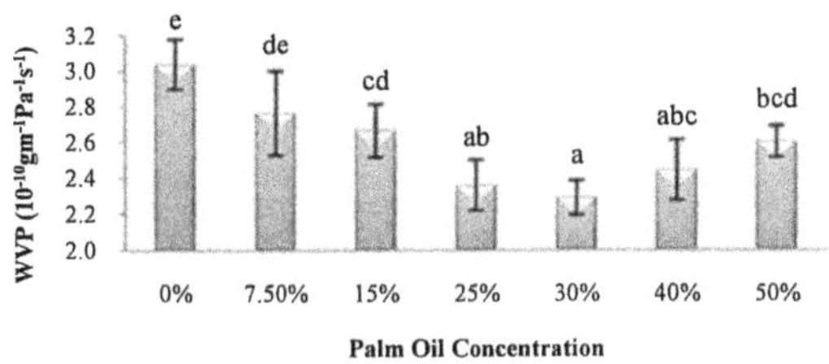

Figure 4: *Water vapour permeability (WVP) of cobia skin gelatin films prepared with different concentration of palm oil*

The cobia skin gelatin film without added palm oil had WVP values approximately half those of mammalian gelatin films. Chiou et al. (2008) reported that bovine and porcine gelatin films showed similar WVP with values 5.17×10^{-10} gm^{-1}s^{-1}Pa^{-1} and 5.15×10^{-10} gm^{-1}s^{-1}Pa^{-1} respectively. It should be noted that the study reported by Chiou et al. (2008) did not include plasticizer in the films, which might affect the comparability in values. However, the film was about twice as permeable to water than that of cod gelatin film (1.20×10^{-10} gm^{-1}s^{-1}Pa^{-1}) (Pérez-Mateos et al., 2009), round scad protein based-film (1.08×10^{-10} gm^{-1}s^{-1}Pa^{-1}) (Prodpran et al., 2007), and catfish skin gelatin (2.33×10^{-10} gm^{-1}s^{-1}Pa^{-1}) (Jiang et al., 2010). Jongjareonrak et al. (2006a) pointed out that the difference could depend on differences in the protein composition, emulsifying protein composition, and emulsifying properties of the gelatins.

Light transmission of cobia skin gelatin-based film containing different palm oil levels for both UV and visible ranges is shown in Figure 5. The results showed that the amounts of oil used were found to affect the light transmission of the film. In general, all films exhibited the low transmission to light in the UV ranges (200–280 nm). For the visible range, gelatin film with 0% concentration of palm oil had the highest transmission and then decreased sharply when 7.5% concentration of palm oil was incorporated. The subsequent incorporation of higher concentration of palm oil resulted in a steady reduction of the transmission in the visible range. In the presence of oil, the films had decreased transmission due to the droplet of dispersed palm oil in cobia skin gelatin-based film, preventing the light transmission through the film (Prodpran et al., 2007). Therefore, the addition of palm oil into the film was an effective means to improve the barrier property to the light (Prodpran et al., 2007). Pérez-Mateos et al. (2009) also suggested that when applied to food systems, oil-containing films could be an excellent barrier preventing UV light-induced lipid oxidation. A decrease of light transmission caused by the lipid components was also reported for cod skin gelatin films with different levels of sunflower oil (Pérez-Mateos et al., 2009), and bigeye and brown stripe red snapper skin gelatin with palmitic acid and stearic acid (Jongjareonrak et al., 2006c).

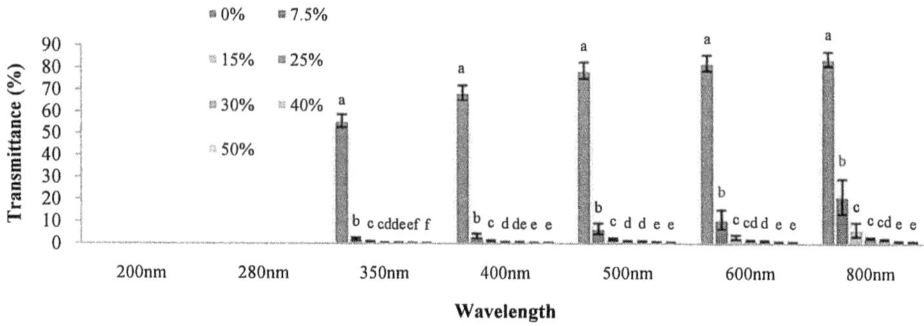

Figure 5: *Light transmission of cobia skin gelatin films with different levels of palm oil at selected wavelengths*

Table 1 shows the colour profiles of the prepared gelatin-palm oil films. In general, the films become less dark when palm oil was incorporated as evidenced by the increase in L^*-value (p<0.05). Whiteness showed a similar increasing trend with lightness with increasing addition of palm oil. The films showed a decrease in a^*-value but increase in b^*-value with increasing palm oil incorporation (p<0.05). This indicated that the film became more yellowish as palm oil was incorporated. However, a^*-value decreased with increase in palm oil incorporation and showed a significant decrease at 25% palm oil concentration and above. Meanwhile, for b^*-value, there were no significant increase at 30% and 50% palm oil concentration.

Table 1: Colour profiles of the prepared cobia skin gelatin films

Palm oil conc.	*L**	*a**	*b**	*Whiteness*
0%	53.82 ± 1.25^d	-0.53 ± 0.05^a	1.22 ± 0.17^e	53.80 ± 1.26^c
7.5%	59.44 ± 2.50^c	-1.93 ± 0.50^b	2.11 ± 0.21^d	59.34 ± 2.46^b
15%	63.29 ± 2.54^c	-2.22 ± 0.19^b	3.26 ± 0.45^c	63.07 ± 2.50^b
25%	73.55 ± 2.01^{ab}	-3.39 ± 0.11^c	6.65 ± 0.42^b	72.51 ± 1.97^a
30%	72.91 ± 1.30^b	-3.40 ± 0.15^c	7.03 ± 0.72^b	71.80 ± 1.36^a
40%	75.12 ± 0.76^a	-3.49 ± 0.03^c	9.21 ± 0.94^a	73.23 ± 0.93^a
50%	76.40 ± 1.77^a	-3.51 ± 0.13^c	10.61 ± 1.28^a	73.85 ± 1.55^a

Different alphabet means significant differences in colour parameters between film formulations.

Similar trend was observed in the study of addition of palmitic acid, stearic acid, and their sucrose ester in the skin gelatin of bigeye snapper and brownstripe red snapper (Jongjareonrak et al., 2006c). They reported that films with palmitic acid added generally had higher b^*-value than those with stearic acid added at same amount. Hence, the cobia skin gelatin film became more yellowish due to the fact that palm oil contains a high percentage of palmitic acid (44.3%).

The difference in color values of films added with various type and amount of hydrophobic substance could be associated with the different characteristics, properties, nature of the lipids, and the compatibility between the lipid and gelatin molecules in the film matrix (Orliac et al., 2003). Besides that, Jongjareonrak et al. (2006) also reported that an increase in solid fat content might enhance the reflection of light at the film surface, leading to the increased L^*-value of film.

Figure 6 shows film solubility of the prepared cobia skin gelatin films. Generally, the solubility of gelatin film with 0% palm oil film was high (around 72%) and adding palm oil at 25% and above increased the solubility ($p<0.05$). Bertan et al. (2005) and Pérez-Mateos et al. (2009) also observed similar trend in their study.

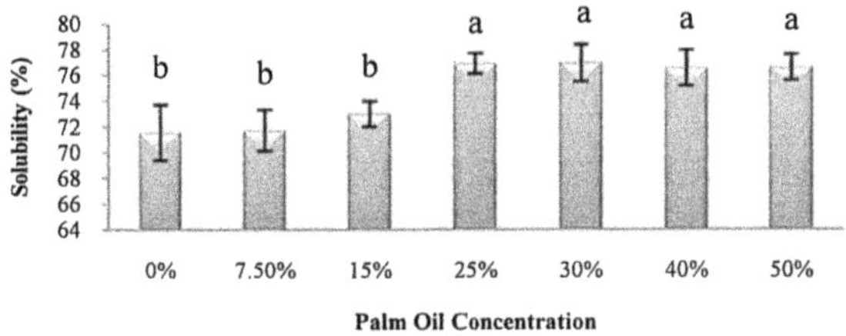

Figure 6: *Solubility of the prepared cobia skin gelatin films*

The biodegradability of the prepared films is shown in Table 2. There was no difference ($p > 0.05$) in the percentage of weight loss between the films with different palm oil concentrations on the 5th day of burial. The subsequent 7th, 10th, 12th and 15th days of burial showed significant different in the weight loss percentage between the film formulations. According to Karnnet et al. (2005), biodegradation is naturally caused by the penetration of microorganisms using water as a medium. Since the films prepared with increased palm oil concentration exhibited increased water solubility, these films exhibited higher biodegradability than those prepared without palm oil.

Table 2: Percentage of weight loss of cobia gelatin films with different concentration of palm oil incorporation after subjected to burial test

Palm oil (%)	% weight loss of cobia gelatin film after subjected to burial test				
	Day 5	Day 7	Day 10	Day 12	Day 15
0	6.44 ± 0.51^{a}	6.51 ± 0.54^{b}	8.16 ± 0.91^{a}	8.30 ± 0.78^{ab}	8.67 ± 0.83^{c}
7.50	6.53 ± 0.51^{a}	7.28 ± 1.07^{ab}	7.82 ± 0.83^{ab}	8.96 ± 0.73^{a}	9.44 ± 0.75^{bc}
15	6.65 ± 0.51^{a}	6.94 ± 0.62^{ab}	6.95 ± 0.90^{b}	9.36 ± 0.52^{a}	10.30 ± 0.64^{a}
25	6.29 ± 0.61^{a}	7.71 ± 0.76^{a}	7.72 ± 0.58^{ab}	8.91 ± 1.12^{ab}	9.72 ± 0.35^{ab}
30	6.69 ± 0.51^{a}	6.96 ± 0.92^{ab}	7.46 ± 0.62^{ab}	7.51 ± 0.91^{b}	9.76 ± 0.58^{ab}
40	6.31 ± 0.82^{a}	7.60 ± 0.77^{ab}	8.05 ± 0.68^{a}	8.59 ± 0.53^{ab}	10.42 ± 0.51^{a}
50	6.57 ± 0.68^{a}	7.29 ± 0.55^{ab}	7.95 ± 0.68^{ab}	7.98 ± 1.18^{ab}	10.01 ± 0.32^{ab}

Different alphabet indicates significant differences in weight loss between film formulations on the same day.

The morphology of cobia skin gelatin film's surface and cross-section micrographs with palm oil incorporated is shown in Figure 7. The film without addition of oil had the smooth and continuous surface without grainy and porous structure. According to Prodpran et al. (2007), such structure indicated that film with ordered matrix was formed. When palm oil was incorporated into film, the surface of film had the irregular surface with distribution of oil droplets. The oil droplets were less intense on the surface as the amount of palm oil increased to 40% and 50%. Those oil droplets were localized further into the film matrix as observed in cross-section view. The film structure as well as the distribution of oil droplets might be associated with the properties of film, particularly the water vapor permeability of the resulting film (Pétroval et al., 2002; Shellhammer and

Krochta, 1997). Moreover, Chambi and Grosso (2006) deduced that protein films with a fibrous structure were responsible for greater tensile strength, greater elasticity and lower water vapor permeability. The compatibility of plasticizers, glycerol and sorbitol, with proteins can be observed in the film morphology images. The almost complete absence of empty spaces in the gelatin film indicated good compatibility of the plasticizers with the gelatin (Chambi and Grosso, 2006). Conversely, when Chambi and Grosso (2006) compared the gelatin film with casein film, they observed that visible empty spaces were present which appeared to be regions occupied by free glycerol. They argued that such phenomenon was related to the phase separation of plasticizer in the polymeric matrix, which generally occurs during film drying when the plasticizer is not compatible with the polymer used.

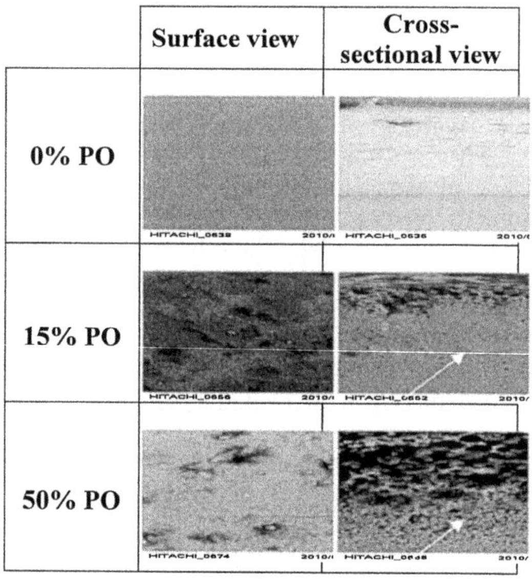

Figure 7: SEM micrograph of cobia skin gelatin films without palm oil incorporation, and 15 and 50% palm oil incorporation. Arrows shows the palm oil distribution of cross section.

3 CONCLUSIONS

This study has shown that an increase in palm oil incorporation into the cobia skin gelatin film resulted in an increase in thickness and yellowness, elongation at break percentage (EAB), water vapor barrier properties, UV and visible light barrier properties, water solubility and biodegradability of the gelatin film. However, it also caused a reduction in tensile strength (TS). Films incorporated with 25% and 30% palm oil exhibited lowest water vapor permeability (WVP) with high elongation at break (EAB) and intermediate tensile strength (TS).

References

M.A. Amiza, and D. Siti Aishah, *Int. Food Res. J.*, 2011,**18**, 159.

R.J. Avena-Bustillos, C.W. Olsen, B. Chiou, E. Yee, E., P.J. Bechtel, and T.H. McHugh, *J. Food Sci.*, 2006,**71**(4): 202-207.

M. Barreto and J.J. Meyer, *South Afr. J. Botany,* 2006, **72**: 521-528.

L.C. Bertan, P.S. Tanada-Palmu, A.C. Siani and C.R.F. Grosso, *Food Hydrocolloids,* 2006, **19**,73.

A. Bigi, G. Cojazzi, S. Panzavolta, N. Roveri and K. Rubini, 2002. *Biomater.*, 2002, **23**, 4827.

H. Chambi and C. Grosso, *Food Res. Int.*, 2006, **39**: 458-466.

B.-S. Chiou, R.J. Avena-Bustillos, P.J. Bechtel, H. Jafri, R. Narayan, R., H.I. Syed, G.M. Glenn and W.J. Orts, *Eur. Polym. J.*, 2008, **44,** 3748.

C.J. Chow. Unpublished data from a report titled ''Utilization and development of by-products of cobia (*Rachycentron canadum*) cultivated in offshore cages (III)" to the National Science Council, Taiwan. 2004.

B. Cuq, B., Gontard, N., Cuq, J.L., and Guilbert, S. *J. Food Sci.* 1998, **60**,1369.

P.G. Dalev, R.D. Patil, J.E Mark, E. Vassileva and S.J. Fakirov, *Appl. Polym. Sci,*. 2000, **78**, 1341.

R.A. de Carvalho and C.R.F. Grosso, *Food Hydrocolloids* 2004, **18,** 717.

A. Gennadios, C.L. Weller, M.A Hanna and G.W. Fronning, *J. Food Sci* 1996, **61**,585.

N. Gontard, S. Guilbert, and J.L. Cuq. *J. Food Sci.*, 1992, 57(1), 190.

M.C. Gómez-Guillén, J. Turnay, M.D. Fernández-Díaz, N. Ulmo, N., M.A. Lizarbe and P. Montero. *Food Hydrocolloids.*, 2002, **16**, 25.

J.D. Gu. *Int. Biodeterior. Biodegrad.,* 2003, **52**, 69.

S. Guilbert, N. Gontard and L.G.M Gorris. *J. Food Sci. Technol.,* 1996, **29**, 10.

G.J. Holt, C.K. Faulk, and M.H. Schwarz, *Aquacul.*, 2007, **268**,181.

M. Jiang, S. Liu, X. Du and Y. Wang. *Food Hydrocolloids.*, 2010, **24,** 105.

A. Jongjareonrak, S. Benjakul, W. Visessanguan and M. Tanaka, M. *Food Hydrocolloids.* 2006a, **20**: 492

A. Jongjareonrak, S. Benjakul, W. Visessanguan and M. Tanaka, M. *Eur. Food Res. and Technol.,* 2006b, **222**, 229..

A. Jongjareonrak, S. Benjakul, W. Visessanguan and M. Tanaka, M. *Eur. Food Res. and Technol.,* 2006c, **222**, 650.

A.A. Karim and R. Bhat. *Food Hydrocolloids.,* 2009, **23**, 563.

S. Karnnet, P. Potiyaraj, and V. Pimpan. *Polym. Degrad. Stab.*, 2005, **90**, 106.

L.M. Kasankala, Y. Xue, Y. Weilong, S.D. Hong and Q. He, Q. *Biores. Technol., 2007,* **98,** 3338.

L.T. Lim, Y. Minc, and M.A. Tung. *J. Agric. and Food Chem.*, 1998, **46**, 4022.

O. Orliac, A. Rouilly, F. Silvestre and L. Rigal. *Ind. Crops and Prod., 2003,***18,** 91.

M. Pérez-Mateos, P. Montero, and M.C. Gómez-Guillén. *Food Hydrocolloids.* 2009. **22**(4), 53.

C. Péroval, F. Debeaufort, D. Despré and A.J. Voilley. *J. Agric. and Food Chem.,* 2002, **50**, 3977.

T. Prodpran, S. Benjakul, and A. Artharn. *Int. J. Bio. Macromol.,* 2007, **41**(5), 605.

T.H. Shellammer and J.M. Krochta, *J. Food Sci.*, 1997, 62, 390.

L. Wang, M.A.E., A. Rau, J.F. Kerry and J.P. Kerr. *J. Food Eng, 2009,* **90**: 11.

CONTROLLED RELEASE OF WATER SOLUBLE VITAMINS FROM CHITOSAN FILMS PREPARED IN ELECTRIC FIELD

M. Marudova[1], T. Yovcheva[1], S. Zhelyazkov[1], G. Zsivanovits[2] and A. Viraneva[1]

[1]Department of Experimental Physics, Plovdiv University, 24 Tzar Assen str., 4000 Plovdiv, Bulgaria, E-mail: margo@uni-plovdiv.bg
[2]Food Research and Development Institute, Agricultural Academy, 154 Vasil Aprilov Blvd., 4000, Plovdiv, Bulgaria

1 INTRODUCTION

Water-soluble vitamins are responsible for many biochemical functions in the human body, and are mainly supplied by the diet. Riboflavin (known as vitamin B_2) plays a key role in energy metabolism, and for the metabolism of fats, carbohydrates, and proteins.[1] It also works as an antioxidant and might have direct medical effect in the treatment of anemia,[2] cataracts[3] and migraine headache[4]. Naturally occurring polymers like chitosan have been suggested as vehicles for encapsulation of active compounds due to their unique properties such as good biocompatibility,[5,6] nontoxicity,[7,8] biodegradability[9,10] and antimicrobial properties.[11] They have also been widely used in the food industry, serve as dietary fibre and are not absorbed in the gastrointestinal tract. It has been reported that chitosan has its own bioactivity and can suppress plasma cholesterol levels.[12] Its capability of gelation in aqueous medium[13] makes it preferable as encapsulation matrix as far as it minimizes the chances of loss of activity or damage of the encapsulated active compounds, protects them against harsh conditions into gastrointestinal tract and enhances their absorption.

The natural polyelectrolytes have more advantages as carriers of active compounds due to their capability of control released. The release of vitamins from polyelectrolyte networks is controlled by the structure of the carrier as defined by the degree of crosslinking, the degree of swelling and related parameters. It also depends on the interactions between the vitamin and the polyelectrolyte, which often have electric origin. Chitosan is a cationic polyelectrolyte and is positively charged at acid conditions.[14] It could be crosslinked by synthetic low-molecular weight compounds like glutaraldehyde or can form polyelectrolyte complexes with anionc polyelectrolytes (for example with sodium alginate).[15] The network properties are governed by the electrostatic forces and can be desirably modified by changing the surrounding electrical environment – pH, ionic strength, applied electric field, etc. The release of vitamin B_2 from chitosan matrices has already been studied by several authors.[16,17] It was shown that the controlled release properties of chitosan films could be improved by creating chitosan/polyanion polyelectrolyte complex. Coating with alginate prolonged riboflavin release noticeably. Building of self-assembly multilayers leads to further delay of the release.

The aim of the present research is to investigate the effect of electrical treatment of chitosan films on the vitamin release properties. The effect of polyelectrolyte complexation on the riboflavin release is also examined.

2 EXPERIMENTAL

2.1 Materials

Chitosan (CH, high viscosity > 400 mPa.s for 1 % in acetic acid at 20 °C), sodium alginate (SA, low viscosity 20 cP for 1 % aqueous solution at 20 °C), riboflavin (vitamin B_2), molecular mass 376.36 and purity 99.7 % were obtained from Sigma-Aldrich and were used without further purification or characterization. Glutaraldehyde was obtained from Merck. All other chemicals were of analytical grade. The double distilled water was used throughout the investigations.

2.2 Preparation of B_2-loaded chitosan films

Chitosan was dissolved in 2 % acetic acid at a concentration of 1 % (w/v). 0.5 % of riboflavin was added to this solution. The solution was stirred thoroughly to ensure complete mixing of the vitamin. Then the films were prepared by casting from a solution and they were left to dry in DC electric field for a day. The electric treatment during the drying process was carried out by using a point-to-plane three-electrode corona discharge system. The experimental set up of the system consisted of a corona electrode (needle), a grounded plate electrode and a grid placed between them. The samples were placed on the grounded plate electrode and were treated until drying under normal relative humidity of (45 % - 55 %), at room temperature of about 25 °C, and atmospheric pressure. Positive or negative 5 kV voltages were applied to the corona electrode. 1 kV voltage of the same polarity as that of the corona electrode was applied to the grid. Etalon films, dried at room conditions (without applying DC electric field), were also prepared. After drying, the films were crosslinked by glutaraldehyde. To improve the controlled release properties of the films, sodium alginate was further coated on the film surface.

2.3 Swelling/degradation of chitosan films

The swelling experiments were carried out in buffers of desired pH (1, 3, 5, and 7) and ionic strength 0.1 M. HCl/KCl, acetate and phosphate buffers were used to cover the pH range. The pre-weighed dry films were immersed in the swelling media and their weight change was monitored at regular time intervals till the films showed constant weight or complete disintegration. The fractional weight change W_t at a particular time t was calculated by using the following expression:

$$W_t = \frac{m_t - m_0}{m_0} \qquad (1)$$

where m_t is the sample mass at time t and m_0 is the initial mass of the sample.

2.4 Riboflavin release study

The amount of riboflavin release during the crosslinking process and later at different time intervals and pH was determined spectrophotometrically at 437 nm.[17] For this purpose the absorbance of the buffer solutions with immersed chitosan/riboflavin films was measured by an Ava-Spec-2048 fibre optic spectrophotometer (USA). The amount of vitamin released was computed by comparing the absorbance with the standard curve prepared for the pure vitamin in the appropriate concentration regions. Since riboflavin is sensitive towards light, the whole study was carried out using glassware with completely blackened surface so as to avoid exposure to light.

3 RESULTS AND DISCUSSION

3.1 Swelling and degradation of chitosan films

The dry chitosan films were allowed to hydrate in buffer solutions with different pH and constant ionic strength 0.1 M. The films, crosslinked with glutaraldehyde were stable for nearly 15 hours at a particular pH and the water uptake was between 200 and 400 (these date are not presented). It was found out that the water uptake depends slightly on the pH and electrical treatment of the films. Hence an assumption could be made that the chitosan/glutaraldehyde crosslinks are stable enough not to change their density due to the pH changes or corona treatment.
The dynamic hydration of chitosan films coated with alginate is shown at Figure 1. There are dramatic changes in swelling behaviour of corona treated and not treated samples at different pH.

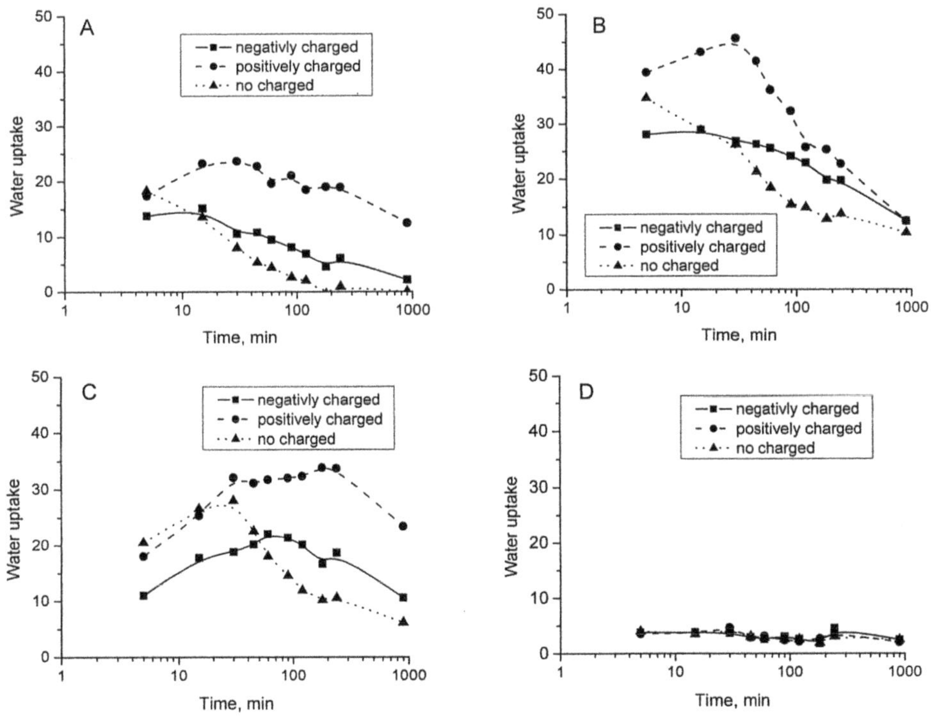

Figure 1 *Dynamic uptake of water as a function of time for the films, coated with alginate at pH 1 (A), pH 3 (B), pH 5 (C) and pH 7 (D).*

The non corona treated films are less stable at a particular pH and begin to disintegrate immediately or in 30 min. The increase of pH leads to enhancement of the stability. This fact could be explained with more intensive electrostatic interactions between protonated

NH_3^+ groups of the chitosan and COO^- groups in the alginate, which interact with electrostatic forces and strengthen the chitosan network.

The corona treatment of polymers leads to 2 effects: 1) increase of surface charge; and 2) orientation and ordination of the dipole molecules. In the case of the chitosan films the surface potential after the charging was nearly 20 V. The second result of orientation of the protonated amino groups may provide more compact structure of the films. Therefore the corona treated samples are more stable and show fewer tendencies to swelling in comparison with non treated ones.

The most stable chitosan films are positively charged. This may be because of the extra surface charge enhances the electrostatic interaction between the chitosan film and the alginate and stimulates the formation of insoluble chitosan/alginate polyelectrolyte complexes.

The minimal water uptake at pH 7 is related to the aggregation and precipitation of chitosan at pH above its pK_a value (pK_a=6.5).[18]

The release of vitamin B_2 during the crosslinking process with glutaraldehyde and coating with alginate was negligible (less than 1%) (Table 1).

Table 1 *Vitamin B_2 release (in %) during crosslinking and coating processes*

Corona treatment	*Negatively charged*	*Positively charged*	*No charged*
Crosslinking with glutaraldehyde	0.582	0.082	0.677
Coated with alginate	1.166	0.666	0.178
Crosslinking with glutaraldehyde and coated with alginate	0.253	0.430	0.448

3.2 Release of vitamin B_2 from chitosan films, crosslinked with glutaraldehyde

Figure 2 describes the release of vitamin B_2 from the chitosan films crosslinked with glutaraldehyde. The following general trends could be observed:

a) The samples treated in negative corona exhibit the slowest vitamin B_2 release, followed by positively charged films. The non corona treated samples possess the fastest release.
b) The increase of pH from 1 to 5 leads to faster release.
c) The release is slower at pH 7, where the fractional release during the first 6 hours is less than 10%.

The negative corona generates primary and secondary electrons, which are injected on the surface and in the volume of the samples.[19] It is well known that vitamin B_2 possesses very good electron acceptor properties.[20] Therefore an assumption could be done that the treatment in negative corona causes charging of the vitamin B_2. The negatively charged vitamin interacts with the protonated NH_3^+ groups of the chitosan and creates complexes. These complexes are more stable at low pH when practically all of the amino groups are protonated. The vitamin B_2 diffusion in the chitosan network is hampered at these conditions and the release is slower. Increase of the pH leads to deprotonation of the amino groups and destruction of the chitosan/vitamin B_2 complexes. The unbound vitamin easily and faster diffuses through the chitosan film. Hence the increase of pH leads to acceleration of the release process. The release is again slower at pH 7 because of the precipitation of the chitosan and formation of an insoluble diffusion barrier.

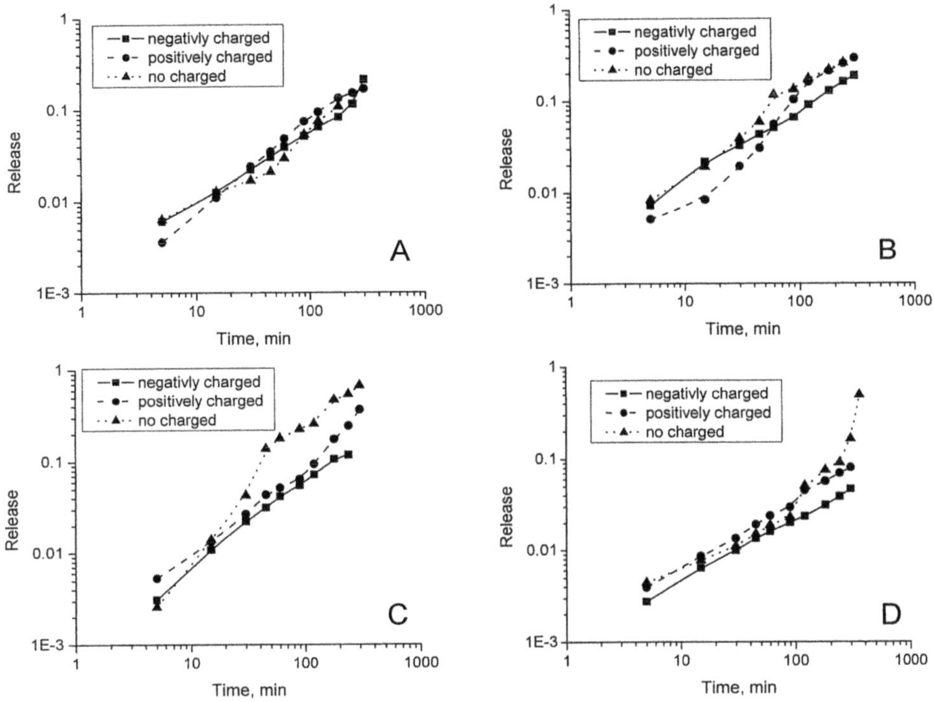

Figure 2 *Release of vitamin B_2 from chitosan films, crosslinked with glutaraldehyde at pH 1 (A), pH 3 (B), pH 5 (C) and pH 7 (D).*

3.3 Release of vitamin B_2 from chitosan films, coated with alginate

After the drying process of chitosan films in or without applied electric field, the films were exposed to alginate solution for 15 min. Then the electrostatic interactions between the protonated amino groups in the chitosan and the ionised carboxyl groups in the alginate cause formation of a polyelectrolyte complex on the surface of the chitosan films. Existence of insoluble layer should prevent the vitamin release process.

The dependence of the fractional release on time at different pH is presented at Figure 3. The following features could be noted:

a) The samples treated in positive corona exhibit the slowest vitamin B_2 release.
b) The fastest release (nearly 100 %) at pH 1 and pH 5 is realized by the non corona treated chitosan films.
c) The fastest release (nearly 100 %) at pH 3 is realized by the chitosan films, which are treated in negative corona.
d) The release of all the samples at pH 7 during the first 6 hours is less than 3 %.

If the corona discharge occurs in air (like in our experimental conditions) the positive corona generates predominantly the following types of cations: $(H_2O)_nH^+$, $(H_2O)_nNO^+$ and $(H_2O)_n(NO_2)^+$. Neutral, chemically active, groups and radicals are also created in the coronas.[21] At the time of charging, these particles are accelerated by the electric field of the corona and interact with the surface of the samples. Some of them generate additional surface trap levels. Hence when positively charged chitosan films are exposed to the alginate solution the additional cations also interact with the negatively charged carboxyl groups and bind an additional amount of alginate that acts as an electrostatic barrier for the vitamin B_2 release.

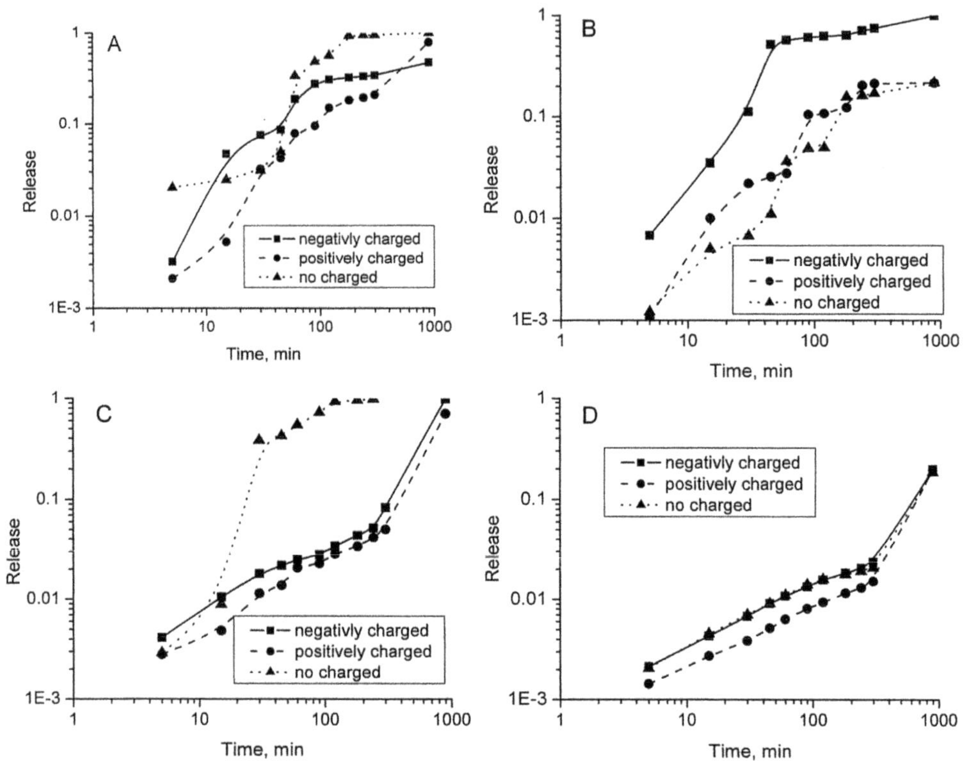

Figure 3 *Release of vitamin B_2 from chitosan films, coated with alginate at pH 1 (A), pH 3 (B), pH 5 (C) and pH 7 (D).*

The slower release of vitamin B_2 for negatively charged chitosan films in comparison to non corona treated ones at pH 1 could be again explained with the vitamin charging and creation of ionic bindings between it and the protonated amino groups in the chitosan.
The polyelectrolyte complexation strongly depends on the pH. Using the dependences between pK_a of weak polyelectrolytes, pH and degree of ionization:

$$\alpha\% = \frac{100}{1+10^{(pK_a - pH)}} \text{ - weak acid} \tag{2}$$

$$\alpha\% = \frac{100}{1+10^{(pH - pK_a)}} \text{ - weak base} \tag{3}$$

and also the fact that the pK_a value of alginate is between 3.38 and 3.65[22], the degrees of ionization of both chitosan and alginate could be estimated. For example, the ionization degree of alginate at pH 3 is about 24 %. Hence the electrostatic interactions between chitosan and alginate are very weak at pH 3 and increasing the negative surface charge on the chitosan films during the corona treatment does not allow polyelectrolyte complex formation. Therefore the release of vitamin B_2 at pH 3 is easier and faster for negatively corona charged films. The alginate ionization degree at pH 5 is nearly 97 %. Both chitosan and alginate are strongly charged at these conditions and an insoluble polyelectrolyte complex may be formed. Considering both processes of vitamin/chitosan binding and chitosan/alginate complexation, the treatment of films in negative corona leads to retardation of the release.

The slower release at pH 7 could be again attributed with the precipitation of the chitosan and formation of insoluble diffusion barrier.

3.4 Diffusion of vitamin B_2 through chitosan films

The release of active compounds from polymer networks is usually interpreted quantitatively in the terms of diffusion coefficients. The integral diffusion at short times, as derived from Ficks first and second law of diffusion is given as:[17]

$$F = 4\sqrt{\frac{Dt}{\pi d^2}} \tag{4}$$

where F is the fractional release, D is the diffusion coefficient, d is the film thickness. The initial diffusion coefficients were evaluated from the slope of the linear plot obtained between F and $t^{1/2}$ values for the initial part of the release processes. The dependences of the diffusion coefficients on the pH are presented at Figure 4.

Figure 4 *Diffusion coefficients of vitamin B_2 through chitosan films, crosslinked with glutaraldehyde (A), coated with alginate (B), crosslinked with glutaraldehyde and coated with alginate (C).*

The relatively lower values for the diffusion coefficients for the samples investigated are evidence of compact structure of the chitosan films crosslinked with glutaraldehyde or with chitosan/alginate polyelectrolyte complex. The further decrease of the diffusion coefficients shown at Figure 4C is a result from the simultaneous effect of glutaraldehyde crosslinking and chitosan/alginate complexation. Based on the diffusion coefficients the release time is estimated to vary in a very broad range – from 400 min to some thousand hours.

4 CONCLUSIONS

The process of vitamin B_2 release from chitosan films is complex and governed by electrostatic interactions between the vitamin, chitosan and crosslinking agents. Therefore it could be modified by treatment of the films in a positive or negative corona discharge system. The negative corona supports the charging of the vitamin B_2, formation of ionic bindings with chitosan and inhibition of the release process. The positive corona increases the positive charge on the chitosan films and helps the formation of crosslinked network. Hence it again inhibits the vitamin B_2 release by creation of diffusion barriers.

Acknowledgements

Authors are thankful to the REGPOT BioSupport project (Grant agreement No 245588) for the financial support.

References

1 J. H. Powers, *The American Journal of Clinical Nutrition*, 2003, **77**, 1352.
2 S. M. Fishman, P. Christian, K. P. West, *Public Health Nutrition*, 2000, **3(2)**, 125.
3 K. A. Head, *Alternative Medicine Review*, 2001, **6(2)**, 141.
4 D. Magis, A. Ambrosini, P. Sandor, J. Jacquy, P. Laloux, J. Schoenen, *Headache*, 2007, **47(1)**, 52.
5 P. Eiselt, J. Yeh, R.K. Latwala, L.D. Shea, D.J. Mooney, *Biomaterials*, 2000, **21**, 1921.
6 Y. Sawayanagi, N. Nambu, T. Nagai, *Chem. Pharmacol. Bull.*, 1983, **31**, 2064.
7 O. Smidsord, G. Skjak-Braek, *Trend. Biotec.*, 1990, **8**, 71.
8 A. Borzacchielo, L. Ambrosias, P. Netter, L. Nicholais, A. Balard, P. Sam, *J. Mater. Sci. Mater. Med.*, 2001, **12**, 861.
9 C.M. Lehr, J.A. Bouwstra, E.H. Schacht, H.E. Junginger, *Int. J. Pharmacol.*, 1992, **78**, 43.
10 R.S. Hermes, R. Narayani, *Trend. Biomater. Artif. Organs.*, 2002, **15 (2)**, 54.
11 H.K. No, N.Y. Park, S.H. Lee, S.P. Meyers, *Int. J. Food Microbiol.*, 2002, **74**, 65.
12 R. M. Kay, A. S. Truswell, In medical aspects of dietary fiber, in: G.A. Spiller, R.M. Kay (Eds.), Plenum Medical, 1980, 153.
13 F. Mi, S. Shyu, T. Wong, S. Jang, S. Lee, K. Lu, *J. Appl. Polm. Sci.*, 1999, **74**, 1093.
14 M. N. V R. Kumar, *Reactive and Functional Polymers*, 2000, **46 (1)**, 1.
15 J. Berger, M. Reist, J. M. Mayer, O. Felt, N. A. Peppas and R. Gurny, *European Journal of Pharmaceutics and Biopharmaceutics*, 2004, **57(1)**, 19.
16 X. Z. Shu, K. J. Zhu, Weihong Song, *International Journal of Pharmaceutics*, 2001, **212 (1)**, 19
17 S.K. Bajpai, Rasika Tankhiwale, *Reactive & Functional Polymers*, 2006, **66**, 1565.
18 A Domard., *Int. J. Biol. Macromol.*, 1987, **9**, 98.
19 L. Loeb, Electrical Coronas Their Basic Physical Mechanisms, 1965, University of California Press.
20 I. Isenberg, and A. Szent-Györgyi, *Proc Natl Acad Sci U S A.*, 1959, **45(8)**, 1229.
21 T. Yovcheva, G. Mekishev, St. Nedev, *Journal of Optoelectronics and Advanced Materials*, 2005, **7(1)**, 237.
22 F. A. Simsek-Egel, G. M. Bondl, J. Stringer, *Journal of Applied Polymer Science*, 2003, **88(2)**, 346.

INTERFACIAL BEHAVIOUR OF CHITOSAN AND CRAYFISH PROTEIN ISOLATE MIXED FILMS

A. Romero[1], T. Verwijlen[2], A. Guerrero[1] and J. Vermant[2]

[1]Departamento de Ingeniería Química, Universidad de Sevilla, C/Prof. García González, 1-41012, Sevilla (SPAIN)
[2]Departament of Chemical Engineering, Katholieke Universiteit Leuven, Willen de Croylaan 46-3001, Leuven (BELGIUM)

1 INTRODUCTION

The interfacial properties of interfaces are of prime importance in a number of technological and biological applications. In fact, most important for emulsions and foams is the actual value of the static/dynamic interfacial or surface tension as a function of the interfacial deformation corresponding to the response of the film against a deformation[1]. In fact, interfacial rheological functions provide fundamental understanding of structure[2]. Therefore, the interfacial rheology of such systems at fluid interfaces has been extensively studied[3-5] although all the used methods show advantages and limitations with respect to different aspects.

Crayfish Protein Isolate (CFPI) and chitosan (CH), both of them being typically produced from by-products of the crayfish or other crustacean food industry, have been used to evaluate the interfacial behaviour of adsorbed films as a function of concentration. An analysis of the surface pressure under compression-expansion cycles was carried out by means of a Langmuir trough. Interfacial rheology has been studied with two sensitive techniques, to avoid the intricate coupling between interfacial and bulk flows: a magnetic rod interfacial stress rheometer (ISR)[6] and double wall-ring geometry in combination with a modern rotational rheometer (Double-wall ring, DWR)[7].

The objective of this work has been to study the A/W and O/W interfacial behaviour of CFPI and chitosan adsorbed films and the influence of the interactions among both biopolymers as a function of CH and CFPI contents. A further objective has been to validate the interfacial rheological analysis through a comparison between two different techniques.

2 METHOD

2.1 Materials

Crayfish flour (CF) was manufactured at pilot-plant scale by ALFOCAN S.A. (Isla Mayor, Seville, Spain). Crayfish meat was separated from the exoskeleton and comminuted to form a meat slurry that was dried at 150-160 °C in a rotator drum dryer, to obtain a low moisture crayfish powder. The flour supplied by ALFOCAN S.A. consisted of 64 wt%, 19 wt% lipids, 13 wt% ashes and 4 wt% moisture.

Chitosan (CH) medium molecular weight (MW 190,000-310,000 Da, 75-85% deacetylated). All chemicals used were of analytical grade purchased from Sigma-Aldrich Chemical Company (St.Louis, USA). Distilled water was used for the preparation of all solutions.

2.2 Preparation of crayfish protein isolate (CFPI)

CFPI was prepared from the crayfish flour by solvent extraction of lipids, alkaline extraction of soluble proteins and isoelectric precipitation, according to the procedure described in a previous paper[8]. The crayfish flour powder was sieved, defatted by hexane, air-dried at room temperature and stored at 4°C until used. The defatted flour was dispersed in water (10% w/v) giving rise to a pH of 6.6, that was adjusted to pH 10.5 with 25% NaOH. The dispersion was stirred at room temperature for 4 hours and centrifuged at 900 x g for 25 min at 4°C in an RC5C Sorvall centrifuge (Sorvall Instruments, Wilmington, DE, USA). The supernatant was then adjusted with 6N HCl to pH 3.4, which corresponds to the actual pI of crayfish protein as determined from zeta potential measurements, and centrifuged at 9 000 x g for 10 min at 4 °C. The pellet was washed and resuspended with distilled water. The protein dispersion was freeze-dried in a Freeze Mivile 3 (VIRTIS, USA).

2.3 Preparation of chitosan solutions

Chitosan solution at 1 wt% was prepared by dissolving the proper amount of chitosan in a buffer solution (10 mM of phosphoric acid, pH = 2.0). The solution was stirred for 3 h at 700 rpm and the system was left to stand for 48 h at 4°C for complete hydration of the polymer and removal of bubbles.

2.4 Chemical composition of crayfish protein isolate

The protein content was determined in quadruplicate as %N x 6.25 using a LECO CHNS-932 nitrogen micro analyzer (Leco Corporation, St. Joseph, MI, USA)[9]. Lipid content was analysed in triplicate by Soxhlet extraction. Moisture and ash content of the isolate was determined in quadruplicate by AOAC approved methods[10]. CFPI, obtained according to the above described isolation procedure, consisted of 90.61±2.14 wt% protein, 0.96±0.08 wt% lipids, 3.98±0.24 wt% ash and 4.45±1.21 wt% moisture.

2.5 Determination of air-water interface isotherms

Measurements of the surface pressure (Π)-surface area (A) isotherms were determined by compression-expansion cycles by the Wilhelmy plate method on an isolated and fully automated Langmuir-type film trough (KSV Instruments, Finland). The trough consisted of two movable barriers with a Pt/Ir Wilhelmy plate as a tension captor. The area of the vessel was filled with each of the buffer systems used for emulsion preparation. Temperature was kept at 20°C by circulating water from a thermostatic bath. The cleanliness of the buffers was checked by a compression and expansion cycle using the buffer without protein injection. The barrier speed was fixed at 6 cm/min and samples were left to stand for 40 min before compression. These measurements were performed at different protein concentration and pH values (2 and 8).

2.6 Determination of interfacial rheology

2.6.1. Magnetic rod interfacial stress rheometry. Sample solutions are placed in a Petri dish with a Wilhelmy balance to control the surface pressure. A magnetized rod is located at the interface and positioned between two microscope slides arranged as a rectangular channel. The physical parameters of the used magnetic rod are: radius=200 μm; length=38 mm and density=1298 kg/m^3. Two coils in Helmholtz condition are positioned at both places and provide a potential well where the magnetic field gradient can be made zero by balancing the current though the coils. From the amplitude ratio and the phase angle has been calculated the interfacial viscoelastic parameters[7].

2.6.2. Interfacial shear rheometry by double wall-ring geometry. Measurements of shear rheology were performed by a double wall ring (DWR). The sample is contained in a trough and a ring is positioned at the liquid/liquid interface and connected to a sensitive stress-controlled rheometer (AR-G2, TA Instruments, USA)[8]. Strain sweep tests were performed at 0.6283 rad/s in order to establish the linear viscoelasticity range. Frequency sweep measurements were carried out at a strain lower than the critical value for linear viscoelasticity.

A numerical analysis of the flow field as a function of the surface viscoelastic properties is performed to evaluate the non-linearities in the surface velocity at a low Boussinesq number[8]. All experiments were carried out, at least in duplicate and the systems were thermostated at 20 ± 0.1 °C by placing the sample holder directly onto the Peltier heated bottom plate.

3 RESULTS

3.1 Air-water interface isotherms

Figure 1 shows the Π-A isotherms obtained for crayfish protein (CFPI), chitosan (CH) and the mixture protein/chitosan (CFPI/CH) as a function of concentration at 20°C by a compression process. These experiments allow the analysis of the behaviour of these systems at a planar air-water interface. It can be observed that surface pressure (Π) is close to zero just before compression and then increased until to reach a final pressure corresponding to the collapse of the interfacial film. In general, for the higher concentrations the pressure increases rapidly. However, in the chitosan system the values of pressure surface are lower and the collapse was not reached in the studied concentration range. Therefore, all these results indicate clearly that crayfish protein spreads more efficiently at air-water interfaces than chitosan. The mixture protein/chitosan shows similar results to the obtained for only crayfish proteins.

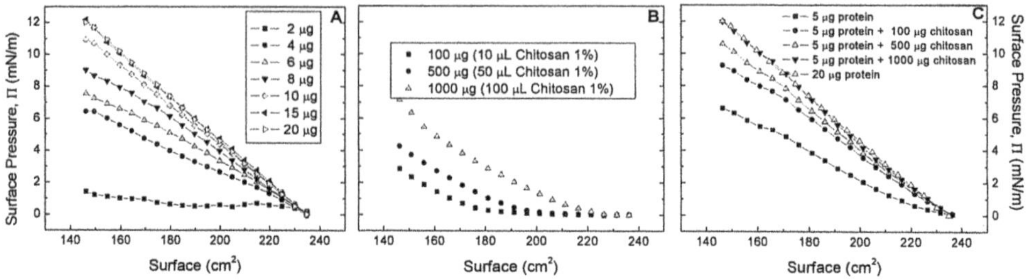

Figure 1 *Surface pressure isotherms (compression curves) at the air-water interface at pH 2 and at 20°C as a function of concentration for: (A) CFPI, (B) CH and (C) CFPI/CH.*

3.2 Air-water interfacial rheology

Strain sweep tests were performed at 0.6283 rad/s for the different systems in order to establish the linear viscoelasticity range (Figure 2). In all the systems, strain sweeps show a predominantly elastic behaviour in the linear regime, especially in systems where chitosan is present. On the other hand, in the systems with proteins the extent of the linear viscoelastic region can be considered higher. As a consequence, frequency sweeps were carried out at strain values between 2 to 3%.

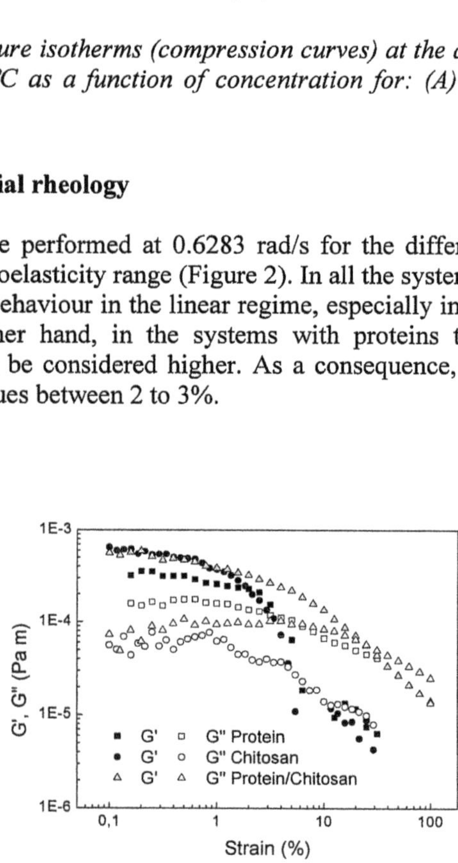

Figure 2 *Interfacial shear strain sweep of air-water interface at pH 2 and at 20°C for CFPI, CH and CFPI/CH.*

The results of the frequency sweep tests for the protein and/or chitosan layers are shown in Figure 3. In general, the adsorbed layer appears predominantly elastic that may be attributed to strong (covalent) bonds present in the film and to the confinement of macromolecules at the interface[11]. As can be seen in Figure 3A and 3B, an increase in the surface concentration leads to higher elastic moduli and, consequently, a decrease in loss tangent (data not shown), which suggests an enhancement of protein network at the air-water interface. However, this increase in concentration tends to a constant value where no differences were found. Moreover, results point out very clearly the more pronounced elastic behaviour of chitosan films as compared to protein films. The mixture protein/chitosan shows an intermediate behaviour. This behaviour has been recently demonstrated by CRYO-SEM microscopy studying the interface of the O/W emulsion stabilized by potato proteins and chitosan[12].

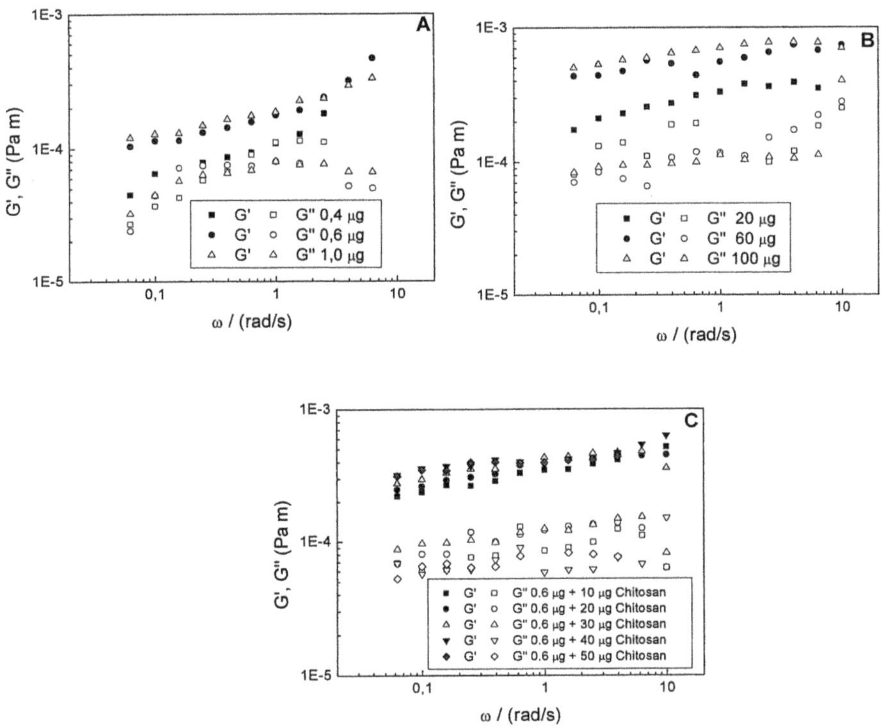

Figure 3 *Interfacial shear frequency sweep of air-water interface at pH 2 and at 20°C as a function of concentration: (A) CFPI, (B) CH and (C) CFPI/CH.*

Figure 4 shows frequency sweeps performed by interfacial stress rheometer (ISR) and by interfacial shear rheometer with a double wall-ring (DWR). In order to compare the rheological measurements, the surface pressure was controlled by a Wilhelmy plate. From all the experimental results those with similar values of surface pressure have been selected. Results put forward a similarity between the interfacial stress and interfacial shear techniques, independent of the studied systems.

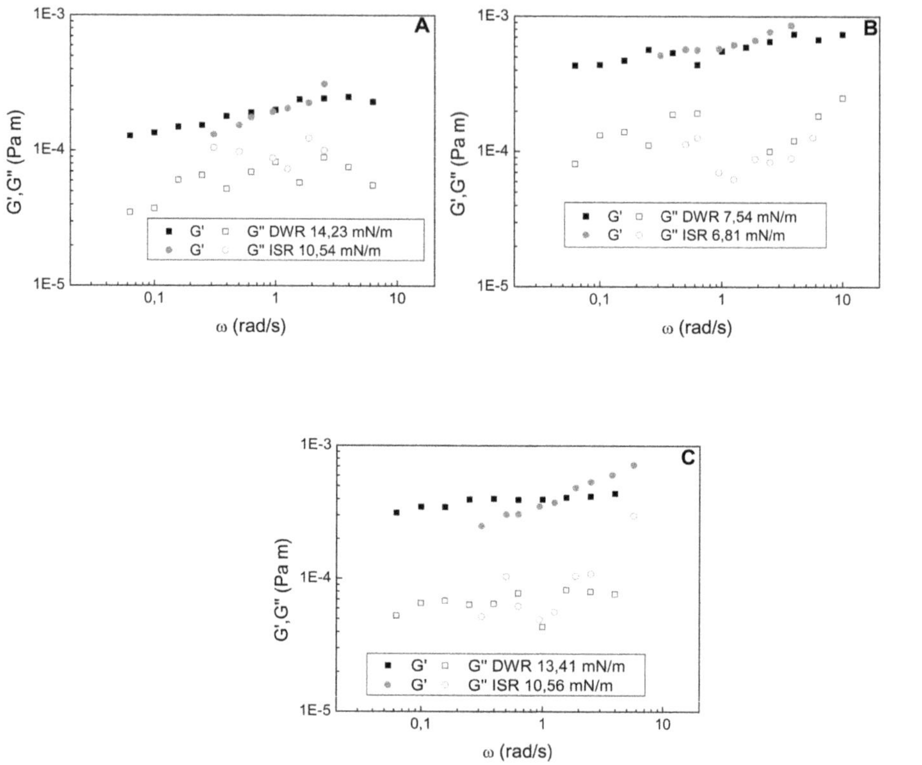

Figure 4 *Frequency sweep of air-water interface at pH 2, at 20°C and at different pressure surface by means interfacial stress (ISR, by magnetic rod) and shear (DWR, by double wall-ring geometry) rheometers: (A) CFPI, (B) CH and (C) CFPI/CH.*

4 CONCLUSION

CFPI and CH confer an elastic dominant behaviour to interfaces showing low interfacial tension values that are dependent of concentration. However, while CFPI show better interfacial tension values with higher surface pressure (related to the adsorption at the interface), CH provide higher elastic interface (related to deformation resistance). A combination of CFPI and CH suggest a good strategy to stabilize systems.
Moreover, two different techniques, interfacial stress rheometry by magnetic rod and interfacial shear rheometer by double wall-ring geometry, allowed study of the rheological properties of the adsorbed layers with similar results.

Acknowledgements

Part of this work has been carried out at Department of Chemical Engineering (K.U.Leuven, Belgium) and by financial support from the University of Seville. The authors also thank the financial support given by the projects P09-TEP-4875 and P09-AGR-4597 from the CICE (Andalousian Government, Spain).

References

1 M.A. Bos, and T. Van Vliet, *Adv. Coll. Int. Sci.*, 2001, **91**, 437.
2 J. Maldonado-Valderrama, M.J. Gálvez-Ruíz, A. Martín-Rodríguez, and M.A. Cabrerizo-Vílchez, *Langmuir,* 2004, **20**, 6093.
3 B.S. Murray. *Proteins at liquid interfaces,* Elsevier, Amsterdam, 1998, pp. 1179-220.
4 E. Dickinson. *Colloids Surf. B*, 1999, **15**, 161.
5 D. Langevin. *Colloid Int. Sci.,* 2000, **88**, 209.
6 S. Vandebril, A. Franck, G.G. Fuller, P. Moldenaers, and J. Vermant, *Rheol. Acta*, 2010, **49**, 131.
7 S. Raynaert, C.F. Brooks, P. Moldenaers, J. Vermant, and G.G. Fuller, *J. Rheol.*, 2010, **52**, 261.
8 R.D. Etheridge, G.M. Pesti, and E. Foster, *Anim. Feed Sci. Technol.,* 1998, **73**, 21
9 AOAC. Official methods of analysis, 1995. 15th Edn., Arlington: Association of Official Analytical Chemists, Washington D.C.
10 W.F. Tan, L.K. Koopal, W.H. Weng, W.H. Riemsdijk, and W. Norde. *Geochim et Cosmochim. Acta,* 2008, **72**, 2090.
11 P. Enri, P. Fischer, and E.J. Windhab. *Langmuir*, 2005, **25**, 10555.
12 N. Calero, J. Muñoz, P.W. Cox, A. Heuer, and A. Guerrero. *Food Hydrocolloids*, In revisión.

INFLUENCE OF THE ADDITION OF A POLYSACCHARIDE TO PROTEIN-BASED BIODEGRADABLE POLYMERIC MATERIALS PROCESSED BY A THERMOMECHANICAL PROCEDURE

L. Zárate-Ramírez[1], A. Romero[1], C. Bengoechea[1], P. Partal[2], A. Guerrero[1]

[1] Departamento de Ingeniería Química, Universidad de Sevilla, 41012 Sevilla (Spain).
[2] Departamento de Ingeniería Química, Universidad de Huelva, 21071 Huelva (Spain)

1 INTRODUCTION

Global plastic market has been growing along the years with an increasing trend. Thus, the world annual consumption of plastics is nowadays higher than 200 million ton with a growth rate of ca. 5%.[1] However, this market will suffer some limitations due to problems related to petroleum reserves, costs and pollution. Extensive studies have been made for exploring the potential use of biobased polymeric materials from renewable resources.

Bioplastics processed from protein/plasticizer blends are among the alternatives that are receiving increasing attention. In particular, wheat gluten (WG), a vegetable protein often obtained as a food industry by-product, has been widely used, being considered as fast degrading polymer. Moreover, gluten/plasticizer blends can be thermoplastic processed to give many kinds of shaped products and thus, may be regarded as an alternative to synthetic plastics for a wide variety of applications.[2]

The addition of polysaccharides to protein/plasticizer blends also merits some research efforts, due to the relevance of non-covalent interactions among protein and polysaccharide that can be used to design and fabricate materials with novel or improved properties. This study uses a combination of protein, polysaccharide and plasticiser to obtain biodegradable polymeric materials. Wheat gluten was selected as the protein component, a mixture of glycerol/water (GLW) as the plasticizer, and Xanthan Gum (XG), as the polysaccharide.

The aim of this work is to analyse the influence of the addition of XG on Dynamic Mechanical and Tensile strength properties of WG/XG/GLW blends subjected to thermoplastic compression moulding, as a function of XG content and pH, comparing the results with those obtained using WG/GLW blends. Water absorption capability and solubility under different extraction media of bioplastic specimens were also evaluated in order to evaluate the effect of the gum on the nature of the interactions taking place in the system. This study would contribute to evaluate the potentials of protein/polysaccharide based bioplastics as absorbent materials exhibiting suitable mechanical properties.

2 EXPERIMENTAL SECTION

2.1 Materials.

Wheat gluten, containing 83 wt% protein, 3 wt% lipid, 1 wt% ash, 8 wt% moisture and 10 wt% starch, was provided by Productos Riba S.A. (Granollers, Barcelona, Spain). Xanthan

Gum was provided by Sigma-Aldrich, (St. Louis, MO, USA). Water and Glycerol, from Panreac Química S.A.U. (Castellar del Vallés, Barcelona, Spain), were used as plasticisers.

2.2 Sample preparation.

WG/GLW blends containing 50 wt% WG, 18 wt GL and 32% water and WG/XG/GLW blends, containing 50 wt% WG/XG with different XG concentration, were mixed in a two-blade counter-rotating batch mixer (Rheomix 3000p; ThermoHaake, Karlsruhe, Germany) at 25°C and 50 rpm under adiabatic conditions. Different pH values were used by adding NaOH or HCl 2M solutions to the 50/18 blend. Bioplastic probes were prepared from dough-like materials obtained after the mixing process by thermomoulding at 9 MPa and 130°C for 10 min. Two types of moulds were used to prepare the probes, a 50 x 10 x 3 mm rectangular shape mould and a type IV probe defined by the Standard Test for Tensile Properties of Plastic (ASTM D638). Rectangular and type IV probes were transferred into recipients at 53% relative humidity (RH) and allowed to reach moisture equilibrium at room temperature for at least 2 weeks before testing. The desired RH value was achieved by means of a desiccator, using a saturated solution of $Mg(NO3)_2$.

2.3 Dynamic Mechanical Thermal Analysis.

DMTA tests were carried out using a RSA3 (TA Instruments, New Castle, DE, USA) by applying dual cantilever bending to rectangular probes. All the tests were carried out at constant heating rate (3 °C min¯1), frequency (1 Hz) and strain (0.01-0.3%, within the linear viscoelastic region). Samples were coated with silicone oil to avoid water loss.

2.4 Tensile Strength Measurements.

Tensile test were performed by using the Insight 10 kN Electromechanical Testing System (MTS, Eden Prairie, MN, USA), according to the ASTM D638 norm. Tensile stress and elongation at break were evaluated from at least three replicates for each product using type IV probes and an extensional rate of 20 mm min^{-1} at room temperature.

2.5 Water absorption capacity.

Water uptake of samples was determined according the ASTM D570 norm, using at least three rectangular specimens immersed in distillate water for 24 h at room temperature.

2.6 Protein extractability

Samples of bioplastic probes were extracted in different extraction media: water; a denaturing agent solution (method A); a solution of denaturing and reducing agents (method B). All the materials were extracted for 2 h at 20 ◦C by magnetic agitation (approximately. 400 rpm). Method A used a 0.086 mol L^{-1} Tris base, 0.045 mmol L^{-1} glycine, 2 mmol L^{-1} EDTA, 10 g L^{-1} Sodium Dodecyl Sulfate (SDS) pH 6 buffer. In method B, proteins were dissolved in the same buffer containing 10 g L^{-1} Dithiothreitol (DTT). Dispersions were centrifuged at 10,000 × g for 10 min at 15°C. The fraction of extractable protein was determined using a modification of the Lowry method.[3] Percentages of soluble protein in the supernatants were calculated.

3 RESULTS AND DISCUSSION

3.1 Dynamic Mechanical Temperature Analysis

Figure 1 shows the values of the storage (E') and loss (E'') moduli and the loss tangent (tan δ) from DMA temperature ramp measurements carried out under dual cantilever bending using 50/18 WG/XG/GLW bioplastics as a function of gum content. All the probes display similar behaviour to that one shown by the reference system (50/18 without gum). Thus, a remarkable decrease in E' and E'' generally took place as the temperature was raised until a plateau value was reached (above 100 °C for E' and even higher for E''). This plateau is followed by an increase in both moduli for those systems containing XG gum, although no apparent change in behaviour was brought about by the presence of gum.

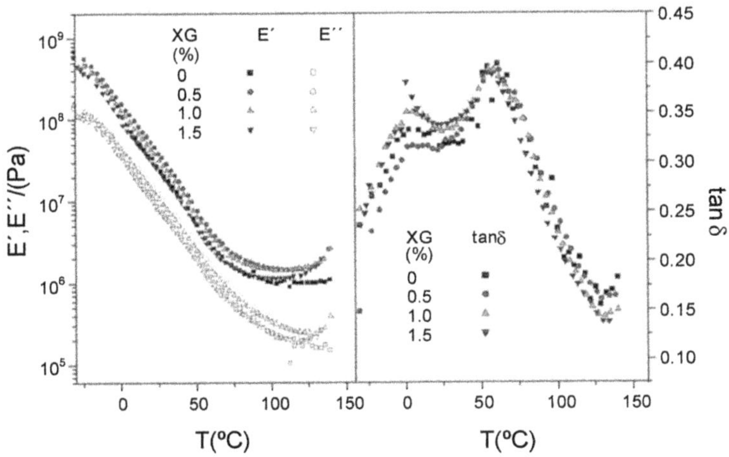

Figure 1 *DMA temperature ramp measurements carried out at constant frequency (1 Hz) and heating rate (5 °C min^{-1}) for WG/XG/GLW bioplastic probes, as a function of gum content*

It should be pointed out that this increase in viscoelastic properties, being a result of the high temperature regime, indicates that the system may be processed at higher moulding temperature in order to achieve an enhanced microstructure (giving rise to higher E' and E'' values) when XG gum is present. In other words, XG-containing biodegradable polymeric materials exhibit a remarkable thermosetting potential. These results also suggest that once protein-protein crosslinking seems to have come to an end (at about 100°C) polysaccharide molecules may still participate in crosslinking formation at higher temperatures. The values of the plateau or minimum temperature (T_{min}) and the elastic modulus obtained at this minimum (E'_{min}) are shown in Table 1 as a function of XG concentration. As may be deduced from figure 1A and Table 1, addition of 0.5% XG led to a remarkable enhancement of viscoelastic properties, particularly above T_{min}. However, a further increase in gum concentration provoked a reduction in E' and E'', so that the system containing 1.5 wt% XG developed similar profiles to those of the reference probe. Parameter E'_{min} also followed the same evolution with increasing XG content, being displaced at lower temperatures.

Table 1 *DMA parameters obtained from E' and tan δ profiles as a function of XG concentration*

XG concentration (wt%)	E'_{min} (MPa)	T_{min} (°C)	$T_{\alpha1}$ (°C)	$T_{\alpha2}$ (°C)
0	1.07	110	6.0	60
0.5	1.51	114	8.9	60
1.0	1.50	110	5.9	58
1.5	1.18	103	4.2	56

As may be observed in Fig. 1B, tan δ profiles at 1 Hz display two well-defined peaks, $T_{\alpha1}$ and $T_{\alpha2}$, whose values are included in Table 1. These two peaks were identified in a previous work for systems containing gluten as the biopolymer and glycerol/water as the plasticiser.[2] The first peak, $T_{\alpha1}$, would be mainly related to the characteristic glass transition of the plasticiser blend affected by some fractions of gluten protein, whereas $T_{\alpha2}$ was attributed to a glass transition of the plasticised gluten.[4-6]

An initial increase in XG up to 0.5 wt% yielded an increase in $T_{\alpha1}$, and a tendency of both peaks to overlap. This behaviour indicates a certain tendency to the compatibility of both biopolymers. However, an evolution towards lower $T_{\alpha1}$ and $T_{\alpha2}$ values may be observed with a further increase in XG concentration. This is a relatively usual behaviour, taking into account that XG is an anionic polysaccharide and bioplastic probes were prepared at conditions near the isoelectric point (IEP). Such conditions typically induce segregative separation through steric exclusion among two biopolymers, one of them charged (XG) and the other uncharged (WG) or similarly charged.[7] The above mentioned compatibility effect obtained at the lowest XG content may be due to the presence of positive charges on some protein residues that interact with polysaccharide anions. In any case, this effect is moderate due to the proximity of the IEP.

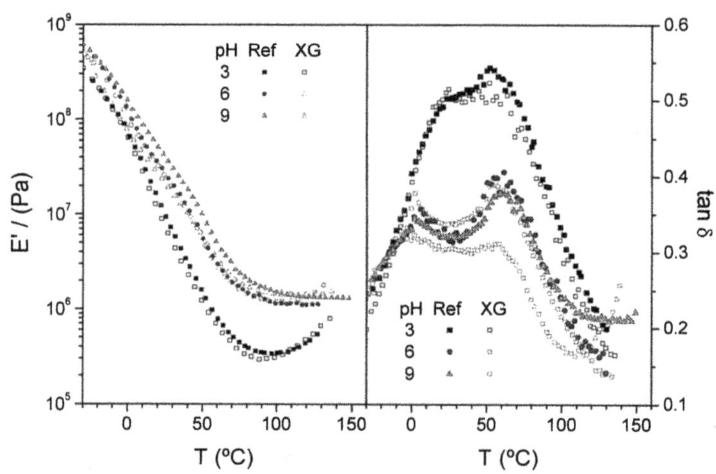

Figure 2 *DMA temperature ramp measurements carried out at constant frequency (1 Hz) and heating rate (5 °C min^{-1}) for WG/GLW and WG/XG/WGL bioplastics, as a function of pH*

Figure 2 shows the effect of pH on the values of E' and tan δ obtained as a function of temperature for bioplastic probes prepared either in absence or presence of XG polysaccharide. Bioplastic specimens prepared at pH 3 exhibited much lower E' values, as well as much higher tan δ, than those prepared at pH 6. Thus, the presence of positive charges on protein surfaces at acidic conditions seems to interfere on bioplastic processing, leading to a softer material with a lower microstructure development and a higher thermosetting potential. Addition of polysaccharide did not induce marked modifications on viscoelastic properties at pH below the IEP, when both biopolymers were oppositely charged, or at the IEP, at which the balance of charges for gluten proteins was zero. Conversely, alkaline conditions, at which both biopolymers were negatively charged, yielded a remarkable reduction (being higher than 50% at low temperature) when XG is present. However, this effect vanishes at the highest temperature region. Figure 2B shows that the pH leading to the most remarkable modifications on tan δ profile is again pH 9.

3.2 Tensile Strength Measurements

Figure 3 shows some typical stress-strain curves obtained after subjecting type IV probes to uniaxial extension tests at constant extensional rate.

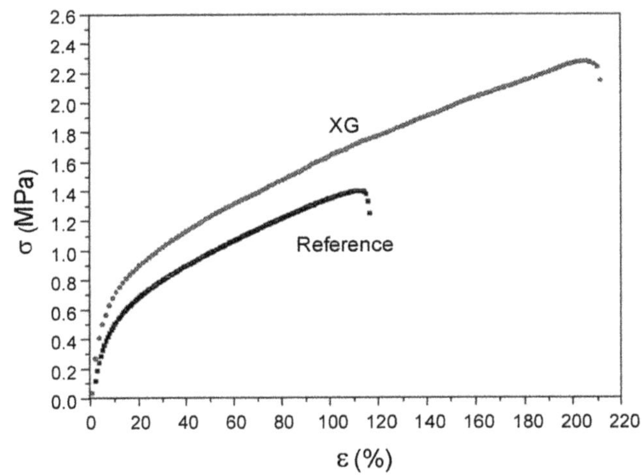

Figure 3 *Tensile Stress-Strain curves for the reference and 0.15 wt% XG-containing bioplastics*

All the materials exhibit a linear elastic region with a marked increase in strength at low strain values, characterized by high values of the Young's Modulus (E), followed by an extensive plastic region with a steady decrease up to a constant stress–strain slope. Finally, after a critical stress (σ_{max}) or strain (ε_{max}), the rupture of samples takes place. As may be observed, addition of XG at pH 6 (near the IEP) resulted in an apparent increase in mechanical properties. Thus, higher values for the maximum tensile stress and strain and for the Young's Modulus may be observed in Figure 3.

Figure 4 shows the evolution of tensile strength parameters for both WG/GLW and WG/XG/GLW bioplastic specimens processed at different pH values. As may be seen, the presence of charges may produce strong modifications on tensile properties of bioplastic probes either in absence or presence of gum. A remarkable increase in both σ_{max}, and E with increasing pH may be observed for WG/GLW specimens, whereas no significant evolution of ε_{max} may be noticed. The effect of adding gum on tensile properties is also strongly dependent on pH. Thus, the enhancement in tensile properties found at pH 6 no longer takes place at pH far from the IEP. In fact, at acidic pH the presence of XG does not seem to alter the specimen response to elongation (only a slight decrease in E may be noticed). On the contrary, the presence of the polysaccharide impairs the mechanical resistance of the probes subjected to tensile strength testing to a high extent, particularly affecting to the Young's Modulus and σ_{max}.

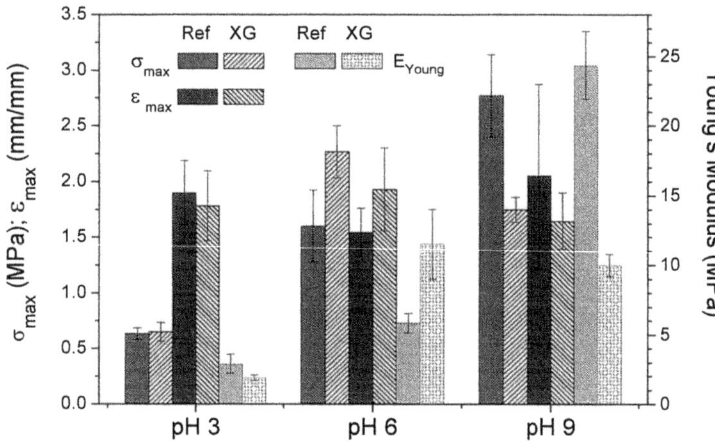

Figure 4 *Parameters from tensile strength measurements (Maximum stress, Maximum strain and Young's Modulus) for the reference and0.15 wt% XG-containing bioplastics, as a function of pH*

3.3 Water absorption capacity

Figure 5 shows the values of water absorption capacity obtained for samples immerse after 2 and 24 h. This figure illustrates how water uptake exhibits a strong dependence on pH either in absence or presence of XG gum. The lowest values always correspond to the IEP, whereas the water absorption capacity is remarkably enhanced at acidic pH, particularly after 24 h. In contrast, only a slight tendency to reduce water uptake capacity of bioplastic probes seems to be produced by adding XG gum, being more evident after 24 h immersion, but remaining always lower than 10 %.

3.4 Protein extractability

Figure 6 shows the values of the percentage of soluble protein over the total amount of protein in different extraction media (water, 10 g L^{-1} SDS buffer, or 10 g L^{-1} SDS and 10 g

L^{-1} DTT buffer) for WG/XG/GLW bioplastic samples with a XG content equal to 1.5% and for different pH values. The addition of XG led to an increase in the solubility of bioplastics for the whole pH range studied. This happens both in water and in the presence of a denaturing agent such as SDS, being only at pH 3 when no significant differences were found after addition of XG. This seems to point out the greater importance of non-covalent interactions in the structure of XG-containing bioplastics when compared to the XG-free reference system.

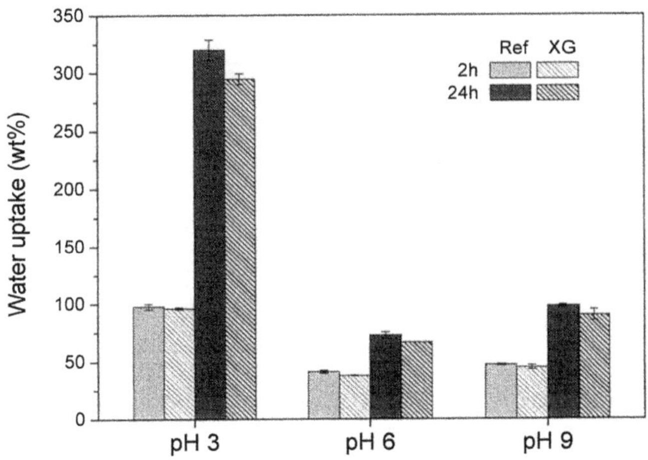

Figure 5. *Results of water uptake tests for WG/CLW and WG/XG/GLW bioplastics as a function of pH*

A decrease in solubility if SDS was added in combination with a dissociating agent such as DTT was found when XG was used at pH 3. This result agrees with those from DMA and Tensile Stress-Strain tests, suggesting that the presence of XG would promote the cross-linking in the protein matrix through interactions with positively charged gluten. However, a greater solubility in the presence of XG was found at pH 6, which may be related to the plasticizing effect of XG in those systems that could lead to a lower cross-linking. This agrees with previous results from DMA and Tensile Stress-Strain tests.

The solubility in the SDS + DTT buffer at pH 9 is independent of the addition of XG. Consequently, crosslinking reactions generating the gluten matrix seem not to be affected by the presence of gum, though phase separation would take place in that matrix. That may be related to the similar charge of gluten and XG at that pH value, and it would be linked to the decrease in E' and E'' in the DMA tests and in the Tensile Strength test parameters.

4 CONCLUDING REMARKS

An increase in XG content produces two effects on WG-based bioplastic probes: (i) an enhancement of the protein matrix that reaches a maximum at low XG concentration and (ii) a continuous increase in its thermosetting potential. The former may be related to a certain degree of interaction and compatibility between both polymers that evolves towards segregative separation with increasing XG content. The later effect suggests that polysaccharide may play a role in crosslinking, particularly if thermomoulding were carried out at higher temperatures.

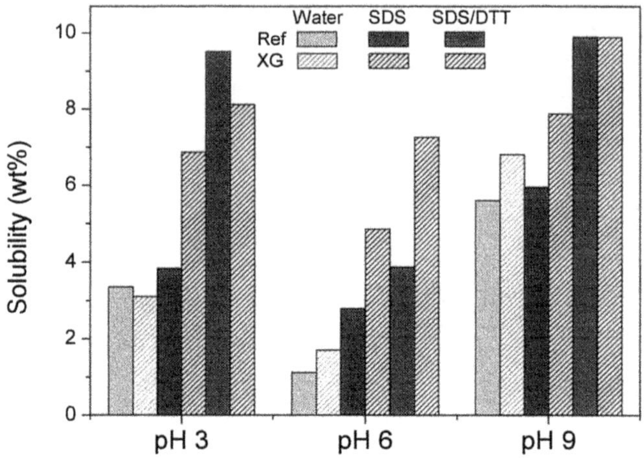

Figure 6 *Extractability profiles in water, 1% SDS buffer and 1% SDS +1% DTT buffer for WG/GLW and WG/XG/GLW bioplastics as a function of pH*

The presence of charges induced strong modifications on DMA and tensile properties of biopastic probes, either in absence or presence of gum. So did on water absorption capacity that exhibited a remarkable enhancement at pH 3. Addition of polysaccharide did not induce marked modifications on viscoelastic properties at pH near or below the IEP, but yielded a remarkable reduction at alkaline pH. Tensile properties were only improved near the IEP, whereas water uptake always underwent a moderate reduction.

Extractability experiments revealed that XG induce an increase in solubility of WG/GLW blends in water, which may be observed even after thermomoulding processing, as well as an increase in the contribution of physical interactions to the bioplastic microstructure, which is independent of pH. As for the contribution of XG to covalent interactions, a strong dependence on pH was observed.

Acknowledgements

Financial support from the MICINN/FEDER through the project MAT2010-21404-C02-02/01 is gratefully acknowledged.

References

1 V. Siracusa, P. Rocculi, S. Romani, M. Dalla Rosa, *Trends Food Sci. Technol.* 2008, **19**, 634.
2 L.S. Zarate, I. Martínez, A. Romero, P. Partal, A. Guerrero, *J. Sci. Food Agric.* 2011, **91**, 625.
3 G.L. Peterson, *Anal. Biochem.* 1979, **100**, 201.
4 A. Jerez, P. Partal, I. Martínez, C. Gallegos, A. Guerrero *Biochem. Eng. J.* 2005, **26**, 131.
5 S. Sun, Y. Song, Q. Zheng Q, *Food Hydrocolloid* 2007, **21**, 1005.
6 S. Sun, Y. Song, Q. Zheng Q, *Food Hydrocolloid* 2008, **22**, 1006.
7 D.J. McClements, *Biotechnol. Adv.* 2006, **24**, 621.

Microstructure and texture

POTENTIAL USES OF HYDROCOLLOIDS AND GUMS IN FOOD TEXTURE STUDIES

M. Peleg[1] and M.G. Corradini[2]

[1] Department of Food Science, University of Massachusetts, Amherst, MA 01002, USA
[2] Instituto de Tecnología, Universidad Argentina de la Empresa, Buenos Aires, Argentina

1 INTRODUCTION

Hydrocolloids and gums have been used in a wide variety of foods, primarily as thickeners, stabilizers and gelling agents. [1-4] Their functionality in these capacities has been intensively studied. It can be attributed to the chemical character of the monomers and the polymer's molecular structure (including the chain length and branching pattern where pertinent). These dictate the polymer's spatial conformation and interaction with water and other food components. They also determine how the hydrocolloid or gum's performance is affected by factors such as temperature and pH, the presence of certain ions, etc. Although the chemistry and physics of hydrocolloids and gums are most interesting research topics in their own right, most of the studies concerning their structure and properties have been primarily motivated by utilitarian considerations. Consequently, there is a large body of knowledge of how to exploit their properties in order to improve existing food products and to develop novel ones, especially with reduced sugars and fat. Information on how to use hydrocolloids and gums in foods (and other consumer products) can be found in books and numerous technical publications and scientific articles. It is also the central topic of the conference to be discussed by several contributors who are experts in this field. The main objectives of this communication are two: to present two unconventional methods to assess texture and to explore the potential use of hydrocolloids and gums as a tool in texture research using these methods. The focus will be on how they can assist in the solution of outstanding problems concerning objective evaluation of selected foods' rheological properties, and how these might be associated with texture as perceived by humans. Although this might seem a marginal subject, it has bearing on quantifying the effects of added hydrocolloids and gums on the texture of two distinct classes of popular foods: semi liquid products and brittle especially cellular cereals and snacks. The emphasis is on the testing methodologies and their results interpretation. The first part will address the assessment of the consistency of semi liquid foods. There are of course many types of viscometers that are used to evaluate the consistency of semi liquid foods. But most, regardless of their degree of sophistication, render results whose interpretation requires special scrutiny. Different questions arise with regards to the mechanical evaluation of brittle solid foams. Many of the fundamental issues concerning rheological measurements and their relation to texture perception are yet to be resolved [5,6] and some are currently a topic of intensive research in several laboratories. What we'll do here is suggest how hydrocolloids could be helpful in this effort and try to highlight the potential practical aspects of the experiments in which they might be involved.

This communication is based on the experimental work of several researchers' at the University of Massachusetts at Amherst whose individual contributions are cited in the text, shown in the figures and listed in the References. Since this article is not a review, and because its focus is on testing methodologies developed in our laboratory, we have made no attempt to cite all the relevant works of published by others. However, the interested reader could find many of them in the references list of the original publications that are cited, or through a search of the pertinent literature.

2 SEMI LIQUID FOODS

2.1 Consistency Measurements

Many semi liquid foods owe their consistency to native hydrocolloid(s), e.g., pectin in tomato juice concentrate and its derivatives, ketchup for example, or added ones, gums in salad dressing, for example. The term 'consistency' in a textural context is intuitively clear but its objective evaluation is not a straightforward task. Unlike the viscosity of Newtonian beverages or pure oil, that of pseudoplastic foods such as ketchup, mustard or yogurt, especially when having a yield stress, is shear rate dependent. Hence their flowability, or resistance to deformation ought to be characterized by two or more rheological parameters, e.g., the "yield stress", "consistency" and "flow index" of the Herschel-Bulkley model. These flow characteristics have been traditionally determined by shear viscometry using a Couette or cone and plate sensor.

There are two serious problems with these methods, long known but rarely satisfactorily solved: (a) many semi liquid foods are basically a suspension and consequently exhibit slip ("plug flow"), and (b) the internal structure, whose integrity is manifested in the food's pseudoplasticity and yield stress, is frequently partially destroyed upon the specimen insertion into the narrow gap of the rheometer's sensor.

The result is that many of the published flow parameters are not of the original foods, but of slipping specimens whose structure has been destroyed to an unknown extent. The solution to slip is to use sensors with serrated sensors and repeat the test with sensors of different gaps and geometries to verify the remedy or correction's efficacy. Using such procedures is frequently judged impractical and hence they are rarely employed in food texture research, let alone routine quality control. A supposedly effective way to avoid the second problem is to allow the specimen to 'rest' in the sensor prior to the shear application. The expectation is that during the resting time the damage to the structure will be fully repaired and its original properties recovered. Unfortunately, the time for the damage repair is unknown and full restoration of the structure is either not guaranteed [7] or simply impossible, as in the case of the casein gel's irreversible disruption in yogurt [8] [Making the yogurt in the sensor itself, although possible, is rarely a feasible option.]

A practical way to avoid both problems, although perhaps not completely, is to replace the conventional shear by lubricated squeezing flow viscometry. The method, which was originally developed for polymer melts was introduced to food research by Edward E. Bagley [9]. The method's main limitations are being only applicable to low deformation rates and in that the there are discrepancies between the theoretical relationships between the elongational and shear viscosities and those observed experimentally. Still even in its 'empirical version' that we have used in our laboratory, it has the great advantage of enabling to test semi liquid foods practically intact. And slip, which is a source of artifacts in shear testing, is not only acknowledged but intentionally introduced for a valid test.[10]

2.2 Lubricated Squeezing Flow

Schematic view of the constant area- constant displacement rate version of the squeezing flow sensor is given in Fig. 1.

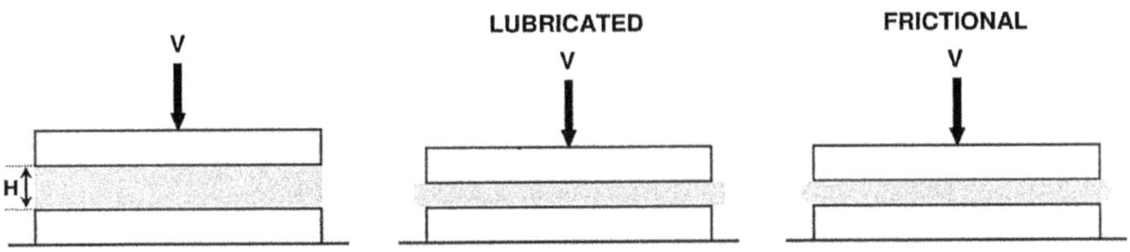

Figure 1 *The sensor in the constant area constant displacement rate version of squeezing flow viscometry. Notice the flat front of the exiting fluid in lubricated squeezing flow.*

The sensor consists of two lubricated or Teflon® made parallel plates to practically eliminate friction. Plates ought to be very large relative to the compressed specimen's height (D>>h) in order to minimize end effects. The food specimen is removed from its container and placed on the bottom plate with a wide spoon or spatula to avoid disruption of its structure – see below. The test starts by lowering the upper plate, attached to the crosshead of a Universal testing Machine, at a constant speed, V, and recording the developing force, F(t). The crosshead is stopped at a chosen height, and before it is withdrawn to allow for a new specimen to be tested, the decaying force is recorded for little longer than 2 min. The logged data of the first stage are converted into and plotted as a log F[H(t)] vs. Log F[H(t)] relationship. Typical plots of these relationships are shown in Fig. 2. The pertinent part to the analysis is the linear part of the log F[H(t)] vs. Log F[H(t)] curve on the left side of the bend. It corresponds to a specimen height of about 1-3 mm reached when squeezing flow has been fully developed. The data in the right side of the bend primarily represent entry and transient flow effect and are therefore discarded. Since the H(t) if measured from the bottom, any initial variability in the specimen's height or shape becomes immaterial.

Theoretically, in ideal lubricated squeezing flow of a Newtonian liquid, the slope of the linear part is supposed to be 1.0 – see figure. In contrast, in fully frictional flow it ought to be 3.0. In ideal pseudoplastic fluids, the slope is expected to be somewhat smaller than these values. Thus an observed value close to 1.0 is indicative of fully developed lubricated squeezing flow while close to 3.0 to frictional flow. Moreover, while the existing fluid in lubricated squeezing flow is expected to be flat, that in frictional flow is expected to be bulging at the center – see Fig. 1. [To watch the differences between ideal lubricated and frictional squeezing flows in animation open the interactive Wolfram Demonstration entitled Lubricated and Frictional Squeezing Flow at http://demonstrations.wolfram.com/LubricatedAndFrictionalSqueezingFlow/]

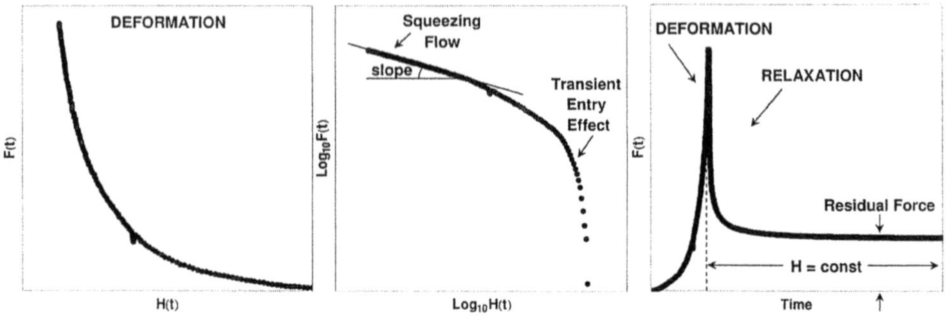

Figure 2 *Typical force-time, log force-log height and force relaxation curves in squeezing flow viscometry. Notice that the residual force (stress) after relaxation can be considered a measure of "solidity"*

Consistency comparison of specimens can be done on the basis of their recorded apparent stress at a given specimen height, 1 mm, say, calculated as $\sigma_{@H=1}$=F(H=1)/πD^2, or two heights, H=1 and H=2 say, for mutual verification. As long as the height is in the fully developed squeezing flow region, the corresponding apparent stress is a very sensitive and reproducible consistency measure. It could be used not only to distinguish between a product's samples made by different manufacturers but also by same manufacturer having different dates. Determination of the standard deviations the $\sigma_{@H=1}$ measurement of various semi liquid foods where very closed to those obtained in testing mineral oils, which have no structure that could be damaged during handling. Since it is very improbable that we produced the almost exactly same damage to successive food specimens, we had to conclude that they were tested practically intact.

Many semi liquid foods exhibit "yield stress". It is manifested in the ability of their specimens to support their own weight, for example. A "yield stress" entails that the food specimen has a certain degree of 'solidity'. The relative magnitude of this degree of solidity could be revealed from the residual stress at a given displacement, or height in our case, after relaxation. In a true liquid, the stress only develops as a response to flow and it dissipates completely and almost instantly when the flow ceases. But since testing machines have a time response (a small fraction of a second) the force decay may not appear instantaneous only as a rapid force drop. A considerable residual stress after 1 or 2 minutes, on the order of 10-20% of the initial stress, is therefore not an instrumental artifact but an indication that the specimen can maintain static stress at least on this time scale. In other words the residual stress may be used as a 'solidity measure' of sort and different from that of the overall consistency. For more on the issue see Corradini et al. [7]

The possibility of obtaining textural parameters of practically intact specimens and in a test where slip is not a source of artifacts opens new ways to assess the structure-consistency relationship in semi liquid foods.

2.3 Evaluation of the Extent of Structural Integrity Loss and its Restoration

A semi liquid food of a single origin can be divided into several lots. One can be tested as such and the results, the apparent stress at a given height and residual stress after a given time, considered measures of the intact food's consistency and degree of solidity. The other lots can be subjected to various extents of shear, by stirring for different times, for example, or by using different stirrers or stirring methods. The sheared samples can then be tested in the same way, which guarantees that hardly any additional damage to the structure will occur. Moreover, the squeezing flow test can be performed not only immediately after the stirring but also after some of the samples were allowed to rest for controlled times. The observed changes in the consistency and solidity parameters, relatively to those of the original intact food could then be used to quantify the amount of damage done under different stirring or handling conditions, whether the original properties could be restored completely or only partially and how much rest will be need for the restoration to occur. Such tests have actually been performed on samples of mayonnaise, mustard and tomato paste. [7] They revealed that for all three foods even 3 hrs at rest were insufficient for complete recovery as judged by the apparent stress at a given height. As to the degree of solidity as judged by the residual force after relaxation, it was almost completely restored in the mayonnaise but not in the mustard and tomato paste.

Experiments of this kind can be used to monitor the effect of added hydrocolloids and gums not only on the consistency of semi liquid foods but also on their structural stability if exposed to shear. The results would indicate whether added gum and its concentration could affect the degree of recovery and the time needed for the structure restoration.

With conventional methods of testing, it will always be uncertain whether the observed changes in the rheological parameters, unless dramatic, solely reflect changes in the food structure's mechanical integrity, or also a different interplay between the artifacts.

2.4 Statistical versus Practically Significant Textural Differences

One of the main purposes of rheological evaluations of foods, in both product development and quality control, is to minimize the use of sensory panels to assess texture and replace them by the 'objective' mechanical testing machine or viscometer. These instruments render results much more rapidly and with very few exceptions are more cost effective to operate. The underlying assumption or rationale of such instruments use is that that there is a correlation between the rheological properties measured and the food's perceived texture. The meaning of such correlations is controversial.[5,6] The controversy arises for various reasons. Among them, apart from the obvious differences between the fate of a food sample deformed and disintegrated in the mouth and in a mechanical testing device regardless of its sensor, are certain ambiguities in the texture vocabulary, the fact that the mouth (or fingers) do not operate in the same manner as a rigid man made machine and that textural properties are rarely if ever perceived in isolation. In addition, the mechanoreceptors that generate the coded signals sent to the brain are spatially distributed in a very different pattern than that of their counterparts in a man made machine and they themselves are viscoelastic and adapting. But suppose, for the sake of the discussion, that both could somehow assess the same rheological properties. If so then one should at least consider the possibility that for some properties the testing machine is more sensitive than humans, i.e., it can detect differences that humans cannot, or vice versa. One would therefore expect that establishing which is the case should precede any attempt to correlate instrumental and sensory data. Surprisingly, the issue has rarely been discussed in food texture literature. This might be partly because many of the reported correlations between instrumental and sensory parameters, especially the early ones, have been obtained with very different foods (or fruits and vegetables at very different stages of maturity) to expand the scale. Suppose that in assessing the consistency of semi liquid foods the testing machine is more sensitive than humans. If so, a statistically significant effect produced by a given treatment might not be perceived as such by humans, be they panelists or the consumer population at large. In the opposite case, finding no statistically difference not detected by the machine could still be significant to the consumer.

To resolve the issue, and avoid interference from factors such even slight differences in taste or appearance, one should compare specimens from exactly the same food who differ only in respect to their texture. The already described squeezing flow experiments designed to assess consistency loss by shearing and the degree and rate of its restoration offer a way to fulfill this requirement almost completely. The only difference is that the sheared samples and those left intact can now be served to a sensory panel to reveal the magnitude of the change that would be detected by the panelists. The sensorily detectable differences, expressed in terms of the apparent stress at the chosen specimen height and/ or the residual apparent stress after relaxation could then be used as criteria in product development and as guideline to establish the margins for quality control. The problem with the procedure as outlined above is that upper limit of the consistency, and that of the degree of solidity, are set by the food's original structure. To expand their ranges, one could use hydrocolloids and gums of different kinds and concentrations, including mixtures to boost the food's consistency to a higher level. Once this has been done, one can produce samples with intermediate consistencies and degrees of solidity within the expanded range by shearing. Again, all the samples would be practically identical as far as taste and appearance are concerned except for their texture, which as before could be assessed without introducing new artifacts. Where an issue, though, the sensory and squeezing flow experiments should be scheduled in such a way that the samples would be at the same degree of recovery.

2.5 Determination of the Practically Significant Difference

Sensory analysis based on triangle tests is probably the most reliable, albeit rarely a convenient option because of logistic considerations. During the evaluations, the panelists are served specimens arranged in triplets, two identical and one different, identified by a random code. For the purpose of consistency differences detection, the panelists were asked to identify the odd sample in each triplet and state whether it was 'thicker' had 'stronger consistency', etc. than the pair having the same texture. The percent correct answers were logged and plotted against the magnitude of difference between specimens' apparent stress at the chosen height and the residual stress after relaxation for a chosen time determined in parallel. Schematic view of the expected relationship is given in Fig. 3. Mathematically, the relationship could be described by the log-logistic model:

$$Y(\Delta\sigma) = Y_0 + \ln\{1 + \exp[b(\Delta\sigma - \Delta\sigma_c)]\} \tag{1}$$

Where Y_0 is the % correct answers before the rise (which could be primarily attributed to chance), $\Delta\sigma$ is the difference in the apparent stress, $\delta\sigma_c$ a marker of the difference in the apparent stress where the % correct answers rise becomes noticeable and *b* the approximate slope of the curve where $\Delta\sigma > \Delta\sigma_c$.

Where $\Delta\sigma_c = 0$, Eq. 1 is reduced to:

$$Y(\Delta\sigma) = Y_0 + b(\Delta\sigma - \Delta\sigma_c) \tag{2}$$

These terms could be used to determine the $\delta\sigma$ level that would correspond to 50% correct identification, for example, or any other level deemed significant in terms of customer satisfaction or other marketing criterion. Statistical analysis of the data can be used to determine the result confidence level.[11,12]

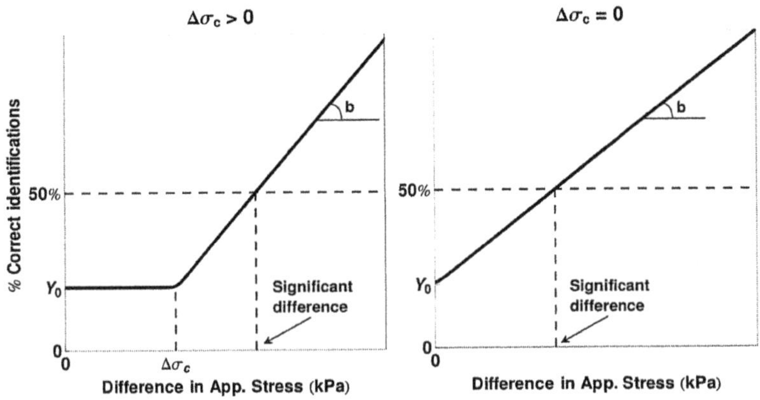

Figure 3 *Schematic view of the expected relationship between the number of correct identifications of the samples with stronger consistency and the corresponding difference in the samples' apparent stress at a given height, Δσ. After Chanasattru et al. (2002)*[13]

Demonstration of the procedure's application to commercial mustard, chocolate pudding and salad dressing, using the apparent stress in kPa at 1mm specimen height as the consistency measure, is given in Fig. 4.

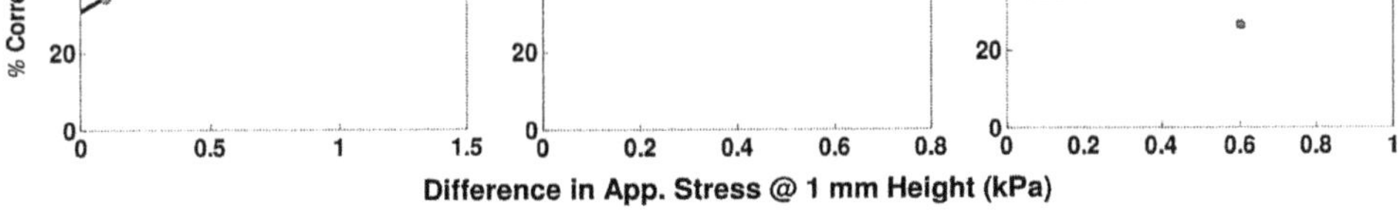

Figure 4 *Experimental correct identifications of the difference in the consistency of sheared samples of commercial chocolate pudding, mustard and salad dressing. From Chanasattru et al. (2002)*[13]

The corresponding practically significant Δσ's were 0.43, 0.74 and 0.5 kPa. This suggests that the practically significant difference for consistency determination might not be the same in different foods. Support for this view comes from that similar results were obtained when the chosen specimen's height was 1.5 mm and when the analysis was performed on the stresses ratio instead of the absolute differences.[13] However, because the conclusion is based on a limited database it still requires independent verification.

In principle, similar relations could be written for the difference in the residual stress after relaxation. However, examination of the sensory data revealed no pattern in any of the three foods. This might have been due to that the residual stresses range was not sufficiently large to establish a pattern. But it might also indicate that it is the overall resistance to deformation that determines a food's consistency and not the particulars of its rheological behavior.

3 BRITTLE CELLULAR SOLIDS

Brittle or "crunchy" breakfast cereals and snacks are popular around the world. Most are made of cereal grains or potato and contain sugar, salt and other additives. Many cereals and snacks are produced by puffing, which makes them solid foams. The mechanics of cellular solids, brittle and soft, natural or man made, is well understood.[14] There are also numerous mathematical models to describe their compressibility and fracture, ranging from purely theoretical to totally empirical. The theoretical models developed for synthetic polymeric solid foams used for cushioning and insulation can provide insight into the general deformation mechanisms in soft cellular solid foods, such as breadcrumb or popcorn. They can also be used to explain such food's compressive force-deformation or stress-strain curves. However, their application to brittle cereals and snacks is hindered for several reasons. Most cereals and snacks are fragile particulates and many have non-uniform or peculiar morphologies. Because of their fragility, shaping them into a convenient specimen for testing, a cylinder or cube for example, is an extremely difficult if not impossible task. Consequently, they ought to be tested in their original shape, or as an assembly, which complicates the interpretation of the recorded force-displacement curve. The particulates brittleness, geometric variability and non-uniform internal structure make their force-displacement curves irregular and irreproducible, examples of which are shown in Fig. 5.

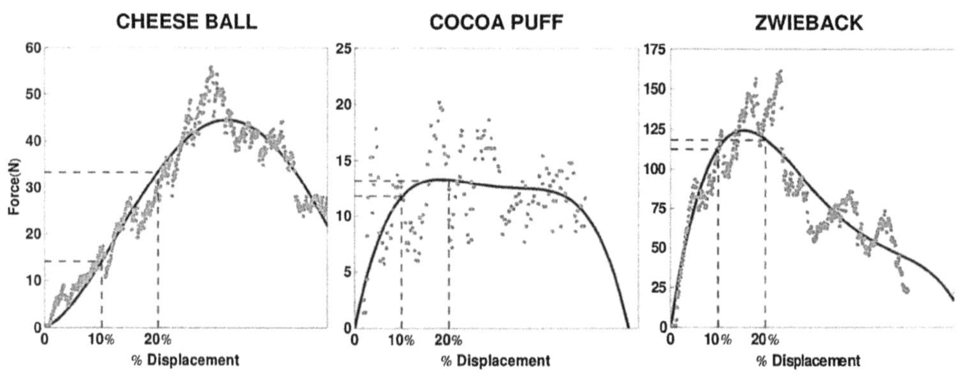

Figure 5 *Typical force-displacement curves of brittle cellular foods. Notice the irregular and irreproducible shape.*

Extraction of useful information from such irregular force-displacement curves can be done in the following way[15]: The recorded (or raw) force-displacement data are fitted with a polynomial or any other convenient empirical model. The force that corresponds to 10, 20 or 30% displacement, say, is calculated from the fitting model equation and used as a measure of the specimen's stiffness. [Where this force represents the area under the curve, it can be also treated as measure of the specimen's toughness.] The curve's degree of jaggedness can be quantified in different ways.[16] Perhaps the most convenient is the apparent fractal dimension, D_f, which can be determined rapidly by different algorithms[17] for mutual verification. Theoretically, the apparent fractal dimension of a curve n a plane can vary between 1.0, that of a smooth Euclidian line, and close to 2.0, that of a curve so jugged that it occupies almost the whole area in which it is imbedded, whose dimension is exactly 2.0. Examples of computer generated fractal curves having $D_f = 1.5$ and $D_f = 1.2$ are shown in Fig. 6. They demonstrate that the degree of jaggedness when expressed in term of a fractal dimension is independent of the curve's morphology and the algorithm that has produced it. [An interactive Wolfram Demonstration entitled 'Line jaggedness visualization with the Mandelbrot-Weierstrass Function' that creates, displays and animates the relationship between a curve's fractal dimension, its jaggedness and the resolution at which it is recorded can be found at http://demonstrations.wolfram.com/LineJaggednessVisualizationWithTheMandelbrotWeierstrassFunct/ .]

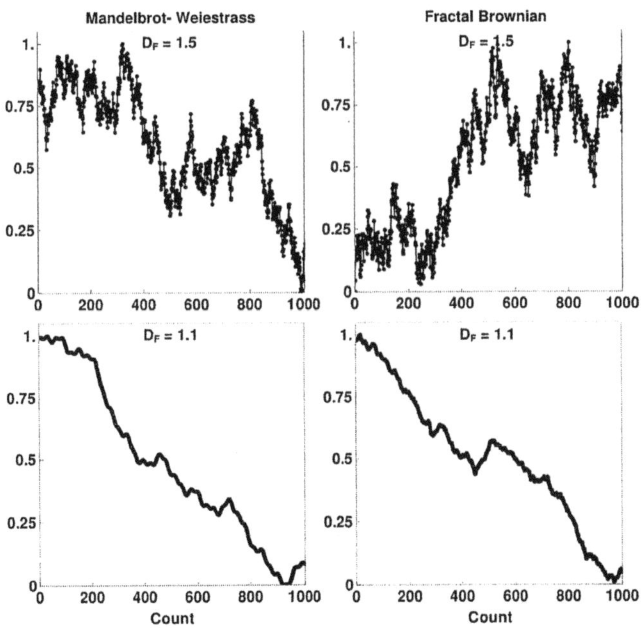

Figure 6 *Comparison of computer generated jagged curves having the same fractal dimension. Notice that the degree of jaggedness is independent of the curve's general shape*

When the above methods to assess stiffness and determine the degree of jaggedness are applied to actual compressive force-displacement curves of cereals and snacks, the scatter in the calculated parameters is considerable. Yet their reproducibility is sufficient to reveal the effect of absorbed moisture on the loss of brittleness, expressed in a decreasing apparent fractal dimension, and the initial toughening of the cereal or snack expressed in the initial rise of the fitted force at 25 or 35% displacement, before it drops when the specimen starts to dissolve. This is demonstrated in Fig. 7, which also shows the corresponding sensory evaluations. The figure shows that both testing machine and people can sense, simultaneously, brittleness or "crunchiness" loss and the rise in stiffness or "hardness", caused by moderate water sorption. [The increase in the samples' stiffness caused by the absorption of moderate amounts of water (a very effective plasticizer) should not come as a surprise. Because failure cannot propagate in the partially plasticized structure as easily as when it is dry and brittle, the specimen can resist higher forces.]

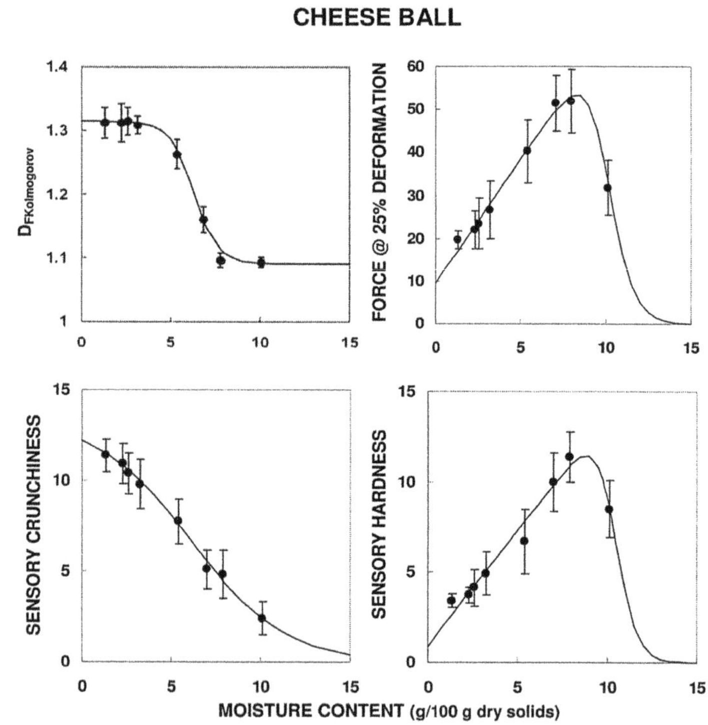

Figure 7 *Examples of simultaneous brittleness loss and initially increased stiffness (toughness) in cellular cereal products caused by moisture sorption. [After Suwonsichon and Peleg (1998)]*[18]

4 POTENTIAL USES OF HYDROCOLLOIDS IN THE STUDY OF CELLULAR SOLIDS MECHANICS

The mechanical and sensory properties of brittle solid foams are primarily determined by the cells walls mechanical properties, thickness and integrity and by the cellular structure, e.g., the cells' shapes and size distribution, whether they are open or closed, etc. It is difficult to separate the roles of the two in real solid foams because in most cases a modification of one will inevitably affect the other. In order to study their roles in isolation one might resort to model systems where different cellular structures can be made with the same cell wall material.

Having the described tools to assess the foams stiffness and brittleness will allow investigating the manifestation of the cell wall material and cellular structure in the foam's mechanical structure. The cellular structure can be characterized by a number of image processing methods, and other properties, such as the foam's density and porosity, by conventional physical methods. Consider freeze-dried gels as a case study.[19, 20] In principle and in practice one can modify the cell wall material by using different or different combinations of gelling agents and by infusing the gel with sugars, starches and other food ingredients. Although yet to be done, one can affect the dried gel's pore size by controlling the freezing rate – slow freezing will produce large ice crystals and hence relatively large pores and fast freezing small ice crystals and subsequently pores. Aeration can be another way to control the structure. Through such means one can create foams with practically the same composition, differing only in their cellular structure. For the particular purpose, gels-based model systems might have an advantage over actual foods produced by puffing, for example. This is because once the gel has been prepared the wall material is unlikely to undergo drastic chemical changes of the kind that occur in an extruder, crosslinking for instance. These are determined by the extrudate's thermal and shear histories, which are affected by the feed's consistency (and in turn by its initial moisture contents). Another advantage of gel-based models is that the specimens can be formed in shapes that are convenient for mechanical testing. An example of the use of freeze-dried gels in the investigation of solid foams' brittleness stiffness, and acoustic signature, is given in Fig. 8.

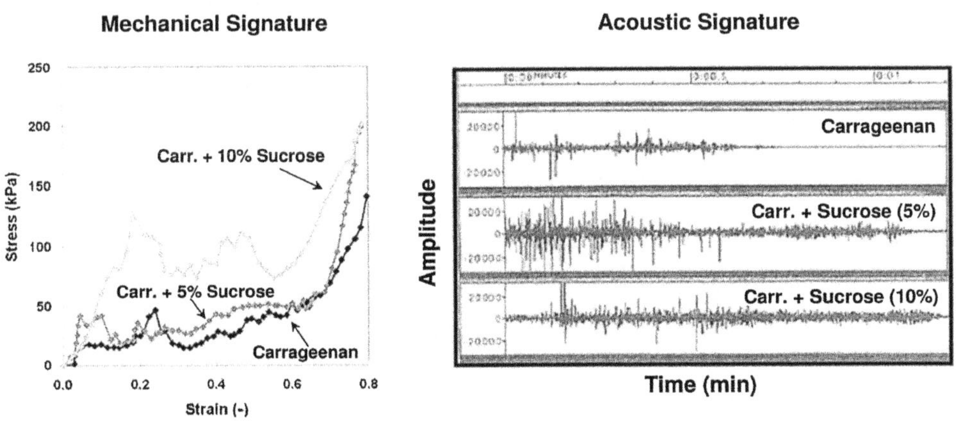

Figure 8 *Typical stress-strain relationships of freeze-dried carrageenan gels and their corresponding acoustic signatures. From Nussinovitch et al. (2000).*[19]

It should be stressed that the testing methodology is not limited to models systems. Hydrocolloids can be and are sometime used in the manufacturing of snacks. Their effect can be measured using the described methods, which were specifically developed for puffed cereals and snacks. However, the application of these methods can be extended to flaked cereals such as cornflakes, and snacks such as potato chips. Since meaningful testing of individual flakes is extremely difficult, the alternative is to test many of them together in a metal cell. To estimate an individual particulate's brittleness and stiffness from the force-displacement curves of its assemblies requires the use of several cells having different dimensions for extrapolation [21] and a simple formula to correct for the jaggedness loss as a result of the averaging effect.[22, 23]

Acknowledgements

Contribution of the Massachusetts Agricultural Experiment Station at Amherst.

References

1 A. Nussinovitch, *Hydrocolloid Applications: Gum Technology in the Food and Other Industries,* Chapman and Hall, London, 1997.
2 A. Nussinovitch, *Water Soluble Polymer Applications in Foods*, Blackwell Science, Oxford, 2003.
3 G.O. Phillips and Williams, P.A., *Handbook of Food Hydrocolloids*, Woodhead Publishing Limited, Cambridge, 2000.
4 A. Imeson, *Thickeners and Gelling Agents for Foods,* Blackie Academics and Professional Publishers, Glasgow, 1992.
5 M. Peleg, *Food Hydrocolloid*, 2006, **20**: 405.
6 M. Peleg and M.G. Corradini, 'Soft Machines Mechanics and Oral Texture Perception', in *Food Oral Processing*, ed., J. Chen, Wiley Blackwell, Oxford, 2011 (In press).
7 M.G. Corradini, R. Engel and M. Peleg, *J. Texture Stud.*, 2000, **31**, 363.
8 T. Suwonsichon and M. Peleg, *J. Sci. Food Agr.*, 1999, **79**, 911.
9 E.M. Casiraghi, E.B. Bagley and D.D. Christianson, *J. Texture Stud.*, 1985, **16**, 281.
10 O.H. Campanella and M. Peleg, *Crit. Rev. Food Sci.*, 2002, **42**, 241.
11 G. Jellinek, *Sensory Evaluation of Food. Theory and Practice*. VCH Publishers, Chichester, 1985.
12 H. Stone and J. Sidel, *Sensory Evaluation Practices*, 2nd. edn., Academic Press. New York, 1993.
13 W. Chanasattru, M.G. Corradini and M. Peleg, *J. Texture Stud.*, 2002, **33**, 445.
14 L.J. Gibson and M.F. Ashby, *Cellular Solids. Structure and Properties*, Cambridge University Press, New York, 1997.
15 M.G. Corradini and M. Peleg, *J. Texture Stud.*, 2006, **37**, 538.
16 M. Peleg, *Food Sci. Technol. Int.*, 1997, **3**, 227.
17 J.C. Russ, *Fractal Surfaces*, Plenum Press, New York, 1994.
18 T. Suwonsichon and M. Peleg, *J. Texture Stud.*, 1998, **29**, 255.
19 A. Nussinovitch, M.G. Corradini, M.D. Normand and M. Peleg, *J. Texture Stud.*, 2000, **31**, 205.
20 A. Nussinovitch, M.G. Corradini, M.D. Normand and M. Peleg, *Food Res. Int.*, 2001, **34**, 871.
21 R. Nixon and M. Peleg, 1995, *J. Texture Stud.*, **26**, 59-69.
22 D. Ulbricht, M.D. Normand, M. Peleg and J. Horowitz, *Powder Technol*, 1994, **81,** 83.
23 D. Ulbricht, M.D. Normand and M. Peleg, *J. Sci. Food Agr.*, 1995, **67**, 453.

STRUCTURING PROPERTIES OF PLANT PARTICLES AS A RESULT OF PHYSICAL TREATMENTS

P Lopez-Sanchez [a], L Bialek[a], R Farr[b], S Schumm[a] and M Langton[c]

[a]Structured materials and process science, Unilever R & D, Olivier Van Noortlaan 120, 3130 AC Vlaardingen, The Netherlands
[b]Structured materials and process science Unilever R&D, Colworth Science Park, Bedford, UK, MK441LQ
[c]Structure and Material Design, SIK-Swedish Institute for Food and Biotechnology, Box 540, 1 402 29 Göteborg, Sweden

1 INTRODUCTION

The World Health Organization (WHO) and the European commission recommend a daily intake of fruit and vegetable of around 600 g, however few people manage to consume this amount, which may reduce the risk for different types of cancer and cardiovascular diseases[1,2]. An increase in micronutrients bioavailability and the consumer's liking of ready-to-eat plant based food products may help achieve the recommended levels of fruit and vegetable consumption. Within this context the main aim of the Healthy Structuring project (European community's sixth programme) was to improve the structural and nutritional qualities of these types of products through the optimisation of all the manufacturing steps from plant growth, harvest, post-harvest storage and processing conditions. Processing could be optimised to make a better use of the natural polymers present in the plant material; this might help design textures using mainly the plant materials as natural texturisers and stabilisers. Furthermore the generated microstructures could be controlled in order to increase the bioavailability of specific micronutrients[3].

Suspensions produced from comminuted plant material are very common in the food industry; they are present in liquid and semi-liquid food products such as soups, drinks and sauces. In general suspensions generated from fresh plant material by conventional processing are highly polydisperse in particle size, shape, and composition [4,5]. They can be described as a combination of a liquid phase or *serum* containing salts, sugars and organic acids and a dispersed phase formed of all the plant insoluble solids mainly from cell wall material [6,7,8]. The insoluble dispersed particles in plant food suspensions are mainly from non-structural parenchyma tissue. The cells in parenchyma tissue are spherical or polyhedral in shape, with diameters of 50-500 μm and they are held together by the pectin of the middle lamella. The individual cell walls have a thickness of 0.1-10 μm depending on plant variety [9] and on the processing conditions. The composition of the plant cell wall depends on plant variety, growing conditions, and ripening stage. In general terms the primary cell wall of higher plants is composed of 70-90 % water and high molecular weight cellulosic and non cellulosic compounds, mainly pectins. The cell walls of carrot, broccoli and tomato in terms of dry matter contain about: 20-30 % cellulose, 10-30 % hemicelluloses and 35-50 % pectin, which is typical for dicotyledonous plants[10-13]. Other compounds such as proteins and phenolic material (lignin) are found in smaller amounts (5 %).

During processing plant tissues undergo external stresses, such as heat and shear, which disrupt the cell membranes reducing cell turgor and leading to more deformable particles. It has been shown that processing conditions influence the way plant tissues are disrupted [14,15]. Failure of the tissue can occur across the cell walls, or by cell-cell separation due to reduction of the inter-cellular adhesion. Based on these different ways of disruption, different particle morphologies can be obtained. The rheological properties and textures of the resulting suspensions are linked to the physicochemical properties and microstructures of the plant tissue components. The rheology of suspensions depends mainly on the rheology of the suspending medium and particle parameters such as particle volume fraction, size and morphology (for example the aspect ratio), the surface morphology and hardness of the particles[16].

Despite the increasing interest in the rheology of food suspensions[17,18,19] a unified treatment of the subject is still lacking. In particular the findings for plant food suspensions are not always conclusive because the particle attributes strongly depend on the specific raw material and processing conditions used. Despite the amount of available models to predict the rheological properties of suspensions, the direct application to plant particle suspensions is not always possible. This is due to the fact that those models make use of particle phase volume, and measuring this parameter for the soft and deformable plant particles is very difficult. Therefore models using particle concentration, rather than phase volume, have been empirically developed[8,19]. Therefore more systematic studies are required in order to understand the complex relationship between microstructure and rheological properties of these types of plant suspensions.

In this work carrot, broccoli and tomato were selected as model plant materials, due to their high content of specific micronutrients (β-carotene, folates and lycopene respectively) and the fact that they form a significant part of the Western diet. Different thermal and mechanical treatments to decompose plant material were utilised to generate particles with different characteristics in order to investigate the influence of particle shape, size, and concentration of the dispersed plant material on the rheology of the resulting suspensions. Furthermore the use of high pressure homogenisation (HPH) was evaluated as a way to further disrupt the plant tissues, compared to conventional blending, in order to generate novel structures.

2 MATERIALS AND METHODS

2.1 Materials

Fresh carrot (Daucus carota, var. Nantes), broccoli (Brassica oleracea, var. Italica) and tomato (S. lycopersicum, salad variety) were purchased from a local supermarket in The Netherlands. They were stored chilled (5 °C) for 1-2 days prior to processing.

2.2 Processing of the plant particle suspensions

Carrot, broccoli and tomato were washed in cold water and cut into pieces of approximately 1x1x2 cm size. The carrots were peeled prior to washing. The large broccoli stem and the core and stem of the tomatoes were removed. The pieces of each material were randomly mixed to reduce variability from individual specimens. They were mixed with deionised water, to enable proper homogenisation in a kitchen blender, in a ratio 1:1 of vegetable: water for carrot and broccoli, while for tomatoes the ratio was 9:1 due to the high amount of unbound water naturally present in this fruit. In the process referred to as 'cold-break' pieces were first mechanically disrupted using a kitchen blender at the maximum speed for 3 min. The resulting slurries were heated. Two thermal treatments were selected: 70 °C for 10 min as a mild cooking, and 90 °C for 40 min as a more extreme cooking step. 'Hot-break' samples were first heated followed by blending. The temperature in the core of the pieces was monitored using two thermocouples attached to a data logger. The blended tomato samples were passed through a fine metal screen (1 mm pore size) to remove seeds and large skin pieces to prevent blockage of the high pressure homogeniser. Selected samples were high pressure homogenised (Panda, Niro Soavi, Parma, Italy used in its standard configuration) via a single pass at 60 MPa. Samples were kept at 5 °C until further analysis.

Suspensions at different concentrations were prepared. In order to have standardised dispersed particles in sufficient amount the carrot and broccoli suspensions were weighed in several large flasks and centrifuged at 3500 rpm equivalent to approximately 1900 g. The supernatants were removed and dilutions were prepared from the sediment. In the initial tomato sample there were enough solids present, therefore the dilutions for tomato were prepared from the initial suspension without centrifugation. Particles were gently mixed with deionised water covering a range from 5 to 90% of sediment (the balance being the added deionised water).

2.3 Methods

2.3.1 Dry mass fraction measurement

Dry mass fraction w, was defined as the mass of a dried sample divided by its initial wet mass. Dry mass fraction was determined by drying 5 ± 0.2 g of each sample in a vacuum oven at 70 °C and a pressure of 4KPa for 5 hours. The dry mass fraction was calculated from the sample weight before and after drying.

2.3.2 Particle size measurement

Particle size distributions were measured by light scattering using a Malvern Mastersizer. Approximately 0.5 mL of each sample were pipetted into a water-continuous diluting accessory (2000 Hydro-S) filled with 1000 mL of deionised water. The water containing traces of particulate is pumped through a measurement window located in the optical path of a laser beam. The laser light is diffracted to detector units at different angles with respect to the incident laser beam. The particle size distribution is calculated from the intensity profile of the scattered light using the instrument software (Mastersizer 2000, version 5.40).

2.3.3 Light microscopy
Light microscopy images were taken using a Zeiss Axioplan microscope (Zeiss, Carl Zeiss Ltd, Hertfordshire, UK) equipped with a video camera. The samples were diluted 10 or 20 times with deionised water and gently mixed. Two droplets were placed on an object holder and covered with a glass slide. At least twelve images were obtained per sample. The samples were observed with a x10 or x40 dry objective lens using either conventional bright field or the differential interference contrast mode (DIC).

2.3.4 Scanning electron microscopy (cryo-SEM)
A drop of the material was placed on top of a rivet-type sample holder and directly plunged frozen in liquid ethane. The samples were stored in N_2 (l) until further analysis. In order to obtain a representative view of the bulk of the sample, the sample was cryo planed in a cryo-ultramicrotome (Ultracut UCT + EM FCS; Leica Microsystems), first with a glass knife, and the last sections with a diamond knife. For SEM analysis, the sample was briefly sublimated (freeze-etched) at -90 °C in the cryo-preparation chamber of the microscope. After sublimation a few nanometres thick layer of Au/Pd was sputtered onto the surface and the sample was transferred into the microscope. The sample was analysed at -125 °C, at 3 kV and a short working distance of about 6 mm using the in-lens secondary electron detector (SEI). A collection of approximately 25 images at different magnifications was taken per sample.

2.3.5 Rheological measurements
Suspending medium (Serum): suspensions were centrifuged at approximately 1900 g for 30 min and the serum (supernatant) was collected and passed through a paper filter (retention of particles larger than 8 μm). The viscosity of the serum phase was measured by increasing the shear rate from 0.1 to 1000 s-1 during 2 min with an AR1000 rheometer (TA Instruments). A cone and plate geometry was used (40 mm diameter, 1.59° angle, with truncation gap 71 μm).

Plant particle suspensions: the experiments were carried out in a stress-controlled rheometer (ARG2; TA Instruments, New Castle, DE, USA). The vane tool (four blades) in combination with a roughened cup was chosen and two gap sizes were used: 1 mm and a wider gap of 8.7 mm. Prior to the measurements the samples were degassed in a vacuum chamber under slow stirring for 15 min. Approximately 50 mL of sample was loaded into the rheometer. A solvent trap was used to prevent drying of the samples. All measurements were performed at 20 ± 0.1 °C maintained with a Peltier system. The suspensions were presheared at 100 s^{-1} for 60 s followed by a resting step of 5 min before starting the oscillatory experiments. To determine the viscoelastic behaviour dynamic small deformation tests were performed: a strain sweep from 0.01 up to 300 % was applied at a frequency of 1 Hz. The flow behaviour of the suspensions was studied by steady shear experiments increasing the shear rate from 0.01 up to 100 s^{-1} or 1000 s^{-1} and back during a total time of 4 min. From these measurements two methods were used to calculate the yield stress a) the yield stress was taken as the maximum value of the viscosity times the stress as shear stress increases and b) by fitting the up and return flow curves to the Herschel-Bulkley model[16].

3 RESULTS AND DISCUSSION

3.1 Microstructures generated

The processing conditions applied in our studies influence the plant tissue disruption, cell turgor and enzymatic activity leading to suspensions with different characteristics. During manufacturing the thermal and mechanical treatments affect the chemical composition of the serum (such as the content of pectin), and the characteristics of the dispersed phase (Figure 1) such as concentration, size, shape and hardness of the plant particles. All these parameters will have an influence on the final rheological properties.

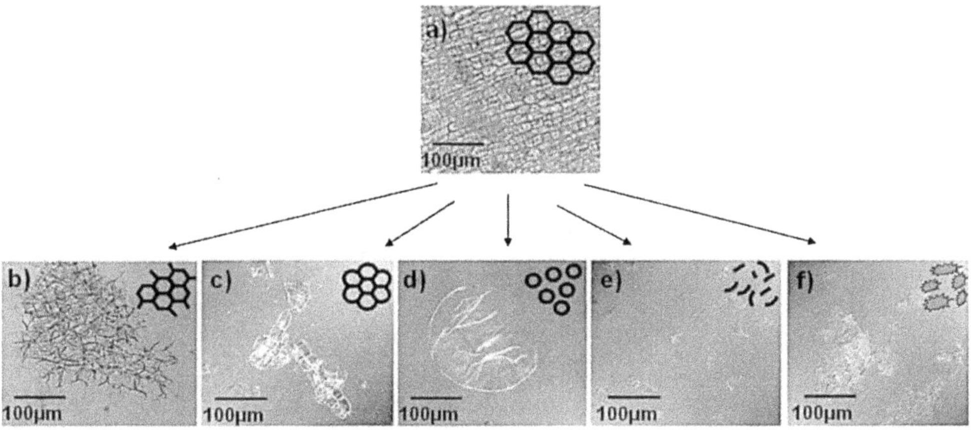

Figure 1 *Light micrographs and schematic representation of the plant particle morphologies generated. a) unprocessed broccoli tissue, b) clusters of cells broken across the cell wall from carrot, c) clusters of cells separated across the middle lamella from carrot, d) individual cell from tomato, e) cell fragments from carrot and f) cell fragments from tomato. Broccoli and carrot showed similar structures.*

3.2 Suspending medium

Our results (Figure 2) showed that the nature of the serum phase, Newtonian with water-like viscosity, did not contribute significantly to the rheological properties of the suspensions. No significant difference in the measured rheological properties of original and serum replaced samples could be observed[20]. This was tested for serums containing approximately 35 g kg^{-1} of soluble solids and suspensions with a total dry mass fraction of 0.05 for carrot, 0.04 for tomato and 0.06 broccoli, so the mass fraction of insoluble solids will be lower. Processing affects the release of cell contents, thus influencing the chemical composition of serum and dispersed phase. Studies on several plant materials such as onion[21], carrot[22], potato[23] and apple[24] have shown a release of water soluble polysaccharides (pectin and also low molecular weight glucose, fructose and sucrose) as a result of thermal treatments. The temperature and duration of heat exposure will affect not only the concentration but also the macromolecular structure of the biopolymers present in the serum[8,25]. Thermal treatments have also an effect on the enzymatic activity of the different enzymes in the plant material, such as pectinmethylesterase (PME) and polygalacturonase (PG) [26,27,28] in turn affecting the rheological properties. These enzymes are still active after a cold break and they can lead to pectin depolymerization of the cell wall during processing. It has been suggested that tomato juices resulting after a hot break are more viscous than from a cold break due to the inactivation of these enzymes[29]. Furthermore it has been observed that the higher the temperature of the hot break the higher the viscosity of the resulting tomato material[30]. In our studies the contribution of the serum phase to the overall rheology seemed to be less important than the properties of the dispersed phase. Nevertheless the serum contribution might be important for lower concentrations of insoluble solids, higher concentrations of soluble solids in the serum or when enzymes are still active with potential to act on polymers from the particles, such as pectin.

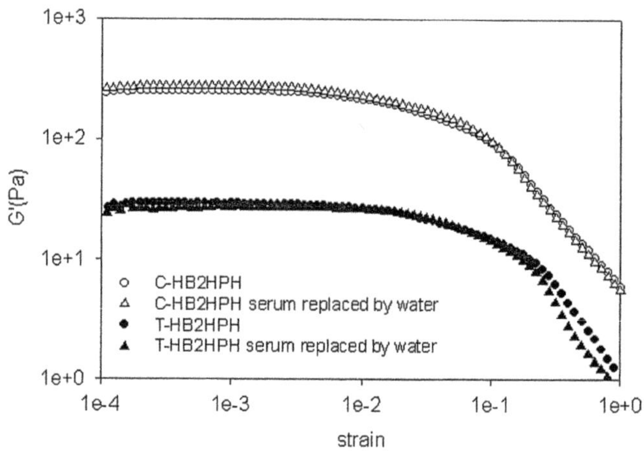

Figure 2 *Curves showing the effect of the replacement of serum by water on the elastic modulus of the plant material suspensions[20]. Carrot and tomato homogenised hot break samples are shown as examples, with dry mass fractions (including soluble solids in the serum) of 0.05 for carrot and 0.04 for tomato. For sample coding see abbreviations section.*

3.3 Dispersed phase: particle size and morphology

3.3.1 Suspensions of broken/smooth cell clusters and compact cell fragments

Cell clusters with broken edges visible at the surface (broken clusters) were obtained from carrot and broccoli (Figure 1a) by a cold break[20]. Similar morphologies were obtained by a shear treatment of unheated carrot in the work of Pickard et al.[31] A second type of cluster (Figure 1b), with smooth surfaces (smooth clusters), could be obtained by a hot break at 90 °C for 40 min indicating that during this treatment the cell separation through the middle lamella was favoured in both carrot and broccoli tissues. This can be explained by softening during heating caused by thermal depolymerisation of pectin[11,14,15]. Enzymatic activity was not detected therefore the pectin degradation is likely occurring via β-elimination which reduces cell adhesion along the middle lamella and influences the mechanical disruption. In contrast the mild hot break at 70 °C for 10 min led to clusters with broken edges; presumably the thermal input was not enough in this case to decrease adhesion and cell rupture was favoured instead of cell-cell separation. In general broken clusters showed larger sizes than the smooth clusters (Figure 1), and a reduction in particle size was observed at increasing temperature[20]. The compact cell fragments were obtained by high pressure homogenisation at 60 MPa (Figure 1d) of the intense heated (90° for 40 min) hot break sample. Moreover the high pressure treatment at 60 MPa further reduced the particle size and size distribution of all samples[20]. Thus carrot and broccoli showed a stepwise breaking: from larger to smaller clusters, to individual cells and cell fragments [20,32]. This behaviour has been reported previously for carrot tissues which have undergone different pre-treatments compared to our study[31].

These generated microstructures were associated to different rheological properties. For instance suspensions of broken clusters had higher yield stress, G' and G" values than the smooth clusters (Figure 3). For carrot suspensions a large yield stress, storage and loss modulus were linked to a larger particle size; however for broccoli no clear correlation to particle size was found. The suspensions of fragments obtained after HPH treatment showed lower viscoelastic properties and yield stress (Figure 3) than the suspensions of clusters (either with rough or smooth surfaces) from which they were generated. This indicates less "solid like" behaviour or the presence of a weaker network in these samples. This decrease was linked to a change not only in particle shape, and presumably in flexibility, but also to a reduced particle size and particle size distribution. Therefore the combination of all these parameters could explain the different rheological properties.

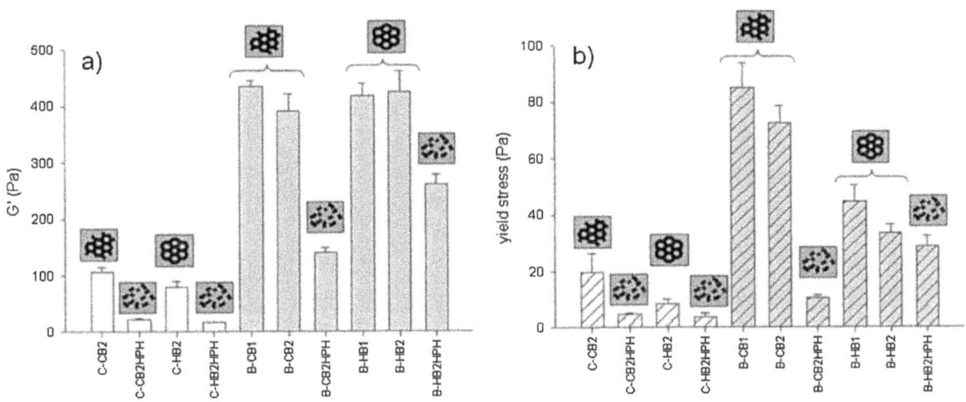

Figure 3 *Suspensions containing clusters and compact fragments with dry mass fraction of 0.05 for carrot and 0.06 for broccoli. a) G' at 0.1% strain and b) yield stress obtained from flow measurements. For sample coding see abbreviations section.*

3.3.2 Suspensions of individual cells and open cell fragments

Individual cells were obtained by simple mechanical disruption of tomato, with or without heat input (Figure 1c). Part of the cells (intact and broken) showed some remains of adjacent cells attached to them. Previous studies have shown that mechanical disruption of tomato followed by heating at 80 °C for 10 min can lead to intact individual cells[33] or broken cells (cell wall debris) when the heating occurs at higher temperatures 85-107 °C even for very short times[30]. In general the largest particle sizes[20] were observed for tomato compared to carrot and broccoli. Cell fragments were easier to obtain from tomato than from the other plant materials studied. A low shear treatment was enough to disrupt some of the individual cells. The more drastic shear treatments using high pressure homogenisation (Figure 1f) led to a dramatic change in the morphology of the tomato fragments. The difference between the structure of the cell fragments from carrot/broccoli and tomato was clearly displayed by cryo-SEM (Figure 4). In the case of carrot and broccoli the cell fragments had a compact cell wall structure. In contrast, tomato fragments showed cell walls considerably swollen with an open structure and the biopolymers from the cell walls expanding into the liquid phase. A closer look at the cell structures revealed some swelling of the tomato cell walls even before homogenisation, likely induced by changes in cell wall composition during the thermal treatment. This swelling was increased by the homogenisation process. Thus, the effect of an intense high-pressure treatment led to a change not only in particle size and morphology but also in the nature of the material composing the tomato particles.

Similar behaviour to the one observed for tomato has been reported for other fruits such as kiwi[33]. This suggests that cell walls from fruits, tomato and kiwi, are easier to disrupt than from vegetables such as carrot and broccoli. This could be due to differences in the chemical composition of the cell walls, in particular the cellulose/pectin ratio, and the extent of crosslinking between the polysaccharides might explain the observed differences in behaviour of tomato compared to broccoli and carrot. An alternative explanation for the higher shear required to break the carrot cells could be the presence of non lignin cell wall phenolics that might cross-link cell wall polymers, contributing to strengthening the cell walls[34].

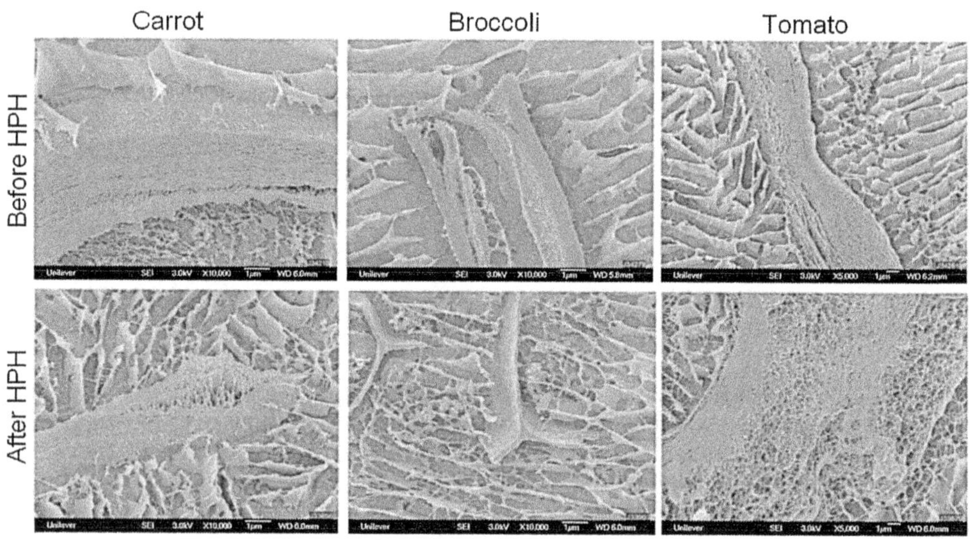

Figure 4 *cryo-SEM micrographs. From left to right carrot, broccoli and tomato. Before HPH (top) and after high pressure homogenisation HPH at 60 MPa (bottom).*

Interestingly in contrast to the compact fragments from carrot and broccoli, the cell fragments with an open structure obtained from tomato showed a higher elastic modulus G' and yield stress than the individual cells (Figure 5), even though a reduction in the particle size was observed. This could be due to the open structure of the tomato cell walls which might lead to increased polymer interactions, via physical entanglements and perhaps other attractive forces such as electrostatic interactions, producing a stronger network and increasing the elastic modulus and yield stress of the suspensions. This enhancement of the rheological parameters is in agreement with the results earlier reported for tomato cell fragments[35], where an increase in the viscosity was related to a drastic change in the particle size distribution and morphology of the tomato material.

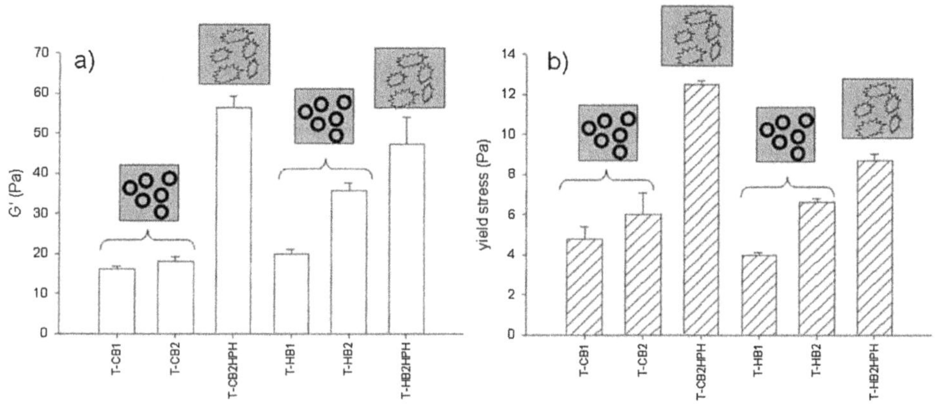

Figure 5 *Rheological properties of tomato suspensions (see abbreviations section) containing individual cells and open fragments. a) G' at 0.1% strain and b) yield stress from flow measurements.*

3.4 Dispersed phase: particle concentration

For all suspensions the apparent yield stress and the elastic moduli increased with particle concentration, expressed as total dry mass w (Figure 6). The dependence of the elastic modulus G' and yield stress values on particle dry mass concentration followed approximately a power-law. The yield stress scaling showed a fitted exponent of 3 ± 0.5 for all cases, independently of the plant origin and particle shape. These values lie within the range of the reported power-law indices for the yield stress as a function of tomato insoluble solids[35]. And similar exponents have been found for flocculated suspensions[36] and wood suspensions[37] investigated as a function of particle volume fraction.

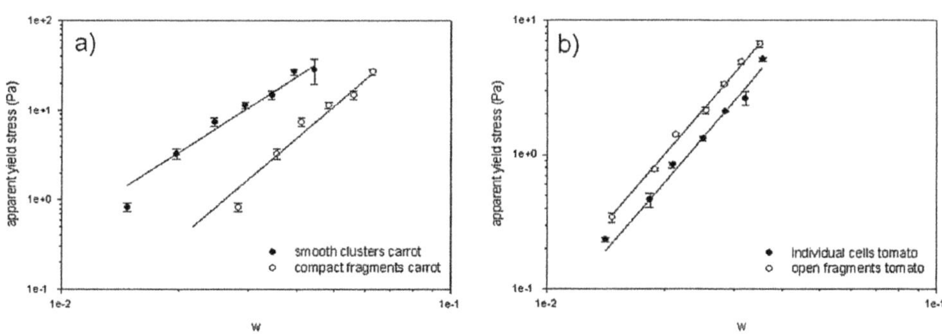

Figure 6 *Apparent yield stress (from fitting Hershel-Buckley to the up-curves) as a function of dry mass fraction w. a) Clusters of cells and compact fragments from carrot. b) Individual cells and open fragments from tomato. Solid lines represent the power law fit.*

In the related literature the rheological properties of plant food suspensions of different botanical origin have been related to the amount of solids they contain. The studies of Sharma et al.[25] showed that the insoluble fraction of the total solids content was more important for the consistency of tomato serum phases than the water-soluble solids (such as fructose and glucose). A high content of insoluble solids has been related to high values of the yield stress[38] and flow properties[25] in tomato dispersions. Similarly the dry matter content of carrot suspensions[39] was related to the final rheological properties of the suspensions. Apple and orange systems showed also a good linear correlation between the insoluble solids level and the consistency[8]. Theoretically at very low particle concentrations, the particles will not be space-filling: they will sediment and the system has no elastic modulus or yield stress, since the serum has neither. This behaviour was observed for suspensions with a dry mass fraction below 0.01 for carrot and tomato and 0.03 for broccoli. At some concentration, a percolated network will form and such a system will yield and flow only when attractive bonds break or sliding friction forces are overcome. If there are asperities on the surface, the particles may withstand static friction, so that they can roll over each other, but not easily slide (like cogwheels in two dimensions). As the concentration is increased above the percolation threshold, the elastic modulus will increase, and the particles are likely to be increasingly deformed, even when there is no imposed strain. Finally, at very high concentrations, the particles will necessarily be highly deformed, in order to fit into the available space. This was observed in the cryo-SEM images (data not shown), where buckled cell walls and crushed cells are clearly evident. In this regime, steric effects alone will be enough to prevent liquid-like flow of the system (although friction and attractive forces may also be present). More particularly, the elastic modulus should be determined purely by the deformation of the particles. Above a certain concentration, which depends on the type of plant material and processing conditions, the insoluble solids are the main contributor to the rheological properties of plant suspensions. Below this concentration the rheology of the suspending medium, due to soluble solids in the liquid phase such as pectic substances, plays also an important role. However, as pointed out by Sanchez et al.[40] the effect of insoluble solids can not be considered just in terms of concentration since the size, shape and elasticity of the particles will also influence the material properties.

4 CONCLUSIONS

Our results show that physical treatments can be used to generate different plant microstructures with specific rheological properties. The control of the plant material disruption and the insight into the microstructure/rheology relationship of plant particle suspensions may be used to generate a new broad range of plant derived microstructures with potential to structure food products in a more natural way, while reducing the use of other texturisers and stabilisers. Furthermore beyond the application for food industries, the improved knowledge on how the material properties arise from plant particles in suspension can be utilised to optimise the processing conditions and rheological properties of other types of plant suspensions for example from cellulose pulp, for paper and biofuel production.

Abbreviations

C, B, T	carrot, broccoli, tomato
raw	unheated comminuted material
CB1	cold break at 70 °C for 10 min
CB2	cold break at 90 °C for 40 min
HB1	hot break at 70 °C for 10 min
HB2	hot break at 90 °C for 40 min
HPH	High pressure homogenisation

Abbreviations for samples may be composed of several of the above terms, for example "B-raw" would mean "unheated, comminuted broccoli", and "T-CB2HPH" would mean "cold break tomato, post heated to 90 °C for 40 min and then high pressure homogenised".

Acknowledgements

The authors would like to thank V. Chapara for helping to carry out the rheological experiments. P. Versluis is kindly acknowledged for helpful discussions. The authors gratefully acknowledge financial support from the European Commission FOOD-2004-T5.4.1.3: Project 023115-Healthy Structuring.

References

1 Singh,P.; Goyal,G.K. Dietary lycopene: Its properties and anticarcinogenic effects. Comprehensive Reviews in Food Science and Food Safety 2008; 7:255-270

2 Steinmetz,K.A.; Potter,J.D. Vegetables, fruit, and cancer prevention: A review. Journal of the American Dietetic Association 1996; 96:1027-1039

3 Aguilera,J.M.; Sanchez,M.C. Why food microstructure? Journal of Food Engineering 2005; 67:3-11

4 Den Ouden,F.W.C.; vanVliet,T. Particle size distribution in tomato concentrate and effects on rheological properties. Journal of Food Science 1997; 62:565-567

5 Yoo,B.; Rao,M.A. Effect of Unimodal Particle-Size and Pulp Content on Rheological Properties of Tomato Puree. Journal of Texture Studies 1994; 25:421-436

6 Barrett,D.M.; Garcia,E.; Wayne,J.E. Textural modification of processing tomatoes. Critical Reviews in Food Science and Nutrition 1998; 38:173-258

7 Hayes,W.A.; Smith,F.G.; Morris,A.E.J. The production and quality of tomato concentrates. Critical Reviews in Food Science and Nutrition 1998; 38:537-564

8 Rao, M.A., Qiu, C. G. Rheological properties of plant food dispersions. In: AnonymousQuality factors of fruit and vegetables. 1989: 149-171

9 Buchanan,B.B.; Gruissem,W.; Jones,R.L. Biochemistry & molecular biology of plants, American society of plant physiologists. Rockville, USA. 2000

10 Massiot,P.; Rouau,X.; Thibault,J.F. Structural Study of the Cell-Wall of Carrot (Daucus-Carota L) .1. Isolation and Characterization of the Cell-Wall Fibers of Carrot. Carbohydrate Research 1988; 172:217-227

11 Greve,L.C.; Mcardle,R.N.; Gohlke,J.R.; Labavitch,J.M. Impact of Heating on Carrot Firmness - Changes in Cell-Wall Components. Journal of Agricultural and Food Chemistry 1994; 42:2900-2906

12 Gross,K.C.; Wallner,S.J. Degradation of cell wall polysaccharides during tomato fruit ripening. Plant Physiology 1979; 63:117-120

13 Muller,S.; Jardine,W.G.; Evans,B.W.; Vietor,R.J.; Snape,C.E.; Jarvis,M.C. Cell wall composition of vascular and parenchyma tissues in broccoli stems. Journal of the Science of Food and Agriculture 2003; 83:1289-1292

14 Waldron,K.W.; Smith,A.C.; Parr,A.J.; Ng,A.; Parker,M.L. New approaches to understanding and controlling cell separation in relation to fruit and vegetable texture. Trends in Food Science & Technology 1997; 8:213-221

15 Ormerod,A.P.; Ralfs,J.D.; Jackson,R.; Milne,J.; Gidley,M.J. The influence of tissue porosity on the material properties of model plant tissues. Journal of Materials Science 2004; 39:529-538

16 Macosko,C.W. Rheology:principles,measurements and applications, VCH publishers. New York, USA. 1993

17 Servais,C.; Jones,R.; Roberts,I. The influence of particle size distribution on the processing of food. Journal of Food Engineering 2002; 51:201-208

18 Gallegos,C.; Franco,J.M.; Partal,P. Rheology of food dispersions Rheology reviews. Rheology reviews 2004; 19-65

19 Genovese,D.B.; Lozano,J.E.; Rao,M.A. The rheology of colloidal and noncolloidal food dispersions. Journal of Food Science 2007; 72:11-20

20 Lopez-Sanchez, P.; Nijsse, J.; Blonk, H.C.G.; Bialek, L.; Schumm, S. and Langton, M. Effect of mechanical and thermal treatments on the microstructure and rheological properties of carrot, broccoli and tomato dispersions. Journal of the Science of Food and Agriculture (2011), 91 (2) 207-217.

21 Lecain,S.; Ng,A.; Parker,M.L.; Smith,A.C.; Waldron,K.W. Modification of cell-wall polymers of onion waste - Part I. Effect of pressure-cooking. Carbohydrate Polymers 1999; 38:59-67

22 Ng,A.; Waldron,K.W. Effect of cooking and pre-cooking on cell-wall chemistry in relation to firmness of carrot tissues. Journal of the Science of Food and Agriculture 1997; 73:503-512

23 Ng,A.; Waldron,K.W. Effect of steaming on cell wall chemistry of potatoes (Solanum tuberosum cv. Bintje) in relation to firmness. Journal of Agricultural and Food Chemistry 1997; 45:3411-3418

24 Vetter,S.; Kunzek,H. The influence of the sequential extractions on the structure and the properties of single cell materials from apples. European Food Research and Technology 2003; 217:392-400

25 Sharma,S.K.; LeMaguer,M.; Liptay,A.; Poysa,V. Effect of composition on the rheological properties of tomato thin pulp. Food Research International 1996; 29:175-179

26 Rodrigo,D.; Cortes,C.; Clynen,E.; Schoofs,L.; Van Loey,A.; Hendrickx,M. Thermal and high-pressure stability of purified polygalacturonase and pectinmethylesterase from four different tomato processing varieties. Food Research International 2006; 39:440-448

27 Sila,D.N.; Smout,C.; Vu,S.T.; Van Loey,A.; Hendrickx,M. Influence of pretreatment conditions on the texture and cell wall components of carrots during thermal processing. Jounal of food science 2005; 70:85-91

28 Verlent,I.; Smout,C.; Duvetter,T.; Hendrickx,M.E.; Van Loey,A. Effect of temperature and pressure on the activity of purified tomato polygalacturonase in the presence of pectins with different patterns of methyl esterification. Innovative Food Science & Emerging Technologies 2005; 6:293-303

29 Goodman,C.L.; Fawcett,S.; Barringer,S.A. Flavor, viscosity, and color analyses of hot and cold break tomato juices. Journal of Food Science 2002; 67:404-408

30 Xu,S.Y.; Shoemaker,C.F.; Luh,B.S. Effect of Break Temperature on Rheological Properties and Microstructure of Tomato Juices and Pastes. Journal of Food Science 1986; 51:399-402
31 Pickardt,C.; Dongowski,G.; Kunzek,H. The influence of mechanical and enzymatic disintegration of carrots on the structure and properties of cell wall materials. European Food Research and Technology 2004; 219:229-239
32 Lopez-Sanchez, P.; Svelander, C.; Bialek, L.; Schumm,S.; Langton, M. Rheology and microstructure of carrot and tomato emulsions as a result of high pressure homogenisation conditions. Journal of Food science 2011; 76:130-140.

33 Redgwell,R.J.; Curti,D.; Gehin-Delval,C. Physicochemical properties of cell wall materials from apple, kiwi fruit and tomato. European Food Research and Technology 2008; 227:607-618
34 Parr,A.J.; Ng,A.; Waldron,K.W. Ester-linked phenolic components of carrot cell walls. Journal of Agricultural and Food Chemistry 1997; 45:2468-2471
35 Bayod,E.; Mansson,P.; Innings,F.; Bergenstahl,B.; Tornberg,E. Low shear rheology of concentrated tomato products. Effect of particle size and time. Food Biophysics 2007; 2:146-157
36 Van der Aershot,E.; Mewis,J. Equilibrium properties of reversible flocculated dispersions. Colloids and surfaces 1992; 69:15-22
37 Bennington C.P.J.; Kerekes R.J.; Grace J.R. The yield stress of fibre suspensions. Canadian Journal of Chemical Engineering 1990; 68:748-757
38 Yoo,B.; Rao,M.A. Yield Stress and Relative Viscosity of Tomato Concentrates - Effect of Total Solids and Finisher Screen Size. Journal of Food Science 1995; 60:777
39 Nyman,E.M.G.L.; Svanberg,S.J.M.; Andersson,R.; Nilsson,T. Effects of cultivar, root weight, storage and boiling on carbohydrate content in carrots (Daucus carota L). Journal of the Science of Food and Agriculture 2005; 85:441-449
40 Sanchez,M.C.; Valencia,C.; Gallegos,C.; Ciruelos,A.; Latorre,A. Influence of processing on the rheological properties of tomato paste. Journal of the Science of Food and Agriculture 2002; 82:990-997

RELATIONS BETWEEN MASTICATION VARIABLES, MECHANICAL PROPERTIES AND SENSORY SCORES OF CHEESE SAMPLES WITH VARIOUS TEXTURES

K Kohyama, F Hayakawa and T Sasaki

National Food Research Institute, National Agriculture and Food Research Organization, Tsukuba, Ibaraki 305-8642, Japan. E-mail: kaoruk@affrc.go.jp

1 INTRODUCTION

Food texture is perceived during mastication and can change with every chew. It is often difficult to characterise texture by mechanical testing because such testing is normally performed on food before it is consumed. Human masticatory measurements involving electromyography (EMG) have recently been introduced in food texture evaluation in order to integrate physical, physiological and psychophysical approaches.[1, 2]

We used cheese of various textures as samples because cheese changes dynamically during mastication with the breakdown of its structure, increase in moisture from saliva and melting at oral temperature. Such temporal changes in texture have not been adequately evaluate by instrumental testing. EMG recordings have been made from the masseter and anterior temporalis muscles while the subjects masticated different cheese samples.[2-5] Because these muscles are typically used to close the mandible (jaw-closing muscle), their EMG activities are useful in analysing the texture of hard foods and the breakdown of their structure.

The hardness of food correlates with the amplitude of the jaw-closing muscles[1-3, 6-10] and mastication of more adhesive foods leads to greater activity of the jaw-opening muscles.[9-11] The suprahyoid musculature, including the anterior belly of the digastric, mylohyoid and geniohyoid muscles, act in the jaw-opening phase.[2, 9-11] We recently measured EMG activities from the masseter and suprahyoid muscles during the mastication of soft gels.[12] EMG activity of the suprahyoid musculature correlated with the compression load of samples under very large strains, although the masseter activity was very minimal because the samples were not chewed by the teeth. The importance of mechanical properties under very large strains has been reported for hard food that requires chewing: the magnitude of masseter EMG activities and the compressive load at large stains were more significantly correlated than any other mastication variable and mechanical parameter measured at smaller strains.[13]

We introduced EMG masticatory measurements of the suprahyoid muscles and masseter muscles for texture evaluation of cheese samples. Most applications of EMG in solid foods have been employed for the analysis of the mastication process.[2-11, 13] We also studied EMG activities after the first swallow while the subjects freely ate a mouthful of the cheese sample. EMG activities of early and late stages of mastication were analysed separately and related to fracture properties and bolus formation. These correspond to the degree of structure and the degree of lubrication, which are the two main factors in the oral processing of foods.[14]

2 METHODS AND RESULTS

2.1 Materials

Four natural varieties of cheese [Caprice des Dieux (C), Mozzarella (M), Beaufort (B) and Parmigiano-Reggiano (P)] and a processed cheese (Y, Snow Brand, Baby Cheese) were cut into blocks measuring 20 × 20 × 10 mm, wrapped in plastic film and stored at 5°C until experiments.

2.2 Methods

The texture of the cheese samples was evaluated with a combination of instrumental tests, human masticatory measurements using surface EMG and sensory evaluation.

Two-bite texture profile analysis (TPA)[14] was perfomed using a Texture Analyser (TA-XT2plus, Stable Micro Systems, Godalming, UK). A block of cheese sample was compressed twice between flat plates at a constant speed of 1 mm/s with an interval of 4 s. Compressive strain was set to 80% of initial thickness. The Young's modulus, fracture force and strain, hardness (maximum force value at the first bite), adhesiveness, cohesiveness and springiness were measured. The Young's modulus was calculated from the slope of the load curve at 5% compressive strain and the fracture point was determined at the first decrease in load value. Cohesiveness and springiness were expressed as ratios of work and compressive length at the second bite to those at the first bite to standardize initial sample size.

The sensory properties of degree of firmness, brittleness, adhesiveness, springiness, melting in the mouth and chewiness were evaluated using a 10-cm unstructured line scale by a trained panel of six members.

EMG experiments were conducted according to the guidelines of the ethics committee at the National Food Research Institute. Nine young adults participated in EMG recording after having given informed consent. Each subject was served mouthfuls (5 g) of cheese samples in random order with two repetitions and ate freely the sample. As cheese P is often consumed in powdered form, shredded P (PP) was compared with a block sample of cheese P. EMG activities were recorded using bipolar surface electrodes (EL503, Biopac Systems Inc., Goleta, CA, USA) from the left and right masseter muscles (LM and RM) as typical jaw-closing muscles and from the suprahyoid muscles (SH) which open the jaw and move the tongue. After filtering (10–500 Hz), removal of 50 Hz noise from the power supply and amplification to 1,000 times the original level using three EMG 100C amplifiers (Biopac Systems Inc.), the EMG signals were saved in a Windows® computer using an MP-150 system (Biopac Systems Inc.) at 1,000 Hz. The subject pressed a button switch on beginning to chew and at every swallow (Figure 1). The data were analysed using wave analysis software (AcqKnowlegde® ver. 3.8.2; Biopac Systems Inc.). Mastication time was divided into two stages: T1, rhythmical masseter activity before the first swallowing signal; T2 from the end of stage T1 until the last swallowing signal. The number of chews was counted based on the number of masseter bursts. Amplitude, duration, cycle time and muscle activity, as integrated EMG amplitudes (Figure 2) were calculated for each masseter muscle. EMGs for each chew, and averages were calculated for the two sides and for each chew. EMG duration and muscle activities were summed for T1 and T2 and total muscle activity until the last swallow was also calculated.

Figure 1 *Example of mastication recording chart for cheese*

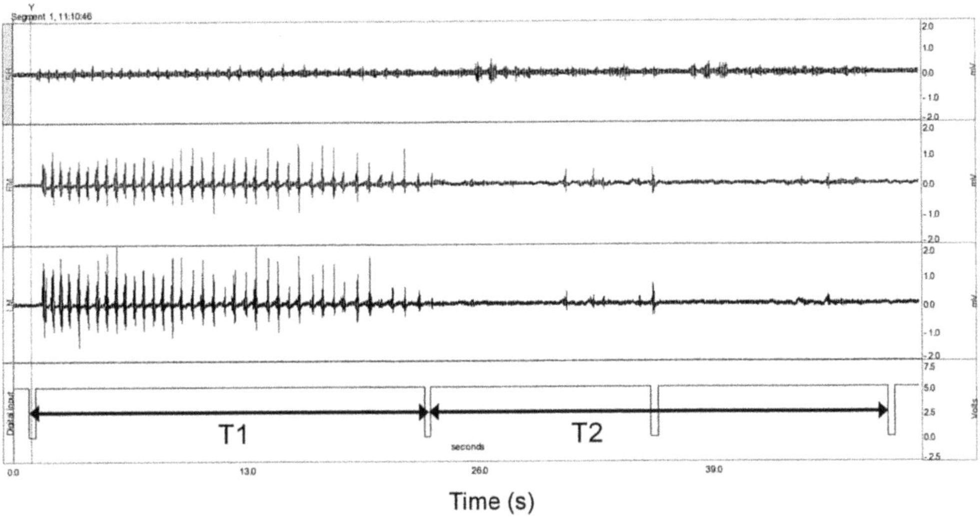

From the top, signals of EMG activities from the suprahyoid (SH), right masseter (RM) and left masseter (LM) muscles and button switch pressed at the beginning to chew and swallows by the subject. Two stages T1 and T2 are before and after the first swallow.

Figure 2 *EMG variables taken from one burst*

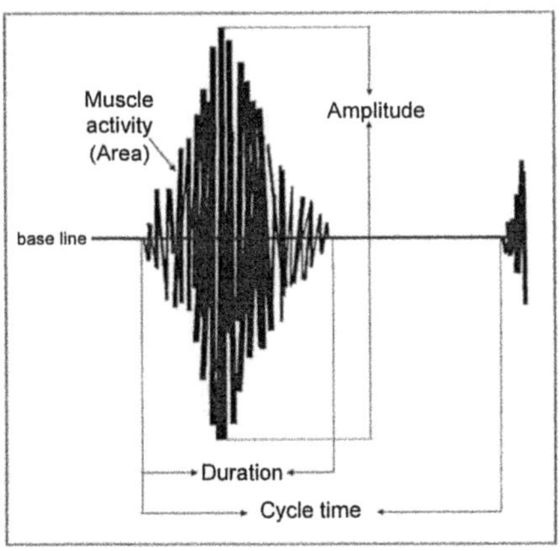

Statistical analyses were conducted using the SPSS® package (ver. 17.0J for Windows®) (SPSS Inc., Chicago, IL, USA), with significance defined at $p < 0.05$. EMG parameters and sensory scores were analysed by repeated-measures analysis of variance (ANOVA), and when the sample difference was significant, the paired *t*-test with Bonferroni's correction was used for all combinations of two samples. The paired *t*-test was used for comparisons between block and shredded samples. One-way ANOVA was performed for instrumental parameters and when appropriate, Tukey's multiple comparison was used to test differences between samples. Mean values of mechanical, EMG and sensory variables were subjected to the principal component analysis.

2.3 Results

Typical TPA curve and mean values of TPA parameters are shown in Figure 3 and Table 1.

Figure 3 *Typical two-bite texture profile curves of five cheese samples*

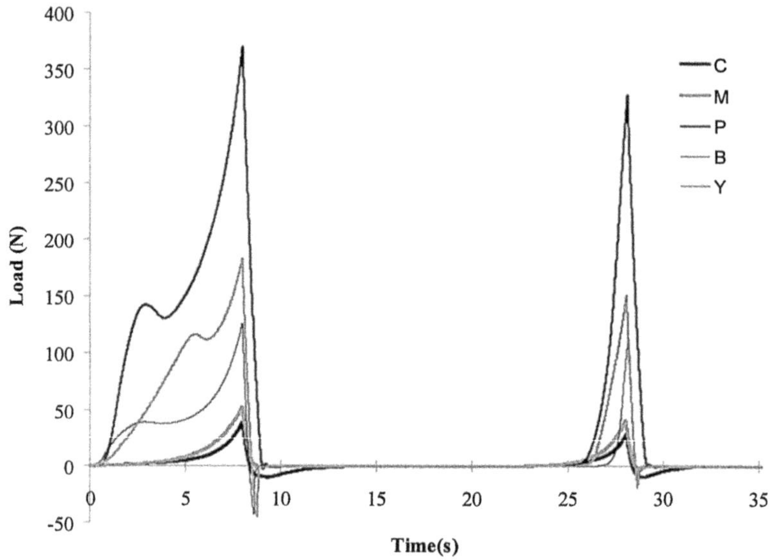

Table 1 *TPA results of five different cheese samples*

Sample	Young's modulus (kPa)	Fracture force (N)	Fracture strain	Hardness (N)	Adhesive-ness (N.mm)	Cohesive-ness	Springiness
C	0.172 a	nd	nd	39.7 a	28.0 c	0.484 b	0.964 c
M	0.096 a	nd	nd	54.1 a	0.67 a	0.636 b	0.800 c
B	6.02 b	42.9 a	0.285 a	136 b	11.1 b	0.129 a	0.257 a
Y	1.90 a	116 b	0.574 b	180 c	10.5 b	0.205 a	0.507 b
P	21.8 c	141 c	0.270 a	370 d	2.8 a	0.226 a	0.480 b

nd; not fractured.

Mean values with a different letter were significantly different by Tukey's multiple comparison test ($p < 0.05$).

As shown in Figure 3 and Table 1, the mechanical properties of the five cheese samples were quite different. Three cheese varieties were broken but two varieties of soft cheese were not. Mechanical testing should be performed under a very large strain in order to obtain a high correlation between masticatory variables;[12, 13] 80% strain seemed enough because the hardness or maximum load value for the first bite was greater than the breaking load for the harder types of cheese. When we conducted a simple compression test until 90% strain, the final load values were of the same order as the hardness values.

Samples C and M were soft, but C was more adhesive than M. Sample P was the hardest but its adhesiveness was low.

Similar characteristics were obtained in the sensory evaluation (Figure 4). Sensory firmness showed a good correlation with instrumental hardness. Brittleness could be sensed as combination of low adherence with high firmness. Springiness was greatest for M and melting in the mouth was greatest for C; both of these were soft varieties of cheese. Harder varieties of cheese were evaluated as being chewier.

Figure 4 *Sensory evaluation results of five cheese samples*

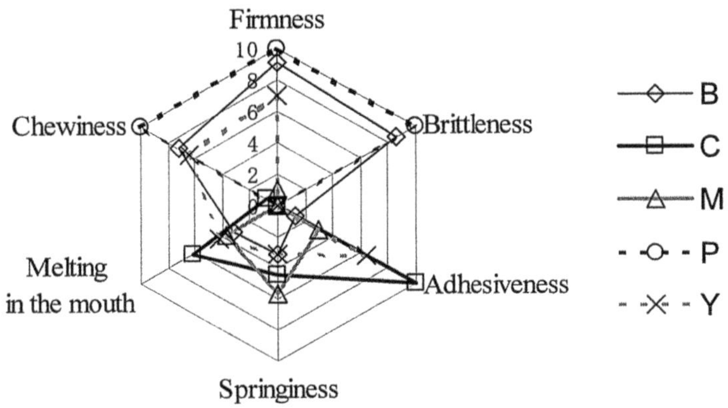

Table 2 *EMG variables of five different cheese samples*

Sample	Number of chews	Time before the first swallowing, T1 (s)	Time after the first swallowing, T2 (s)	Total muscle activity (mVs)	EMG duration / chew (s)	Cycle time / chew (s)	Amplitude / chew (mV)	Muscle activity / chew (mVs)
C	29.6 b	13.6 b	21.5 b	0.80 b	0.285 a	0.680 ab	1.28 b	0.027 b
M	19.1 a	7.9 a	15.0 a	0.42 a	0.290 a	0.642 a	1.01 a	0.022 a
B	41.5 c	20.1 c	26.9 c	1.46 c	0.300 b	0.709 b	1.52 c	0.035 c
Y	32.4 b	16.3 b	20.6 b	0.93 b	0.303 b	0.678 ab	1.18 b	0.029 b
P	49.9 d	22.8 d	29.4 c	2.04 d	0.319 b	0.680 ab	1.66 c	0.041 d

Mean of RM and LM muscle EMGs.
Values of mean of nine subjects with different letters are significantly different ($p < 0.05$).

Table 2 shows EMG variables which exhibited significant sample effects. As the masseter muscles act to close the mandible, masseter EMG activity per chew was greater for harder varieties of cheese. A longer chewing period and a greater number of chewing strokes before the first swallow were found in harder samples. In the jaw-opening phase, the suprahyoid muscles worked alternately with the masseter muscles in the first stage of mastication (T1, Figure 1). The rhythmical pattern of the masseter and suprahyoid muscles disappeared and suprahyoid activity became predominant over masseter activity in the second stage of oral processing (T2). This suggests that breaking the food structure with the teeth is the main act in the first stage of mastication and bolus formation using the tongue is important in the second stage.

In this study T1 and T2 were of the same order and suprahyoid muscle amplitude was not significantly different among the samples. No new suprahyoid muscle variables relating to difficulty in bolus formation were found, though T2 and the sum of muscle activity during the second stage could be derived from the masseter EMGs.

Table 3 *Electromyographic variables of a 5 g block and shredded Parmigiano-Reggiano cheese*

Sample P	Number of chews	Time before the first swallowing, T1 (s)	Total mastication time (s)	Total muscle activity (mVs)	First EMG duration (s)	First muscle activity (mVs)
Block	49.9	22.8	52.2	2.04	0.487	0.051
Shredded	35.6	15.6	45.7	1.42	0.366	0.041

Average values of nine subjects.
Other EMG variables were not significantly different.

We compared block of cheese P with a shredded sample of cheese P as shown in Table 3. The shredded sample required a significantly shorter mastication time, with smaller number of chews. This suggests that shredding a hard cheese makes it easy to chew. Muscle activity per chew and duration of the first chew for shredded cheese were significantly smaller than values for the block sample, but averaged EMG amplitude, duration, cycle time and muscle activity were not significantly different. This may be explained by the brittleness, low cohesiveness and low breaking strain of sample P.

2.4 Relation between EMG Variables and Cheese Characteristics

We did not conduct sensory evaluation of cheese texture at the same time as EMG measurement because we considered it possible that the subjects would not masticate in the normal way during a difficult task such as texture evaluation. Our previous work on bite force measurement of cheese samples revealed that bite speed was relatively slow during sensory texture evaluation.[15-17] The fracture force increased and fracture strain decreased with bite speed, because cheese shows viscoelastic characteristics. Sensory evaluation was conducted separately to characterise cheese texture in this study. The sensory characteristics were essentially consistent with the instrumental parameters. Cheese with high sensory firmness had great values in hardness and Young's modulus in the instrumental test. Although degree of melting in the mouth and chewiness are evaluated after mastication is started, these values related with hardness measured in the instrumental test.

Total masseter muscle activity can be used as be an index of chewing difficulty as it is related to muscle effort until the food is swallow.[10] We investigated the correlations between total muscle activity and all variables from the instrumental and sensory texture tests. It is confirmed that Kendall's correlation coefficient between the total masseter muscle activity and sensory score of chewiness was significantly high (Table 4). Total muscle activity was also highly correlated with Young's modulus, compressive load values at a fixed strain ranged less than 50% and 60–80% and fracture strain in the mechanical test. To compare samples C and M, which had similar hardness levels, the more adhesive C sample required greater mastication effort.

Combination of the sensory scores and EMG or mechanical parameters with significantly high correlation is listed in Table 4. Highest correlation with sensory firmness was found with compressive load at 90% strain (r = 1.000). It suggests that cheese firmness is sensed when it is compressed completely by the teeth rather than at smaller deformation. Fracture load may not influence the sensory attributes. Brittleness, adhesiveness and melting in the mouth were not significantly correlated with any of EMG and mechanical parameters. Because high chewiness and low springiness were observed in hard cheese, we could not analyse the influence of resilience separately from firmness.

Table 4 *Significantly high correlation between sensory attributes and EMG or mechanical parameters*

Sensory parameters	Correlated parameters
Firmness	Compressive load at 90% strain
Brittleness	
Adhesiveness	Melting in the mouth
Springiness	Chewiness, Number of chews, Total mastication time, T1, Total EMG duration, Total muscle activity, Muscle activity/chew, Duration at early stage, Muscle activity at early stage, Initial amplitude, Young's modulus, Fracture strain, Compressive load at 5–80% strain
Melting in the mouth	Sensory adhesiveness
Chewiness	Springiness, Number of chews, Total mastication time, T1, Total EMG duration, Total muscle activity, Muscle activity/chew, Muscle activity at early stage, Initial amplitude, Young's modulus, Fracture strain, Compressive load at 5–40 and 60–80% strain

Average values were used to calculate Kendall's correlation coefficient.
Springiness showed negative correlation.

We performed principal component analysis using the all parameters of mechanical testing, EMG variables and sensory scores. The first principal component (78.2%) related with many EMG, mechanical and sensory variables. Factor loading values of fracture strain, cohesiveness and springiness in the instrumental test and springiness and melting in the mouth in the sensory test were negative but those of many other parameters such as load at a defined compressive strain and hardness in the mechanical test and firmness, brittleness, chewiness in the sensory test were positive. The second principal component (13.7%) showed a large positive factor loading value in fracture force and a large negative value in TPA adhesiveness. Fracture properties had little effect on masticatory and sensory evaluated texture.

3 CONCLUSION

Using EMG variables recorded at each mastication stage, we related mechanical properties and sensory scores of textural characteristics. In general, EMG activities of both masseter muscles were greater in the first stage of mastication and lesser in the second stage. Contrary to the masseters, the suprahyoid muscles exhibited greater activity in the late stage of mastication.

Cheese samples with great hardness values in the mechanical test were evaluated firmer in sensory test. They had long mastication time, high EMG amplitudes and high muscle activity per chew. When the varieties with similar hardness were compared, those with higher adhesiveness and springiness exhibited greater EMG activities.
The hardness of commercial varieties of cheese correlated with many variables measured in this study, and it is difficult to choose a mechanical property, that is independent of hardness and still influences mastication. A further study using cheese models with a precisely defined texture is being planned.

Acknowledgment

This study was partly supported by the Grant-in-Aid for Scientific Research (B) 21300274 from the Japan Society for the Promotion of Science and by the Research and development projects for application in promoting new policy of Agriculture Forestry and Fisheries 22026.

References

1 J. Chen, *Food Hydrocolloids*, 2009, **23**, 1.
2 E. A. Foegeding, C. R. Daubert, M. A. Drake, G. Essick, M. Trulsson, C. J. Vinyard, F. van de Velde, *J. Texture Stud.*, 2011, **42**, 103.
3 F. R. Jack, J. R. Piggott and A. Paterson, *J. Food Sci.*, 1993, **58**, 1313.
4 R. González, S. Sifre, J. Benedito and V. Nogués, *J. Dairy Res.*, 2002, **69**, 151.
5 A. Terrega, C. Yven, E. Sémon and C. Salles, *Int. Dairy J.*, 2008, **18**, 849.
6 R. González, I. Montoya and J. Cárcel, *Food Sci. Technol. Int.*, 2001, **7**, 461.
7 K. Kemsley, M. Defernez, J. C.Sprunt and A. C. Smith, *J. Electromyogr. Kinesiol.*, 2003, **13**, 197.
8 A. Woda, K. Foster, A. Mishellany and M. A. Peyron, *Physiol. Behav.*, 2006, **89**, 28.
9 K. Kohyama, K. Ohtsubo, H. Toyoshima and K. Shiozawa, *J. Texture Stud.*, 1998, **29**, 101.
10 H. Sakamoto, T. Harada, T. Matsukubo, Y. Takaesu and M. Tazaki, *Agric. Biol. Chem.*, 1989, **53**, 2421.
11 K. Kohyama, H. Sawada, M. Nonaka, C. Kobori, F. Hayakawa and T. Sasaki, *Biosci. Biotechnol. Biochem.*, 2007, **71**, 358.
12 S. Ishihara, M. Nakauma, T. Funami, T. Tanaka, K. Nishinari and K. Kohyama, *J. Texture Stud.*, 2011, **42**, in press.
13 K. Kohyama, T. Sasaki and F. Hayakawa, *Biosci. Biotechnol. Biochem.*, 2008, **72**, 1690.
14 J. B. Hutchings and P. Lillford, *J. Texture Stud.*, 1988, **19**, 103.
15 M. C. Bourne, *Food Texture and Viscosity*, 2nd Edn., Academic Press, San Diego, 2002, Chapter 4, p. 107.
16 H. Dan, F. Hayawaka and K. Kohyama, *Appetite*, 2008, **158**, 158.
17 H. Dan and K. Kohyama, *Arch. Oral Biol.*, 2007, **52**, 455.
18 H. Dan, M. Okamoto, Y. Wada, I. Dan and K. Kohyama, *Physiol. Behav.*, 2007, **92**, 601.

THE TEXTURE OF LOW FAT YOGHURT PREDICTED FROM ITS CONTENT OF STARCH, PECTIN AND GELATINE

L. Knarreborg and H.L. Nielsen

CP Kelco, Ved Banen 16, DK-4623 Lille Skensved, Denmark

1 INTRODUCTION

Since it is very expensive and troublesome to train and maintain a sensory panel, there is a need for mechanical measurements that correlates with sensory yoghurt attributes. Several studies indicate limited correlations between sensory attributes and rheological parameters.[1-6] There is, however, some disagreement regarding which rheological methods to use. The shear rates ranging from relatively low values, 2 – 10s^{-1}, up to relatively high shear rates from 100 – 241s^{-1}, are all stated to correlate to mouth thickness, oral viscosity or oral perception.[1,2,3,5,6] But also elastic parameters as G' seem to correlate to oral viscosity and mouth coating attributes.[4,5] The general observation is that it is difficult to find good mechanical methods that describe or predict sensory attributes of low fat stirred yoghurt. In this paper correlations between rheological parameters and sensory attributes will be presented as well as correlations between yoghurt texturizer content, sensory attributes and rheological parameters.

2 MATERIAL AND METHODS

2.1 Yoghurt formulation

The study consists of 19 low fat yoghurt productions with various combinations of texturizers including 4 identical samples (Table 1). The yoghurt formulations were selected to represent formulations ranging from thin consistency (typical European style) to more heavy consistency (typical American style). Pasteurized, homogenized 0.5% fat milk, 2% skimmed milk powder (3.9% protein in final yoghurt), 7% sugar and IFF strawberry flavour were used. The yoghurt was fermented with YF-L903 from Chr. Hansen.

Starch, gelatine and pectin were selected as they are commonly used as yoghurt texturizer agents. Thermtex (**MS**) (modified starch) and Novation 2300 (**NS**) (native starch) are both derived from waxy maize. The starches are from National Starch and recommended for yoghurt application. Native starch is typically used in organic yoghurt. Starch is normally not added above a level of 2%. PBG 55 (230-bloom gelatine) (**GT**) is from PB Gelatine. Gelatine was dosed at 0 or 0.5%, which is a typical gelatine dosage level in yoghurt.

Table 1 *Composition of yoghurt samples.*

Sample #	Texturizer composition	Sample #	Texturizer composition
1	0.2% PA	9	1.25% MS
2	0.3% PA	10	1.75% MS
3	0.2% PA + 1.25% MS	11	1.25% NS
4	0.2% PA + 1.75% MS	12	0.75% MS + 0.5% GT
5	0.2% PA + 1.25% MS + 0.5% GT	13-16	1.25% MS + 0.5% GT (4 identical samples)
6	0.2% PB	17	1.75% MS + 0.5% GT
7	0.2% PB + 1.25% NS	18	0.5% GT
8	0.2% PB + 1.75% NS	19	0.5% fat

GENU® pectin type A (**PA**) is amidated low ester pectin and GENU® pectin type B (**PB**) is newly developed non-amidated low ester pectin. Both pectin types are designed for yoghurt application. Pectin is normally not dosed above 0.3% in yoghurt.

2.2 Yoghurt process

The dry ingredients were mixed with the milk, then hydrated for ½ hour at 20°C. The mixture was homogenized at 50°C at 50 bars, pasteurized at 85°C for 15 min, fermented at 42°C to pH 4.55, cooled to 20°C and filled and stored at 5°C until rheological and sensory measurements on day 2.

2.3 Rheological measurements

Different rheological methods were used to characterize the samples including non-destructive tests like low strain oscillation and destructive tests using back extrusion on Texture Analyzer.

Table 2 *Rheological readings: MCR301 Anton Paar Rheometer with bobcup table (CC17); Texture caption from Stable Micro Systems with back extrusion geometry with a 40 mm disc (A/BE40).*

<table>
<tr><th colspan="2">Creep and creep-recovery</th><th>Extracted Parameters</th></tr>
<tr><td colspan="2">20 points, stress = 0.25Pa
20 points log distr., stress=0Pa</td><td>G2 (secondary elastic component) (Pa)
(Figure 1)</td></tr>
<tr><th colspan="2">Strain sweep</th><th>Extracted Parameters</th></tr>
<tr><td>Time sweep</td><td>time=300s, f=1Hz, strain=0.2 %</td><td></td></tr>
<tr><td>Strain sweep</td><td>f=1Hz, strain=0,1...100 % log distr.</td><td>Calculated Yield Strain (CYS) (-)
(Rheoplus software)</td></tr>
<tr><th colspan="2">Consistency measurement</th><th>Extracted Parameters</th></tr>
<tr><td>Back extrusion</td><td>speed=0.5mm/s, distance=30mm, post speed=10mm/s</td><td>Firmness (F) (g),
Cohesiveness (C) (g)</td></tr>
</table>

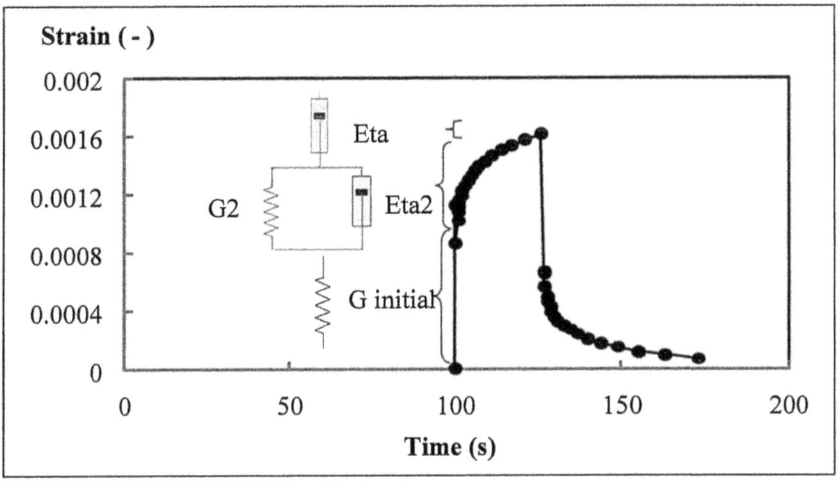

Figure 1 *The rheological parameters that are extracted from the creep and creep-recovery measurements: G initial (Pa), G2 (Pa), Eta2 (Pas), Eta (Pas), by fitting the curve to a Maxwell and Kelvin elements for low fat stirred yoghurt sample (1.25% MS + 0.5% G).*

Although multiple parameters were extracted four parameters have been selected to be presented in this paper, listed in Table 2.

Consistency measurements were made by filling 100 g yoghurt in a beaker, leaving it undisturbed for 1 hour and then measuring on Texture Analyzer. Yoghurt samples for the rheometer were loaded into the geometry and left for 5 minutes before the measurement was done. The yoghurt samples had a temperature of 13°C when measured.

2.4 Sensory evaluation

The sensory evaluation was performed by the Sensory Laboratory at Department of Sensory Science, Faculty of LIFE Sciences, University of Copenhagen. The panel consisted of 10 panel members. The panel trained for one week evaluating the same type of yoghurt as was evaluated during the study. The purpose of the training was to have the panel select sensory attributes to describe the yoghurt as well as to make the panel familiar with the attributes and the use of the scale. Each product was evaluated three times by each panellist. The temperature of the yoghurt was 13°C. All evaluations were made on a 15 cm unstructured line scale, where 0 was low and 15 high.

The panel came up with 32 different attributes which they found relevant to characterize the yoghurt. The attributes could be categorized into 7 groups: Visual, Physical evaluation, Texture in Mouth, Flavour, After Taste, Physical evaluation of texture after stirring, Texture in Mouth after stirring. Out of these 6 sensory attributes were selected for presentation in this paper (Table 3). They were selected because they represent important characteristics when evaluating the performance of texturizers in yoghurt. Each of these 6 attributes correlates with many of the attributes which are not presented here.

Sensory results are subjected to a number of sources of variation due to for example difference between judges in sensitivity for certain attributes and differences in scale use. This source of variation is reduced by averaging the sensory scores obtained from all assessors for each sample and attribute.

Table 3 *Selected sensory attributes and their sensory category*

Sensory category	Sensory attributes
Physical evaluation	Sharpness of edge: How sharp is the edge in the yoghurt when a spoonful is taken?
Texture in mouth	Oral Viscosity: How thick does the sample feel in the mouth?
	Smoothness: How smooth does the sample feel in the mouth?
	Stickiness: How sticky does the sample feel in the mouth?
Physical evaluation of texture after stirring	Stirred Sharpness of Edge: How sharp is the edge left in the sample when a spoonful is taken after stirring yoghurt10 revolutions with a plastic spoon?
Texture in mouth after stirring	Stirred Oral Viscosity: How thick does the sample feel in the mouth after stirring the yoghurt10 revolutions with a plastic spoon?

2.5 Data analysis

MODDE v.8.0 from Umetrics was used for analysing sensory and rheological data. Multiple linear regressions (MLR) were applied and the regression coefficient (R^2) provides the goodness of fit of the model. The plots displayed in Figures 2 and 3 show the scaled and centred coefficients of the selected response when changing the factor from 0 to high, with a 95% confidence interval as error bars. The models are listed in tables 4, 5 and 6.

3 RESULTS AND DISCUSSION

3.1 Link between yoghurt texturizer content and rheology results

The texturizers influence the rheological parameters differently, making it a great puzzle to tailor the texture of low fat yoghurt; however, it also provides a great number of possibilities to alter the rheological properties by changing the texturizer composition.

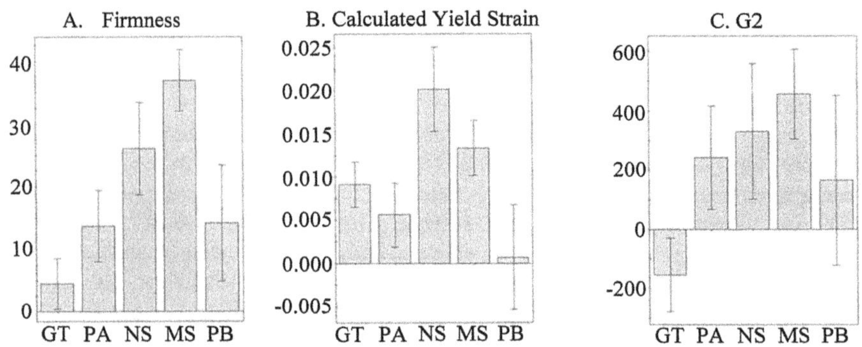

Figure 2 *Scaled & centred coefficients for* ***A****. Firmness,* ***B****. Calculated Yield Strain and* ***C****. G2 (secondary elastic modulus).*

Table 4 *Equations showing the links between texturizer content (%) of yoghurt and rheology parameters and the regression coefficients (R^2). Insignificant factors are underscored.*

Rheological parameter	Model for predicting rheology parameters from yoghurt texturizer content	R^2
Firmness (g)	19 [GT] + 91.3 [PA] + 94.7 [PB] + 29.8 [NS] + 42.2 [MS] + 26.3	0.959
CYS (-)	0.037 [GT] + 0.037 [PA] + 0.005 [PB] + 0.023 [NS] + 0.015 [MS] + 0.022	0.938
G2 (Pa)	- 616 [GT] + 1618 [PA] + 1109 [PB] + 377 [NS] + 522 [MS] + 290	0.826

Three different rheological parameters have been selected to show how the texturizer composition influences these.

Firmness is the maximum force measured when pressing a piston down into the yoghurt, and as seen in Figure 2A the Firmness is mainly driven by addition of starch, but also pectin provides a significant contribution to the Firmness of yoghurt. Gelatine, however, only provides a small contribution and is barely significant.

The Calculated Yield Strain (CYS) is less affected by pectin (Figure 2B). Gelatine plays a major role together with starch in providing stability when subjected to deformation, as larger deformations (CYS) can be applied before rupture.

The secondary elastic component (G2) from the creep experiment (Figure 1) can be interpreted as the rigidity holding together the structure as the sample is subjected to force as when the tongue is applying pressure on the yoghurt. G2 seems to be negatively affected by gelatine whereas both pectin and starch contribute to the parameter (Figure 2C). Contrary to starch and pectin, gelatine provides a weak structure which easier collapses when force is applied to the yoghurt. The models linking yoghurt composition and rheological parameters are listed in table 4.

3.2 Link between yoghurt texturizer content and sensory yoghurt attributes

All the tested texturizers have an impact on "Sharpness of Edge"; however, modified starch and gelatine have the largest impact (Figure 3A).

Pectin B has more effect than Pectin A on "Sharpness of Edge". Starch and pectin have an impact on "Stirred Sharpness of Edge" (Figure 3B) while gelatine has a small negative but significant influence on this attribute. Interestingly, the structure which gelatine creates in the yoghurt is lost when the yoghurt is stirred. The yoghurt consumer will recognize this when eating a gelatine containing yoghurt as the yoghurt becoming visually thinner (less "Sharpness of Edge") the more the consumer has stirred or scooped into the yoghurt during consumption. Starch and pectin are more shear resistant than gelatine and they lose only part of their structure when stirred.

All texturizers investigated have an impact on the "Oral Viscosity" of yoghurt (Figure 3D). The largest impact comes from modified starch. In practice this means that if the yoghurt producer is seeking a high oral viscosity then starch in most cases will be one of the yoghurt texturizers added. Pectin A contributes less to "Oral Viscosity" than Pectin B. Gelatine contributes more to "Oral Viscosity" than pectin.

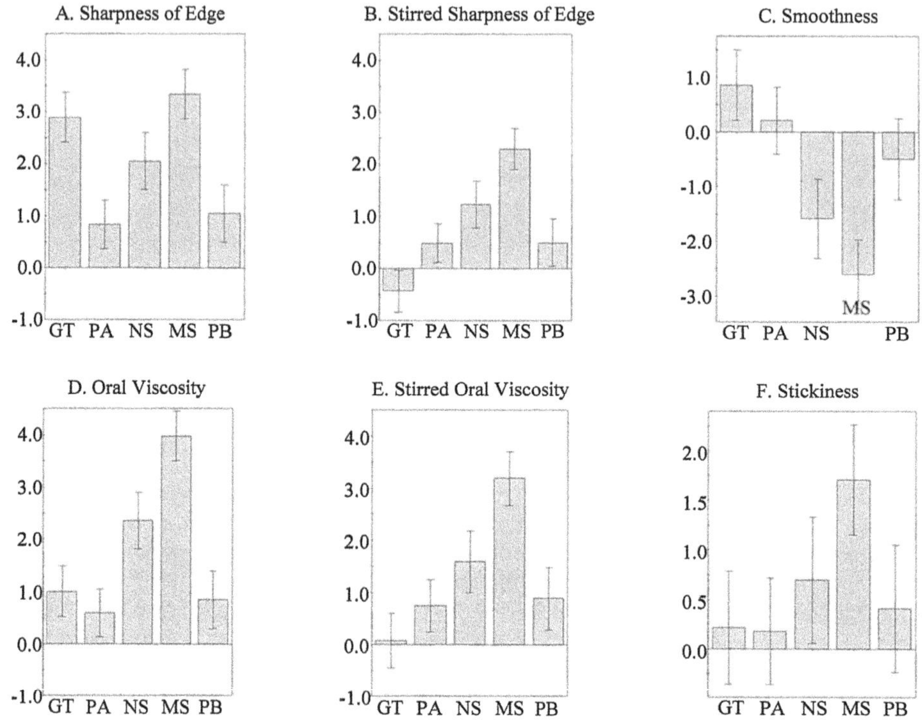

Figure 3 *Scaled & centred coefficients.* ***A****. Sharpness of Edge,* ***B****. Stirred Sharpness of Edge,* ***C****. Smoothness,* ***D****. Oral Viscosity,* ***E****. Stirred Oral Viscosity and* ***F****. Stickiness.*

It is seen that the effect of gelatine on "Stirred Oral Viscosity" becomes small and insignificant after stirring the sample (Figure 3E), indicating that the structure provided by gelatine is broken down after mechanical stirring of the sample, while starch and pectin stabilized yoghurt are more shear stable.

Starch lowered the smoothness of the yoghurt samples evaluated significantly (Figure 3C). Gelatine has a significant, however small, positive influence on smoothness. Pectin does not have any significant impact on smoothness.

Only starch has an impact on the yoghurt's stickiness and particularly modified starch increased the stickiness of the yoghurt (Figure 3F).

Often a yoghurt producer needs to combine the benefits of several texturizers to fulfil multiple needs which besides texture enhancement include improving storage and transport stability as well as syneresis control. Individually these texturizers provide certain attributes, but using them together can create even more possibilities. Starch is very good at increasing the oral viscosity of yoghurt, but it also increases the stickiness and reduces the smoothness of the yoghurt. It is also well known that starch is masking the flavour of the yoghurt particularly at higher dosages. These findings have been confirmed by our work, however, this is not shown here. Oral viscosity enhancement above a certain point could be obtained with the addition of other texturizers than starch, like gelatine or pectin, in order to counteract the negative impact of starch.

Table 5 *Equations showing the links between texturizer content (%) of yoghurt and sensory attributes and the regressions coefficients (R^2). Insignificant factors are underscored.*

Sensory attribute	Models for predicting sensory attributes from the texturizer content	R^2
Sharpness of Edge	11.4 [GT] + 8.2 [PA] + 14.0 [PB] + 3.8 [NS] + 4.7 [MS] – 0.5	0.975
Oral Viscosity	3.9 [GT] + 5.8 [PA] + 11.3 [PB] + 4.4 [NS] + 5.5 [MS] + 2.0	0.969
Stickiness	0.8 [GT] + 1.8 [PA] + 5.4 [PB] + 1.3 [NS] + 2.4 [MS] + 3.5	0.789
Smoothness	3.4 [GT] + 2.1 [PA] - 6.6 [PB] - 2.9 [NS] - 3.6 [MS] + 11.6	0.884
Stirred Oral Viscosity	0.3 [GT] + 7.3 [PA] + 11.8 [PB] + 2.9 [NS] + 4.5 [MS] + 1.7	0.936
Stirred Sharpness of Edge	-1.7 [GT] + 4.8 [PA] +6.7 [PB] + 2.3 [NS] + 3.2 [MS] + 0.5	0.931

Gelatine influences the sensory yoghurt attributes differently than starch and pectin, as it impacts the gel structure ("Sharpness of Edge") and mouthfeel ("Oral Viscosity") significantly. However, it is much less shear stable than starch and pectin. The two pectin products, Pectin A and B, perform differently in yoghurt. Pectin B provides the yoghurt with a higher physical structure ("Sharpness of Edge") and a higher "Oral Viscosity".

The models linking the sensory attributes with the yoghurt texturizer composition are listed in table 5 where high R^2 scores are seen.

3.3 Link between sensory yoghurt attributes and rheology results

As it is time-consuming and expensive to maintain a sensory panel to evaluate yoghurt, it would be beneficial to be able to predict sensory attributes from rheological parameters.

The maximum force of pressing a plunger down into the yoghurt may very well be a good estimate of the tongue moving through the yoghurt and pressing against the palate, and both visual viscosity (Stirred Sharpness of Edge) and oral viscosity (Oral Viscosity and Stirred Oral Viscosity) are predicted very well from Firmness (Table 6).

Firmness alone, however, was not able to predict "Sharpness of Edge", as the gelatine containing samples clearly differed from the other samples (Figure 4). It makes sense that Firmness was not able to predict this effect from gelatine seen in Figure 2A, as gelatine contributes only slightly to Firmness, whereas CYS is able to detect the effect from gelatine (Figure 2B) and was able to predict "Sharpness of Edge" quite well. Gelatine provides yoghurt with a structure observed as "Sharpness of Edge", but this structure is not to the same degree detected when the yoghurt is evaluated in the mouth observed as "Oral Viscosity". This is also supported by Figures 3A and 3D which show that gelatine affects "Sharpness of Edge" much more than "Oral Viscosity".

Figure 4 shows that the observed "Sharpness of Edge" and "Oral Viscosity" is reduced for all samples when sheared, however, the gelatine containing samples particularly lose their "Sharpness of Edge" when sheared.

"Stickiness" correlates with the cohesiveness from TA.XT measurements which determines the maximum negative force when pulling the piston backwards after being submerged in the yoghurt, similar to pulling the tongue back from squeezing the yoghurt between the palate and the tongue.

Table 6 *Models showing the links between rheological parameters and sensory attributes and the regression coefficients (R^2). CYS: Calculated Yield Strain; F: Firmness TA.XT; C: Cohesiveness; G2: Secondary elastic component.*

Sensory attribute	Model for predicting sensory attributes from rheological parameters	R^2
Sharpness of Edge	259.3 [CYS] – 5.35	0.82
Stirred Sharpness of Edge	0.071 [F] – 1.93	0.89
Oral Viscosity	0.34 [F] – 0.0014 [F]^2 – 8.37	0.95
Stirred Oral Viscosity	0.19 [F] – 0.00058 [F]^2 – 4.15	0.97
Stickiness	0.038 [C] + 2.73	0.86
Smoothness	- 0.015 [G2] + 4.997 [G2]^2 + 15.07	0.85

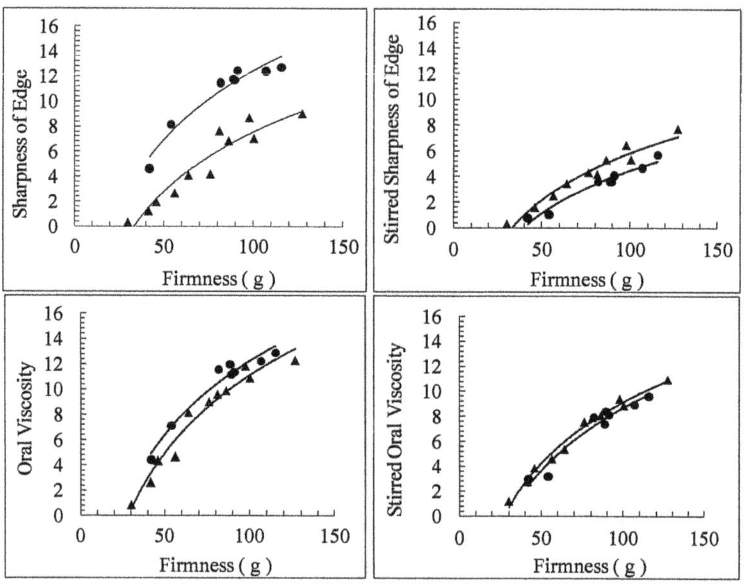

Figure 4 *"Sharpness of Edge" and "Oral Viscosity" before and after stirring as a function of Firmness (TA.XT). Gelatine containing yoghurt samples are marked with circles, whereas yoghurt samples not containing gelatine are marked with triangles.*

"Smoothness" is best predicted from G2, indicating that the lower the secondary elastic modulus, the smoother the yoghurt appears in the mouth. G2 is a measure of how much the structure is kept intact when force is applied.

4 CONCLUSION

Models linking the yoghurt texturizer content to rheological parameters (Table 4) as well as sensory attributes of low fat yoghurt (Table 5) were developed with very high correlation, explaining 89 – 98% of the variation in data. These models make it possible to predict rheological parameters and sensory attributes of low fat stirred yoghurt when composition is altered among the texturizers (gelatine, pectin and starch).

Furthermore, models were generated with high prediction rate linking rheological parameters to 6 sensory attributes, providing a great tool for estimating sensory parameters from mechanical measurements (Table 6). It is clear that one rheological parameter is not enough for predicting all sensory attributes of low fat stirred yoghurt. The choice of texturizer will influence different attributes and rheological parameters in various ways. This provides a toolbox for tailoring the best texture solution for the consumer.

Results show that instrumental measurements are a useful alternative to sensory analysis and can greatly facilitate quality control of commercial yoghurt as well as making feasible high-throughput screening of novel products.

Acknowledgements

It is a pleasure to acknowledge our colleagues at CP Kelco's technology group for the yoghurt preparation and rheological measurements as well as providing useful discussions along the way.

The Sensory Laboratory at Department of Sensory Science, Faculty of LIFE Sciences, University of Copenhagen is acknowledged for their sensory measurements.

References

1 N. Cutler, E. R. Morris and J. L. Taylor, *J. of Texture Studies*, 1983, **14**, 377.
2 D. M. Folkenberg, P. Dejmek, A. Skriver, H. S. Guldager, and R. Ipsen, *International Dairy J.*, 2006, **16**, 111.
3 T. Janhøj, C. Blangsted Petersen, M. B. Frøst and R. Ipsen, *J. of Texture Studies*, 2006, **37**, 276.
4 W. –J. Lee and J. A. Lucey, *J. of Dairy Science*, 2006, **89**, 2374.
5 A. Skriver, J. Holstborg, and K. Qvist, *J. of Dairy Research*, 1999, **66**, 609.
6 F. W. Wood, in *Psychophysical studies on the consistency of liquid foods. In SCI Monograph No. 27. Rheology and Texture of Foodstuffs.* Society of Chemical Industry, London, 1968, p. 40.

Food applications

CONTRIBUTIONS OF THE PARTICULATES AND SOLUBLE MATERIALS TO THE VISCOSITY BEHAVIOUR OF TOMATO PUREE

Rachael Abson, Bettina Wolf, Sandra Hill and Tim Foster

Division of Food Sciences, Sutton Bonington Campus, University of Nottingham, Loughborough, Leics LE12 5RD, UK

1 INTRODUCTION

Tomato puree is a suspension of particulates within a serum prepared by breaking down the fruit tissue during a pureeing process. During ripening of tomato fruits, many changes occur in the cell walls and these reflect on tissue strength. The cell walls are a composite structure of cellulose, hemicelluloses and pectin, [1] and individual cells are adjoined to each other with a pectin rich layer, the middle lamella. The firmness of the tissue is determined in part by the strength of the cell wall composite as well as the intracellular adhesion. During the production of puree, the tissue is broken down into particulates which may be clusters of cells, individual cells and cell fragments. These are found suspended in a serum of soluble pectin and simple sugars released from the tomato tissue during pureeing. Both firmness and viscosity are affected by the genetic make-up of the plant and stage of ripeness of the fruit. In unripe tomato fruit, the cell walls are relatively rigid and cell to cell adhesion is strong. It has been well established that during ripening, cell wall disassembly, and breakdown and dissolution of the middle lamella occur due to enzymes and ultimately the genes that control the functioning of these enzymes.[2-4] Ripening of tomato fruit is also associated with major changes in the pectin of the cell walls, and this has previously been shown to have an effect on the rheological properties of tomato purees.[2]

The effect of tomato ripeness on the material properties of the particulates and the soluble material and how these consequently affect the viscosity of the puree is not fully understood. The aim of this research was to establish the link between fruit firmness and viscosity of the puree. The material properties of the soluble and non-soluble fractions in purees prepared from tomatoes at different stages of ripening were characterised and the contribution of these fractions to tomato puree viscosity quantified.

2 MATERIALS AND METHODS

Commercial tomatoes were donated by Evesham Vale Growers (Evesham, UK). They were separated by colour into four groups namely; mature green, breaker (turning from green to slightly pink), pink and red ripe. Fruit colour was then measured using the Minolta Chroma meter CR-400 (Konica Minolta, Osaka, Japan) with the a* chromaticity coordinate indicating a colour change with values negative to positive expressing the colour change from green to red . To assess firmness of the fruit, maximum load values were recorded using the Ametek Lloyd LF PLUS texture analyser (Fareham, UK) equipped with a 10 N load cell on a 6 mm transverse section cut from each fruit (ten fruits at each stage of ripeness, two replicates per fruit). A 1.6 mm diameter cylindrical probe was moved 4 mm into the pericarp at 10 mm/min and load recorded throughout. The maximum load represents the greatest load (N) required to cause failure of the tissue integrity, with a higher maximum load indicating a firmer fruit. Tomatoes were then washed, halved to remove seeds and locular tissue and stored frozen at -80 °C until puree could be made.

To make the puree, the tomatoes were peeled frozen and the pericarp tissue was cut into approx 1cm^3 cubes and then microwaved at 800 W. The temperature was checked to be 95 °C +/- 2 °C and enzymic activity assumed inhibited by this treatment. Samples were cooled to room temperature, and pureed with a hand held blender for three minutes.

The total solids of the puree samples were determined by oven drying for 18 hours at 105 °C. Brix was measured using an Atago PAL-1 digital pocket refractometer (Atago, Tokyo, Japan). A Rapid visco analyser RVA Super 4 (Newport Scientific, Sydney, Australia) was used to obtain a stirred viscosity value at 25 °C on 25 g of sample at 180 rpm which equates to a characteristic shear rate across the stirrer of 60 s^{-1} [6]. A rotational rheometer (MCR301, Anton Paar, Graz, A) was used to obtain viscosity values at 25°C and 10 s^{-1} using a serrated parallel plate geometry (diameter 25 mm, gap height 1mm) and a double concentric cylinder geometry for the pellet and the serum, respectively. Results are shown as relative viscosity, η_r. Particle size distributions of the puree samples were measured using a laser diffraction particle sizer (Beckman Coulter LS-13320, Meritics, Dunstable, UK) and the data analysed with the instrument's software choosing the Fraunhofer diffraction model. To separate the particulates from the serum, tomato puree was centrifuged at 2700 g and 25 °C for 60 minutes. The supernatant was carefully poured out to separate from the pellet. The particulate to total sample volume ratios were calculated from this. All measurements were carried out at least in duplicate. A one way ANOVA using Design Expert (Version 6.011 Stat-Ease, Inc Minneapolis, USA) was used for the statistical analysis.

3 RESULTS

Properties of whole tomato and purees were measured (**Table 1**). It is evident that the stage of ripeness, quantified by a* values, had a major effect on the firmness of the fruit tissue indicated by the tissue maximum load. The stirred viscosity of the puree also decreased on ripening. The total solids content and pH (titratable acidity) have been reported to alter as the fruit ripens [7,8], but in this study the levels were not clearly related to the ripeness of the fruit. Total solids also do not correlate to stirred viscosity of the puree or firmness of the fruit. As the tomato ripens the water soluble content (often assessed as Brix) increases[8] and this was observed, from mature green through to red ripe.

Purees from breaker and red ripe tomatoes were diluted with water to a range of total solid levels and stirred viscosity measured (**Figure 1**). The results indicate that at the same total solids the viscosities were greater for the breaker tomato than the red ripe. Ripeness was therefore shown to affect viscosity, and the solids level had some influence on viscosity. This indicates that the molecular assemblies of hydrocolloids within the total solids may be different during the ripening stage, and has an effect even after the tomato tissue is macerated.

Table 1 *Properties of the tomato samples before and after pureeing with (standard deviations) shown*

Ripeness	**Mature green**	**Breaker**	**Pink**	**Red ripe**
a* (colour of tomato skin)	-10.3 (1)	-3.2 (2.6)	7.6 (1.9)	19.8 (2.6)
Fruit tissue texture, Max. Load N	2.3 (0.3)	1.7 (0.1)	0.9 (0.4)	0.5 (0.11)
Puree stirred viscosity cP [x]	663 (42)	475 (66)	254 (41)	191 (22)
Puree total solids %	5.4 (0.09)	5.7 (0.01)	5.0 (0.06)	5.7 (0.05)
Puree brix %	4.3	5.0	4.5	5.5
Puree pH	4.6	4.3	4.3	4.3
Particulate: total sample volume ratio	0.28 (0.03)	0.24 (0.02)	0.18 (0.01)	0.30 (0.01)

[x] Samples shown to be significantly different p>0.05

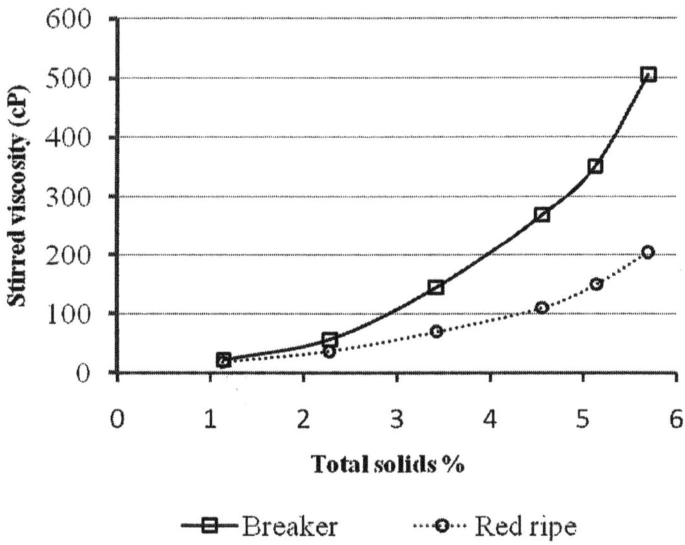

Figure 1 *Effect of total solids and stage of ripeness on stirred viscosity of tomato puree, diluted with water*

Table 2 *Effect of liquid phase (68% by weight) on stirred viscosity of pellet (32% by weight) obtained from red ripe tomato puree*

Liquid phase	**Total solids of liquid phase%**	**Stirred viscosity cP (standard deviation)***
Serum	4.4	212 (4.8) [a]
Sucrose solution	4.4	211 (10.4) [a]
Water	0.0	193 (2.0) [b]

* Samples with same letter are not significantly different p>0.05

Tomato puree contains particulates and a serum of soluble material (small sugars and hydrocolloids). To observe the effect of the serum on tomato puree viscosity the pellet fraction was spun out of the red ripe tomato puree by centrifugation. This pellet was resuspended at a ratio of 32:68 % by weight pellet: liquid (serum) phase, with either 4.4% w/w sucrose solution, water or the original serum. The concentration of the sucrose solution was chosen to match the solids content of the serum. The stirred viscosity of these three samples was measured (**Table 2**). The fact that stirred viscosity is independent of whether the suspension medium constitutes sucrose solution or serum phase shows that little contribution to the viscosity is coming from the soluble hydrocolloids in the tomato sera. The viscosity was lower in the samples resuspended in water, a fluid of lower osmotic strength then the other two suspending fluids, indicates that the osmotic strength of the suspending medium does have an influence. It is hypothesised that the lower viscosity is due to contraction of the particles.

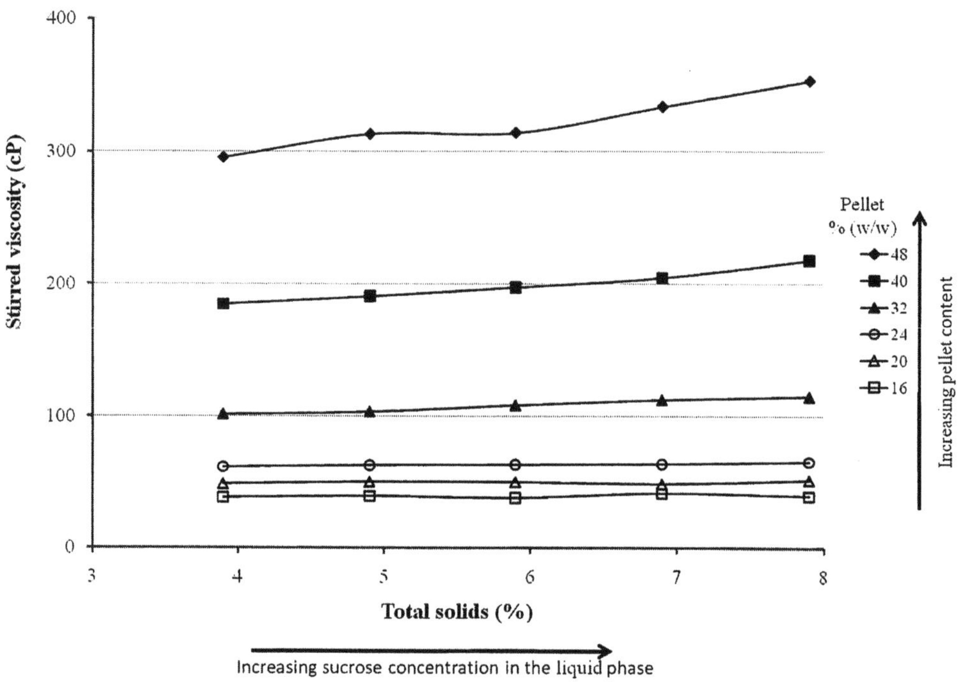

Figure 2 *Effect of total solids, concentration of pellet (16-48%) and sucrose concentration in the sera on stirred viscosity*

To gain further understanding of the effect of particulate (pellet) concentration and interaction between particles and soluble solids, the pellet from the red ripe tomato was diluted with sucrose solutions. The concentration of sucrose and amount of pellet were adjusted to maintain the overall total solids of the samples between 3.9 and 7.9%. Stirred viscosity of the resulting samples was measured (**Figure 2**). For lower sucrose levels contributing to the total solids content, and increasing particle concentration (> 24% and most noticeable at the highest pellet content of 48%), suspension viscosity was affected by the sucrose content of the serum. However, with all sucrose concentrations measured, the amount of pellet was the main source of the viscosity.

The particulates are obviously key to the viscosity of the puree. As viewed under a light microscope (**Figure 3**), it is observed that the particles are mainly single and fragmented cells, with the mean diameter of a single tomato cell at approximately 400 μm in diameter. Quantitative data for size distribution (**Figure 4**) show that as the tomato ripened from green to red, particles were smaller on average and of a broader size distribution. In the red ripe sample more of the particles were broken cell fragments, perhaps due to the disassembly of polysaccharide structures in the cell wall making them more susceptible to breakdown during pureeing. The size distribution for the mature green puree extends to sizes much larger than the diameter of a single cell (~ 400 μm) indicating that some clusters of cells remained in the puree due to strong cell to cell adhesion of the middle lamella pectins. In a previous study using plant based suspensions similar to tomato puree [9], a large particle size distribution has been associated with a lower viscosity, whereas larger particles caused increased viscosity. The results shown in this chapter indicate a similar trend.

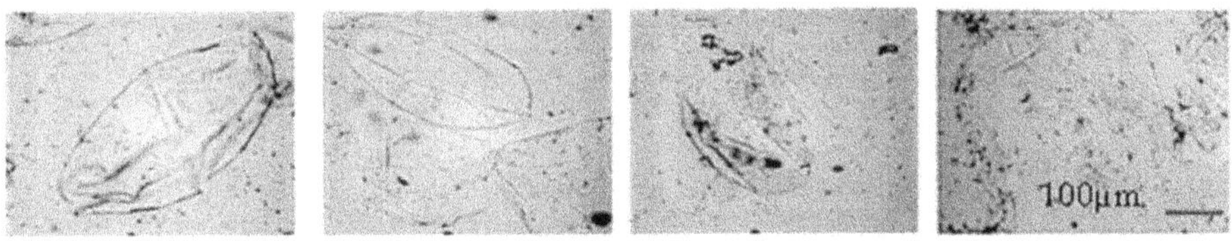

Figure 3 *Micrographs of mature green, breaker, pink and red ripe tomato puree (left to right)*

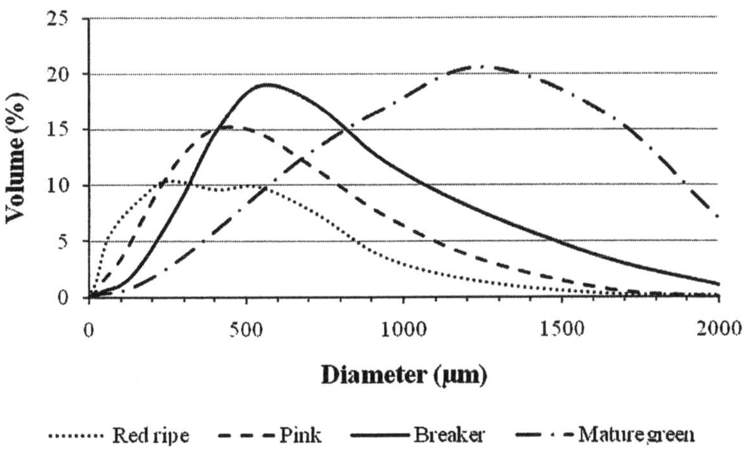

Figure 4 *Particle size distributions of pure samples as determined by laser light scattering*

Apart from their size and shape, it is important to characterise the packing and deformability of the particles. Information about these can be obtained from centrifugation measurements and by calculating the ratio of particulate volume relative to the total sample. The trend for the particulate to sample volume ratio **(Table 1)** does not directly reflect the viscosity decrease with ripening of the fruit. This could suggest that the viscosity of the puree is not wholly dependent on the particle volume. Similar results were also observed by Errington[2] where viscosity values and particle volume of genetically modified tomatoes with suppressed ripening related pectic enzymes were reported. Another reason for this variation could be that the sedimentation of particulates is dependent on several factors such as liquid viscosity, particle size and density, fraction of dissolved solids and gravitational force applied to hydrated particles.

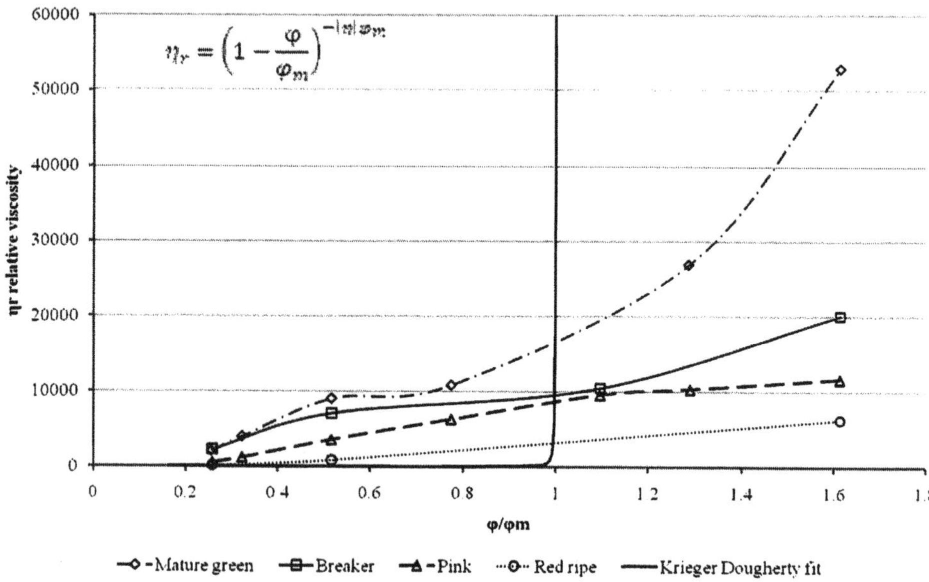

Figure 5 *Relative viscosity η_r of pelleted material from tomato puree compared with the Krieger Dougherty fit (see insert in graph for equation)* $\eta_r = \left(1 - \frac{\varphi}{\varphi_m}\right)^{-|\eta|\varphi_m}$ *for solid spheres whereby with $\varphi_m = 0.62$ and $[\eta] = 2.5$. Note that relative particle phase volume for pelleted material was calculated with $\varphi_m = 0.62$.*

However, viscosity will be affected by the mechanical properties of the particles especially as the volume fraction of the dispersed phase increases towards the possible maximum. This was assessed by preparing suspensions of increasing particle volume fraction and determining relative viscosity at 10 s^{-1} (parallel plate geometry) (**Figure 5**). It should be noted that the viscosity data were to some extent influenced by the choice of gap height in the parallel plate geometry. However, the effect was not of several orders of magnitude to invalidate the finding that the puree particles are deformable and behave quite differently from what would be expected for a solid sphere suspension. The relative viscosity behaviour for a suspension can be modelled with the Krieger Dougherty [10] equation (see insert in Figure 5), and the behaviour for solid spheres assuming a maximum packing volume of $\varphi_m = 0.62$ and for which the intrinsic viscosity = 2.5 has been included with the experimental data for tomato puree based suspensions. The differences are striking and much higher relative viscosities as well as packing fractions as indicated by the packing fraction ratio φ/φ_m exceeding values of unity achieved. Similar results were obtained using a vane geometry (data not shown). At low particulate concentrations, pellets from different sources of tomatoes seem to fit within the same range of viscosities, but at higher concentrations the values are very different, this may imply that it is a change in the robustness/ solidity of the particles that is the biggest change through ripening, which reflected the firmness of the flesh before pureeing.

4 CONCLUSIONS

The contribution of the particulates and soluble material to tomato puree viscosity has been studied with reference to the varying composition of solids and firmness of tomato fruit at four stages of ripeness; mature green, breaker, pink and red ripe. When purees from the red ripe and the breaker fruit were diluted to a range of total solids content, both samples decreased in viscosity. On further investigation, the viscosity of the puree was shown to be affected only to a small degree by the viscosity of the serum. The absence of soluble sugars within the (serum) liquid phase caused a slight decrease in viscosity, possibly due to osmotic potential on the particles (**Table 2**). Most dominant for viscosity was the particulate fraction. The particulate fraction was affected by the maturity of the fruit which in turn was associated with changes in texture, particulate size and size distribution of the puree. Addition of more of the particulate (pellet) fraction increased viscosity, but the particles do not behave as rigid spheres, as shown by the large departure from the corresponding Krieger Dougherty fit (**Figure 5**); ripened fruit deviating the furthest. This suggests the more ripened fruit has softer particles, which was shown to have softer tissue before the pureeing process. This may also have made the tissue more susceptible to rupture into smaller particles during pureeing.

Consumers prefer the flavour and colour of ripened tomatoes, but pureed green unripe fruit have been shown to offer more potential as an effective natural viscosifier for foods such as soups and sauces. New varieties of tomato can be developed with the study of fruit from the *Lycopersicon pennellii* introgression library where cultivated tomatoes have been crossed in a highly specific way with a wild type tomato, *Lycopersicon pennellii* [11]. Fruit displaying desirable characteristics (e.g. increased viscosity) when ripe can then be further analysed to determine the responsible gene(s). Future studies in this line of research will therefore focus on tomato with variations in their genes rather than ripeness.

References

1 N.C. Carpita and D.M. Gibeaut, *Plant J.*, 1996, **3**, 1.
2 N. Errington, G.A. Tucker and J.R. Mitchell, *J. Sci. Food Agric.* 1988, **76**, 515.
3 D.A. Brummell and M.H. Harpster, *Plant Mol. Biol.*, 2001, **47**, 311.
4 G.B Seymour, I.J. Colquhoun, M.S. DuPont, K.R. Parsley and R.R. Selvendran, *Phytochemistry*, 1990, **29**, 725.
5 A.L.T. Powell, M.S. Kalamaki, P.A. Kurien, S. Gurrieri and A.T. Bennett, *J. Agric. Food Chem.*, 2003, **51**, 7450.
6 K.P. Lai, J.F. Steffe and P.K.W. Ng, *Cereal Chem.*, 2000, **77**, 714.
7 G.W. Winsor, J.N. Davies and D.M. Massey *J. Sci. Food Agric.*, 1962, **10**, 304.
8 A.A. Kader, A. Stevens, M. Albright-Holton, L.L. Morris and M. Algazi, *J. Amer. Soc. Hort. Sci.*, 1977, **6**, 724.
9 A.C.K. Sato and R.L. Cunha, *J. Food Eng.*, 2009, **91**, 566.
10 I.M. Krieger and T.J. Dougherty, *Trans Soc Rheology* 1959, 3, 137.
11 Y. Eshed and D. Zamir, *Genetics*, 1995, **141**, 1147.

INCLUSION OF OAT FLOUR IN THE FORMULATION OF REGULAR SALTED DRIED NOODLES AND ITS EFFECTS ON DOUGH AND NOODLE PROPERTIES

Mahsa Majzoobi*, Behnaz Layegh, Asgar Farahnaky

Department of Food Science and Technology, College of Agriculture, Shiraz University, Shiraz, Iran. *Tel. +987116138356; Fax +98711-6289017; e-mail: majzoobi@shirazu.ac.ir

1 INTRODUCTION

In many Asian countries, noodles have been used as a staple food for many years. The main ingerdients of noodles are wheat flour, water and salt. Today, most of the noodles are produced by machine, while hand-made noodles are still available and common in many parts of the world. In both methods (i.e. using hand or machine) the process usually involves mixing of the ingredients to make a uniform dough, sheeting and cutting the dough to different lengths. Depending on the type of noodles, dough strands can be consumed fresh or they may be dried, steamed, boiled, frozen, steamed and deep fried and steamed and hot air dried. Accordingly, the type of noodle processing before consumption has been used as a method for classification of the noodles. The formulation and appearance of the noodles varies in different parts of the world due to regional eating habits, taste preference and advances in technology which has made it difficult to classify noodles. Based on the salt content, noodles are classified as regular salted and alkaline noodles. In Iran, noodles are gerenally produced from white wheat flour, 2-8% salt and water and can be classified as dried regular salted noodles with flat a surface. These type of noodles are also very common in Japan, China and Korea.[1] Normally, flours of low extraction rate are used in noodle production since the appearance of dark bran specks in the noodles is undesirable. Such flours contain very low dietary fiber and hence noodles are considered as low fiber foods. Therefore, increasing the fiber content of the noodles can improve their nutrition value and have positive effects on human health.

Amongst different cereals, oats (*Avenal sativa* L.) contains a high amount of nutritive components such as dietary fiber (especially β-glucan), proteins, unsaturated fatty acids, vitamins, minerals, and photochemicals. Therefore, the use of oats in the human diet is increasing. Oat β-glucan, the main non-starch polysaccharide of the grain, is well recognized as a functional component. The amount of β-glucan in different varieties of oats varies between 2.3 to 8.5%. Consumption of foods rich in β-glucan significantly reduces serum cholesterol levels and low-density lipoproteins (LDL) as well as reducing the risk of colon cancer. Because of numerous health benefits of β-glucan, the Food and Drug Administration

(FDA 2001), has allowed health claims for foods that contain more than 0.75 g of oat-derived β-glucan for each serving portion and can reduce the risk of heart disease (Wood 1994).[2] In recent years, a number of research studies have been performed to include oat flour or its derivatives in production of different types of noodles.[3,4,5] The main objectives of this study were to include oat flour in production of dried regular salted noodles, determine the appropriate level of oat flour and to study the effects of different levels of oat flour on the physical properties of the dough and regular salted noodles before and after cooking in distilled and salted water.

2 MATERIAL AND METHODS

Oat grains were gifted from Isfahan University of Technology, Isfahan, Iran. Wheat flour with extraction rate of 75% (weight of flour obtained from 100 units of clean wheat kernels) and average particle size of about 130 μm was obtained from Sepidan Milling Factory, Zarghan, Iran. Iodine free table salt (NaCl) was purchased from local market. Other chemicals used in this study were purchased from Merck (Darmstadt, Germany).

2.1 Production of oat flour

In order to produce oat flour, first the grains were cleaned by hand to remove impurities. The grains were then heated in a jacket pan at atmospheric pressure at 75 °C for 30 min to facilitate de-husking. The grains were de-husked using a de-husking machine (Strong-Scott, 17809S-S, Germany) followed by sifting to remove husks. The de-husked grains were heated at 100 °C for 1 h in a jacket pan at atmospheric pressure to inactivate enzymes and to stabilized the produced oat flour. Then they were milled using a laboratory mill (Swantech, SJ500, France) followed by sieving to obtain particle size of about 280μm. Chemical composition of wheat and oat flours including moisture, ash, fat and total carbohydrate content were determined.

2.2 Determination of the water absorption and farinograph parameters of the noodle dough

The amount of water required to make noodle dough was determined using a Brabender farinograph (model FE022-NK, Germany) with bowl capacity of 50g. Oat flour was replaced with wheat flour at different levels of 0, 10, 15, 20, 30 and 40% (w/w, wheat flour basis) and mixed well in a farinograph mixer for 2 min at 30 °C and then water was gradually added to make a dough with consistency of 900±20 BU. This consistency was the average consistency of commercial noodle dough commonly produced in local noodle factories. The commercial doughs were made with wheat flour (same flour used in this study), salt (2% w/w, flour basis) and water (50%, flour basis) according to the Institute of Standard and Industrial Research of Iran (No 2018). The amount of water was recorded and used to make the noodle dough. Farinograph parameters of the dough including dough development time, dough stability time and dough softening (12 min after peak) were determined from the farinograms.

2.3 Dough and noodle preparation

To produce noodles, flour containing different levels of oat flour and 2% NaCl were mixed at 45 rpm for 15 min in a laboratory mixer (Kenwood, A701A, UK) while water was added gradually. The dough was allowed to rest for 15 min at room temperature (22 °C±1), then sheeted using a laboratory noodle machine (Marcato, Super Iusso ISO, Italy) into a 3 mm sheet dough. The dough was then allowed to rest for 15 min at ambient temperature (25±2 °C) and then was cut into strands of an average width, thickness and length of 7 mm, 1.6 mm and 20 cm, respectively. The strands were spread in a plastic tray to dry at room temperature for 24 h. The moisture content of the dried noodles was 8% as determined by oven drying. The mean width, thickness and length of the dried noodles were 6.0 mm, 1.3 mm and 20 cm, respectively.

2.4 Rheological properties of noodle dough as determined by a Texture Profile Analyzer

Rheological properties of noodle dough (before sheeting) including elastic and viscous components and maximum force required to deform the dough were determined using a Texture Profile Analyser (TPA) (Texture Analyser, TA Plus, Stable Micro Systems, Godalming, Surrey, England). To obtain a constant shape of the dough, a hollow and open-ended plastic cylinder with very smooth inside wall with diameter and height of 1 and 3 cm, respectively was filled with constant weight of the dough. Then the shaped dough was removed from the cylinder and left to rest for 5 min at room temperature. The dough was subjected to a constant strain of 15% using a cylindrical plunger with diameter of 2.5 cm moving at speed of 2 mm/s. Force-time curve was recorded. The maximum height of the force-time curve was taken as the maximum force and from elastic recovery and permanent deformation, viscous and elastic components were obtained.

2.5 Color of the dried noodles

Color parameters of the dried noodles were evaluated using a modified method of Afshari and Farahnaky (2011). The measured L-value is an indication of the lightness (brightness) ranging from 0 (black) to 100 (white). The a-value is the function of the redness-greenness. Positive a-value indicates redness while negative value indicates greenness. The b-value indicates blueness-yellowness and positive b-values related to yellowness while negative values indicate blueness.

2.6 Textural properties of the dried noodles

Texture of the dried noodles was studied using the Texture Analyser by performing three point bending test. From force-distance graphs and Equation 1, Young's moduli of the samples were calculated

$$E=4Fa^3/3dD^4 \qquad (1)$$

Where E is the Young's moduli (N/m^2); F is force (Kg); a is distance between two supports (m); d is deflection (m) and D is sample distance (m).

2.7 Cooking characteristics of the noodles

To determine optimum cooking time of samples, 10 g of the noodles was cooked in 300 g distilled or salted water (containing 2% NaCl), separately until the centre white core could no longer be observed when a small piece of noodle was crushed between two glass plates.
To determine cooking loss, the cooking water obtained after cooking of the samples for their optimum cooking time was collected and evaporated in an air oven at 105 °C to constant weight. The residue was weighed and reported as a percentage of the original noodle sample according to the Equation 2. For the sample cooked in salted water, the added salt was taken into account.
Cooking loss (%) = [(weight of drained residue in cooking water) / weight of dried noodles] × 100 (2)
Water absorption of the noodles cooked in distilled or salted water (2% NaCl) was determined as follows; 100 g noodles was placed into 300 g of distilled or salted water, separately and cooked in a cooking pan at 97 °C for 18 min. During cooking, for the first 5 min, every 30 sec, 5 g of the sample was removed and drained for 2 min in a sieve. Between 5 to 12 min, each sample was removed every 1 min, and between 12 to 18 min every 2 min the sample was tested as explained and the water absorption of the sample at each cooking time was determined according to Equation 3.

%Water absorption = [(wt of cooked noodle – wt of dried noodle) / wt of dried noodle] × 100 (3)

2.8 Textural properties of cooked noodles

Texture of the cooked noodles was evaluated using the Texture Analyser. Resistance to cutting of cooked noodle samples was compared using a Cutting Shear device (Warner-Blatzler Blade, Stable Micro Systems, Godalming, Surrey, England) determining cutting shear forces. After cooking, the strands were put in a closed container and left on a bench at room temperature to cool down (for 10 min), followed by cutting shear force test. Test speed of 1 mm/s and strain of 80% were used for the experiment. Maximum cutting force and slope of cutting force-distance curve were obtained as indicators of resistance to cutting.
Texture profile analysis (TPA) was performed on the samples cooked in distilled or salted water by performing a two bite compression test at pretest speed of 5 mm/s, test speed of 0.25 mm/s, time interval of 10 s and strain deformation of 25%. The noodle rig held the strands in a way to avoid the movement of them when the plunger was returning upward. From the force-distance curve, the maximum force on the first compression cycle (first bite) was defined as hardness. The ratio of the positive force area under the second to the first compression was defined as cohesiveness. The negative force area of the first bite represented the work necessary to pull the compressing plunger away from the sample and was taken as adhesiveness.

2.9 Sensory Evaluation

Noodles in Iran are cooked in a meat or vegetable gravy before consumption. Therefore, noodles were cooked in a vegetable gravy containing carrot, potato and tomato extract, for 13±1 min (the average optimum cooking time obtained for the noodles cooked in the soup). Then 200 g of the samples were poured in transparent plastic dish, coded with three digits random numbers and presented to 20 in-house semi-trained panelists. The panelists evaluated the samples in terms of texture, color, stickiness and taste using a ranking test and scored them from 0 to 5 (0-1 for very bad; 1-2 for bad; 2-3 for fair; 3-4 for good and 4-5 for excellent).

2.10 Statistical Analysis

The experiments were performed in a completely randomized design. All experiments were conducted in triplicate and the mean values and standard deviations were calculated. Analysis of variance (ANOVA) was performed and the results were separated using the Multiple Ranges Duncan's test ($\alpha < 0.05$) using statistical software of SPSS 16 (SPSS, Inc., New Jersey, USA). The variables were levels of oat flour (0, 10, 20, 30 and 40%) and salt in cooking water (0 and 2%).

3 RESULTS AND DISCUSSION

Chemical composition of wheat flour and oat flour is given in Table 1. The results showed that the oat flour significantly had higher fat, ash, protein, total, soluble and insoluble fiber content than wheat flour.

Table 1. *Chemical composition of wheat and oat flours used in this study based on dry weight basis*.*

Flour type	Moisture content (%)	Protein (%)	Lipid (%)	Ash (%)	Total Dietary fiber (%)	Soluble fiber (%)	Insoluble fiber (%)
Wheat flour	12.30^{a}	11.17^{b}	1.64^{b}	0.60^{b}	5.4^{b}	2.6^{b}	2.8^{b}
Oat flour	7.25^{b}	14.19^{a}	6.42^{a}	1.03^{a}	14.3^{a}	5.2^{a}	9.1^{a}

**Values are the average of triplicates. Different letters in each column indicate significant statistical difference between the values ($p<0.05$).*

Table 2 shows that the water absorption of the dough increased with increasing the oat flour percentage. The results showed that the time required to make the dough with consistency of 900 BU reduced with increasing the oat flour level. Dough stability time is an indiacation of dough strength. The results also showed that dough stability time decreased, while dough softening increased with increasing the percentage of the oat flour in the formulation. With increasing oat flour level, gluten content decreased and the dough became softer. Increasing the lipid content of the dough as a result of oat flour addition, could soften the dough as well. As a result, less force was required to compress the dough (Table 3). The results also showed that the dough became less elastic and more viscous as the oat flour level increased.
The color of the regular salted white noodles is preferred to be white to slightly creamy. Table 4 showed that with increasing the oat flour up to 30%, the L- and b-values of the samples did

not change significantly. Further increasing of the oat flour level caused L- and b-values to decrease and a-value to increase. The results may indicate that the samples became slightly darker and less greener as the oat flour level increased to more than 20%. The ash content of the flour increased with increasing oat flour which has a negative correlation with noodle lightness.

Table 2. *The farinograph parameters of the noodle dough containing different levels of oat flour*.*

Oat flour (%)	Water absorption (%)	Dough development time (min)	Dough stability time (min)	Dough softening (BU)
0	50.6±0.7^{c}	2.05±0.09^{a}	2.15±0.15^{a}	313.3±11.5^{c}
10	52.4±0.0^{b}	1.70±0.39^{b}	2.10±0.12^{a}	340.3±22.6^{b}
20	52.7±0.5^{b}	1.60±0.36^{b}	1.80±0.35^{b}	373.0±11.5^{b}
30	53.5±0.1^{a}	1.32±0.13^{bc}	1.25±0.09^{c}	406.7±11.5^{a}
40	53.6±0.0^{a}	1.25±0.09^{c}	1.12±0.06^{d}	413.3±11.5^{a}

**Values are the mean of triplicates ± standard deviation. Different letters in each column indicate significant statistical difference between the values (p<0.05).*

Table 3. *Rheological properties of noodle dough containing different levels of oat flour as determined by a Texture Analyser*.*

Oat flour level (%)	Maximum force (N)	Elastic component	Viscose component
0	3.551±0.002^{a}	0.160±0.002^{a}	0.841±0.002^{d}
10	3.511±0.003^{b}	0.150±0.002^{b}	0.850±0.003^{c}
20	3.264±0.002^{c}	0.145±0.003^{c}	0.855±0.003^{c}
30	3.164±0.003^{d}	0.137±0.004^{d}	0.863±0.002^{b}
40	3.012±0.002^{e}	0.128±0.002^{e}	0.872±0.002^{a}

**Values are the mean of triplicates ± standard deviation. Different letters in each column indicate significant statistical difference between the values (p<0.05).*

Table 4. *Color parameters of dried noodles containing different levels of oat flour including lightness (L-value); redness-greenness (b-value) and blueness-yellowness (b-value)*.*

Oat flour (%)	L-value	a-value	b-value
0	60.2±2.5^{a}	-4.0±0.4^{c}	33.8±1.0^{a}
10	58.3±2.0^{a}	-3.8±0.4^{bc}	33.5±1.3^{a}
20	57.0±1.8^{a}	-3.5±0.5^{bc}	33.3±1.0^{a}
30	56.2±1.5^{a}	-3.1±0.3^{b}	32.5±2.0^{a}
40	51.2±2.0^{b}	-1.5±0.2^{a}	31.1±1.1^{b}

**Values are the mean of triplicates ± standard deviation. Different letters in each column indicate significant statistical difference between the values (p<0.05).*

The results of three point bending test (Figure 1) showed that the Young's moduli of the dried noodles decreased significantly, as a result of oat flour addition. This may indicate that the noodles became more fragile as the level of oat flour increased, therefore, less force was required to break the dried noodles.

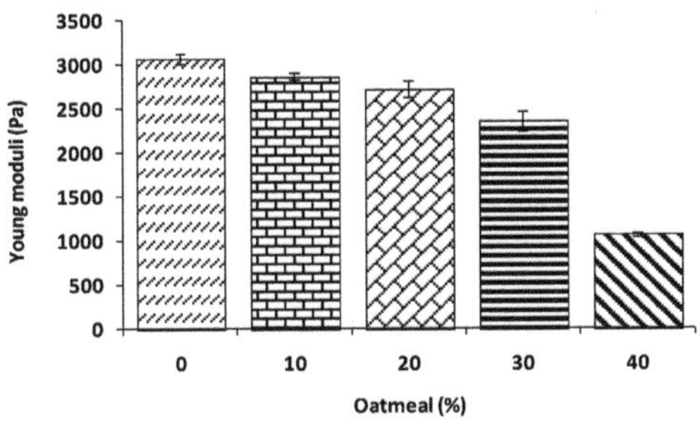

Figure 1. *Young's moduli of the dried noodles containing different levels of oat flour.*

Figure 2 shows that with increasing oat flour level, the optimum cooking time of the samples increased. Determination of the thermal properties of oat and wheat starches has showed that the oat starch had significantly higher onset, peak and conclusion temperatures and enthalpy compared to wheat starch. Cooking of the samples in salted water further increased the optimum cooking time.

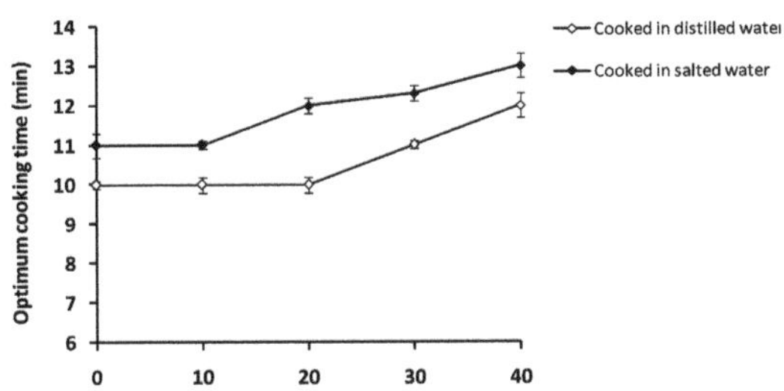

Figure 2. *Optimum cooking time of the noodles containing different levels of oat flour cooked in distilled or salted water.*

Figure 3 shows that the water absorption of the noodles was positively affected by the cooking time. Increasing the water absorption of the noodles during cooking is mainly due to the water uptake of gluten and starch granules during gelatinization. Increasing the oat flour level reduced the water uptake of the samples. Comparing Figure 3-A and 3-B it was found that addition of salt to the cooking water enhanced water absorption of the samples.

With increasing the oat flour percentage in the noodle recipe, the cooking loss of the samples increased, significantly (Figures 4-A and -B). The cooking loss varied from 1.1 to 3.5% for noodles cooked in distilled water and from 0.5 to 2.3% for those cooked in salted water. The results also revealed that cooking of the samples in salted water reduced the cooking loss. Salt can support the gluten network, which can prevent soluble material diffusing out of the noodle readily.

Textural properties are important characteristics of noodles. Both starch and proteins play major roles in governing textural properties of noodles. However, TPA parameters better reflect the contribution of proteins to textural properties than the contribution of starch to texture.

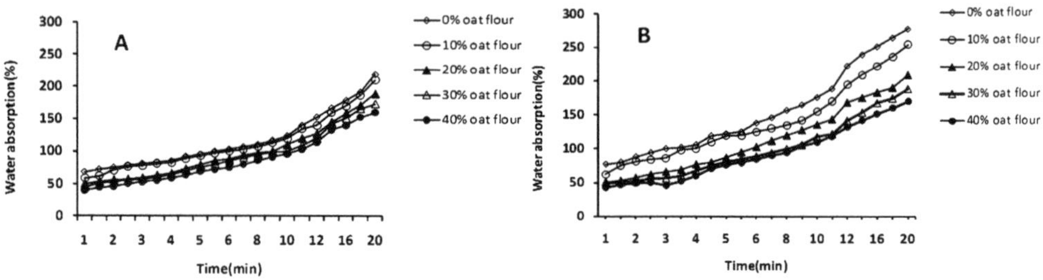

Figure 3. *Water absorption of the noodles containing different levels of oat flour cooked in distilled (A) and salted (B) water.*

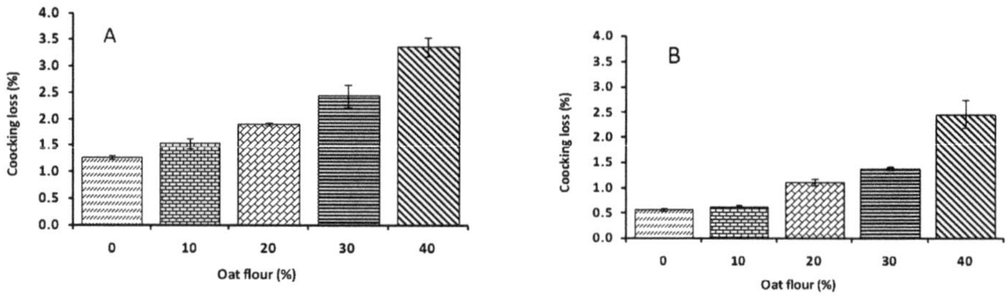

Figure 4. *Cooking loss of the noodles containing different levels of oat flour cooked in distilled (A) and salted (B) water.*

Determination of the hardness of the cooked noodles (Figures 5-A and -B) showed that the hardness of the samples decreased as the cooking time increased. Figs. 5-C, D show that with increasing of the cooking time, cohesiveness of the samples increased. Adhesiveness is an indication of stickiness of the noodles. The results showed (Figures 5 E, F) that with increasing the cooking time, the adhesiveness decreased while with increasing the oat flour in the noodle recipe, this parameter increased.

4 CONCLUSION

Some physicochemical changes may occur when wheat flour is replaced by oat flour. However, addition of up to 30% oat flour caused no considerable problem with dough and noodle production. After drying of the noodles, they became more fragile and hence special care is required during handling and packaging of these noodles. The color of dried oat flour noodles became darker with inclusion of more than 20% oat flour. Nevertheless, after cooking of the samples in soups (as usual way of noodle consumption), darkness of the noodles was not noticable by the panelists. Increasing of the oat flour level also increased the softness and cooking loss of the cooked samples. In general, replacement of maximum 30% of the wheat flour with oat flour is recommended for production of oat noodles.

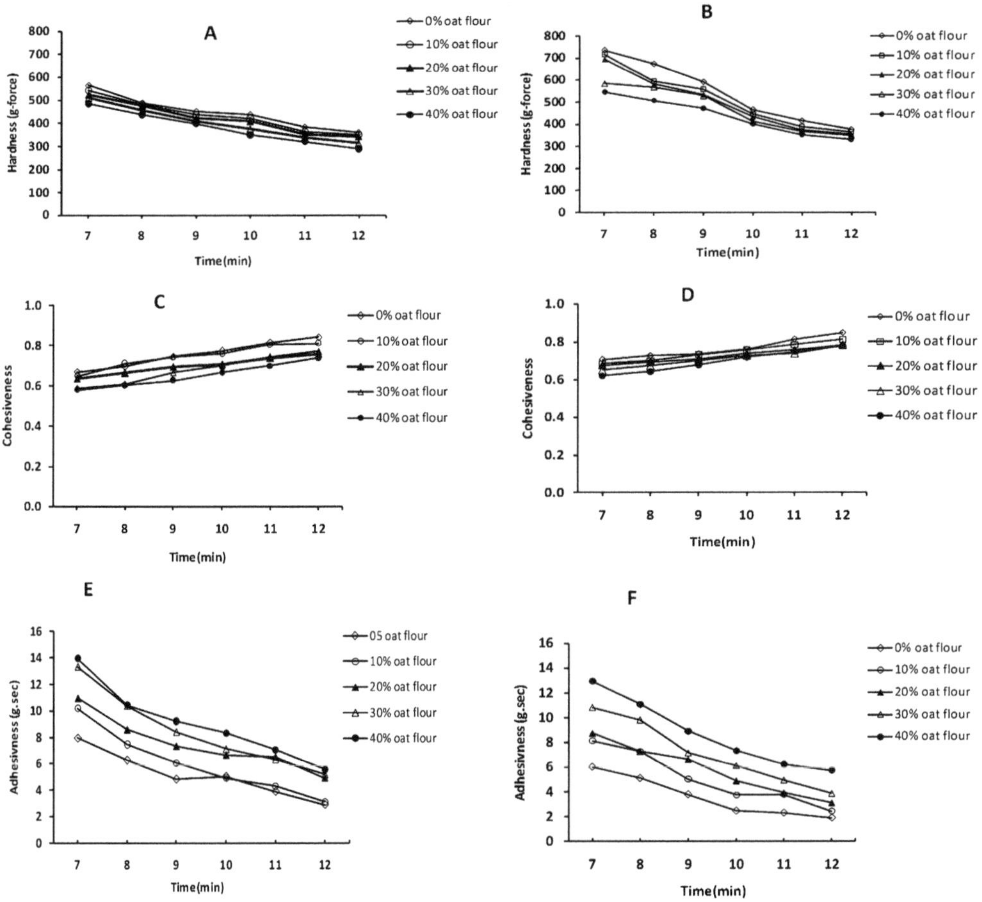

Figure 5. *Textural properties of cooked noodles containing different levels of oat flour. A & B: Hardness of samples cooked in distilled and salted water, respectively; C & D: Cohesiveness of the noodles cooked in distilled water and salted water, respectively; E & F: Adhesiveness of the noodles cooked in distilled and salted water, respectively.*

REFERENCES

1 B.X. FU, Food Res. Int., 2008, **41,** 888.
2 J. WAN, W. HUANG, J. ZHONG, L. HUANG, P. RAYAS-DUARTE, B. LIU, Cereal Chem., 2011. **88,** 153.
3 F. WANG, W. HUANG, Y. KIM, R. LIU, T. MICHAEL, J. Cereal Sci., 2011.1-7 (In Press).
4- P.J. WOOD, Carbohydr. Polym., 1994. **25,** 331.

FUNCTIONALITY OF POLYOLS AS SUCROSE REPLACERS IN SPANISH MUFFINS

S. Martínez-Cervera[1], T. Sanz[1] & A. Salvador[1]

[1] Instituto de Agroquímica y Tecnología de Alimentos (CSIC), Avda. Agustín Escardino, 7, 46980 Paterna (Valencia), Spain.

1 INTRODUCTION

Consumer interest in sucrose reduced or sucrose-free food products is increasing continuously. In Spanish muffins, apart from conferring sweet taste, sucrose has an important role in the development of the final texture. In particular, the role of sucrose in the starch gelatinization process is of special importance.

An appropriate sucrose replacer in bakery products requires two important properties. On one side it must provide sweetness and on the other side must provide the sucrose physical functionality, which is essential to achieve the desired texture.

Different sugar alcohols or polyols have been used as sucrose replacers in bakery products in which the volume and texture are important such as biscuits and cakes[1]. They are able to provide the required sweetness and to assist in product structure development.

From the point of view of nutrition, polyols do not promote the development of dental caries, have a low glycemic response (turning the meat product suitable for diabetics) and have lower calories than sucrose[1]. Sucrose provides 4 Cal/g and the polyols used in this work provide between 0.2 to 2.6 Cal/g.

The aim of the present work was to evaluate the functionality of the polyols sorbitol, maltitol, isomalt and erythritol as sucrose replacers in Spanish muffins. To do so the influence of the polyols on the starch gelatinization process and on the muffin batter rheological properties was studied and the results related to the muffins' texture.

2 METHODS AND RESULTS

2.1. Batter and muffin preparation

Five muffin formulations were prepared, one control formulation and four formulations by replacing all of the sucrose with erythritol, sorbitol, isomalt or maltitol. The Spanish muffin ingredients were wheat flour, pasteurized white egg, pasteurized yolk, refined sunflower oil, whole UHT milk, sucrose or the corresponding polyol, sodium bicarbonate, citric acid and salt. The samples were identified as control, 100% erythritol, 100% sorbitol, 100% maltitol and 100% isomalt.

The batter was prepared in a professional mixer and 45 g of batter was dosed exactly into each paper mould. 12 moulds were arranged in three rows of four muffins on a baking tray and baked for 17 minutes at 175°C in a conventional electric oven. All the formulations were prepared at least in duplicate.
The muffins were left to cool at room temperature for one hour. The twelve muffins baked on the same baking tray were packed in a polyethylene bag, hermetically sealed and stored at 20°C for one day until the texture determinations.

2.2 Measurements in batter muffins

The linear viscoelastic properties of the muffin batters were studied using oscillatory tests carried out with an AR G2 controlled-stress rheometer (TA-Instruments). The batters were all kept at 25°C for 60 min after batter preparation before the rheological test. The samples were allowed to rest in the measurement cell for 5 min as equilibration time. A 40-mm diameter plate–plate sensor geometry with a serrated surface and a 1-mm gap was employed. To simulate the effect of heating in the batter structure, temperature sweeps were performed from 25°C to 95°C at a heating rate of 1.0°C/min and a strain amplitude of 0.0005. The strain applied was selected to guarantee the existence of a linear viscoelastic response according to previous stress sweeps carried out at 25°C and 95°C. Additionally, mechanical spectra in the linear region from 10 to 0.01 Hz at 25 °C were also recorded in separate tests. The storage modulus (G'), the loss modulus (G'') and the complex modulus $|G^*|$ were recorded.

2.3 Measurements in muffins

The instrumental texture measurements of the muffins were made with a TA.XT2 Texture Analyser provided with Texture Expert software. The muffins were cut horizontally at the height of the mould. A double compression test (TPA) was performed with a 75mm diameter flat-ended cylindrical probe and compression to 50% of the initial height, at a speed of 1mm/s with a 5s waiting time between the two cycles. The parameters obtained from the curves were hardness, springiness, cohesiveness, chewiness and resilience.

2.4 Calorimetric measurements in flour, flour/sucrose and flour/polyol mixtures

Differential scanning calorimetric (DSC) measurements were performed with a Q2000 calorimeter (TA-Intruments). Measurements were performed in the flour and in 1:1 mixtures of flour with sucrose or polyol. Ingredients were weighed and distilled water added at a 1:3 (w/w) sample to water ratio in high volume DSC pans. The samples were heated from 25 to 130 °C at 10 °C/min. The enthalpy was expressed in J/g of flour.

2.5 Influence of sucrose and polyols in starch gelatinization

In order to evaluate the influence of sucrose and the different polyols in the flour starch gelatinization process, DSC measurements were performed. All the samples showed an endothermic peak around 70 °C, which corresponds to starch gelatinization (table 1). As expected, the presence of sucrose increased the starch gelatinization temperature significantly (a 4°C increase). All the polyols showed a similar effect to sucrose and also delayed starch gelatinization. Maltitol showed the highest starch gelatinization temperature.

Table 1 Effect of sucrose and the different polyols in the temperature and enthalpy of starch gelatinization

	T (°C)	ΔH (J/g of flour)
Flour	68.07a (0.24)	5.86a (0.06)
Flour-sucrose	71.73bc (0.06)	4.94a (0.49)
Flour-erythritol	70.17ab (0.66)	5.19a (0.34)
Flour-sorbitol	71.46bc (0.12)	4.73a (0.53)
Flour-isomalt	72.66bc (0.22)	5.18a (0.85)
Flour-maltitol	73.04c (1.57)	4.05a (0.47)

The effect in the enthalpy was also similar among sucrose and the different polyols. In all the cases a reduction in enthalpy was found, although the differences were not significant. Maltitol showed the lowest enthalpy value.

2.6 Viscoelastic properties

The frequency dependence of G' and G'' in the linear region at 25°C is shown in Figure 1. In all the batter samples, G' values were greater than G'' values and both of them showed a smooth dependence with frequency, revealing a typical rheological behaviour of soft gels. The control muffins batter (no sucrose replacement), the maltitol batter and the sorbitol batter showed a very similar behaviour: G' values at 1 Hz were, 56.77 Pa, 50.37 Pa and 55.10 Pa, respectively. The erythritol and the isomalt batters showed a different group with higher values of both the elastic and the viscous modules: values of G' at 1 Hz were 135.50 Pa and 150.80 Pa, respectively.

To simulate the structural changes that take place in the muffin batter during heating in the oven, the linear viscoelastic properties during a temperature sweep from 25 °C to 95°C were studied. Values of the complex modulus G* versus temperature are shown in Figure 2. In all the batters, the initial increase in temperature produced a decrease in the values of the G*. A similar temperature profile was found among the control batter and the sucrose replaced by polyols batters. The polyols, similarly to sucrose increase the temperature at which the inflexion in $|G^*|$ occurred.

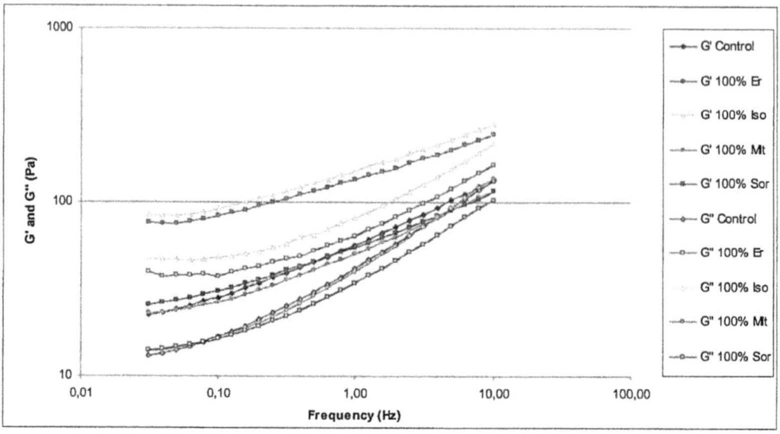

Figure 1 Effect of sucrose replacement by the different polyols in the frequency dependence of the muffin batters at 25°C.

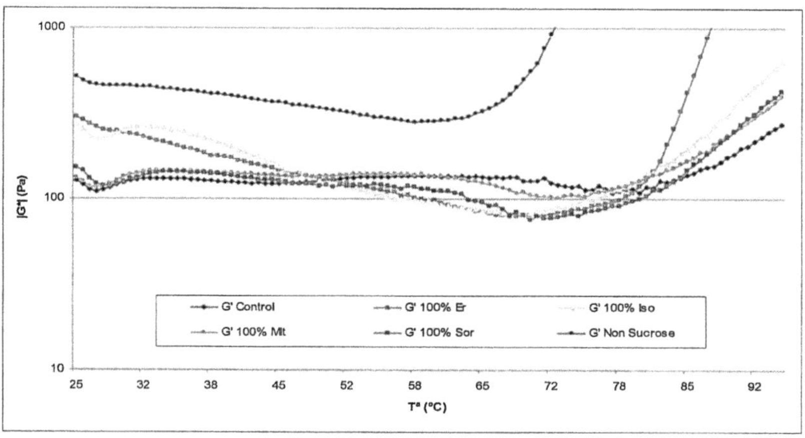

Figure 2 Complex modulus, |G*|, as a function of increasing temperature for the control batter, a non sucrose batter and the sucrose replaced by polyols batters

2.7 Instrumental texture and height of muffins

A texture profile analysis test was carried out to evaluate the instrumental texture of the control muffins and of the 100% sucrose replacement by polyols muffins. Values of hardness, springiness, cohesiveness, chewiness and resilience, together with the muffins' height values are shown in Table 2. The hardness values of muffins with isomalt, maltitol or sorbitol were lower than the hardness of control muffins, but muffins with sorbitol were the softer. On the other hand, muffins with erythritol were the hardest. The maltitol and sorbitol muffins showed similar values of springiness than the control while erythritol and isomalt muffins showed lower springiness. Significant differences were found in cohesiveness values, being the control muffins the ones with the highest cohesiveness and the erythritol muffins the ones with the lowest cohesiveness. Chewiness is a secondary texture parameter that is associated with difficulty in chewing and forming a bolus before swallowing. The chewiness values followed the same trend as the hardness values (Table 2). The erythritol muffins showed the highest chewing difficulty and the sorbitol muffins the lowest. All the polyols decrease resilience values.

All the polyols reduced significantly muffin height in comparison to the control. Once more the effect was more noticeable for erythritol, which showed significantly the lowest height.

Table 2 Mean height and TPA values (H: hardness; S: springiness; Ch: cohesiveness; Cw: chewiness; R: resilience) (N=8) of the control and the sucrose replaced muffins by different polyols.

	Height (mm)	TPA				
		H (N)	S	Ch	Cw (N)	R
Control	48.15a (2.14)	29.02ab (0.88)	0.85a (0.01)	0.60a (0.00)	14.69b (0.52)	0.24a (0.00)
100%Er	38.88c (3.12)	104.65a (5.64)	0.78d (0.01)	0.36d (0.02)	28.87a (1.85)	0.13e (0.01)
100%Iso	44.24b (1.27)	26.83bc (1.79)	0.82c (0.01)	0.47c (0.01)	10.43c (0.78)	0.18d (0.00)
100%Mlt	44.93b (1.53)	23.99cd (1.95)	0.84ab (0.01)	0.53b (0.01)	10.70c (0.99)	0.20b (0.00)
100%Sor	44.69b (1.84)	20.37d (2.90)	0.83bc (0.01)	0.47c (0.01)	7.92d (1.30)	0.18c (0.01)

3 CONCLUSIONS

The different polyols affected the muffin texture properties differently. Erythritol provided the most negative effect in texture with a high significant increase in hardness and chewiness and a decrease in springiness. Contrary to erythritol, the other polyols showed a small decrease in hardness and no difference in springiness in comparison with the control.

The effect of polyols in the flour starch gelatinization properties and in the batter linear viscoelastic behaviour can not explained the texture differences observed among polyols. In a future study the role of bubbles formation and stability during baking will be studied to try to understand the differences among polyols.

References

1 M. Kroger, K. Meister and R. Kava, *Comp Rev Food Sci F*, 2006, **5**, 35.

REPLACING SUCROSE IN SPANISH MUFFINS WITH ERYTHRITOL AND XANTHAN GUM.

S. Martínez-Cervera[1], E. de la Hera[2], T. Sanz[1], M. Gómez[2] and A. Salvador[1]

[1] Instituto de Agroquímica y Tecnología de Alimentos (CSIC), Avda. Agustín Escardino, 7, 46980 Paterna (Valencia), Spain.
[2] Food Technology Area. College of Agricultural Engineering, University of Valladolid, Avd./ Madrid, nº44, 34071 Palencia, Spain.

1 INTRODUCTION

Spanish muffins are sweet, high-calorie baked products which are highly appreciated by consumers due to their good taste and soft texture. The principal problem is that their high fat and sucrose content makes them a high calorie product.
Due to the complex structural functionality of sucrose in baked products, obtaining good quality low-sucrose products is a difficult task. Sugar alcohols are typical sucrose replacers in baked goods[1] and their properties have been studied by many authors[2,3,4]. Erythritol (Er) plays a similar role to sucrose, providing some of its properties in bakery products[5]. Erythritol is a 4-carbon sugar alcohol with about 60% to 80% of the sweetness of sucrose[6] which is obtained via a natural fermentation process[5]. It is suitable for people with diabetes because blood glucose and insulin levels do not increase when erythritol is administered orally to healthy male subjects[7]. Erythritol can replace sucrose, reducing the calorie content of the product to less than 0.5 kcal/g[8].
A strategy to improve the physicochemical properties of baked products is to add hydrocolloids to the batters. Hydrocolloids change the viscosity of batters and aid air incorporation during batter mixing, thus achieving an improvement in the height and volume of the final product[9].

2 METHODS AND RESULTS

2.1. Batter and muffins preparation

Fourteen muffin formulations were prepared by replacing part or all of the sucrose with erythritol (Er) or erythritol and xanthan gum (ErXan), or reducing or eliminating it without adding a replacement (SR). The Spanish muffin ingredients were wheat flour, pasteurized whole egg, refined sunflower oil, whole UHT milk, sucrose, erythritol, xanthan gum, tap water, sodium bicarbonate, citric acid and salt. The samples were identified as control, controlXan, 25%SR, 50%SR, 75%SR, 100%SR, 25%Er, 50%Er, 75%Er, 100%Er, 25%ErXan, 50%ErXan, 75%ErXan and 100%ErXan.
The batter was prepared in a professional mixer and 45 g of batter was dosed exactly into each paper mould. 12 moulds were arranged in three rows of four muffins on a baking tray

and baked for 25 minutes at 190°C in modular electric oven. All the formulations were prepared at least in duplicate.

The muffins were left to cool at room temperature for one hour. The twelve muffins baked on the same baking tray were packed in a polyethylene bag, hermetically sealed and stored at 20°C for one day until the determinations were conducted.

2.2 Measurements in batter muffins

The specific gravity of the batter (SG) was measured as the ratio of the weight of a standard container filled with batter (W2) to that of the same container filled with water (W1) (specific gravity =1 g/ml) as follows:

$$SG = W2/W1$$

The bubbles in batter samples were imaged using a DM750 microscope with a magnification of 40x. A linked EC3 video camera provided images using LAS EZ software.

The linear viscoelastic properties of the muffin batters were studied using an AR G2 controlled-stress rheometer. The batters were all kept at 25°C for 60 min after batter preparation before the rheological test. The samples were allowed to rest in the measurement cell for a 25 min equilibration time. A 40-mm diameter plate–plate sensor geometry with a serrated surface and a 1-mm gap was employed. To simulate the effect of heating in the batter structure, temperature sweeps were performed from 25°C to 95°C at a heating rate of 1.0°C/min and a strain amplitude of 0.0005. The strain applied was selected to guarantee the existence of a linear viscoelastic response according to previous stress sweeps carried out at 25°C and 95°C.

2.3 Measurements in muffins

The muffins were numbered by marking the underside of the mould and were weighed before (W3) and after baking and cooling (W4). The weight loss upon baking was calculated as follows: Weight loss = (W3-W4)*100/W3

The muffin height was measured with a calliper from the highest point of the muffin to the bottom of the paper mould after cooling for 1h at room temperature.

Muffin volume was determined using a laser sensor with the BVM-L 370 volume analyzer. This procedure involves an instrument that spins the sample round and has a sensor laser and software to calculate the volume of the whole muffins.

The instrumental texture measurements of the muffins were made with a TA.XT2 Texture Analyser provided with Texture Expert software. The muffins were cut horizontally at the height of the mould. A double compression test (TPA) was performed with a 75mm diameter flat-ended cylindrical probe and compression to 50% of the initial height, at a speed of 1mm/s with a 5s waiting time between the two cycles. The parameters obtained from the curves were hardness, springiness, cohesiveness, chewiness and resilience.

Crumb muffins bubbles were captured using a DSC-W300 super steady shot camera. Image processing was performed using Image Pro-Plus 4.5 software. The images were converted to an 8-bit grey scale and segmented separately using a histogram-based segmentation that was defined individually for each image.

2.4 Viscoelastic behaviour during heating

The initial G' and tan delta values (values at 25°C) varied among the different batters. Elimination of sucrose from the control batter progressively increased the initial G' values

and decreased tan delta, meaning that the sucrose decreased the viscoelasticity of the batter system (Figure 1A and 1B). As with sucrose, although to a lesser extent, erythritol decreased the G' values and increased the viscoelasticity, thus at 25°C the viscoelastic effect of erythritol in the batter lay between that of sucrose and that of its absence.
In all the batters, the initial increase in temperature produced a decrease in the values of the elastic moduli. This decrease was more evident for 100%SR, 50%SR and 100%Er than for the control and 50%Er batters.
In the control batter the inflexion point occurred at approximately 80°C, while 50% sucrose reduction decreased it to approximately 67°C and 100% reduction to 60°C. In this regard the erythritol was very successful in mimicking the effect of sucrose, as it increased the temperature at which the inflexion point occurred. The resulting delay, mainly in starch gelatinization, is crucial to avoid earlier thermosetting and give enough time for appropriate air and vapour expansion during baking.
The effect of adding xanthan gum and a double quantity of leavening agent on the erythritol batter's viscoelastic properties during heating can be seen in Figure 1C. The temperature at which the elastic modulus (G') started to increase was not altered by the addition of xanthan gum and a double quantity of leavening. The only difference compared to erythritol alone was a slight decrease in the initial G' values (25°C) in this sample, which achieved the values reached by the control batter. This decrease in G' may be related to the double amount of leavening agent, which reduced the batter density by increasing the air content of the batter.

2.5 Specific gravity (SG) and microstructure of the batter

Sucrose reduction significantly increased the SG values (Table 1), implying a reduction in the amount of air. This reduction in the amount of air may explain the increase in the elastic properties found with sucrose reduction, due to an increase in batter density. In the batters where sucrose was replaced by erythritol (Table 2) the same trend was observed, although less noticeably. The SG values were in agreement with the light micrography images of the batters. As the sucrose in the batter was reduced, a progressive reduction in the number and size of air bubbles was found. This may partly be associated with the lower elasticity of the control batter, as offering less resistance to the shearing forces applied during mixing would favour air entrapment in the batter. The micrographs of the batters in which sucrose was replaced by erythritol also reflected a decrease in the number and size of air bubbles as the sucrose was reduced. However, this decrease was slightly lower than in the sucrose reduction batters, which implies that erythritol had an effect on the batter's air-holding capacity. Indeed, the number and size of bubbles in the 25%Er and 50%Er batters was very similar to those of the control formulation. In the present case, the increase in specific gravity correlates to the light micrographs of the batter, which show a reduction in both the number and size of bubbles in the 75%Er and 100%Er formulations.
The addition of xanthan gum and doubling of the leavening agent resulted in an increase in the total amount of air in the batter compared to the control and erythritol muffin batters, as reflected by the significant decrease in the SG of all the xanthan gum batters in comparison with the control (Table 3). Observation of the batter microstructure revealed that adding xanthan gum and doubling the leavening increased the size of the air bubbles compared to the control batter. The xanthan gum batters contained medium and large size bubbles while those without xanthan gum had smaller bubbles or, at high replacement levels, the air bubbles disappeared.

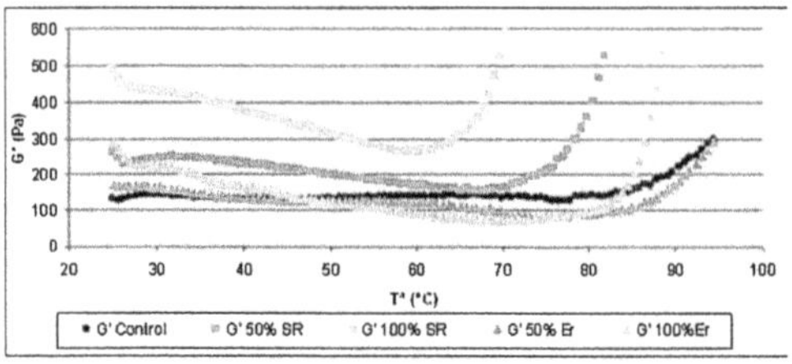

1A

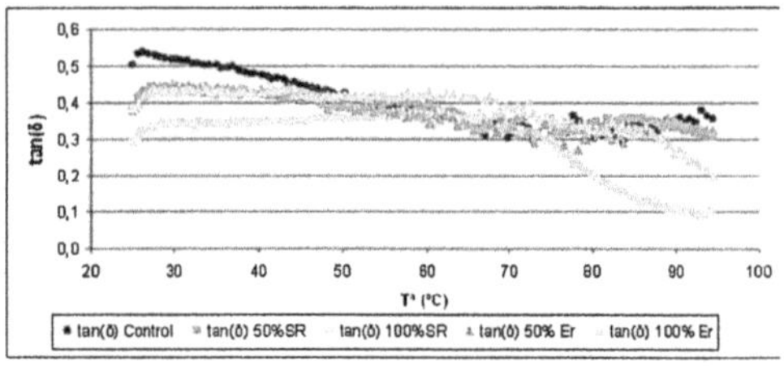

1B

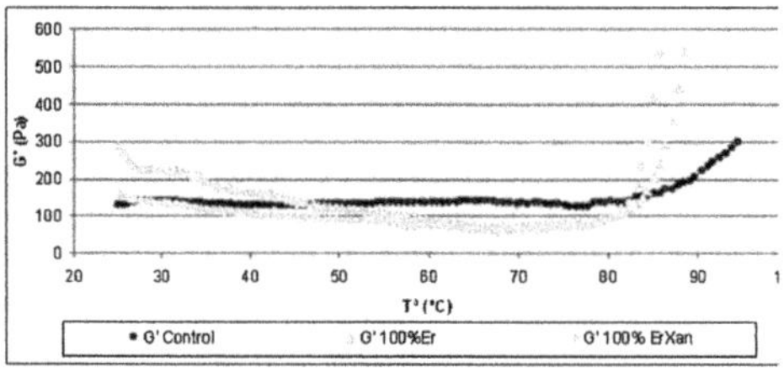

1C

Figure 1 Elastic modulus (G') (A) and Tan delta. (B)as a function of increasing temperature in the Spanish muffin control batter, sucrose reduction batters and batters with sucrose replaced by erythritol. (C) Elastic modulus (G') as a function of increasing temperature in the Spanish muffin control batter and in batters with sucrose replaced by erythritol alone or erythritol with xanthan gum.

2.6 Bubbles in the crumb and muffin height, volume and weight loss

The control crumb possessed large and small air bubbles, which conferred an appearance of springiness. The elimination of sucrose caused larger bubbles to appear. The 100%SR muffin crumb had very few small bubbles and many large bubbles. These large bubbles formed tunnels from the base to the surface, so some air was lost during the formation of the crumb structure. The decrease in air bubble expansion was associated with a significant decrease in height and volume as the sucrose was reduced (Table 2).

Sucrose plays an important role in water retention in muffins, avoiding greater moisture loss during baking. As the sucrose was eliminated the weight loss progressively increased; the differences compared to the control muffin were significant for the 75% and 100% sucrose reduction levels.
The effect of erythritol can be seen Table 2 (muffin height, volume and weight loss). The erythritol samples also differed from the control sample, with the bubbles becoming larger as the amount of sucrose replacement increased; the 75%Er and 100%Er samples contained larger bubbles than the control sample. However, the crumb structure bore a greater resemblance to that of the control than in the sucrose reduction samples, implying that erythritol was conferring some of the structural effects provided by sucrose. The height and volume of the muffins also increased in the presence of erythritol in comparison with the equivalent sucrose reduction muffins. Erythritol also reduced the weight loss during baking. In the erythritol muffins, the weight loss was lower than in the control muffins. It would appear that erythritol works as a moisturizer and prevents water from leaking out during baking.

Table 1 Specific gravity (SG) of batters, muffin weight loss during baking (WL), muffin height, muffin volume and mean TPA values (H: hardness; S: springiness; Ch: cohesiveness; Cw: chewiness; R: resilience) of the control muffin and of muffins prepared with decreasing quantities of sucrose.

	Batter	Spanish muffin							
	SG	WL	Height	Volume	TPA				
	(g/L)	(g)	(mm)	(cm^3)	H (N)	S	Ch	Cw (N)	R
Control	0.95a (0.00)	14.82a (0.68)	48.24a (0.86)	114.50a (5.98)	16.20a (0.57)	0.88a (0.01)	0.56a (0.00)	7.93a (0.24)	0.28b (0.01)
25% Er	0.96ab (0.00)	15.07a (0.53)	48.63a (1.47)	117.50a (5.88)	14.46a (0.64)	0.89a (0.01)	0.57a (0.01)	7.23a (0.25)	0.30a (0.01)
50% Er	0.97ab (0.00)	13.96b (0.54)	46.08b (1.19)	115.25a (3.54)	26.58b (1.47)	0.85ab (0.01)	0.49b (0.01)	10.90b (0.66)	0.23c (0.01)
75% Er	0.98bc (0.01)	12.98c (0.43)	42.00c (2.13)	106.38b (5.66)	31.58c (2.32)	0.83b (0.07)	0.48b (0.04)	12.50c (1.10)	0.23c (0.02)
100% Er	1.02c (0.01)	12.88c (0.65)	36.09d (1.53)	93.88c (1.64)	74.88d (5.19)	0.81b (0.01)	0.39c (0.02)	23.62d (1.76)	0.18d (0.01)

Table 2 Specific gravity (SG) of batters, muffin weight loss during baking (WL), muffin height, muffin volume and mean TPA values (H: hardness; S: springiness; Ch: cohesiveness; Cw: chewiness; R: resilience) of the control and the formulas in which sucrose was replaced by erythritol

	Batter	Spanish muffin							
	SG (g/L)	WL (g)	Height (mm)	Volume (cm^3)	TPA H (N)	S	Ch	Cw (N)	R
Control	0.95a (0.00)	14.82a (0.68)	48.24a (0.86)	114.50a (5.98)	16.20a (0.57)	0.88a (0.01)	0.56a (0.00)	7.93a (0.24)	0.28b (0.01)
25% Er	0.96ab (0.00)	15.07a (0.53)	48.63a (1.47)	117.50a (5.88)	14.46a (0.64)	0.89a (0.01)	0.57a (0.01)	7.23a (0.25)	0.30a (0.01)
50% Er	0.97ab (0.00)	13.96b (0.54)	46.08b (1.19)	115.25a (3.54)	26.58b (1.47)	0.85ab (0.01)	0.49b (0.01)	10.90b (0.66)	0.23c (0.01)
75% Er	0.98bc (0.01)	12.98c (0.43)	42.00c (2.13)	106.38b (5.66)	31.58c (2.32)	0.83b (0.07)	0.48b (0.04)	12.50c (1.10)	0.23c (0.02)
100% Er	1.02c (0.01)	12.88c (0.65)	36.09d (1.53)	93.88c (1.64)	74.88d (5.19)	0.81b (0.01)	0.39c (0.02)	23.62d (1.76)	0.18d (0.01)

Table 3 Specific gravity (SG) of batters (N=4), muffin weight loss during baking (WL) (N=14), muffin height (N=8), muffin volume (N=8) and mean TPA values (H: hardness; S: springiness; Ch: cohesiveness; Cw: chewiness; R: resilience) (N=8) of the control and the formulas in which sucrose was replaced by erythritol and xanthan gum

	Batter	Spanish muffin							
	SG (g/L)	WL (g)	Height (mm)	Volume (cm^3)	TPA				
					H (N)	S	Ch	Cw (N)	R
Control	0.95a (0.00)	14.82a (0.68)	48.24b (0.86)	114.50ab (5.98)	16.20b (0.57)	0.88b (0.01)	0.56c (0.00)	7.93c (0.24)	0.28b (0.01)
Control Xan	0.84b (0.00)	15.49ab (0.61)	51.72a (1.89)	131.63c (3.93)	6.52d (0.52)	0.89bc (0.01)	0.60d (0.02)	3.50a (0.23)	0.31c (0.01)
25% ErXan	0.85b (0.00)	15.93b (0.60)	48.05b (2.63)	118.25b (5.87)	6.24d (0.50)	0.90c (0.01)	0.61d (0.02)	3.40a (0.27)	0.34d (0.01)
50% ErXan	0.85b (0.00)	15.56ab (0.73)	45.67b (2.34)	117.88b (6.62)	7.51d (0.75)	0.89bc (0.01)	0.58cd (0.02)	3.87a (0.36)	0.33cd (0.01)
75% ErXan	0.84b (0.01)	15.01a (0.81)	42.38c (3.57)	121.50b (4.78)	12.85c (1.68)	0.87b (0.02)	0.53b (0.02)	5.87b (0.51)	0.28b (0.02)
100% ErXan	0.87b (0.02)	14.74a (1.05)	35.67d (2.22)	108.13a (5.00)	25.05a (2.42)	0.82a (0.01)	0.44a (0.03)	9.07d (1.09)	0.22a (0.02)

When xanthan gum was added and the leavening agent doubled, the crumb structure of the erythritol muffins became more similar to that of the control. For all the replacement levels, the samples contained bubbles of different sizes, evenly distributed. The increase in bubble stability associated with the presence of xanthan gum was also associated with an increase in muffin volume (Table 3). Xanthan gum increased the batter viscosity and therefore decreased the mobility of the air bubbles, so less coalescence occurred. In the presence of xanthan gum and double quantities of leavening agent, no significant differences were found between any of the erythritol-xanthan gum muffins and the control muffin. The muffin height also increased in the control muffin but no improvement in muffin height was found in the erythritol muffins. Adding xanthan gum to the control muffin and doubling the leavening agent slightly increased the weight loss during baking. This can be related to the increased volume of the muffins, which means that a greater surface area is available for water evaporation. In the erythritol muffins, the addition of xanthan gum and a double quantity of leaving agent conferred a similar weight loss to the control muffin, so this combination counterbalances the decrease in weight loss attributed to the erythritol.

2.7 Muffin texture

The elimination of sucrose had a considerable effect on the hardness of the muffins (Table 1). 50% reduction of sucrose gave samples with a significant increase in hardness. The same behaviour applied to chewiness. Springiness and resilience were not affected by sucrose reduction but cohesiveness decreased significantly when more than 75% of sucrose was removed.

The erythritol muffins also showed significantly different hardness values to the control (Table 2). It was only for 25% sucrose replacement with erythritol that no significant differences were found. Chewiness followed the same trend as hardness. On the other hand, springiness, cohesiveness and resilience decreased when up to 50% of the sucrose was replaced by erythritol, reflecting a more compact crumb.

When xanthan gum was added and sucrose was replaced by erythritol the hardness and chewiness values were significantly higher than for the control, but when the formulation was improved the hardness and chewiness values were lower than those of the control muffins (Table 3). Cohesiveness showed similar values to the control muffins. The same behaviour was observed with the springiness values. The resilience values were lower in all the formulations except 75%ErXan, where the values were similar to the control muffins, and 25%ErXan and 50%ErXan, which obtained higher values.

3 CONCLUSIONS

Erythritol was able to partially simulate sucrose functionality in Spanish muffins, producing an increase in the air content of the batter and in muffin height and number of bubbles in comparison with the corresponding low sucrose muffins. The positive effect of erythritol on muffin properties can be associated with the fact that erythritol, like sucrose, reduces the decrease in elasticity during heating and raises the thermosetting temperature. However, the effect of the erythritol was not enough to improve the volume and texture of the muffins.

The combination of erythritol with xanthan gum and double quantities of leavening agent was effective in improving the texture properties of the muffins, reducing their hardness;

this was associated with a significant increase in batter air content and in muffin height, volume and number of bubbles in comparison with the employment of erythritol alone.

References

1 P.M. Olinger and V.S. Velasco, *Cereal Food World*, 1996, **41**, 110.
2 B.S. Kamel and V.F. Rasper, *J Texture Stud*, 1988, **19**, 307.
3 M.R. Baeva, V.V. Terzieva and I.N. Panchev, *Nahrung/Food*, 2003, **47**, 154.
4 F. Ronda, M. Gómez, C.A. Blanco and P.A. Caballero, *Food Chem*, 2005, **90**, 549.
5 F. Haji, *Agro Food Ind Hi Tec*, 2009, **20**, 36.
6 J. Goossens and H. Roeper, H, *Food Sci Technol Today*,1994, **8**, 144.
7 K. Noda, K. Nakayama and T. Oku, *Eur J of Clin Nutr*, 1994, **48**, 286.
8 T. Oku and K. Noda, *Nutr Res*, 1990, **10**, 987.
9 F. Ronda, M. Gómez, P.A. Caballero, B. Oliete and C.A. Blanco, *Food Sci Technol Int*, 2009, **15**, 193.

LINEAR VISCOELASTIC PROPERTIES OF SHORT DOUGH ENRICHED WITH RESISTANT STARCH

L. Laguna[1], M. J. Hernández[2], A. Salvador[1] T. Sanz[1]

[1] Instituto de Agroquímica y Tecnología de Alimentos (CSIC), Avda. Agustín Escardino, 7, 46980 Paterna (Valencia), Spain.
[2] Dpto. de Física de la Terra i Termodinàmica, Facultats de Física i Farmacia, Universitat de València, Burjassot 46100, Valencia, Spain.

1 INTRODUCTION

Resistant starch (RS) refers to the sum of starch and starch degradation products not absorbed in the small intestine of healthy individuals[1]. The fact that RS escapes digestion confers many positive physiological health effects[2]. RS is, therefore, a dietary fibre, as stated in the latest approved dietary fibre definition by the AACC International. In comparison to more traditional fibres, RS confers better appearance, texture and mouth feel and possesses the advantage of its white colour[3]. Due to its positive health effects and technological advantages, different foods have been enriched with RS.
The objective of the present work was to investigate the effect of RS incorporation in the structure of short dough biscuits by means of the study of the dough linear viscoelastic rheological properties. Oscillatory and creep and recovery tests were performed in a short dough biscuit formula without RS (control) and with 20%, 40% and 60% wheat flour replacement by RS.

2 METHODS AND RESULTS

2.1. Short dough biscuit preparation

Four formulations were prepared using the same quantity of all the ingredients except the wheat flour and RS. The proportions of these two ingredients were 100:0 (control), 80:20, 60:40 and 40:60. The dough biscuit ingredients were: a) soft wheat flour, a source of RS (Hi-maize 260, National Starch Food Innovation), shortening 30g/100g, sugar 15g/100g, milk powder 0.3 g/100g, salt 0.1 g/100g, g, sodium bicarbonate 0.1 g/100g, ammonium hydrogen carbonate and tap water 4.44 g/100g.
The shortening was creamed in a mixer for 4 minutes to obtain a homogeneous cream. After this, the sugar was added and mixed in for 2 minutes at speed 4. The milk powder, previously dissolved in all the water, was added and mixed in for 2 minutes at the minimum speed. Finally, the flour (or flour/RS), sodium bicarbonate and ammonium hydrogen carbonate were mixed in together at minimum speed for 2 minutes. The dough was sheeted 4 mm and frozen during 24h hours.

2.2. Rheological experiments

Oscillatory and Creep and Recovery tests were performed at 25°C in a controlled stress rheometer (AR-G2, TA Instruments, Crawley, England). A 20mm roughened parallel plate with a gap of 1.5 mm was employed.
Frequency sweep test within the linear region were carried out at a frequency from 10 to 0.01 Hz. Values of G', G'' and tan δ were recorded.
Creep and recovery tests were carried out by applying instantly a constant stress within the linear viscoelastic region, σ_0, which was maintained for 1200s, then removed, allowing the sample to relax for 1200s. The system deformation per unit stress, called compliance: $J(t) = \gamma(t)/\sigma_0$, was measured.

2.3. Frequency sweep test

The frequency sweep obtained in the linear region for the different biscuit dough formulae is shown in Figure 1.In addition in Table 1 values of G' and tg δ at 1Hz are shown.
The rheological behaviour found in all the doughs is typical of soft gels with values of G' higher than G'' and with a soft dependency of both moduli with frequency. The incorporation of RS increases the values of G', although no significant differences were found among the control dough and the 20RS dough. The values of tan δ reveal that no significant differences were found among the different samples, so incorporation of RS does not affect significantly the viscoelastic properties.

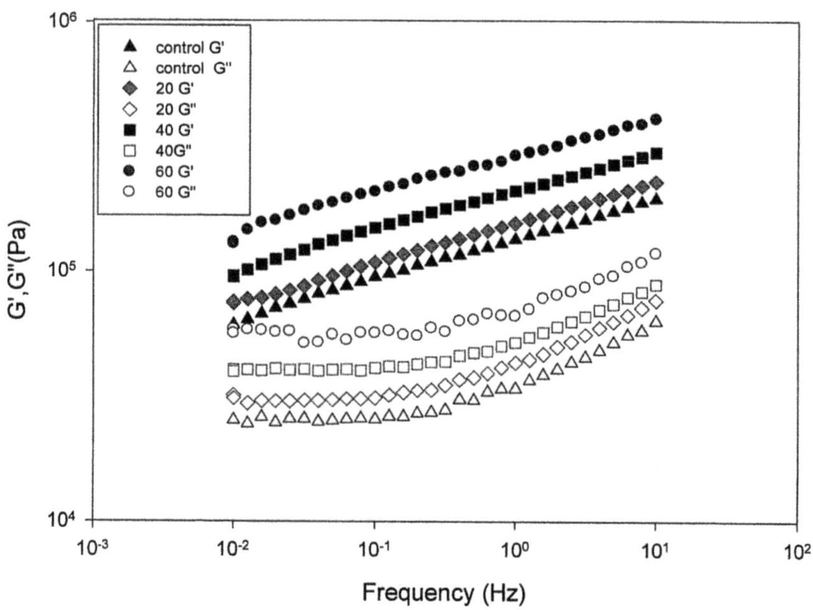

Figure 1 Influence of RS in the frequency sweep.

Table 1 Influence of RS in G' and tan δ at 1 Hz.

Type of formula	*G' (Pa)*	*tan δ*
control	107000a	0.27a
20RSRI	123933.3a	0.29b
40RSRI	162400b	0.26a
60RSRI	217866.7c	0.257a

2.4. Creep and Recovery tests

The values of compliance (J) as a function of time for the creep and recovery tests are shown in Figure 2.
The creep data were well adjusted to the Burger model (equation 1).

$$J(t) = \frac{1}{G_0} + \frac{1}{G_1}\left[1 - \exp\left(\frac{-t\,G_1}{\eta_1}\right)\right] + \frac{t}{\eta_0} \qquad (1)$$

The vales of G_0, η_0, G_1 and η_1 are reported in Table 2.

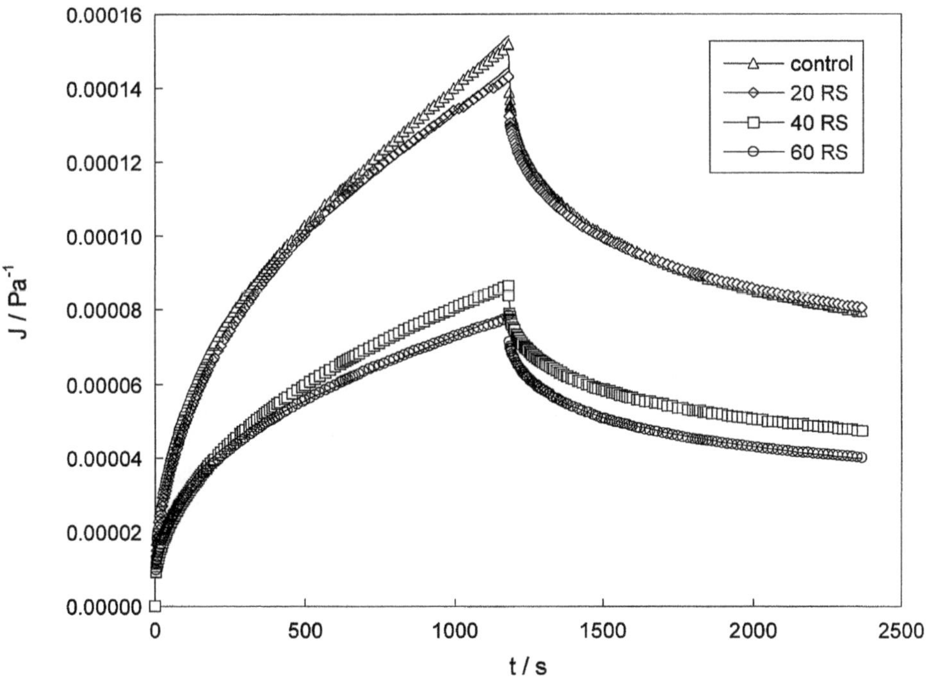

Figure 2 Influence of RS in the creep and recovery tests

Table 2 Creep parameters

Type of dough	$G_0 \times 10^{-4}$ (Pa)	$\eta_0 \times 10^{-7}$ (Pa s)	$G_1 \times 10^{-4}$ (Pa)	$\eta_1 \times 10^{-6}$ (Pa s)
control	6.50ab	1.34a	1.95a	1.70a
20RSRSI	5.60a	1.57b	1.84a	2.78a
40RSRI	9.80bc	2.57c	3.23b	5.14b
60RSRI	9.90c	2.94d	3.41b	4.39b

The replacement of wheat flour by RS at levels of 40% or higher reinforced the structure, incrementing significantly all the parameters of the Burger model.
The values of J as a function of time during the recovery test were adjusted to the following equation (equation 2)

$$J(t) = J_\infty + J_{KV}.\exp(-B.t^C) \qquad (2)$$

The values of J_{SM} were calculated as

$$J_{SM} = J_{MAX} - (J_\infty + J_{KV})$$

and the final percentage recovery as

$$\%R = [(J_{MAX} - J_\infty)/ J_{MAX}] \times 100$$

The values of B and C were very similar in all the dough formulations with mean values of B= 0.045±0.005 and C= 0.43±0.01. The other parameters are shown in Table 3.
The recovery parameters were significantly affected by incorporation of RS, however the differences among the control and the 20% RS samples were non significant. The % R was practically the same in all the samples implying that incorporation of RS in the dough does not affect the type of structure of the control dough.

Table 3 Recovery parameters

Type of dough	$J_{max} \times 10^5$ (Pa^{-1})	$J_\infty \times 10^5$ (Pa^{-1})	$J_{kv} \times 10^5$ (Pa^{-1})	$J_{sm} \times 10^6$ (Pa^{-1})	%R
Control	15.2d	4.4b	10.2b	5.3c	70.50a
20 RSRI	14.3c	3.9ab	10.0b	4.1b	73.00a
40 RSRI	8.6b	2.6ab	5.8a	2.9a	70.50a
60 RSRI	7.8a	1.9a	5.6a	2.5a	75.00a

References

1 N.G. Asp and I. Björck. *Trends Food Sci. Tech.*, 1992, **3**, 111.
2 P. Yue and S. Waring. *Food Australia*, 1998, **50**, 615.
3 R. Baixauli, A. Salvador, G. Hough and S.M. Fiszman, S. M. *Food Qual. Prefer.*, 2008, **19**, 628.

Health related aspects

THE APPLICATION OF CELLULOSE ETHERS AS RHEOLOGICAL MODIFIERS IN A CONTROLLED DROPLET DELIVERY SYSTEM

G.H. Zhao[1], N. Kapur[2], J. T. Guthrie[1]

[1]Department of Colour Science, School of Chemistry, University of Leeds, Leeds, UK
[2]School of Mechanical Engineering, University of Leeds, Leeds, UK

1 INTRODUCTION

Cellulose ethers are popular rheological modifiers in the food, cosmetics, paints, inks and pharmaceutical industries because of their interactive behaviour and their biodegradability, when required. The majority use of this type of additive is required to ensure the effective formulation, delivery and stabilisation of delivered components. Additionally they can assist in the controlled release of active components under specific environments.[1-3]

Attention has been given to the application of cellulose ethers especially, hydroxypropyl cellulose, (HPC), and sodium carboxymethyl cellulose, (CMC-Na) in liquid-dose formulations. There are still concerns with respect to the use of cellulose ethers as the rheological modifiers. These concern the quality and the consistency of the cellulose sources, the reliability of their industrial production and potential interactions with other components.[4,5] Therefore, in their use, it is necessary to understand the physical and chemical nature of cellulose ethers and their composite formulations. In this regard, the characterisation, manufacturability and deliverability of additives in appropriate formulations is vital. The goal of this study was to establish the contribution made by two selected polymer types, Na-CMC and HPC, the samples being selected on the basis of their molecular weight and molecular weight distribution and on the basis of their degree of substitution, with respect to the properties they give to their composite solutions and with respect to their "delivery" and their "delivery process"

2 MATERIALS AND METHODS

2.1 Materials

HPC in three different grades, identified as (L, SL and SSL) was supplied by Nisso, Germany. Three grades of CMC denoted (7M 8SF, 12M 8P and 7H OF) were supplied by IMÇD, ltd, UK, and manufactured by Ashland, Wilmington, USA. Deionised water was used to make up to the appropriate solutions.

2.2 Sample preparation

Several concentrations of the solutions were prepared for the various studies. For the rheological characterisation, the loading of the CMCs was 2% (w/w). The samples were denoted 7H CMC 2, 7M CMC 2 and 12M CMC 2. These samples were characterised by rotational rheometry and by capillary extensional breakup rheometry, (CaBER).

Previous studies have shown that by selecting different loadings for the three grades of the HPCs, namely, HPC-L, HPC-SL and HPC-SSL, it is possible to give a similar shear viscosity of the three solutions, W L 6% (w/w); W SL 8% (w/w) and W SSL 15% (w/w). This then allows meaningful extensional rheological behaviour to be studied.

These solutions were assembled by dissolving the relevant amount of the polymer in deionised water. The process of dissolving was optimised by the use of magnetic stirring, for 48 hours, at 25°C.

2.3 Methods

Size exclusion chromatography (SEC)

Solutions of the CMC samples and the HPC samples were analysed by size exclusion chromatography using a Suprema Linear-M, Mixture Bed column, S/N: 7041112. To support the determination of the average molecular mass and polydispersity, an Optiab DSP interferometric refractomer (Wyatt Technology Corporation, USA) was used. The GPC was characterised using mono-dispersed poly(ethylene oxide) in water, and had a delay volume of 0.225 mL. The eluent contained 0.1 M sodium nitrate and 0.001 M sodium azide, in distilled water. The reasons for adding the sodium nitrate and the sodium azide were to screen the charge of electrolyte and to prevent bacterial growth, respectively. The specific refractive index increment of the polymer solution was obtained using an Optiab DSP interferometric refractomer (Wyatt Technology Corporation, USA) was 0.137 cm^3/g at 25°C. This value is in close agreement with that obtained by Pramanik.[6]

Rheological characterisation (controlled shear approach)

The rheological properties of the assembled hydrogels were assessed using a AR1500ex Rheometer (TA Instruments, Crawley, UK). This unit gives either a controlled shear rate or a controlled shear stress, in a cone and plate rheometry format. A 60 mm cone (1° angle geometry) was used for all of the measurements, to provide uniform shearing to the sample. Samples were transferred to the plate with the minimum of shearing. The equilibration time before measurement was 2 minutes. The temperature was kept at 25±0.1°C for each measurement.

CaBER evaluations

The extensional rheological behaviour of the CMC solutions was evaluated using Capillary Break-up Extensional Rheometry (CaBER), (Haake, Thermo Electrical C. Germany). In this, the sample was placed between two plates (diameter 4mm) with a distance of 2mm. The Hencky strain range used during the measurements was between 1.04 and 1.56. The typical stretch time applied for all of the measurements, was 40ms. The conditions were chosen according to the pre-experimental condition of the samples. Every measurement was repeated 10 times in order to ensure consistency in behaviour.

Trimaster evaluations

A capillary thinning device that was used in the study, was the Trimaster unit, (Department of Chemical Engineering and Biotechnology, University of Cambridge, Cambridge, UK). The mechanism and the operation of the Trimaster unit has been explained by Vadillo.[7] The Trimaster unit operates with the initial distance between two plates, L_o, of 600±10 μm. The diameter of the plate, D_o, was 1.2mm. A final stretch distance of 1.6mm for each side of the plate was used in the characterisation of the formulations. The aspect ratio of the measurements was 2.7. The high speed camera, which was used to record the thinning process, was a Photron Fastcam 1024 PCI, (Photorn, Bucks, UK). A frame rate of 6000 frame per second (fps), a shutter time of 3 μs, and a solution of 512×256 pixels were chosen in the recording of the processes. The camera was synchronized with the movement of the plates. Thus, the high speed camera recorded the whole process from stretching and thinning, to breaking up. There were ten repeat measurements for each sample.

3 RESULTS AND DISCUSSION

3.1 Molecular weight identification

The results from the determination of the average molecular weight and the polydispersity of the HPC samples and the CMC samples are listed in Table 1. For the HPC samples, the average molecular weight increases from SSL, SL to L, whilst the polydispersity and the molar substitution shows a similar pattern. For CMC-Na, 7H CMC has double the average molecular weight of 7M CMC, while the values of molar substitution are similar. 7M CMC possesses a similar average molecular weight to that of 12M CMC, while the molar substitutions varied.

The information relating to the molar substitution values was obtained from the suppliers. The characterisation of the molecular weight distribution of cellulose ethers is fundamental to any effective investigation of their other physical properties, such as the interactions that take place within the polymer solutions and as observed in studies of rotational viscometry rheology, and extensional rheology.

The shear viscosity profiles of the solutions of the three CMC samples, at 2% (w/w), in the shear rate range of 0.1 to 1000s^{-1}, have been determined and are shown in Figure 1 (a). 7H CMC 2 gives a steeper shear thinning profile than that of 7M CMC 2 and that of 12M

Table 1 *Average molecular weight, polydispersity and molar substitutions of sample HPC and CMC*

	Average M_w (g/mol)	Polydispersity	Molar substitution
HPC-SSL	36,000	1.56	3.3
HPC-SL	76,000	1.45	3.3
HPC-L	98,000	1.56	3.3
7M CMC	204,000	2.11	0.7
12M CMC	296,000	1.81	1.2
7H CMC	450,000	3.55	0.7

3.2 Shear rheological characterisation

CMC 2, due to its relatively high molecular weight and polydispersity. In addiiton, the viscosity of 7H CMC 2, at the shear rate of 0.1 to 100 s^{-1}, is greater than that of 7M CMC 2 and that of 12M CMC 2, due to the greater molecular weight. The effect of the degree of substitution on the shear thinning profile of 7M CMC and of 12M CMC is also illustrated in the shear rate range of 0.1 to 100 s^{-1}, in Figure 1 (a).

In addition, at the loading of 2% of 7M CMC, 12M CMC and 7H CMC, the solutions gave shearing thinning behaviour, indicating that the CMC molecules, in solution, presented complex structures. These stuctures were caused by entanglements within the polymer chains in the solution and are the result of the high molar mass and the high polydispersity. This structure is relatively weak, and can be destroyed by moderate shearing at 100 s^{-1}. The viscosities of these three CMC solutions possessed a value after the shear rate of 100 s^{-1}. The more substituted 12M CMC has a stonger network, which means that there are more interactions between the substituents than exist in the corresponding aqueous solution of 7M CMC.

The shear viscosity profiles of the three HPC samples, at the identified loadings, in the shear rate range of 0.1 to 1000s^{-1}, were determined and are shown in Figure 1 (b). All of the three HPC solutions behaved as a Newtonian fluid, due to the relatively low average molecular weight at the relatively high molar substitution value. This indicates that interactions between the polymer chains and the solvent dominate. Thus, three loadings were chosen, according to the average molecular weight, to give a similar viscosity value for each of the all aqueous solution, W L 6, W SL 8 and W SSL 15, Figure 1 (b). This approach assisted in establishing the effect of molecular weight of the HPC in the solutions on the extensional rheological properties, by avoiding any variations in the flow behaviour as encountered in extensional behaviour, (stretching).

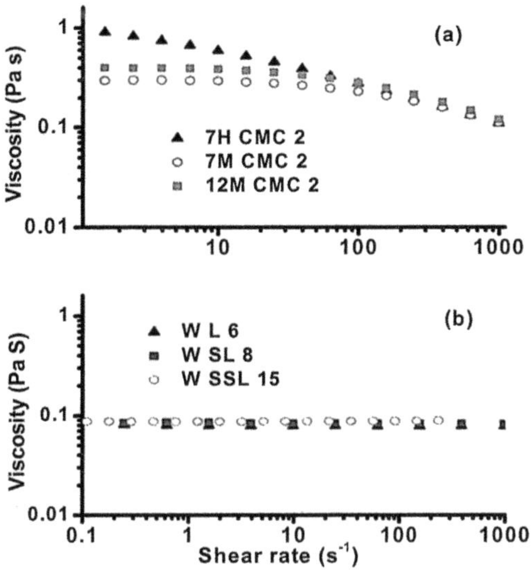

Figure 1 *The variation of the viscosity versus the variation of the shear rate of the identified samples, (a) Identified CMC samples; (b) Identified HPC samples*

3.3 CaBER

The extensional rheological behaviour of solutions of each of 7H CMC 2, 7M CMC 2 and 12M CMC 2 was measured on the CaBER unit, under Hencky strains of 1.04 and 1.46. A comparison of the breakup time of solutions of these three samples is illustrated in Figure 2. During a typical CaBER measurement, the elastic forces, the viscous forces and the surface tension forces attempt to hold the shape of the filament at the beginning of the break up process. The elastic properties of the fluid dominate. However, gravitational forces and surface tension forces act on the filament to try to break it. Near the end of the break up process, a Newtonian response was observed. The break-up time was increased by increasing the elastic properties, the viscous properties and the surface tension. Figure 2 shows greater differences with respect of the break up time than are seen in studies of the shear viscosity. The relaxation time of the molecular chains of 7H CMC 2 was much longer than that of 7M CMC 2 and that of 12M CMC 2, due to the increase in the average molecular weight. In addition, the difference in the break up time caused by a difference in the molar substitution has an effect on the extensional behaviour, see the break up time for 7M CMC 2 and 12M CMC 2, Figure 2. The greater the molar substitution of the polymer, the longer were the breakup times. However, the influence of the molar substitution on the extensional viscosity was less than the influence of the average molar mass. Only two Hencky strain values were applied during the CaBER experiments.

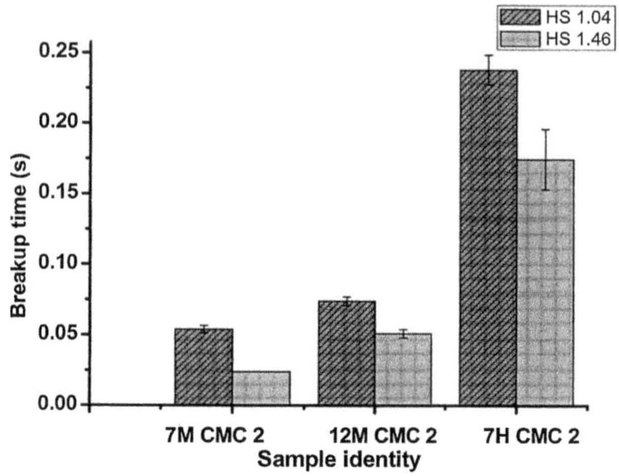

Figure 2 *The breakup time of the identified CMC samples with the Hencky strains of 1.04 and 1.46*

3.4 Trimaster

Aqueous solutions of three HPC samples, denoted W L 6, W SL 8, and W SSL 15 (the number indicates the (w/w %) loading of the polymer in the aqueous solutions) were evaluated using a capillary thinning device, the Trimaster. The mechanism and the operation of the Trimaster in measurement are described elsewhere.[7,8] Solutions of these three samples were formulated to give the same viscosity of 80mPa s in an investigation of the effect of only the elasticity of the solution on the behaviour of break up. Figure 3 gives the experimental results of the break up processes of the three solutions as provided by a high speed camera, at the frame rate of 6000fps. The break up time of W SL 8 was slightly longer than that of W SSL 15, even through W SL 8 and W SSL 15 had a similar viscosity. W L 6 gave the longest break up time among these three solutions. The effect of the elasticity on the filament breakup behaviour is shown in Figure 3.

The extensional rheological behaviour of three solutions is shown in Figure 4, by using the relationship between the Trouton ratio and the Hencky strain. The Trouton ratio is the ratio of the extensional viscosity to the shear viscosity. The Hencky strain is defined as the ratio of the thickness of the filament to the original thickness of the filament.[7] Thus, Figure 4 gives the extensional viscosity increase during the extension process, (during the increase of the Hencky strain) until filament breaking up ooccurs. The extensibility of these three solutions is relatively weak. The extension becomes broken at the Hencky strain of 3. Solutions of HPC-L showed a relatively greater extensibility than solutions of HPC-SL and solutions of HPC-SSL. This indicates that the solutions of HPC-L showed have longer break up length than those of HPC-SL and of HPC-SSL, during the drop delivery. It also increases the probability that satellite drop formation will occur and reduces the degree of control that can be provided during the delivery.

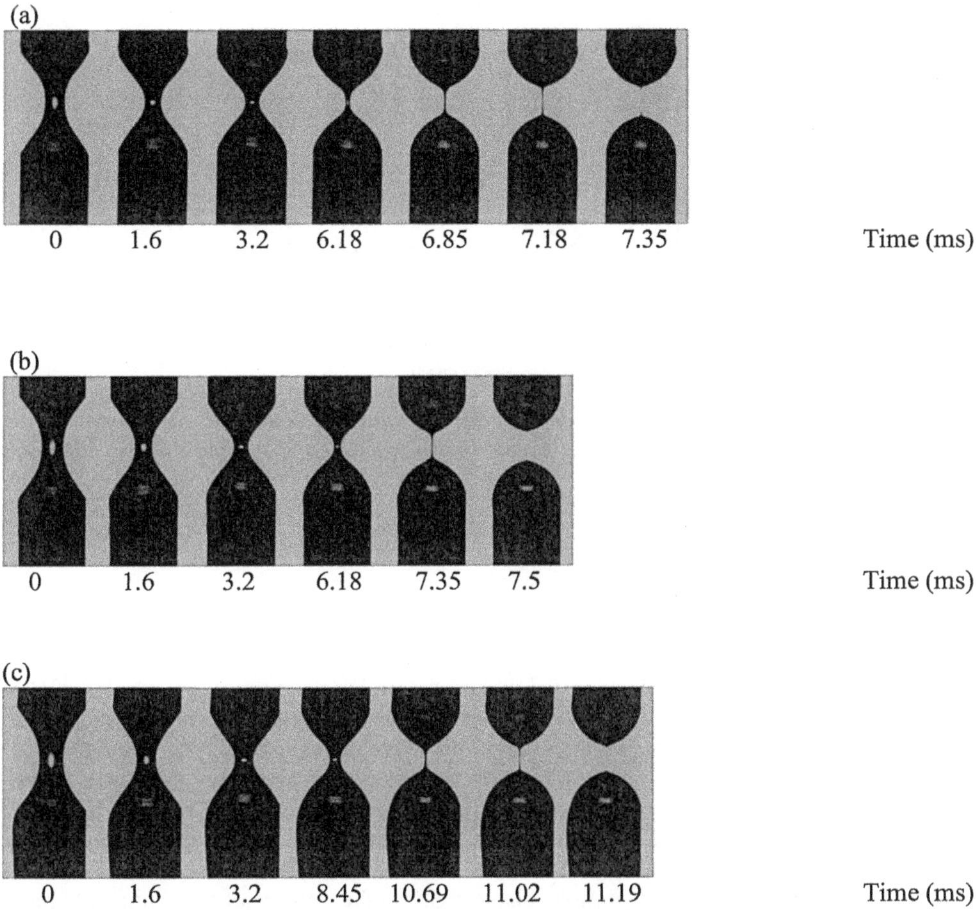

Figure 3 *The high speed images from measurements of the extensional viscosity of the (a) aqueous solutions W SSL 15; (b) aqueous solutions W SL 8 and (c) aqueous solutions W L 6*

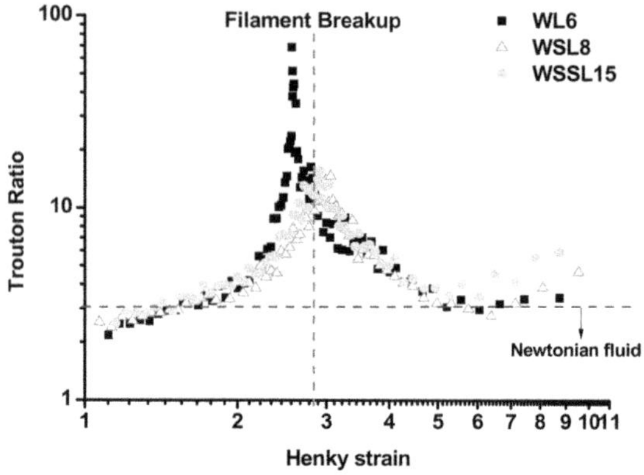

Figure 1 *Extensional rheological behaviour of the identified fluids as shown by the variation of the Trouton ratio versus the Hencky strain of the identified samples*

4 CONCLUSION

The effects of the average molecular weight and the molar substitution of samples of HPC and of CMC on the shear rheological behaviour and on the extensional rheological properties of their respective aqueous solutions were investigated. The average molar weight plays an important role with respect to the presence of complex structures in their aqueous solutions. For the solutions studied here, these complex structures can be broken down at the shear rate of $100s^{-1}$. The influence of the molar substitution of the polymers on the shear rheology is felt in the interactions within between the polymer chains. In addition, the effect of the average molecular weight on the process of the filament breakup was greater in the extensional rheology, (the relaxation time of the molecular chains), than on the shear rheology. Attention should always be given to the behaviour of the solution when interpreting the data relating to the properties of the chosen polymer/solvent combinations for controlled-delivery.

Acknowledgement

The authors acknowledge the valued contributions made by the EPSRC (Dorothy Hodgkin award for G.H.Z).

References

1. Hamman, J.H. and Tarirai,C. *Functional excipients.* Chimica Oggi-Chemistry Today, 2006. **24**(5): p. 57
2. Chang, C.Y. and Zhang, L.N. *Cellulose-based hydrogels: Present status and application prospects.* Carbohydrate Polymers, 2010. **84**(1): p. 40-53.
3. Kamel, S., et al., *Pharmaceutical significance of cellulose: A review.* Express Polymer Letters, 2008. **2**(11): p. 758-778.

4. Pifferi, G., Santoro, P., and Pedrani, M., *Quality and functionality of excipients.* Farmaco, 1999. **54**(1-2): p. 1-14.
5. Pifferi, G. and Restani, P. *The safety of pharmaceutical excipients.* IL Farmaco, 2003. **58**: p. p: 541-550.
6. Pramanik, A. and Chowdhur.Pk, *Degration of Na amylose Xanthate in dilute caustic soda solution druing ripening.* Journal of Polymer Science Part a-1-Polymer Chemistry, 1969. **7**(4PA1): p. 1055-&.
7. Vadillo, D.C., et al., *Evaluation of the inkjet fluid's performance using the "Cambridge Trimaster" filament stretch and break-up device.* Journal of Rheology, 2010. **54**(2): p. 261-282.
8. Anna, S.L. and McKinley, G.H., *Elasto-capillary thinning and breakup of model elastic liquids.* Journal of Rheology, 2001. **45**(1): p. 115-138.

PECTIN-ZEIN HYDROGELS FOR THE DELIVERY OF DRUGS AND NUTRIENTS

Z. K. Muhidinov, Kh. I. Teshaev, G. F. Kasimova, A. S. Nasridinov, and L. S. Liu*

Chemistry Institute of Tajikistan Academy of Sciences, Dushanbe 734063, Tajikistan
*Easter Regional Research Centre, ARS, USDA, 600 East Mermaid line, Wyndmoor, PA 19038, USA

1 INTRODUCTION

Polysaccharide/protein coacervates are widely used for oral drug delivery. In comparison with other drug formulations, those of polysaccharide/protein have several advantages, such as easy preparation, only simple and inexpensive apparatus are required, the reaction can be conducted at room temperature and under environmentally friendly conditions, etc[1]. Furthermore, both polysaccharide and protein are naturally occurring biopolymers; if the polysaccharide and protein used are food-grade materials, the products are considered to be biocompatible. The procedure to obtain FDA approval may also be relatively simple. The main challenge to broaden the use of polysaccharide/protein coacervates is how to enhance drug encapsulation efficiency. This is typically critical with drugs of low molecule weight, where usually only a small portion of the drug can be encapsulated.

In the present study, we explored a new approach to prepare polysaccharide/protein coacervates in aqueous solution. Piroxicam (PX) was chosen as a model drug. This weakly acidic organic compound is a non-steroidal anti-inflammatory drug with poor water solubility. The dissolution and diffusion of PX from the polysaccharide/protein hydrogel beads was also evaluated against the changes in composition and ratios of the biopolymers in coacervates. Drug formulation conditions were optimized in an attempt to obtain higher efficiency of drug encapsulation and to prolong the duration of drug release.

2 METHOD AND RESULTS

2.1 Isolation and Identification of Corn Zein and Different Pectin

Zein proteins from corn were prepared according to methods previously reported[2-3]. The proteins were further isolated and purified, and used as the hydrophobic components in hydrogel formulation in the following study. Two types of low methoxyl pectins from citrus (GENU LM-12CG; GA, 69.0%; DE, 38%; Mw, 138KD) and (GENU L/200-CP Kelco) GA, 69,6 %; DE, 9 %; Mw, 466KD) were obtained from CP Kelko (Wilmington, DE) and used as the hydrophilic components in the complex preparation.

2.2 Turbidimetric titration method for investigation of ionotropic gelation process between pectin and zein

The formation of a complex hydrogel from hydrophilic pectin and hydrophobic zein in water-ethanol solution containing two diferent concentrations of Ca^{2+} ions was determined by monitoring the changes in solution turbidity of the mixtures, using a 2100 ANIS (HACH, USA). Data obtained was expressed as $\Delta T_2/\Delta\gamma$ as described in refrerence[3]. Zein solutions of different concentrations was titrated with 0.1% pectin (GENU LM-12CG) solution. After each addition of a fraction of pectin solution, the turbidity of the mixture was dertermined using a nephelometric unit (NTU). The solution turbidity (T_0) of 0.1% pectin solution was 3.3

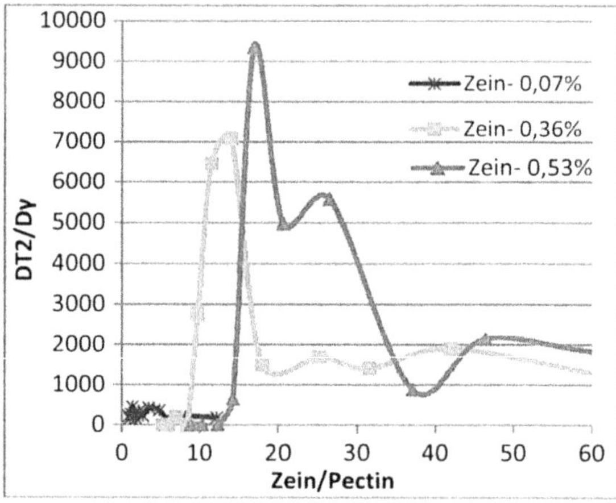

Figure 1 *Differential curve of the turbidimetric titration of corn zein by 0.12 % pectin (GENU LM-12CG) in presence of 0.0028% $CaCl_2$*

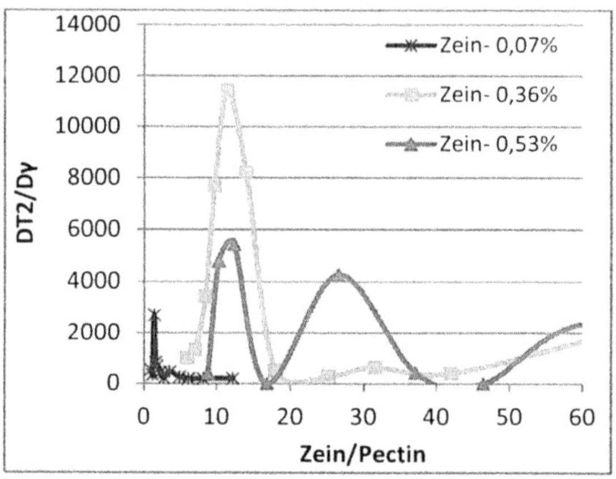

Figure 2 *Differential curve of the turbidimetric titration of corn zein by 0.12 % pectin (GENU LM-12CG) in presence of 0.028% $CaCl_2$*

NTU; T_0 of the zein solutions increased from 6.8NTU to 151 NTU as zein concentrations increased from 0.07% to 0.53%.

The results of the turbidity titration are shown in Figures 1 and 2. At the lowest zein concentration (0.07%), the solution turbidity was very low at both low and high calcium concentration and hence complex formation under the current experimental conditions was not significant. When the zein concentration was increased to 0.36%, the solution turbidity increased with the addition of pectin, indicating pectin/zein complex formation. The titration curves showed a multi-modal distribution characteristic. This, probably, relates to the formation of two types of complex: one for pectin- Ca^{2+} zein and the second for pectin-zein complexes. The third peak could be attributed to particle association. Further increase of the zein concentration to 0.53% resulted in an increase in peak intensity, regardless of the content of calcium ions. The increase of zein concentration also had an effect on peak position. At lower Ca^{2+} concentration (0.0028%), the peak position shifted to a higher zein/pectin ratio (Figure1), because more hydrophobic polymer was required for phase separation to occur within the hydrogels. In contrast, at a higher Ca^{2+} concentration (0.028) the peak shifted to a lower zein/pectin ratio (Figure 2).

2.3 Preparation of Pectin/zein hydrogel and its characterisation

Pectin/zein complexes were formed by adding pectin (GENU LM-12CG) solution (2 %) drop-wise into a 75 % ethanol solution containing zein and piroxicam with varying concentrations, as well as Ca^{2+}. The amount of metal ions, Ca^{2+} used was calculated as corresponding to the maximum gelling point of calcium pectate (30 mg Ca^{2+} per gram pectin at room temperature). The composition and ratios of the three components are shown in Table 1. The resultant complexes in the form of hydrogel beads were washed with distilled water, and with 70 % ethanol to remove non-reacted components, and then air-dried at room temperature until no weight change could be detected.

Table 1 *Complex yields, swelling degree and piroxicam entrapment percentage in pectin/zein hydrogel of different ratio in presence of calcium ion.*

P/Z ratio	Complex yields, %	Swelling degree	Piroxicam, %
6.0	18.57	139.0	12.4
3.0	17.50	179.0	9.8
1.0	40.00	49.0	37.0
0.7	47.06	3.2	57.0
0.5	58.67	19.4	72.6
0.2	66.67	7.5	93.8

The hydrogel beads thus obtained were characterized for yield, swelling behaviour, and drug encapsulation rate. The results are shown in Table 1. Both swelling behaviours and drug loading efficiency were found to be dependent on the pectin/zein ratio. The more zein in the complex, the less the beads swelled. The presence of zein suppressed beads swelling. Whereas, drug loading efficiency increased as zein fraction increased in the complex. This fact, possibly, is connected with the nature of the complex and the drug. The driving force of encapsulation is considered to be an enthalpic effect. The hydrophobic piroxicam is likely dispersed in the zein matrix.

Table 2 *Comparison of beads output, swelling and PX loading capacity for formed beads with calcium or zinc ions.*

P/Z ratio	Me^{++} in complex,%		Complex yield, %	Swelling degree	PX in Complex, %
	Ca^{2+}	Zn^{2+}			
3.0	3.56		64.20	29.10	81.80
1.0	2.44		76.30	10.33	66.00
0.7	2.02		46.60	21.00	95.20
0.5	3.70		66.70	13.69	75.80
3.0		1.85	61.20	31.10	86.00
2.0		1.08	28.60	19.50	52.80
0.7		1.27	45.00	16.00	80.60
0.5		1.08	34.50	5.80	80.60

In the following experiment, we compare the effect of Ca^{2+} and Zn^{2+}on complex formation and drug loading. A low DE pectin (GENU L/200) and double amount of PX were used. The results were summarized in Table 2. In contrast with using calcium ions, the use of Zn^{2+} in gel formation produced upside-down bell curves when plotting pectin/zein ratio vs. gel yield (or swelling degree, or drug loading). At the pectin/zein ratio of 2.0, all parameters showed the lowest value.

2.4 Drug release investigation from pectin/zein composites

The release of PX from pectin/zein hydrogel beads[4] was investigated using a set of buffer solutions that mimic the conditions of the gastrointestinal (GI) tract: A 0.2M NaCl/KCl buffer at pH 1.2, followed by a 0.2 M phosphate buffer at pH 6.4, and then a 0.2 M phosphate buffer at pH 6.8 containing pectinase. All experiments were conducted according to the USP standard method in a standard dissolution apparatus at 37 °C[5]. 10 g pectin/zein beads were suspended in 60 ml of dissolution medium. At predesignated time intervals, 3 ml sample solutions from each test were withdrawn using a syringe equipped with a Millex Millipore filter (pore size, 0.45μ). 3 ml of fresh medium were added back to the vial to maintain a constant volume. The amount of drug released was calculated based on the UV readings (at 335 nm for sample solutions at pH 1.2 and at 355 nm for sample solutions at pH 6.4). Standard calibration curves were plotted using PX solutions of known concentrations at pH 1.2 and pH 6.4, and used for calculating PX amount in solutions. The quantity of entrapped drug in the beads M_o and released at time M_t was determined.

The results (Figures 3 and 4) showed a faster release (80% in 6 hours) for the samples with high pectin fraction and a slower release (40% in more than 24 hours) for the samples of high zein fraction. The high release rate is likely to be due to the pectin swelling under conditions where the pectin chain more expanded. This means that PX is mainly dispersed within the calcium pectin gel rather than the zein cavities. The zein fraction may only retard the drug by steric hindrance on the pectin surface. However the increase in the zein fraction significantly changed the release mechanism. As mentioned above, PX was mainly entrapped in zein zones. The two step release profile were viewed as being due to the two different mechanisms.

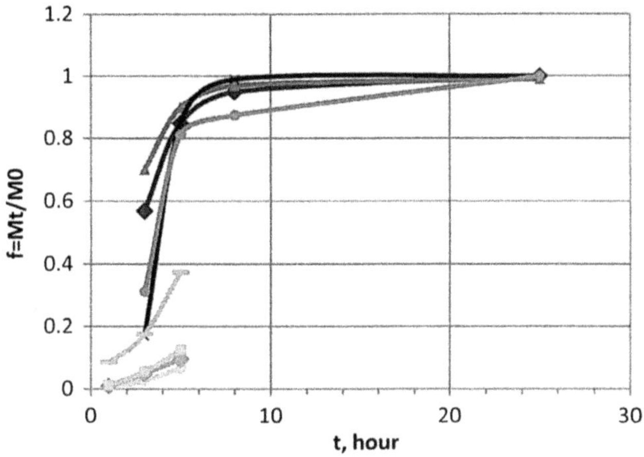

Fig 3 *Piroxicam release rate in the buffer solutions mimicking GI tract, pH 1,2 and 6,4, from the beads formed by different LM citrus pectin/zein ratio containing different level of Ca^{2+} and PX, mg: Ca^{2+} -0,04 , PX-50 (◊); Ca^{2+} -0,02, PX - 20 (Δ) ; Ca^{2+} -0,04, PX 50(X) and Ca^{2+-}0,04, PX 20 (•)*

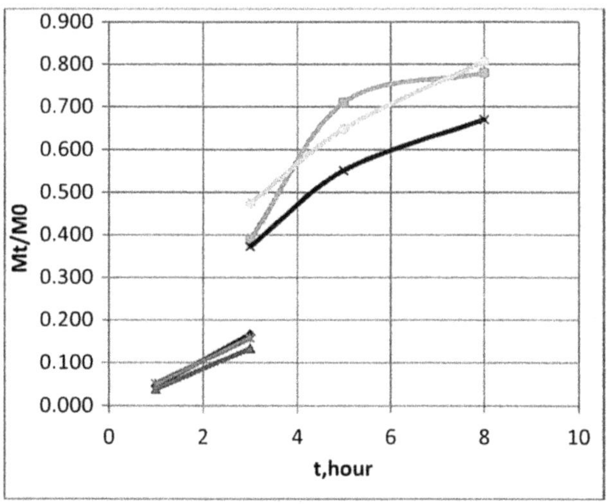

Figure 4 *Piroxicam release rate in the buffer solutions mimicking GI tract, pH 1,2 and 6,4 from the beads of different LM citrus pectin/zein ratio containing 0.04 mg Zn^{2+}:P/Z 3.0 (□); P/Z 1.5 (X); P/Z 0.5 (•).*

Furthermore, comparing Ca^{2+} cross-linking (Figure 3) and Zn^{2+} cross-linking (Figure 4), it was clear that at the lower pH, PX released from Zn^{2+} cross-linked beads had a profile that was similar to that obtained from Ca^{2+} cross-linked; both exhibited near zero-order kinetics. However, at the higher solution pH, PX released from Zn^{2+} cross-linked beads showed a much slower release rate than that from Ca^{2+} cross-linked. Indeed, as has been reported earlier[6], the superior cross-linking effect of Zn^{2+} over the traditional Ca^{2+} ions is due to their spatial geometry and orientation. Zinc ions can form a compact matrix with pectin molecules, with zein helices between homogalacturonan backbones. The zein protein body may form different cylinder like aggregates with diameters of 6 to 20 nm.

The data obtained are of interest, for preparation of complexes that are stable at the upper gastrointestinal site, and show that there is a required quantity of pectin per zein molecules, which retards drug release. The pectin fraction in the complex with zein is stabilized by calcium cross-linking and may protect the protein from protease attack. The zein fraction, in turn, suppressed the degree of pectin polyelectrolyte swelling favoring drug release. The results show that hydrogels prepared with the higher pectin/zein ratio exhibited a lower drug diffusion coefficient. This indicates that the drug diffusion is mainly controlled by the zein fraction and its interaction with pectin in the hydrogel.

3 CONCLUSION

The relevant relationship existing between the hydrogel bead composition and drug release behavior can be exploited to develop unique drug\nutrient delivery devices with specifically defined release properties. The data obtained are of interest, for preparation of complexes that are stable at the upper GI tract and can be degraded in the lower GI tract, and show that there is a minimum quantity of pectin per zein molecule, which retards drug or nutrient release and protects the protein fraction from pepsin attack. The pectin fraction stabilises the zein clusters through a calcium crosslinked network and protects protein from protease attacks. The zein fraction, in turn, suppressed pectin polyelectrolyte swelling favouring drug/nutrient release.

References

1. Dickey L.C., Dallmer M.F., Radewonuk E.R., Parris N., Kurantz M., Craig J.C.- Cereal Chem, 1998, v 75, p 443-448.
2. Liu LS.,Kost J.,Fishman M.L.,Hicks K. A (2008). Review: Controled Release Systems for Agricultural and Food Applications, in New Delivery Systems for Controled Drug Release from Naturally Ocuring Materials. Ed. N.Parris, LS Liu et al. ACS Symposyum series 992, 265-281.
3. Kasimova G.F., Bobokalonov J.T., Khalikova M.Kh., Muhidinov Z.K., Kukaniev M.A. The news of Tajikistan Academy of Science. 2007, 127(2), p. 42-50.
4. Muhidinov Z.K.,Kasimova G.F., Bobokalonov J.T., Teshaev Kh.I., Khalikova M.D., Liu LS. Pectin-Zein microspheres as a Drug Delivery Systems. Pharmaceutical Chem. 2010, v. 44(10), p. 20-24
5. Pillay, V.; Danckwert, M. P.; Muhidinov, Z. K.; Fassihi, R. Drug Dev. & Ind. Pharm. 2005, 31, 191-207.
6. US Pharmacopea

TEXTURE DESIGN OF DYSPHAGIA FOODS USING POLYSACCHARIDES

Takahiro FUNAMI, Sayaka ISHIHARA, Makoto NAKAUMA

Texture Design Division, San-Ei Gen F.F.I., Inc., 1-1-11, Sanwa-cho, Toyonaka, Osaka 561-8588 JAPAN

1 INTRODUCTION

Texture is important not only for food palatability but also for the safety of eating for the elderly, where the number of people with mastication and swallowing difficulty is increasing. Accordingly, the market of food products for nursing-care purpose is expanding. Textural properties of nursing-care foods should be considered from rheological, colloidal, and tribological aspects so that food products can be masticated and swallowed easily even by these patients.

Food texture is often modified by the addition of polysaccharides in the food industry. Polysaccharides exhibit multiple functions in food systems, including thickening, gelling, foaming, emulsifying etc, and have been used as a texture modifier in almost every kind of processed foods by changing the structural and geometrical properties of foods. There is still need for further progress in polysaccharide engineering for the texture design of nursing-care foods because textural attributes required for these foods are different to those required for other existing foods.

Swallowing deficiency, namely dysphagia, is commonly managed by texture-controlled foods. These foods are usually designed to moderate the flow speed of bolus through the pharynx by thickening or gelling. Thin liquids like water and tea are sometimes unfavorable for dysphagia patients, showing the delay in the swallowing reflex and in the closure of the epiglottis. These liquid foods pass quickly through the pharynx and can go into the lungs of the patients, causing aspiration and sometimes pneumonia. Aspiration pneumonia is now one of the leading causes of the death for elderly people in Japan. With such a social background, texture design of dysphagia foods is now focused in the food industry and many commercial dysphagia products of advanced functionality and diversity are available in Japan.

Dysphagia thickeners often contain xanthan gum as an essential ingredient. Although many polysaccharide thickeners are available in the food industry, why is xanthan gum so favoured by food manufactures and consumers? There are scientific reasons for the popularity of xanthan gum in this application. The most important one is textural characteristics of xanthan gum. When thickened with xanthan gum, liquid foods flow as one coherent bolus without dispersing through the pharynx and without adhering onto the pharynx during swallowing. This physiological flow behaviour may lead to the sensation of swallowing ease. It is recognized in the food industry that systems should have "weak gel" characteristics to produce a coherent bolus that is swallowed in one go. Xanthan gum is thus expected to be more than thickener.

In the initial part of this paper, various advantages of xanthan gum as a main ingredient of dysphagia thickeners are presented along with historical review of the development of dysphagia thickeners in Japan. In the second part, *in-vivo* acoustic analysis of swallowing sound by the authors' research group is presented. The analysis is aimed to describe physiological flow behavior of xanthan gum through the pharynx and to know the relationship with rheological and microstructural properties of the gum. To find objective parameters that represent perceived swallowing ease is also discussed from the correlation with sensory testing.

2 DYSPHAGIA THICKENERS

Dysphagia thickeners make liquid foods easier to swallow for the patients by increasing the viscosity (called *toromi* in Japanese). In Japan, the first generation of commercial products was starch-based, particularly instant starches due to their swelling and thickening capabilities without heating. Recently, non-starch polysaccharides have gained more popularity as thickening agents; guar gum for the second generation, followed by xanthan gum for the third generation. Now is the stage of diversity and specialization of the products using multiple polysaccharides to build up synergistic and complementary systems.

The usage of dysphagia thickeners by the end users is very simple. In the most cases, whole amount of the thickener is put into liquids (3.0 g in 100 ml liquid for example) and stirred manually using a spoon or spatula for 30 sec or so. Thickened liquids are prepared in a short time and ready to drink. Requirements for dysphagia thickeners include;

1) no lumping even with gentle and weak agitation
2) rapid viscosity increase without heating
3) stability of viscosity during storage
4) applicability to the wide range of liquid foods
5) textural non-adhesiveness and easiness to form a bolus
6) no off-flavor or color

There is no doubt at the moment that xanthan gum is the best choice for dysphagia thickeners because the gum satisfies all these functionalities to a considerable degree.

2.1 Conformational and structural features of xanthan gum

Xanthan gum is an extracellular polysaccharide produced by a microorganism *Xanthomonas campestris* and is widely used in the food and non-food industries. The primary structure, functionalities, and applications are described in detail elsewhere[1]. In particular, conformational and structural features of xanthan gum are important in explaining its functional properties. Xanthan gum forms a double helical structure at ambient temperature, elucidated well by the finding that the molecular weight of the gum measured in NaCl aqueous solutions was almost double the measurement using cellulose-solvent cadoxen[1]. Conformation of xanthan gum ranges from rigid rod to stiff worm-like molecules in aqueous systems without heating, depending on the molecular weight of the gum and the ionic strength of the systems. Due to these conformational features, xanthan gum forms microgels as a result of the interhelical associations by the increase in concentration and/or ionic strength, as visualized by atomic force microscopy (Figure 1). The presence of microgels is responsible for the weak gel behavior of the gum. Xanthan gum is characteristic in its high stability against heating (including freezing), acid, salt, and human digestive enzymes in comparison with other polysaccharide thickeners, which may be also related to these conformational and structural features.

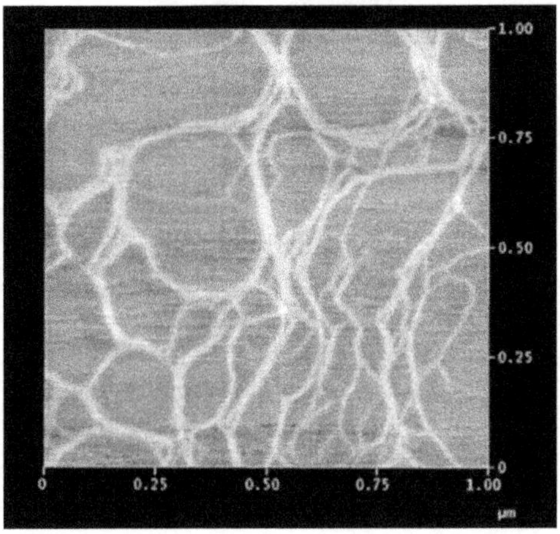

Figure 1 *Topographical atomic force microscopy image of xanthan gum interhelical associations on freshly-cleaved mica surface in air. Concentration of xanthan gum was 1 µg/ml in test solutions prior to deposition, and 2 µl of the solutions was deposited onto the surface.*

2.2 Hydration and swelling behavior of xanthan gum

It has been confirmed experimentally that xanthan gum hydrates and swells more easily in water at an ambient temperature (e.g., 20 °C) than guar gum, enhancing viscosity rapidly. Hydration speed of xanthan gum was estimated to be 6-7 times faster than that of guar gum[2]. This relates to the requirements no. 2 and 3 for dysphagia thickeners.

For most polysaccharide thickeners, the relationship between hydration (i.e., viscosity enhancement) and dispersibility (i.e., the degree of lumping) is interdependent. The faster the hydration of a polysaccharide, the lower the dispersibility is (the easier the lumps form) and vise versa. Lumps look like fish-eyes, where only the periphery of the particle swells with centre still remaining unhydrated (Figure 2). Once formed, it is extremely difficult to eliminate lumps. It is an actual case from the author's experience that xanthan gum forms lumps more easily than guar gum. A good balance between these opposite attributes is necessary for the functional design of dysphagia thickeners. This relates to the requirement no. 1 for dysphagia thickener.

One of the solutions for the good balance is granulation of polysaccharide thickeners to a granule form in the presence of bulking agents. Apparatus for granulation, its operating conditions, and the composition of granules should be optimized by considering mechanical strength of the granules, which is associated with the tolerance against disintegration during manufacturing and transportation, and reproducibility of the production in addition to the balance between hydration and dispersibility. That is, the blending knowledge of polysaccharide thickeners and the granulatation technique should be fused together for good products.

Figure 2 *Lumping or fish-eye formation of polysaccharide thickeners. Powder causes lumping when added to water with weak agitation (left), which can be prevented by granulation (right).*

2.3 Rheological properties of xanthan gum

Xanthan gum in solution can be described as a 'weak-gel'. This rheological behavior of the xanthan gum solution is represented by the frequency-dependence of dynamic viscoelastisities (i.e., mechanical spectra); both the storage modulus G' and the loss modulus G'' increase slightly with increasing frequency with G' being continuously larger than G'' throughout the frequency range within a realistic time scale (e.g., 0.1-100 rad/s) (Figure 3a). This is in contrast to the behavior of the guar gum solution for which both G' and G'' increase with increasing frequency, showing a crossover point between these moduli at a certain frequency (Figure 3b). The weak gel behavior of the xanthan gum solution is attributed to the formation of microgels, enhancing the perception of solid-like texture rather than that of viscous texture. This relates to the requirement no. 5 for dysphagia thickeners.

Rheological behavior of xanthan gum is less affected by external conditions, including the addition of salt (to 3 %), the pH decrease (to 3.5), and temperature than that of other polysaccharide thickeners. This relates to the requirement no. 4 for dysphagia thickeners.

2.4 Variety of xanthan gum product source

There is a variety in the product grade for commercially available xanthan gum, depending on the particle size, transparency, acid and salt tolerance, rheological flow behavior, and compatibility with galactomannans etc. In particular, transparent grade is much more favorable than standard turbid grade since transparency does not deteriorate the palatability of thickened foods in terms of flavor, color, and appearance. This relates to the requirement no. 6 for dysphagia thickeners.

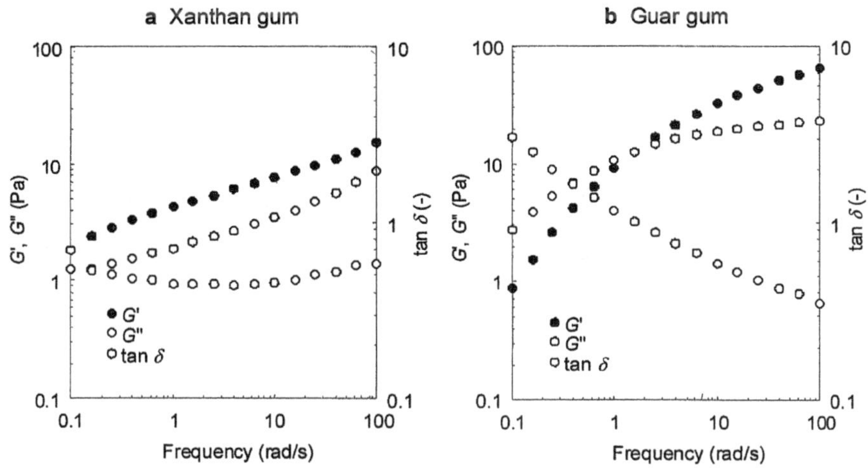

Figure 3 *Frequency-dependence of dynamic viscoelasticities for xanthan gum (a) and guar gum (b) in solutions. Solutions were prepared by dispersing each gum in water at 80 °C for 30 min, followed by quenching to 20 °C. Gum concentration: 1%, Measurement temperature: 20 °C, G': storage modulus, G": loss modulus, tan δ: mechanical loss tangent.*

3 PERCEIVED SWALLOWING EASE OF XANTHAN GUM FROM IN-VIVO ACOUSTIC ANALYSIS

Physiological flow behavior of liquid foods during swallowing depends on the thickener used even when compared at an equivalent viscosity. The authors' trial[3] is to use *in-vivo* acoustic analysis of the swallowing sound to describe the physiological flow behavior through the pharynx and to find objective parameters that represent perceived swallowing ease from the correlation with sensory test.

Swallowing sound is collected using a throat microphone (i.e., vibration sensor) (Figure 4). Representative acoustic profile of the swallowing sound (in the case of water) is illustrated (Figure 4). The acoustic profile is divided into three parts, each of which is assigned to the closure of the epiglottis, the bolus flow, and the opening of the epiglottis in order of occurrence[4]. The duration of each oral event; t_1, t_2, and t_3 were determined.

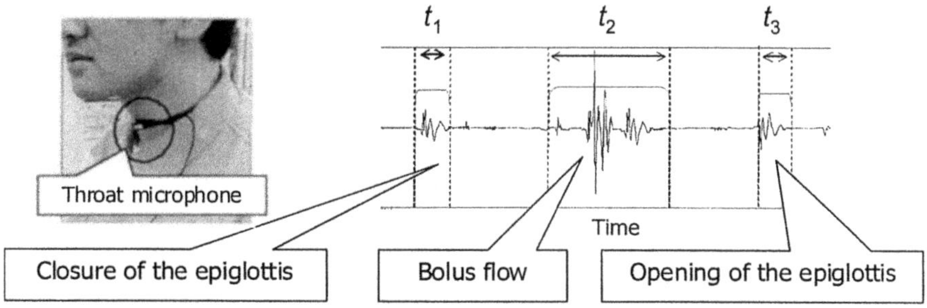

Figure 4 *Location of throat microphone for collection of the swallowing sound and representative acoustic profile when water is swallowed.*

Xanthan gum and locust bean gum were used as samples in this experiment. Each polysaccharide was dispersed in de-ionized water at 85 °C for 10 min with mechanical stirring, deaerated under vacuum, and quenched to 20 °C. The dispersions were left standing at 20 °C overnight to attain equilibrium before use. Concentrations of each polysaccharide were 0.3, 0.45, 0.6, 0.75, and 0.9 w/v% for xanthan gum and 0.5, 0.6, 0.7, 0.75, and 0.8 w/v% for locust bean gum. These concentrations correspond to a shear viscosity at 10 s^{-1} of 0.3, 0.6, 0.9, 1.2, and 1.5 Pa·s in the increasing order of concentration for each polysaccharide.

Three males and two females ranging from 29 to 37 yrs of age (mean age 33.4 yrs) participated as subjects. All subjects were healthy and free from mastication or swallowing problems. Fifteen ml of each sample was fed to subjects at 20 °C. Subjects were asked to put the sample into the mouth and to swallow the whole amount once without mastication by teeth or squeezing between the tongue and the hard palate. The same amount of water was used as control.

Two sensory attributes were evaluated, including cohesiveness (i.e., bolus formation) and swallowing ease perceived in the pharynx. Evaluations were carried out using a visual analogue scale (VAS), where the subjects were asked to indicate the score of each attribute by marking a position on a 100 mm-length scale. The larger the length (i.e., the larger the VAS value), the greater the sensation is for each attribute. Each sample was ranked by the length on the scale.

Xanthan gum solutions showed smaller t_2 than locust bean gum solutions of equivalent shear viscosity and water, indicating that the time required for the bolus to transfer through the pharynx is shorter (Figure 5). Duration t_2 for xanthan gum decreased with increasing concentration despite the viscosity increase, whereas that for locust bean gum was much less concentration-dependent and was consistently larger than that for xanthan gum when compared at an equivalent shear viscosity at 10 s^{-1} (Figure 6). Duration t_2 correlates well with sensory perceived cohesiveness and swallowing ease (Figure 7). Results indicate that xanthan gum solutions flow as one coherent bolus through the pharynx with smaller variation of flow velocity than locust bean gum solutions, leading to a greater sensation of cohesiveness and swallowing ease.

In summary, duration t_2 can serve as an indicator of perceived swallowing ease. The usefulness of xanthan gum as a base ingredient for dysphagia thickener is also evidenced. In the future, the authors should confirm the validity of t_2 as the indicator using a wide range of liquid foods of different viscosity profiles. Also, to carry out similar experiments involving elderly people with different swallowing characteristics compared to young subjects should be another area of research interest.

4 CONCLUSION

Texture in nature is complex and many research areas are associated with texture studies, including psychology, oral physiology, and physicochemistry. For a better understanding of texture and thus the development of new food products, collaborative work among scientists of different research backgrounds should be necessary.

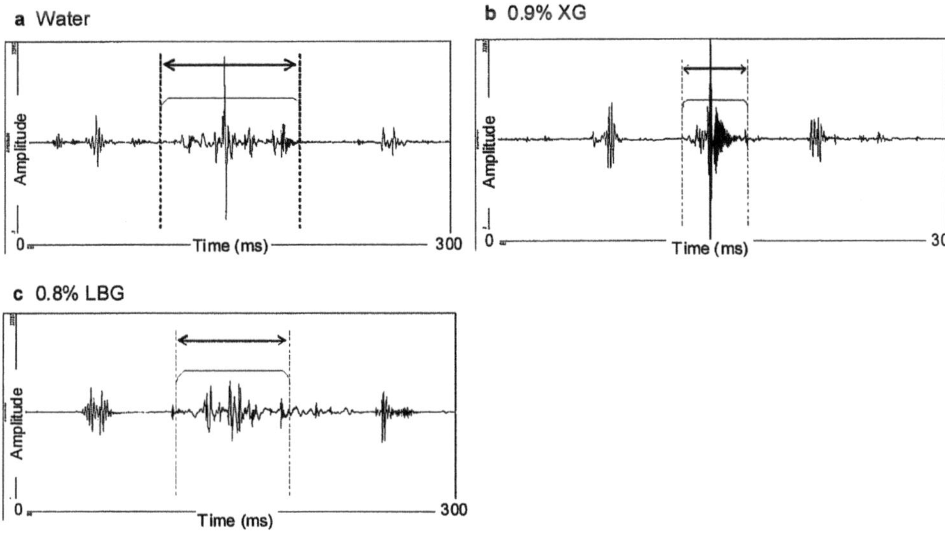

Figure 5 *Representative acoustic profile of water (a), 0.9% xanthan gum solution (b), and 0.8% locust bean gum solution (c). The solutions (b) and (c) are equivalent in shear viscosity at 10* s^{-1} *(approx. 1.5 Pa·s).*

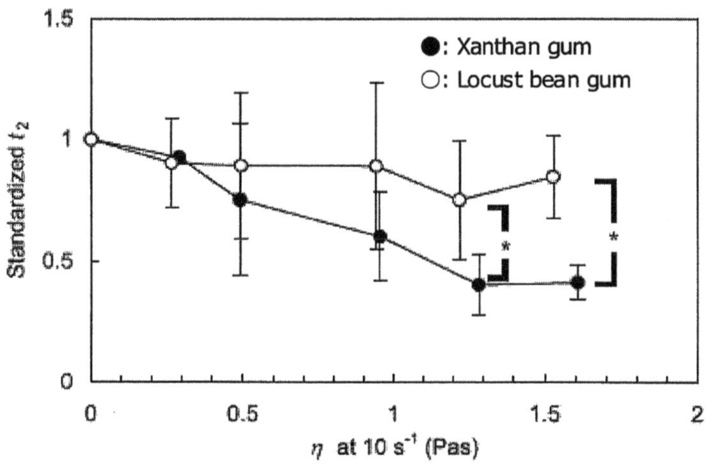

Figure 6 *Duration* t_2 *during swallowing for polysaccharide solutions as a function of shear viscosity at 10* s^{-1}. *Duration* t_2 *for each polysaccharide solution was standardized with that for water. Data with * are different at* $p < 0.05$.

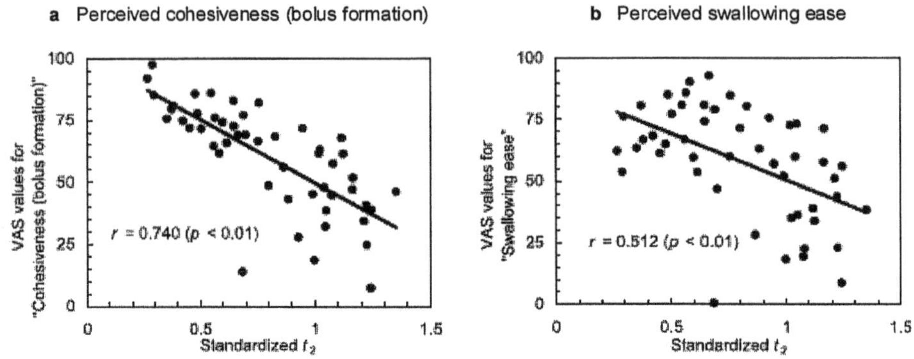

Figure 7 *Correlation between duration t_2 and perceived cohesiveness (a) and perceived swallowing ease (b) during swallowing for polysaccharide solutions.*

References

1 V.J. Morris, 'Bacterial Polysaccharides' in *Food Polysaccharides and Their Applications*, eds., A.M. Stephen, G.O. Phillips and P.A. Williams, CRC Press, London, 2006, Chapter 12, pp. 413-454.
2 T. Funami, T. Tsutsumino and K. Kishimoto, *Journal of Cookery Science of Japan*, 2006, **39**, 233 (in Japanese).
3 M. Nakauma, S. Ishihara, T. Funami and K. Nishinari, *Food Hydrocolloids*, 2011, **25**, 1165.
4 S.L. Hamlet, R.L. Patterson, S.M. Fleming and L.A. Jones, L.A. *Dysphagia*, 1992, **7**, 160.
5 International conference on food oral processing - physics, physiology and psychology of eating, The Royal Society of Chemistry, London, 2010.

PROBING SPECIFIC INTERACTIONS BETWEEN PECTIN-DERIVED GALACTOOLIGOSACCHARIDES AND THE MAMMALIAN LECTIN GALECTIN 3 – TESTING A MODEL FOR PECTIN BIOACTIVITY

V.J. Morris, R.J.M. Bongaerts and A.P. Gunning

Institute of Food Research, Norwich Research Park, Colney, Norwich NR4 7UA, UK

1 INTRODUCTION

Pectin is a structural component of non-graminaceous plant cell walls that provide a source of both soluble and insoluble fibre. Pectin extracts are added to processed foods as gelling agents or as stabilisers for acidic milk drinks. In addition to the health benefits associated with a role as dietary fibre new aspects of pectin bioactivity are starting to emerge. There is a growing interest in the role that natural polysaccharides present in the diet may play in moderating cancer progression and metastasis[1-2]. One of the best characterised bioactive roles for pectin is the potential use of a modified form of citrus pectin (MCP) as an anti-cancer agent[1-2]. The accumulated evidence in the literature on the anti-cancer activity of MCP suggests that it is effective for a range of cancers at several stages of development.

Such a broad range of activities would seem to imply that the effects arise due to a simple generic mode of action. A testable molecular model has been proposed as a hypothesis for explaining this type of pectin bioactivity[2]: it has been suggested that the modification of pectin causes the release, or results in the release during digestion, of pectin fragments that are capable of binding specifically to, and thus inhibiting the various biological roles of the mammalian lectin galectin 3 (Gal3). Gal3 is an important biomarker for cancer and is recognised as a target for therapy[3-5]. The ability to bind to and inhibit the behaviour of Gal3 is important because Gal3 regulates a number of key stages in the development and spread of cancers[3-6]: these include, for example, the adhesion of cancer cells to tissue, the growth of blood vessels to tumours (angiogenesis) and several factors which enhance the survival of cancer cells, such as self-association (emboli formation) and resistance to programmed cell death (apoptosis).

Gal3 is one, albeit an important member of a large family of highly conserved lectins found in a range of species from lower invertebrates to mammals. Gal3 is literally a galactose-binding lectin containing a ~14 KD carbohydrate recognition domain (CRD) with an affinity for β-galactoside residues[7-8]. The molecular interactions through which Gal3 regulates key stages in the development and spread of cancers[3-6] involves binding of molecular species into the Gal3 CRD.

Pectin extracts from plant cell walls are very complex structures[9]. The pectin extracts are likely to be complex mixtures rather than simple individual molecules and the structure of the mixtures is still incompletely understood. However, these extracts have been shown to contain two main types of structure relevant to their functional properties. These structures comprise linear homogalacturonan backbone (smooth) regions, largely responsible for their gelling ability, and branched (hairy) rhamnogalacturonan 1 (RGI) regions[9], which are the likely origins of their bioactivity. Figure 1 shows a simple pictorial illustration of these two main components.

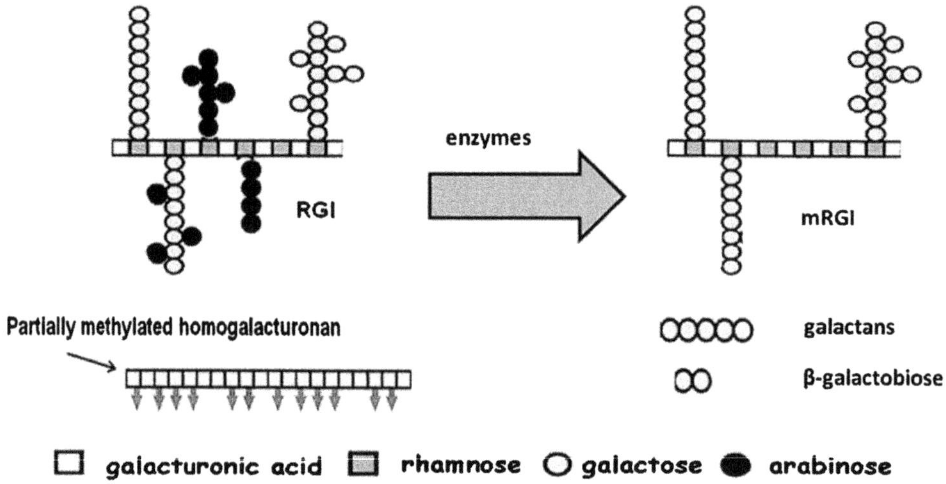

Figure 1 *Schematic diagram illustrating the two main structural components of pectin extracts plus a variety of galactan-enriched fragments that can be obtained by enzymatic treatment of purified pectin extracts.*

The homogalacturonans are composed of (1→4) linked α-D-galacturonic acid residues (Gal*p*A) in which some of the carboxyl groups are methyl-esterified. The RGI units consist of a repeating disaccharide [→4)-α-D-Gal*p*A-(1→2)-α-L-Rha*p*-(1-)] of galacturonic acid and rhamnose (Rha*p*) residues. A fraction of the Rha*p* residues are substituted with neutral sugar sidechains that, depending on the source and mode of extraction, can include arabinans, galactans and arabinogalactans. In terms of the bioactivity of pectin the most interesting types of neutral sidechains are linear (1→4) β-D-galactans and branched (1→3,6)-β-D-galactans linked to C4 of the backbone rhamnose residues. The (1→4) β-D-galactans may be linear unsubstituted chains or may contain predominately (1→3)-α-L-Ara*f* (arabinose) sidechains. All of the arabinogalactans contain terminal galactose residues and the simplest candidates for binding to Gal3 are linear (1→4) β-D-galactans: the simple linear structures might be expected to better facilitate binding within the Gal3 CRD.

Studies on MCP as an anti-cancer agent have used a modified form of citrus pectin prepared by alkaline treatment of commercial citrus pectin extracts[10-11]. Alkali treatment[12] will cause depolymerisation and de-esterification of the homogalacturan regions, generating lower molecular weight more soluble oligomers of polygalacturonic acid (PGA), plus RGI regions attached to polygalacturonic acid oligomers.

The potential importance of arabinogalactans or galactans is supported by published evidence for the anti-cancer activity of arabinogalactans[13-15]. Recent comparative studies of pectins from different dietary sources have shown that the galectin inhibitory behaviour is enhanced in pectin polysaccharide extracts with higher arabinose and galactose contents, indicating the importance of arabinogalactans in the inhibition of Gal3-mediated cellular interactions[16].

Recent studies on purified RGI from citrus pectin[17] have shown that a combination of endo-1,5-α-L-arabinanase, endo-1,4-β-D-galactanase, α-L-arabinofuranosidase and β-D-galactosidase will debranch the RGI sample, suggesting the presence of β (1→4)-D-arabinogalactan and/or galactan sidechains. Citrus pectin RGI, or modified RGI regions, are not available commercially. However, potato RGI regions are available commercially in reasonable quantities for experimental studies. An enzymatic analysis[18] of the structure of potato RGI shows the presence of α-(1→5) linked arabinans and an abundance of β (1→4)-D-galactans poorly substituted with arabinose. Thus these samples provide a route to examining the potential importance of the β (1→4)-D-arabinogalactan/galactans as candidates for binding to Gal3. The arabinans and the arabinose sidechains can be removed with an α-L-arabinofuranosidase to produce a modified RGI (mRGI) consisting of a rhamnogalacturonan backbone substituted with essentially linear (1-4) linked β-D-galactans.

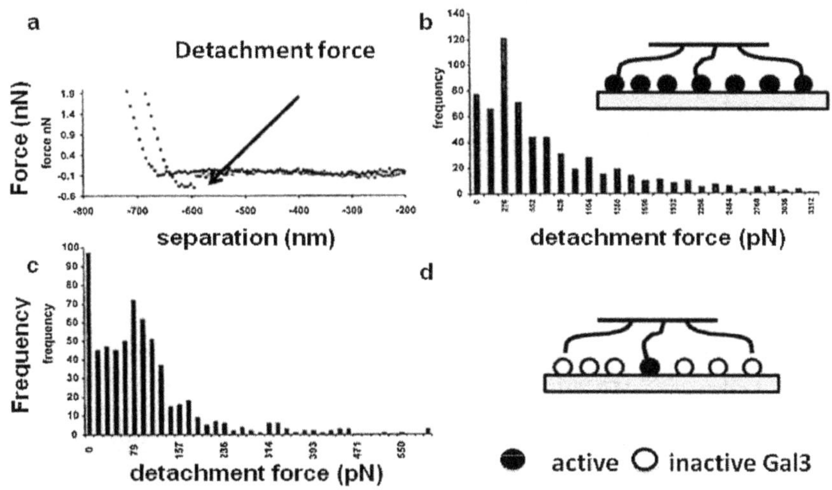

Figure 2 *Force spectroscopy measurements of the interaction of mRGI with Gal3[19-20]. (a) force distance curves showing the detachment of mRGI from Gal3. (b) Histogram of the measured detachment forces with an inset indicating multiple binding of the galactan chains in mRGI to Gal3. (c) Histogram showing the measured detachment forces after addition of the inhibitor lactose to inactive the majority of the Gal3 molecules. The inset (d) shows how the competitive inhibition restricts binding to individual galactan chains and Gal3.*

Biophysical studies using microscopy, flow cytometry and force spectroscopy have been used to probe the specific binding of mRGI, RGI and PGA fragments (Figure 1) to Gal3[19-20]. The microscopy and flow cytometry studies showed that PGA does not bind specifically to Gal3, RGI binds specifically but weakly to Gal3, and mRGI binds specifically to Gal3. The detachment forces determined by force spectroscopy, for interactions of mRGI with Gal3 (Figure 2), were characteristic of carbohydrate-lectin interactions, suggesting binding of the pectin-derived galactans to Gal3[19-20]. This article reports studies to extend these investigations by probing the binding of the purified galacto-oligosaccharide (β-galactobiose) to Gal3.

2 MATERIALS AND METHODS

2.1 Galactobiose and Gal3

The β-galactobiose PGA samples were purchased from Megazyme International Ireland Ltd (Ireland). Recombinant human Gal3 was purchased as a Gal3-lactose complex from Sigma Chemicals Ltd (UK). All other materials were purchased from Sigma Chemicals Ltd (UK) unless otherwise stated.

2.2 Preparation of activated substrates

Two types of substrates were used: silica beads and glass slides. Silica beads (mean diameter 6.8 μm) were purchased from Bangs Laboratories (Fishers, IN, USA). The procedures used to functionalise the beads are described in detail elsewhere[19, 21]. Basically the beads were acid washed to regenerate surface hydroxyl groups and then coated with a layer of 3-mercaptopropyltrimethoxy silane (MTS). For force spectroscopy studies the silanised beads were further functionalised with a carbohydrate coupling agent 4-(4-*N* maleimidophenyl) butyric acid hydrazide hydrochloride (MPBH, Pierce, USA) and then glued onto the end of the AFM cantilevers (long thin NP; Veeco Inc., USA) using a 2-part epoxy resin (40 min cure time; Permatex Inc., USA): the tip was dipped into the liquid resin and then pressed down onto the bead. Once the epoxy resin had hardened the cantilever-bead assemblies were incubated with β-galactobiose solutions (0.1 mg ml^{-1}, 22°C, phosphate buffered saline (PBS), 1 h), rinsed in PBS and then incubated for 30 min in a 10 mg ml^{-1} glucose solution in PBS to cap any unreacted hydrazide sites on the bead, followed by rinsing in PBS. Fully functionalised tips were inserted into the liquid cell of the AFM and maintained under PBS solution.

MTS coated glass slides for force spectroscopy studies were functionalised with Gal3-lactose complexes attached via a heterobifunctional polyethylene glycol (PEG) spacer molecule (O - [N - (6-Maleimidohexanoyl) aminoethyl] - O′ - [3-(N-succinimidyloxy) - 3 - oxopropyl] polyethylene glycol 3,000, Sigma Chemicals, Poole, UK) to facilitate binding of the Gal3 to the small disaccharide molecules. One end of the PEG contains an n-hydroxy succinimide (NHS) ester which covalently binds primary amines at neutral pH and the other end of the PEG contains a maleimido group which binds to the surface adsorbed thiol groups on the MTS coated glass slide. Selected regions of the functionalised slides were incubated with the Gal3-lactose complexes, washed in PBS, glycine-treated to cap any unreacted succinimide groups, rinsed in PBS and inserted into the AFM liquid cell.

2.3 Force spectroscopy

The force spectroscopy data presented in figure 2 was carried out on a Lumina AFM (Veeco Instruments, CA, USA) and that shown in figures 3-5 were obtained using an MFP-3D AFM (Asylum Research, CA, USA). In both instruments the measurements were made by ramping the AFM tip towards and then away from the sample surface under buffer solution (either pure PBS or lactose solution in PBS) at a controlled speed. This process was repeated typically several hundred times for each tip-sample and solution combination to generate statistics on the frequency and magnitude of adhesion. By attaching functionalised silica beads to the end of the AFM cantilevers and pressing them against functionalised glass surfaces molecular interactions could be observed as negative peaks in the force versus separation plots. In the MFP-3D instrument the measurement was further controlled by setting a maximum loading force to minimise damage of the functionalised bead on the end of the AFM cantilever. Furthermore, in the MFP-3D data, a dwell period of 0.3 seconds was imposed on the tip-sample contact phase of the measurement to increase the chances of capturing binding events.

2.4 Analysis of force spectroscopy data

Tip-sample adhesion data were analysed by quantifying the magnitude of the most negative adhesive peak seen upon retraction of the functionalised AFM tip from the functionalised sample surface. The resulting data was then assigned to bins in order to generate histograms depicting the distribution of values.

3 RESULTS AND DISCUSSION

Force spectroscopy studies were made to investigate potential specific binding of β-galactobiose to Gal3. Figure 3a shows a schematic diagram illustrating the attachment of the β-galactobiose-activated sphere to the tip-cantilever assembly and the attachment of Gal3 to the glass substrate via PEG spacer molecules. The substrate was rinsed with PBS to remove lactose from the attached Gal3 complexes and thus to activate the Gal3 molecules. Figure 3b shows an example of approach and retract curves revealing a specific interaction between β-galactobiose and Gal3: the curves show hysteresis between the advance and retract curves revealing evidence of a binding event on retraction. The detachment events shown in figures 2 & 3 differ in nature due to the presence of the PEG spacer molecules. The retract curve in figure 3b shows a region corresponding to extension of the PEG spacer prior to the breakage of the β-galactobiose–Gal3 linkage.

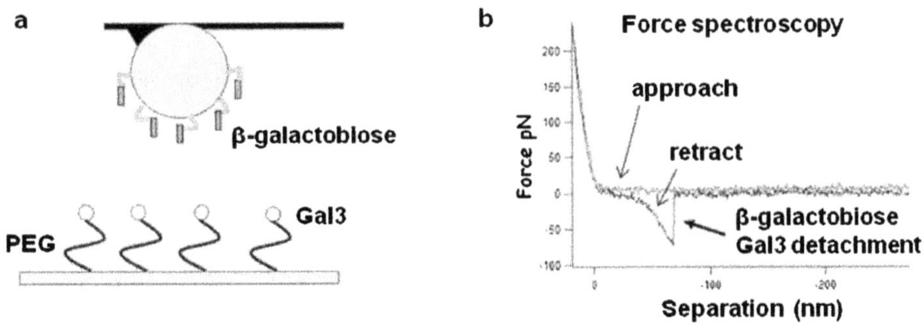

Figure 3 *Force spectroscopy studies for Gal3 and β-galactobiose. (a) Schematic diagram showing the activated substrate and activated spheres attached to the AFM cantilever. (b) Approach and retract curves showing the detachment of the β-galactobiose from Gal3.*

The presence of the PEG spacer facilitates binding of Gal3 to β-galactobiose and also helps to distinguish between specific and non-specific binding events. On approach the cantilever deflection remains constant until the sphere and substrate come into contact. On retraction non-specific interactions will occur at this point of contact. Specific binding events will be shifted along the horizontal axis, by a distance that depends on the contour length of the PEG spacer. The curves shown in figure 3b suggest specific binding of β-galactobiose to Gal3.

The experimental study yields a range of detachment forces shown as a histogram of measured binding events in figure 4a. The calculated modal value of the detachment force was found to be 92 pN. The flexibility of the PEG spacer enables binding to occur but will promote the occurrence of multiple binding events. It is desirable to limit the interactions to single β-galactobiose-Gal3 molecules. Controlled addition of the inhibitor lactose was used to inactivate a high proportion of the immobilised Gal3 molecules. This has the effect of effectively increasing the separation between active Gal3 molecules (Figures 4b & d). Figure 4c shows the effect of such inhibition on the histogram of measured detachment forces. Fewer binding events were observed during the scan and the modal detachment force was reduced to 35 pN. This suggests that the binding between β-galactobiose and Gal3 has been largely restricted to the binding of single carbohydrate and protein molecules. The fuller extent of adhesion events was recovered upon rinsing with PBS to remove lactose.

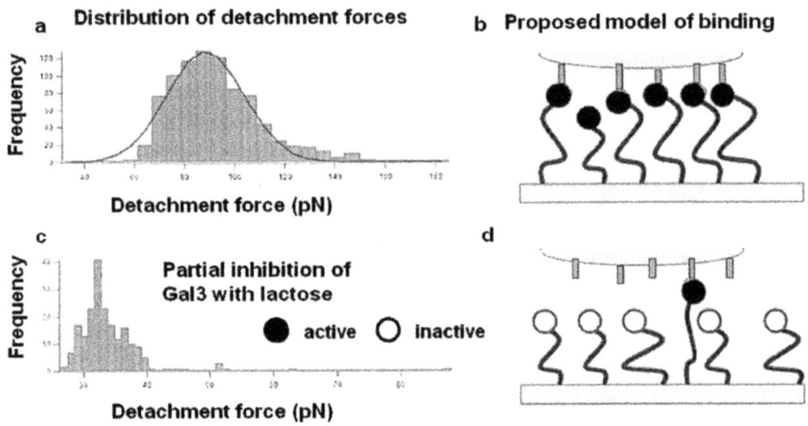

Figure 4 *Detachment force data for β-galactobiose–Gal3 interactions. (a) Before, and (c) after, the addition of lactose to partially deactivate Gal3 molecules. The figure shows schematic models for the interactions between β-galactobiose and Gal3 in the absence (b) and presence (d) of lactose.*

The modal value of the measured detachment force is typical of that measured for other carbohydrate–lectin interactions[22-24], suggesting that there is a specific interaction between mRGI and Gal3 which involves binding into the CRD. Touhami and co-workers[23-24] report a value of 96±55 pN for single carbohydrate interactions between glucose and Concanavalin A and a similar value (121±53 pN) for single interactions between glucose and individual cell-surface-expressed lectins on live yeast cells. Reported studies on lactose–galectin 1 interactions[22] showed that the detachment force varied with loading rate and a measured detachment force of 34±6 pN was reported at a loading rate of 3 nN s^{-1}. The dependence of the detachment force on loading rate means that it is not possible to easily use the relative values of detachment forces to gauge the relative inhibitory effects of different oligosaccharides.

The dependence of the detachment force on loading rate arises because of the kinetic nature of the binding between carbohydrates and lectins.[22, 25] At very slow loading rates the lectin-carbohydrate interaction will have time to dissociate and the limiting detachment force will be zero. By extrapolating the measured detachment forces to zero loading rate it is possible to determine a lifetime for the carbohydrate-lectin interaction which can be used to compare the inhibitory effects of different oligosaccharides. The data shown in figure 4a & c was collected at a constant approach speed of 2 μm s^{-1}. Because of the complicated geometry of the attached carbohydrates and proteins different individual interactions may lead to detachment at different loading rates. Figure 5 shows a comparison of two individual force-distance curves from the set of data collected for the β-galactobiose–Gal3 study. For the two curves shown the measured detachment force decreased with decreased loading rate on detachment. A detailed analysis of the data is needed to establish whether this is true for all the measured curves. Should this prove to be true then it may provide an explanation for the observed spread of measured values of the detachment force. Of more importance is that it could provide a basis for extrapolating the data, and thus calculating lifetimes for the β-galactobiose–Gal3 interaction that could then be used to compare inhibitory values of different oligosaccharides.

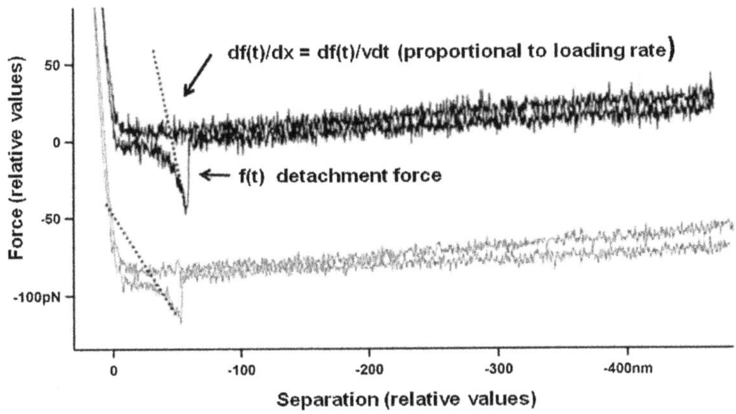

Figure 5 *Comparison of isolated force-distance curves for β-galactobiose–Gal3 interactions. These two curves illustrate a decreased detachment force for a decreased loading rate.*

4 CONCLUSIONS

The present studies confirm the specific binding of the galactooligosaccharide β-galactobiose to Gal3. The experimental data is consistent with the proposed molecular hypothesis for the anti-cancer action of modified pectin extracts.

Acknowledgement The research described in this article was supported by the BBSRC core strategic grant.

References

1 P. Nangia-Makker, J. Conklin, V. Hogan and A. Raz, *Trends Mol. Med.*, 2002, **8**, 187.
2 V. Glinsky & A Raz. Carbohydrate Research 344 (2009) 1788-1791 .
3 A. Krzelška and A. Lipińska, *Cellular & Mol. Biol. Letts.*, 2004, **9**, 305.
4 P. Demetter, N. Nagy, B. Martin, A. Mathieu, P. Dumont, C. Decaestecker and I. Salmon, *J. Pathol.*, 2008, **215**, 1.

5 A. Nakahara and A. Raz, *Anti-Cancer Agents in Med. Chem.*, 2008, **8**, 22.
6 Y. Takenaka, T. Fukamori and A. Raz, *Glycoconj. J.*, 2004, **19**, 543.
7 S.H. Barondes, V. Castronovo, D.N.W. Cooper, R.D. Cummings, K. Drickmer, T. Feizi, M.A. Gitt, J. Hirabayashi, C. Hughes, K. Kasai, H. Leffler, F.T. Lui, R. Lotan, A.M. Mercurio, M. Monsigny, S. Pillai, F. Poirer, A. Raz, P. W.J. Rigby, J.M. Rini and J.L. Wang, *Cell,* 1994, **76**, 597.
8 S.H. Barondes, D.N.W. Cooper, M.A. Gitt and H. Leffler, *J. Biol. Chem.*, 1994, **269**, 20807.
9 B.L. Ridley, M.A. O'Neil and D. Mohnen, *Phytochem.*, 2001, **57**, 929.
10 K.J. Pienta, H. Naik, A. Akhtar, K. Yamazaki, T.S. Replogle, J. Lehr, T.L. Donat, L. Tait, V. Hogan and A. Raz, *J. National Cancer Institute*, 1995, **87**, 348.
11 D. Platt and A. Raz, *J. National Cancer Institute*, 1992, **84**, 438.
12 C.M.G.C. Renard and J-F. Thibault, *Carbohydr. Res.*, 1996, **286**, 139.
13 J. Beuth, H.L. Ko, V. Schirrmacher, G. Uhlenbruck and G. Pulverer, *Clin. Exper. Metastasis*, 1998, **6**, 115.
14 G. Uhlenbruk, J. Beuth, K. Oette, W. Roszkowski, H.L. Ko and G. Pulverer, *Naturwissenschaften*, 1986, **73**, 626.
15 B. Hagmar, W. Ryd and H. Skowmedal, *Invasion & Metastasis*, 1991, **11**, 348.
16 U.V. Sathisha, S. Jayaram, M.A.H. Hayaka and S.M. Dharmash, *Glycoconj. J.*, 2007, **24**, 497.
17 B.M. Yapo, P. Lerouge, J-F. Thibault and M.C. Ralet, *Carbohydr. Polym.*, 2007, **69**, 426.
18 J. Obro, J. Harholt, H.V. Scheller and C. Orfila, *Phytochem.* 2004, **65**, 1429.
19 A.P. Gunning, R.J.M. Bongaerts and V.J. Morris, *FASEB J.*, 2009, **23**, 415.
20 V.J. Morris, R.J.M. Bongaerts and A.P. Gunning, *Gums and Stabilizers for the Food Industry 15*, Royal Society of Chemistry, Cambridge, UK, 2010, 395.
21 S.K. Bhatia, L.C. Shriverlake, K.J. Prior, J.H. George, J.M. Calvert, R. Bredhorst and F.S. Ligler, *Anal. Biochem.*, 1989, **178,** 408.
22 W. Dettmann, M. Grandbois, S Andre′, M. Benoit, K. Angelika, A.K. Wehle, H. Kaltner, H-J. Gabius, and H.E. Gaub, *Archiv. Biochem. & Biophys.*, 2000, **383**, 157.
23 A. Touhami, B. Hoffmann, A. Vasella, F. Denis and Y.F. Dufrene, *Langmuir*, 2003, **19**, 1745.
24 A. Touhami, B. Hoffmann, A. Vasella, F. Denis and Y.F. Dufrene, *Microbiol.*, 2003, **149**, 2873.
25 V.J. Morris, A.R. Kirby, A.P. Gunning, *Atomic Force Microscopy for Biologists.* Imperial College Press, UK, 2nd Edition. 2010, chapter 9, 356.

SATIETY INDUCTION WITH NOVEL FOOD GRADE METHYL CELLULOSE

Roland Adden[1], Kerr Anderson[2], Wim Calame[3], Stuart Clegg[4], Britta Hübner-Keese[1], Matthias Knarr[1], Roberta Re[4], Martin Wickham[4]

[1] Dow Wolff Cellulosics, August-Wolff-Str.13, 29699, Bomlitz, Germany
[2] The Dow Chemical Company, Larkin Laboratories, Midland MI 48674, USA
[3] StatistiCal BV, Consultancy in Statistics and Experimental Design, Strandwal 148,2241 MN Wassenaar, The Netherlands
[4] Leatherhead Food Research, Randalls Road, Leatherhead, Surrey KT227RY, UK

1 INTRODUCTION

Overweight and obesity, as well as consequent cardiovascular disease, are primarily driven by over-availability of food and an increasingly sedentary lifestyle. One approach to treatment is to manipulate appetite and reduce food intake through control of satiety. It is believed that reduction of casual snacking between meals and the size of the meals ingested by controlling appetite will have a major impact on the prevention and reduction of obesity.

Materials incorporated into the diet that form a gel mass in the stomach, such as alginate and pectin, have been shown to enhance satiety[1], by, it is thought distending the stomach wall. However, these require an external chemical trigger, such as divalent metal ions or pH, which complicates the formulation of foodstuffs.

We have developed a modified, food grade, methyl cellulose (Satisfit^{TM}LTG) which gels at temperatures below body temperature but which is unaffected by pH. Here we have reported a clinical trial with healthy human volunteers demonstrating testing the satiety effect of this novel food grade methyl cellulose.

2 METHODS AND RESULTS

2.1 Study Design and Measurement

Healthy human volunteers (32) tested the novel methyl cellulose in comparison to control samples in a William's squared randomised double-blind cross-over design. The subjects have eaten the test products as a mid morning snack. All test diets were flavored.

During the test period, satiety feelings were scored by subjects using a Visual Analogue Scale (VAS) on hand-held computers.

After consumption of an *ad libitum* meal, the remaining food portions were weighed and consumed calories were calculated.

2.2 Clinical Study Results

The Visual Analogue Scale indicates that appetite recovery was slower with novel methyl cellulose than with control products.
In analysis of energy intake, we have also seen a significant reduction consumed calories in comparison to control products.

3 CONCLUSION

We have reported a trial with healthy human volunteers which demonstrates clear perception of greater satiety. Analysis of results from the Visual Analogue Scale indicates that appetite recovery is slower with novel gelling methyl cellulose than with control products. We have also seen a significant reduction in energy intake which might be explained by the gelation of the methyl cellulose in the stomach. The satiety effect has been shown to last for at least two hours after ingestion of the product, suggesting a role in control of snacking and meal size.

Reference

1 C. L. Hoad, P. Rayment, R. C. Spiller, L. Marciani, B. de Celis Alonso, C. Traynor, D. J. Mela, H. P. F. Peters and P. A. Gowland, *J. Nutr.*, 2004 , **134**, 2293.

SUBJECT INDEX

Acetate level, 151-153
Acetylated glutenin, 115-122
Acetylation, 116-122
Acoustic analysis, 411-414
Acoustic signature, 312-316
Adhesiveness, 332-338, 373
AFM imaging, 221-226
Afzelia africana, 27-44
Agarose microgels, 166-173
Air-water interface, 288-292
Alginate, 79-85
Anisotropic phase, 77-83
Arabinoxylan, 59-66
Atomic force microscopy, 209-216, 408, 409

Ball milling, 69-76
Batchelor's equation, 168,169
Batter preparation, 369-373, 375-383
Bile salts, 220
Bimodal, 136
Bioactivity, 415-421
Biodegradable films, 115, 257-267, 269-277
Biodegradable materials, 295-302
Bioplastics, 295-302
Biosurfactants, 219-225
Brachystegia eurycoma, 27-44
Bradford assay, 134
Brittleness, 312-316, 332-338, 373
Broccoli, 320-327
Bubbles, 378-383
Bulk rheology, 99-106
Burger model, 387

CaBER, 3993-399
Caesalpinia pulcherrima, 27-44
Calorimetric measurement, 370-373
Carboxymethylcellulose, 391-399
Carboxymethylcellulose, acid gel 175-182
Carrabiose, 19-25
Carrageenan structure, 19-25
Carrageenan, gel, 183-190
Carrageenans, hybrid, 22
Carrageenases, 20-24
Carreau model, 100
Carrot, 320-327
Caseinate emulsions, 221
Caseinate, 133-139
Cell clusters, 322
Cell fragments, 322
Cellulose ethers, 45-57, 391-399
Cellulose, 67-76
Cellulose, amorphous, 74

Characteristic time, 101
Cheese, 331-338
Chitosan films, 279-286
Chitosan mixed films, 287-292
Chlorella biopolymers, 123-129
Cloud point, 48-57
CLSM micrographs, 232
Coacervates, 401-406
Cobia, 269-277
Coil-helix transition temperature, 153
Cola milleni, 27-44
Colipase, 220
Complex modulus, 371
Complex viscosity, 103
Compression modulus, 179
Confocal laser scanning microscopy, 142- 147
Conformational structure, 67-76
Consistency index, 92
Consistency measurement, 306, 340
Controlled release, 279-286
Corn fiber gum, 59-66
Corn zein, 401-406
Crayfish protein, 287-292
Creep tests, 387-388
Creep, 340-347
Critical coil overlap concentration, 33
Crosslink density, 177-182
Crosslinking, 239
Crossover point, 192
Cryo-SEM, 320-327
Crystallinity, 71,72
Cured gels, 144

Deformation, 308
Degradation, 280-286
Degradation, thermal, 202
Degree of substitution, 47
Demethoxylation, 200
Depletion interactions, 215
Detarium microcaspum, 27-44
Dextran fibres, 248
Differential scanning calorimetry, 48-57, 154-159, 203-206, 370-373
Diffusion coefficients, 285-286
Digestion, 222
Dough, 115, 359-368, 385-388
Drug delivery, 401-406
Dynamic mechanical thermal analysis, 296-302
Dynamic surface pressure, 104
Dynamic viscoelastic measurements, 51-57
Dynamic viscosity, 30
Dysphagia, 407-414

Einstein equation, 168,169
Elastic deformation, 171
Elastic modulus, 80, 81,153-159, 378-383
Electromyographs, 331-338
Electrospinning, 237-245, 247-256
Electrostatic gels, 141-148
Elongation tests, 259-267, 271-276
Elongation, 244
Emulsifying activity, 118
Emulsion interface, 219-225
Emulsion preparation, 220
Emulsions, 227-233
Emulsions, caseinate, 221
Emulsions, β–lactoglobulin, 221
Encapsulation, 247-256, 279
Endosperm, 12
Endosperm-seed ratio, 29-32
Endothermic peak, 49-57
Enthalpy of gelatinisation, 371-373
Enzymatic treatment, 221
Enzyme modification, 19-25
Enzyme, analysis using, 19-25
Enzymes, in biosynthesis, 23-25
Erythritol, 375-383
Extensional rheology, 393-399
Extrusion, 69-76
Exudate polysaccharides, 99

Farinograph parameters, 360-368
Fedor's equation, 30
Fibres, dextran, 248
Film forming, 74
Film thickness, 259-267, 271-276
Films, biodegradable, 115
Films, chitosan, 279-286, 287-292
Films, edible, 257-267
Films, gelatin, 269-277
Films, protein, 221-226
Firmness, 332-338, 373
Flour, oat, 359-368
Flour, wheat, 359-368
Flow behaviour, 100-106
Flow properties, 90-96
Fluid gels, 183-189
Fluorescence microscopy, 135-139
Fluorescent brightener, 134
Foam stability, 120,121
Foaming capacity, 118
Force displacement, 312-316
Force microscopy, 209-216
Force relaxation, 308
Force spectroscopy, 417-421
Force-distance curves, 214
Fractal dimension, 143,146,147
Frequency dependence, 230
Frequency sweep, 186, 386-388
Friction coefficient, 185
Frictional squeezing, 307-316
FT-IR microscopy, 11-19, 178-182

Galactans, 19-25
Galactobiose, 418-421
Galactolipid, 219
Galacto-oligosaccharides, 415-421
Galacturonic acid backbone, 199
Galacturonic acid, 416
Galectin -3, 415-421
Gel fraction, 176-182
Gel hardness, 147,148
Gel permeation chromatography, 30,40
Gel preparation, 192
Gel strength, 238
Gel, CMC, 175-182
Gelatin fibres, 237-245
Gelatin films, 269-277
Gelatin, 339-347
Gelatin, fish, 237-245
Gelatinisation, 370-373
Gelatinisation, enthalpy of, 371-373
Gelation, pectin, 191-197
Gelled network, carrageenan, 183-190
Gelled particles, 183
Gelling process, 195
Gels, cured, 144
Gels, fluid, 183-189
Gels, quiescent, 183
β glucan, 133-139
Glutaraldehyde, 239, 279
Gluten, 115
Gluten, wheat, 295-302
Glutenin,115-122
Granulation, 409-414
Guar gum, 32
Gum karaya, 99
Gum, corn fiber, 59-66

Hardness, 331, 373
Hershel-Buckley, 325
HPLC system, 64
Hydration profile, 90-96
Hydration, 409-414
Hydrid carrageenans, 22
Hydrocolloid market, 3-8
Hydrogels, 401-406
Hydroxypropyl cellulose, 391-399
Hydroxypropylmethylcellulose, 45-57

Image analysis, 143
Immunostimulatory biopolymers, 123-129
Interface, air-water, 288-292

Interface, emulsion, 219-225
Interfacial behaviour, 287-292
Interfacial films, 210-216
Interfacial rheology, 289-292
Interfacial shear rheology, 210-216
Interfacial tension, 210-216
Intrinsic viscosity, 30
Isotherms, 259-267, 288-292
Isotropic phase, 79

Karaya, 99
Konjac glucomannan, 151-161
Krieger-Dougherty, 168, 169, 356

Lactobacillus johnsoni, 250-256
Lactococcus lactis, 250-256
β lactoglobulin / xanthan, 141-148
β lactoglobulin emulsions, 221
Lacunarity measurements, 143- 147
Langmuir-Blodgett, 210-216
Lectin, 415-421
Light micrographs, 354, 355
Light microscopy, 319-327
Lipase, 219
Lipid content, 63-66
Lipolysis, 222-226
Liquid crystalline phase, 77-83
Lubricated squeezing flow, 306-316
Lubricating behaviour, 183-190
Lysine content, 117-119

Maize starch, 11-19
Market overview, 3-8
Mastication, 331-338
Mechanical properties, 331-338
Methylcellulose, 45-57, 421, 422
Microalgae, 123-129
Microgel suspension, 165-173
Microgels, agarose, 166-173
Microrheology, 183
Microstructure, 145,146, 377
Modulus, complex, 371
Modulus, compression, 179
Modulus, particle, 165-173
Molecular weight, 30, 40-42, 393-399
Morphological properties, 111,112
Muffins, 369-373, 375-383

Nanofibres, gelatin, 237-245
Nematic phase, 79
Newtonian plateau, 91
Nigerian plants, 27-44
Noodles, 359-368

Oat flour, 134, 359-368
Obesity, 421
Oil-in-water emulsions, 227-233
Oil-water interface, 212
Osmotic deswelling, 172
Ozonisation, 107-112

Packing fraction, 168
Particle size distribution, 228-233
Particle size measurement, 319
Particle tracking, 183
Particles, plant, 317-327
Pasting properties, 109-112
Pectin gelation, 191-197
Pectin, 321, 339-347, 415-421
Pectin, citrus, 199-206
Pectin, sugar beet, 209-216
Pectin-zein hydrogels, 401-406
Peltophorum pteracarpus, 27-44
Pendant drop, 210-216
Phase behaviour, 133-139
Phase concentration, 79, 81
Phase diagram, 80, 135-139
Phase separation, 49, 79-86
Phase volume, 165-173
Phenolic acid analysis, 60
Piroxicam, 401-406
Plant particles, 317-327
Polarimetry, 184
Polarised light miscroscopy, 12
Polyol, 369-373
Polysaccharide-protein coacervates, 401-406
Pore size measurements, 143-148
Potato protein, 227-233
Power Law, 91, 102
Prices, hydrocolloid, 5
Processability, 67-76
Prosopis Africana, 27-44
Prosopis grandulosa, 27-44
Protein content, 62-66
Protein displacement, 221,224
Protein films, 221-226
Protein interactions, 219-225
Protein, crayfish, 287-292
Protein, potato, 227-233
Protein-polysaccharide gels, 141-149
Protein-polysaccharide mixtures, 133-139
Pseudoplastic region, 91
Pullulan, 256
Pyruvate level, 151-153

Quemada equation, 168,169
Quiescent gels, 183

Raman microscopy, 11-19
Regioselective substitution, 45-57
Relaxation spectrum, 145
Relaxation time, 104
Relaxation, force, 308
Resistant starch, 385-388
Rheological characterisation, 228-233
Rheological measurements, 193, 320-327, 340-347, 386-388
Rheological modifiers, 391-399
Rheological properties, 32, 89-96, 361
Rheology, 48-57, 142, 166-173, 184
Rheology, bulk, 99-106
Rheology, interfacial, 289-292
Rheology, microgel suspension, 165-173
Rheology, surface, 99-106
Rhodamine B, 134
Riboflavin, 279-286
Ridgelimeter method, 193
Ripeness, 353

SAG method, 191-197
Sago starch films, 258-267
Satiety, 421,422
Scanning electron microscopy, 68-76, 111,112, 200, 201, 241-245, 260-267, 276, 320-327
SDS-PAGE, 118-121
Seaweed, red, 183
Seed components, 31
Sensory attributes, 410-414
Sensory evaluation, 341, 362-368
Sensory panel, 339
Sensory scores, 331-338
Shear flow, 230-233
Shear viscosity, 252-256, 393-399
Size exclusion chromatography, 392-399
Solubility, 117
Space occupancy, 40
Specific gravity, 377-383
Springiness, 332-338, 373
Starch gelatinisation, 370-373
Starch ozonisation, 107-112
Starch suspensions, 165
Starch swelling, 80
Starch, 339-347
Starch, 67-76
Starch, maize, 11-19
Starch, resistant, 385-388
Starch, wheat, 107-112
Starch-gelatin films, 257-267
Steady shear flow, 230-233,
Steady shear viscosity, 30-44
Sterculia apetala, 99-106
Sterculia striata, 99-106
Storage and loss moduli, 32-34, 52-57
Strain sweep, 340
Stribeck experiment, 185
Structure formation, 194
Structuring properties, 317-327
Succinoglycan, 89-96
Sucrose palmitate, 257-267
Sucrose replacers, 369-373, 375-383
Sugar beet pectin, 209-216
Sulphurylases, 23-25
Surface characterisation, 103
Surface potential, 221-226
Surface pressure, 221-226
Surface rheology, 99-106
Surface tension, 100
Swallowing, 409-414
Swelling behaviour, 409-414
Swelling degree, 403
Swelling starch, 80
Swelling, film, 280-286
Syneresis, 143,146
Synergistic interaction, 151-161

Tensile strength, 244, 259-267, 271-276, 296-302
Texture design, 407-414
Texture profile analysis, 143
Texture properties, 143, 361-368
Texture studies, 305-316
Texture, 331-338, 339-347
Texture, muffin, 382
Thermal analysis, 203,204
Thermal degradation, 202
Thermal properties, 109-112
Thermogelation, 46
Thermo-rheological study, 45-57
Thickeners, dysphagia, 407-414
Tilapia, 237-245
TNBS method, 117
Tomato puree, 351-357
Tomato, 320-327
Transparency index, 261
Treculia africana, 27-44
Tribology, 184
Trifluoroacetic acid, 134
Trimaster, 393-399
Turbidimetric titration, 402
Turbidity, 48

Vegetation zones, 28
Viscoelastic behaviour,
Viscoelastic properties, 39, 94, 371-373, 376-383, 385-388
Viscoelasticity, 144
Viscosity behaviour, 351-357

Viscosity measurements, 135-139
Viscosity, 90-96
Viscosity, zero shear, 189
Visual analogue scale, 422
Vitamin B2, 279-286

Water absorption capacity, 296-302
Water absorption, 117, 176-182, 360-368
Water binding properties, 147
Water holding capacity, 118-121
Water solubility, 117
Water sorption isotherms, 259-267
Water vapour permeability, 259-267, 271-276
Waxy maize starch, 67-76
Wheat flour dough, 115
Wheat flour, 359-368
Wheat gluten, 116-122, 295-302
Wheat starch, 107-112
Whey protein, 141-148
Williamson model, 30

Xanthamonas campestris, 77
Xanthan gum / β lactoglobulin, 141-148
Xanthan gum blend, 295-302
Xanthan gum modification, 77-85
Xanthan gum rheology, 89-96
Xanthan gum, 375-383, 407-414
Xanthan, acetate level, 151-153
Xanthan, pyruvate level, 151-153
Xanthan, with konjac mannan, 151-161
X-ray diffraction, 68-76

Yield stress, 306, 308, 325
Yoghurt, 254-256, 339-347
Young's modulus, 334, 361-368

Zein hydrogels, 401-406
Zero shear viscosity, 33, 101, 189
Zeta potential, 229-233

www.ingramcontent.com/pod-product-compliance
Ingram Content Group UK Ltd.
Pitfield, Milton Keynes, MK11 3LW, UK
UKHW021815271025
464415UK00005B/34

9 781849 733588